AF755569

Geometry and Algebra of Multidimensional Three-Webs

Mathematics and Its Applications (*Soviet Series*)

Managing Editor:

M. HAZEWINKEL
Centre for Mathematics and Computer Science, Amsterdam, The Netherlands

Editorial Board:

A. A. KIRILLOV, *MGU, Moscow, U.S.S.R.*
Yu. I. MANIN, *Steklov Institute of Mathematics, Moscow, U.S.S.R.*
N. N. MOISEEV, *Computing Centre, Academy of Sciences, Moscow, U.S.S.R.*
S. P. NOVIKOV, *Landau Institute of Theoretical Physics, Moscow, U.S.S.R.*
M. C. POLYVANOV, *Steklov Institute of Mathematics, Moscow, U.S.S.R.*
Yu. A. ROZANOV, *Steklov Institute of Mathematics, Moscow, U.S.S.R.*

Volume 82

Geometry and Algebra of Multidimensional Three-Webs

by

Maks A. Akivis

Moscow Institute of Steel and Alloys,
Moscow, Russia, C.I.S.

and

Alexander M. Shelekhov

Tver State University,
Tver, Russia, C.I.S.

Translated from Russian by
Vladislav V. Goldberg
New Jersey Institute of Technology,
Newark, N.J., U.S.A.

SPRINGER SCIENCE+BUSINESS MEDIA, B.V.

Library of Congress Cataloging-in-Publication Data

Akivis, M. A. (Maks Aĭzikovich)
Geometry and algebra of multidimensional three-webs / by Maks A. Akivis and Alexander M. Shelekhov.
p. cm. -- (Mathematics and its applications (Soviet series) ; v. 82)
Translated from the Russian.
Includes bibliographical references and index.
ISBN 978-94-010-5059-3 ISBN 978-94-011-2402-7 (eBook)
DOI 10.1007/978-94-011-2402-7
1. Webs (Differential geometry) I. Shelekhov, A. M. II. Title. III. Series: Mathematics and its applications (Kluwer Academic Publishers). Soviet series ; 82.
QA648.5.A35 1992
516.3'6--dc20 92-7748

ISBN 978-94-010-5059-3

All Rights Reserved
© 1992 Springer Science+Business Media Dordrecht
Originally published by Kluwer Academic Publishers in 1992
No part of the material protected by this copyright notice may be reproduced or utilized in any form or by any means, electronic or mechanical, including photocopying, recording or by any information storage and retrieval system, without written permission from the copyright owner.

SERIES EDITOR'S PREFACE

'Et moi, ..., si j'avait su comment en revenir, je n'y serais point allé.'
Jules Verne

The series is divergent; therefore we may be able to do something with it.
O. Heaviside

One service mathematics has rendered the human race. It has put common sense back where it belongs, on the topmost shelf next to the dusty canister labelled 'discarded nonsense'.
Eric T. Bell

Mathematics is a tool for thought. A highly necessary tool in a world where both feedback and nonlinearities abound. Similarly, all kinds of parts of mathematics serve as tools for other parts and for other sciences.

Applying a simple rewriting rule to the quote on the right above one finds such statements as: 'One service topology has rendered mathematical physics ...'; 'One service logic has rendered computer science ...'; 'One service category theory has rendered mathematics ...'. All arguably true. And all statements obtainable this way form part of the raison d'être of this series.

This series, *Mathematics and Its Applications*, started in 1977. Now that over one hundred volumes have appeared it seems opportune to reexamine its scope. At the time I wrote

> "Growing specialization and diversification have brought a host of monographs and textbooks on increasingly specialized topics. However, the 'tree' of knowledge of mathematics and related fields does not grow only by putting forth new branches. It also happens, quite often in fact, that branches which were thought to be completely disparate are suddenly seen to be related. Further, the kind and level of sophistication of mathematics applied in various sciences has changed drastically in recent years: measure theory is used (non-trivially) in regional and theoretical economics; algebraic geometry interacts with physics; the Minkowsky lemma, coding theory and the structure of water meet one another in packing and covering theory; quantum fields, crystal defects and mathematical programming profit from homotopy theory; Lie algebras are relevant to filtering; and prediction and electrical engineering can use Stein spaces. And in addition to this there are such new emerging subdisciplines as 'experimental mathematics', 'CFD', 'completely integrable systems', 'chaos, synergetics and large-scale order', which are almost impossible to fit into the existing classification schemes. They draw upon widely different sections of mathematics."

By and large, all this still applies today. It is still true that at first sight mathematics seems rather fragmented and that to find, see, and exploit the deeper underlying interrelations more effort is needed and so are books that can help mathematicians and scientists do so. Accordingly MIA will continue to try to make such books available.

If anything, the description I gave in 1977 is now an understatement. To the examples of interaction areas one should add string theory where Riemann surfaces, algebraic geometry, modular functions, knots, quantum field theory, Kac-Moody algebras, monstrous moonshine (and more) all come together. And to the examples of things which can be usefully applied let me add the topic 'finite geometry'; a combination of words which sounds like it might not even exist, let alone be applicable. And yet it is being applied: to statistics via designs, to radar/sonar detection arrays (via finite projective planes), and to bus connections of VLSI chips (via difference sets). There seems to be no part of (so-called pure) mathematics that is not in immediate danger of being applied. And, accordingly, the applied mathematician needs to be aware of much more. Besides analysis and numerics, the traditional workhorses, he may need all kinds of combinatorics, algebra, probability, and so on.

In addition, the applied scientist needs to cope increasingly with the nonlinear world and the extra mathematical sophistication that this requires. For that is where the rewards are. Linear models are honest and a bit sad and depressing: proportional efforts and results. It is in the nonlinear world that infinitesimal inputs may result in macroscopic outputs (or vice versa). To appreciate what I am hinting at: if electronics were linear we would have no fun with transistors and computers; we would have no TV; in fact you would not be reading these lines.

There is also no safety in ignoring such outlandish things as nonstandard analysis, superspace and anticommuting integration, *p*-adic and ultrametric space. All three have applications in both electrical engineering and physics. Once, complex numbers were equally outlandish, but they frequently proved the shortest path between 'real' results. Similarly, the first two topics named have already provided a number of 'wormhole' paths. There is no telling where all this is leading - fortunately.

Thus the original scope of the series, which for various (sound) reasons now comprises five subseries: white (Japan), yellow (China), red (USSR), blue (Eastern Europe), and green (everything else), still applies. It has been enlarged a bit to include books treating of the tools from one subdiscipline which are used in others. Thus the series still aims at books dealing with:

- a central concept which plays an important role in several different mathematical and/or scientific specialization areas;
- new applications of the results and ideas from one area of scientific endeavour into another;
- influences which the results, problems and concepts of one field of enquiry have, and have had, on the development of another.

A web is a collection of d foliations in a general position of the same codimension. For example, one has the case of 3-webs of curves in the plane (the first interesting case, and the subject of numerous early studies). The French and German names for the concept are respectively 'tissu' and 'Gewebe', which mean tissue, fabric, texture (in its original meaning of woven fabric), web; and these words with these meanings convey a good initial intuitive picture of the kind of geometric structure involved. An easy example of a 3-web of curves in the plane is given by the three systems of lines $x =$ const., $y =$ const., $x+y =$ const.. By definition a codimension-one foliation is locally like a set of parallel hyperplanes. However, whether several foliations, i.e. a web, can be (locally) 'straightened out' simultaneously is a much tougher question. For example the trivial 3-web of lines just mentioned has the following *closure property*. Take a point 0 and draw the three leafs of the three foliations through that point. Take a neighboring point A on one of these leafs and walk around along the leafs and the foliations as indicated in the figure below:

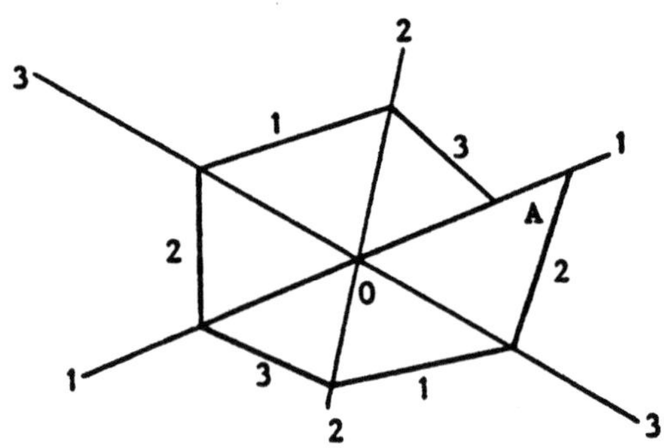

In the case of the given trivial foliation one finishes back in the point A after traversing a hexagon. This turns out to be a necessary and sufficient condition for a 3-web of curves in the plane to look locally like our trivial example, i.e. to be *parallellisable*.

There are, of course, a good many places in mathematics where multiple foliations come up naturally, and where such results are thus important. For example in control theory, although, as far as I know, this particular potential application of web theory remains unexplored.

A web is linearizable if it is equivalent to one consisting of straight lines. It is an old theorem (1924) that a linearizable web in the plane consists of the tangents to a curve of degree 3 in the projective plane, and thus the question arises whether every web arises in some such algebraic manner.

This also provided a rather unexpected link with algebraic geometry, one of the first of many interrelations of the theory of webs with various parts of the theories of symmetric spaces, differential equations, algebraic geometry, integral geometry, singularities, and holomorphic mappings, quasigroups, etc.

Thus far, this part of this series editor's preface is practically indentical to the corresponding part of my preface to V.V. Goldberg, *Theory of Multi-Codimensional* $(n+1)$*-Webs*, KAP, 1988. That volume mainly concerns $(n+1)$ foliations of codimension r on an nr dimensional manifold for general n. The word 'practically' in the one but last sentence refers to a now corrected inaccuracy.

This present book is about 3-webs, i.e. 3 foliations of codimension r. Here 3 is important. It is the first case that there are local invariants.

In 1981 there appeared an 81 page multigraphed text by Akivis and Shelikhov on multicodimensional 3-webs. This has had a very limited circulation; basically, only to cognoscenti who could read Russian. And for those it was a considerable boon. In his preface to his own book, V.V. Goldberg expressed the hope that one day an expanded translation would be available in English. Well, here it is, and not just an expanded translation but a whole new completely up-to-date 350 page unique book on the topic by the very authors who kept the subject developing in the years of neglect and, who together with others (Chern, Goldberg, Griffiths, ...) are responsible for the revival of, and current strong interest in, webs.

The shortest path between two truths in the real domain passes through the complex domain.

J. Hadamard

La physique ne nous donne pas seulement l'occasion de résoudre des problèmes ... elle nous fait pressentir la solution.

H. Poincaré

Never lend books, for no one ever returns them; the only books I have in my library are books that other folk have lent me.

Anatole France

The function of an expert is not to be more right than other people, but to be wrong for more sophisticated reasons.

David Butler

Bussum, 10 February 1992

Michiel Hazewinkel

Table of Contents

CHAPTER 4

The Bol Three-Webs and the Moufang Three-Webs

CHAPTER 5

Closed G-Structures Associated with Three-Webs

CHAPTER 6

Automorphisms of Three-Webs

CHAPTER 7

Geometry of the Fourth Order Differential Neighborhood of a Multidimensional Three-Web

CHAPTER 8

d-Webs of Codimension r

APPENDIX A

Web Geometry and Mathematical Physics (E.V. Ferapontov)

Preface

1. In 1927–1928 the papers of Wilhelm Blaschke, who was a well-known geometer by that time, and his student T. Thomsen appeared (see [Bl 28] and [T 27]). The papers were devoted to a new topic in differential geometry – web geometry. Many of Blaschke's previous papers were connected with applications of ideas of Felix Klein to differential geometry. These ideas were formulated in Klein's famous "Erlangen program" (1872). According to this program, every geometry studies properties of geometric figures that are invariant under transformations composing a certain group. From this point of view, the foundations of affine differential geometry and conformal differential geometry were presented in Blaschke's books, published in the 1920's.

Blaschke extended the same group-theoretic approach to web geometry. He took the group (more precisely, pseudogroup) of all differentiable transformations of a manifold, where a web is given, as a fundamental group, and studied local web invariants relative to transformations of this group.

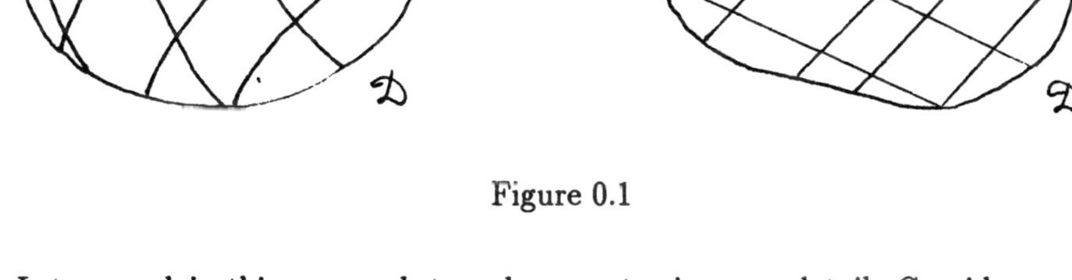

Figure 0.1

Let us explain this approach to web geometry in more detail. Consider a regular family of smooth curves in a two-dimensional domain D. By means of an appropriate differentiable transformation, the domain D can be transferred into a domain D' of an affine plane A^2 in such a way that the family of lines given in D will be transferred into a family of parallel lines of D'. This shows that a family of smooth curves in D does not have local invariants. Two regular families of smooth curves, that are in general position in D, also do not have local invariants since one can always find a diffeomorphism that transfers them into two families of parallel lines of a domain D' of an affine plane A^2 (see Figure 0.1). Thus, the structures, defined in D by one or two families of curves, are locally trivial.

Consider now three regular families of smooth curves in D, that are in general position in D, i.e. form a three-web in the domain D. Such a structure is no longer locally trivial. In fact, in a neighborhood of each point p of the domain D, where the web is given, one can construct a family of hexagonal figures as shown in Figure 0.2.

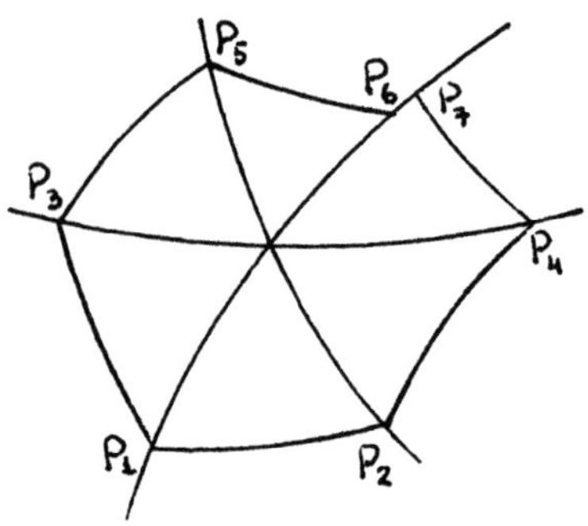

Figure 0.2

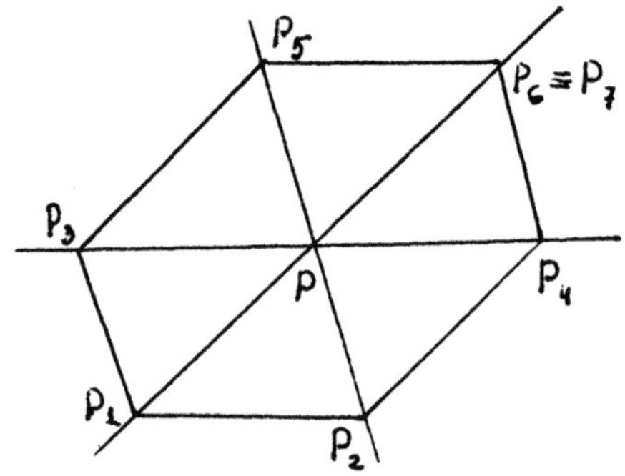

Figure 0.3

In this figure the lines p_1p_6, p_3p_5, p_2p_4 belong to the first family, the lines p_2p_5, p_1p_3, p_4p_7 belong to the second family, and the lines p_3p_4, p_1p_2, p_5p_6 belong to the third family. In the general case, the points p_6 and p_7 of these figures do not coincide, i.e. hexagonal figures are not closed. However, there exist three-webs, on which all hexagonal figures are closed, i.e. on these webs the points p_6 and p_7 coincide. Such three-webs are called hexagonal. For example, the web, formed by three families of parallel lines in an affine plane A_2, is hexagonal (see Figure 0.3). It is remarkable that the converse theorem also holds: any hexagonal three-web in a plane admits a differentiable mapping on a three-web, formed by three families of parallel lines, i.e. such a web is parallelizable. This theorem was first proved by Thomsen in the above mentioned paper [T 27] (see also the book [Bl 55]).

Since the fundamental group in web geometry is the group of arbitrary differentiable mappings, Blaschke refers the theory of webs as one of "topological questions of differential geometry" (Topologische Fragen der Differentialgeometrie). Numerous papers of Blaschke, his students and co-workers as well as the well-known monograph *Web Geometry* (*Geometrie der Gewebe*) [BB 38], written jointly by Blaschke and Bol and published in 1938, were devoted to this subject.

Blaschke noticed the connections of the web theory with many branches of geometry, in particular, with projective geometry. In a projective plane P^2 it is natural to consider a three-web formed by three one-parameter families of curves, each enveloping a curve X_α, $\alpha = 1, 2, 3$. In this case, one straight line from each family passes through a point p of a domain $D \subset P^2$ (Figure 0.4). By a correlative transformation of the plane P^2, these families of straight lines will be transferred into three curves X_α^* in a dual plane P^{2*}, and to each point p there corresponds a straight line p^* (see Figure 0.5). To straight lines of the original rectilinear three-web there correspond pencils in the plane P^{2*} with centers on the curves X_α^*. For this configuration the following remarkable theorem X_α in a plane P^2 is hexagonal if and only if the triple of curves X_α^* in the plane P^{2*}, which is dual to this web, belongs to a cubic curve. This theorem is called the Graf–Sauer theorem and follows from the Chasles' theorem on plane cubic curves. Its proof can be found in the book [Bl 55].

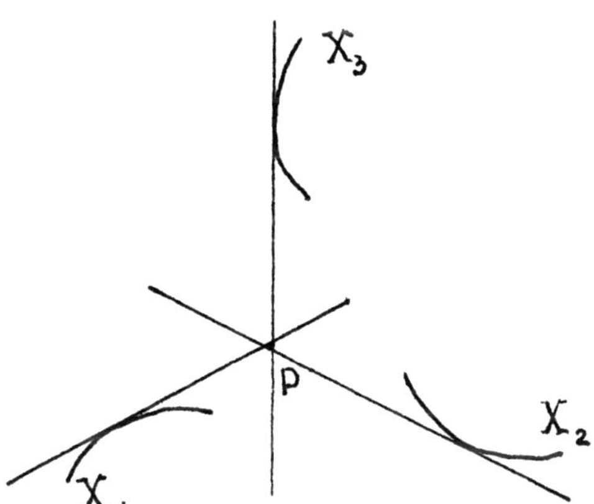

Figure 0.4

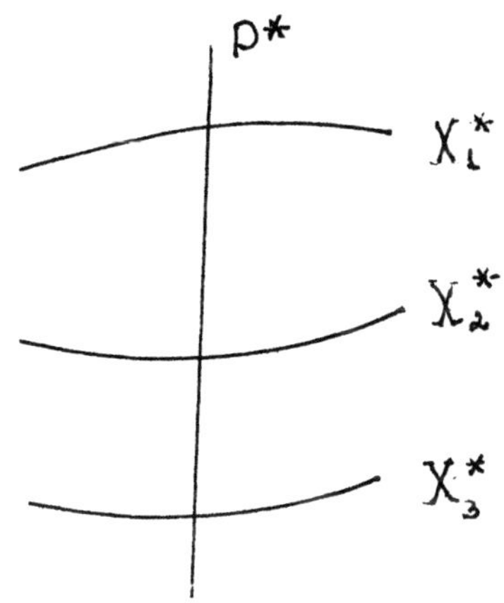

Figure 0.5

At the very beginning of the development of web theory its connection with algebra was discovered. In 1928 and 1932 the papers of K. Reidemeister [R 28] and H. Kneser [Kn 32] were published, in which it was established that a structure that has many of the properties of a three-web, can be constructed from an arbitrary group G. Namely, consider a cartesian square $X = G \times G$, whose elements are pairs (x, y) of elements of G. Call these pairs elements of the considered "three-web". Consider subsets $(x, b), (a, y), \{(x, y) | x \cdot y = c\}$ of the set X, where a, b, c are constants and x, y are variable elements of the group G, and call them lines of this "three-web". In this case, through each point of the web, there pass three lines from different families, and two lines from any two different families intersect at a point. The last two properties of this three-web allow us to consider it as an incidental structure consisting of points and lines. Because of this, the theory of such abstract webs is developed in the same way as the abstract theory of projective planes [Pi 75]. The only difference is that instead of the Pappus and Desargues configurations, which are considered in projective planes, in web theory the Reidemeister and Thomsen figures arise. We will consider these configurations in detail in Chapter 2.

Later on, Bol in his paper [B 37] observed that for construction of the abstract three-web of this type it is possible to use a groupoid of more general kind than a group. At present this kind of groupoid is called a quasigroup. Bol introduced more general closure figures on an abstract three-web, which are called Bol figures. He found that these figures are connected with certain weakened associativity conditions in a quasigroup generating a three-web.

However, it turned out that all these closure conditions cannot be geometrically realized by means of two-dimensional geometric webs, which were considered in the papers of Blaschke and Thomsen: on such webs the hexagonality condition – the weakest among the classical closure conditions – implies the web parallelizability and consequently the realization of all other closure conditions on the web.

Nevertheless all closure conditions indicated above can be realized on multidimensional geometric webs. This was a stimulus for their study. The first works in theory of multidimensional webs were the Bol paper [B 35], where a three-web on a four-dimensional manifold was considered, and the Chern paper [C 36a], where the theory of invariants of a three-web of dimension $2r$ was constructed. These two papers were published in Germany in the 1930's, and after them, for a long time, there were no publications devoted to theory of multidimensional three-webs.

Only at the end of 1960's, was the study of multidimensional three-webs continued. In 1969 the first author of this book published the paper [A 69b], and this paper was followed by an extensive series of his papers as well as his students' papers. At approximately the same time M.A. Akivis' geometry seminar at Moscow Insitute of Steel and Alloys began. Until this day, all of the most important problems of web theory are presented and discussed in this seminar.

Due to the effort of the second author, since 1981 Kalinin State University has regularly published collections of papers under the title *Webs and Quasigroups*. As a result, extensive material in the theory of multidimensional three-webs has been accumulated. The main problems of this theory as well as its connections to related problems of geometry and algebra are presented in the current monograph.

However, because of the size restriction, we cannot include everything we wanted to discuss in this book. Some of the results are given in the book without proofs. Other results are listed only in bibliographical notes and in the Bibliography.

2. We will make a few general remarks for readers of this book. First of all, note that the book is intended for graduate students whose field is differential geometry, and for mathematicians and teachers conducting research in this subject. This book can also be used for a few special courses for graduate students in Mathematics.

In our presentation of material we use the tensorial methods in combination with the methods of exterior differential forms and moving frames of Élie Cartan. The reader is assumed to be familiar with these methods as well as with the basics of modern differential geometry. Many notions of differential geometry are explained briefly in the text and some are given without any explanation. As references, the books [KN 63], [St 83], [CaH 67] and [V 87] are recommended.

All functions, vector and tensor fields and differential forms are considered differentiable sufficiently many times. All variables are assumed to be real, although in many cases this does not exclude a possibility of considering complex variables. For this reason we use the notation $\mathbf{GL}(n)$ here instead of $\mathbf{GL}(n, \mathbf{R})$ or $\mathbf{GL}(n, \mathbf{C})$ as often done.

The book consists of eight chapters. Sections, formulas and figures are numbered within each chapter. Each chapter is accompanied by a set of problems and notes with remarks of historical and bibliographical nature. A sufficiently complete bibliography, a list of notations and index are given at the end of the book.

We asked E.V. Ferapontov to write an appendix devoted to applications of web geometry to some problems of Mathematical Physics. We hope that this supplement enlargens the circle of the readers of our monograph and attracts physicists to the problems considered in the book.

In conclusion, we express our sincere gratitude to V.V. Goldberg, with whom we discussed different aspects of this book in the process of our work on the book, to V.F. Kirichenko and V.V. Timoshenko, who carefully read the manuscript and made many useful suggestions, and to L.V. Goldstein and V.B. Lazareva for their invaluable assistance in preparing the manuscript for publication.

Moscow, C.I.S. *Maks A. Akivis*

Tver, C.I.S. *Alexander M. Shelekhov*

Chapter 1

Three-Webs and Geometric Structures Associated with Them

1.1 G-Structures, Fibrations and Foliations

1. Let X be a differentiable n-dimensional manifold of class $C^s, s \geq 3$, let p be its arbitrary point and $T_p \equiv T_p(X)$ be the tangent space to the manifold X at the point p. Let us denote by $x^u, u, v, w = 1, 2, \ldots, n$, local coordinates in a neighborhood U of the point p and by $\frac{\partial}{\partial x^u}$ and dx^u respectively the natural basis of vector fields and the corresponding co-basis of differential forms in U. If the vectors $e_u = \tilde{x}^v_u \frac{\partial}{\partial x^v}$ form a vector frame, then the corresponding co-frame is formed by the differential forms

$$\omega^u = x^u_v dx^v, \tag{1.1}$$

where $\tilde{x}^v_u$ and x^u_v are matrices mutually inverse to each other. The basis $\{e_u\}$ and the co-basis $\{\omega^u\}$ are connected by the relation $\omega^u(e_v) = \delta^u_v$ (δ^u_v is the Kronecker symbol).

Let us denote by $\mathcal{R}_p(X)$ the set of all vectorial frames in the tangent space T_p. The union $\mathcal{R}(X) = \bigcup\limits_{p \in X} \mathcal{R}_p(X)$ is called the *bundle of frames* on the manifold X. It is a manifold of class C^{s-1} and dimension $n+n^2$, where $n = \dim X$, with local coordinates x^u and x^u_v, which are called the *principal and secondary parameters* respectively.

The forms ω^u defined by the equation (1.1) depend only on the differentials of the principal parameters and are defined in the bundle of frames of the manifold X. Exterior differentiation of these forms gives the exterior quadratic equations

$$d\omega^u = \omega^v \wedge \omega^u_v, \tag{1.2}$$

where d is the symbol of exterior differentiation and $\wedge$ is the symbol of exterior multiplication. The forms ω^u_v contain the differentials of both, principal and secondary parameters. The latter determine the location of the frame $\{e_u\}$ in the tangent space T_p.

If we fix a point p in X, i.e. set $\omega^u = 0$, we obtain the forms $\pi_v^u = \omega_v^u\big|_{\omega^u=0}$, defining an infinitesimal displacement of the frame $\{e_u\}$ in the space T_p:

$$\delta e_u = \pi_u^v e_v. \tag{1.3}$$

Here the symbol δ denotes differentiation with respect to secondary parameters x_v^u and the forms π_v^u are invariant forms of the general linear group $\mathbf{GL}(n)$ acting in the space T_p. These forms satisfy the structure equations of this group

$$\delta\pi_v^u = \pi_v^w \wedge \pi_w^u. \tag{1.4}$$

To equations (1.4) on the manifold X there correspond equations of more general form:

$$d\omega_v^u = \omega_v^w \wedge \omega_w^u + \omega^w \wedge \omega_{vw}^u, \tag{1.5}$$

which become (1.4) if $\omega^u = 0$. Equations (1.5) can be obtained from equations (1.2) by application of the standard procedure of differential prolongation, i.e. by exterior differentiation and application of Cartan's lemma.

Note that the forms ω^u, ω_v^u and ω_{vw}^u admit a global definition in the bundles of frames of appropriate order (see, for example, [La 66]). However, this is not essential for our further considerations since all problems, which will be discussed in the book, are of local nature.

In what follows the notion of a G-structure will play an important role. Let G be a ρ-dimensional subgroup of the general linear group $\mathbf{GL}(n), \rho \leq n^2$. A *$G$-structure* X_G on the manifold X is a subbundle of the frame bundle with the same base manifold X and the structure group G.

For example, the section $X \to R(X)$ gives a frame in each tangent space T_p of the manifold X. This gives a G-structure on the manifold X, and its structure group is $G = e$. This G-structure is called an *e-structure*.

If one takes as a structure group G the orthogonal group $O(n)$, which preserves a scalar product in the space T_p, then the corresponding $O(n)$-structure defines the *Riemannian structure* on the manifold X.

Let an r-dimensional distribution Δ be given on the manifold X. Denote by G the group of linear transformations of the space T_p leaving invariant the subspace Δ. The corresponding G-structure X_G is called the *Pfaffian structure*. Other examples of G-structures can be found in [St 83] as well as in this book.

We will find structure equations of the G-structure. Denote invariant forms of the Lie group G by θ^α. They satisfy the well-known Maurer–Cartan equations:

$$d\theta^\alpha = c_{\beta\gamma}^\alpha \theta^\beta \wedge \theta^\gamma, \quad \alpha, \beta, \gamma = 1, 2, \dots, \rho. \tag{1.6}$$

where $c_{\beta\gamma}^\alpha$ is the *structure tensor* of the group G. This tensor is skew-symmetric in lower indices β and γ and satisfies the *Jacobi identity*:

$$c_{\beta\gamma}^\epsilon c_{\epsilon\delta}^\alpha + c_{\gamma\delta}^\epsilon c_{\epsilon\beta}^\alpha + c_{\delta\beta}^\epsilon c_{\epsilon\gamma}^\alpha = 0.$$

Since the group G is a subgroup of the group $\mathbf{GL}(n)$, then the invariant forms π_v^u of the latter are expressed in terms of the invariant forms θ^α of the group:

$$\pi_v^u = c_{v\alpha}^u \theta^\alpha, \tag{1.7}$$

where the coefficients $c_{v\alpha}^u$ are constant. Relations (1.7) are satisfied provided that the principal parameters are fixed, i.e. if $\omega^w = 0$. Because of this, on the whole manifold X the following equations hold:

$$\omega_v^u = c_{v\alpha}^u \sigma^\alpha + \tilde{a}_{vw}^u \omega^w, \tag{1.8}$$

where σ^α are certain new differential forms satisfying the condition $\sigma^\alpha\Big|_{\omega^u=0} = \theta^\alpha$. Substituting the expansion (1.8) into the equation (1.2), we obtain the structure equations of the G-structure X_G:

$$d\omega^u = c_{v\alpha}^u \omega^v \wedge \sigma^\alpha + a_{vw}^u \omega^v \wedge \omega^w. \tag{1.9}$$

The quantities $a_{vw}^u = \tilde{a}_{[vw]}^u$ form the so-called the *first structure object* of the considered G-structure.

2. The frame bundle and the G-structures are particular cases of a more general structure called a fibration.

A *fibration of class* C^s is a triple $\lambda = (X, B, \pi)$, where X and B are manifolds of class C^s and dimensions n and r respectively and $\pi : X \to B$ is a projection of X onto B satisfying the following conditions:

a) For each point $b \in B$ the set $\pi^{-1}(b) \subset X$ is an $(n-r)$-dimensional submanifold diffeomorphic to a manifold $\mathcal{F}$, which is called a *typical fibre*; and

b) For each point $b \in B$ there exists a neighborhood $U_b \subset B$ such that $\pi^{-1}(U_b)$ is diffeomorphic to the direct product $U_b \times \mathcal{F}$. This means that the fibration is locally trivial.

The manifold X is the *space* of the fibration λ, or its *total space*, the manifold B is its *base*, and $\pi^{-1}(b) \stackrel{\text{def}}{=} \mathcal{F}_b$ is a *fibre* of λ.

There exists a unique fibre $\mathcal{F}_b$ of the fibration λ passing through any point $p \in X$, where $b = \pi(p)$.

The simplest fibration is the *standard trivial fibration*, i.e. the triple $(B \times \mathcal{F}, X, \pi)$, where π is the natural projection $B \times \mathcal{F} \to X$. In the frame bundle $\mathcal{R}(X)$ and in the G-structure X_G the base is the manifold X, and the structure group is the group $\mathbf{GL}(n)$ or the group G respectively. Another example is the tangent bundle $T(X) = \bigcup_{p \in X} T_p(X)$, where a fibre is a tangent space T_p. Other examples of fibrations can be found in [KN 63] and [St 83].

In the tangent space T_p of a point p of an n-dimensional total space X of the fibration $\lambda = (X, B, \pi)$ we define a subspace T_p' of codimension r, which is tangent to the fibre $\mathcal{F}_b$, passing through the point p. In a neighborhood U of the point p we consider a local frame $\{e_i, e_a\}, i, j = 1, 2, \ldots, r; a = r+1, r+2, \ldots, n$, such that the vectors $e_a\Big|_p$ belong to T'_p. The corresponding dual co-frame is composed of the forms ω^i, ω^a, which are also locally defined in the neighborhood U and $\omega^i(e_a) = 0$, i.e. the fibres $\mathcal{F}_b, b \in B$ of the fibration λ are integral manifolds of the system of Pfaffian equations $\omega^i = 0$. This system is completely integrable, since there exists a unique fibre $\mathcal{F}_b$ through any point $b \in U$. Therefore, according to the well-known Frobenius theorem [CaH 67], the exterior differentials of the forms ω^i are expressed through these forms as follows:

$$d\omega^i = \omega^j \wedge \omega^i_j. \tag{1.10}$$

Conversely, any system of forms ω^i, which are defined in U and satisfy equations of the form (1.10), gives in U a fibration of codimension r. Note that the forms ω^i are defined in the frame bundle $\mathcal{R}(B)$ of the base B of the fibration λ and equations (1.10) are the structure equations of the base.

In adapted local coordinates, the fibration λ is defined by the equations

$$F^i(x^u) = c^i = \text{const}, \quad \text{rank}\left(\frac{\partial F^i}{\partial x^j}\right) = r, \quad i, j = 1, \ldots, r; \quad u = 1, \ldots, n.$$

After differentiation we find the corresponding forms ω^i:

$$\omega^i = dF^i = \frac{\partial F^i}{\partial x^u} dx^u.$$

In this case the equations (1.10) have the form $\omega^i = 0$.

We say that we are given a *foliation of codimension* r on C^s-manifold X if

(a) through each point $p \in X$ there passes a smooth C^s-submanifold (leaf) of codimension r, and

(b) there exists a neighborhood U of each point $p \in X$ such that inside U the leaves form a trivial fibration with an $(n - r)$-dimensional base.

Thus, locally a foliation is a fibration, although globally it is not.

In general, a foliation does not have a base. (For example, an irrational winding of a torus forms a foliation; it does not form a fibration since a base does not exist for it.) However, in a sufficiently small neighborhood a foliation becomes a fibration, whose base is the set of leaves. Thus, we can speak about a local base of the foliation. Taking liberty in expression, we will simply call this local base the *base of a foliation*.

1.2 Three-Webs on Smooth Manifolds

1.

Definition 1.1 Let X be a C^s-manifold of dimension $2r, r \geq 1, s \geq 3$. We say that a 3-*web* $W = (X, \lambda_\alpha), \alpha = 1, 2, 3$ is given in X if

(a) three foliations λ_α of codimension r are given in X; and

(b) three leaves (of λ_α) passing through a point $p \in X$ are in general position, i.e. any two of the three tangent spaces to the leaves at the point p intersect each other only at the point p.

There exists a neighborhood of each point p of the web $W = (X, \lambda_\alpha)$, [1] where the foliations λ_α are fibrations. Therefore, from a local point of view, a three-web can be considered as formed by three fibrations. We denote the bases of these fibrations by X_α.

Example 1.2 Consider in an affine space A^{2r} of dimension $2r$ three families of parallel r-dimensional planes, which are in general position. They form a three-web, which is called a *parallel* three-web. We will denote it by W_0.

Example 1.3 Let $X_\alpha, \alpha = 1,2,3$ be three smooth hypersurfaces in a projective space P^{r+1}, and let p be a straight line that intersects them at the points x, y and z respectively. A three-web arises in a neighborhood of the line p on the Grassmannian $G(1, r+1)$ of all straight lines of the space P^{r+1} ($\dim G(1, r+1) = 2r$). The leaves of this web are bundles of straight lines with vertices located on the hypersurfaces X_α. Such a web is called the *Grassmann* three-web and denoted by GW. Our definition of the Grassmann three-web is essentially of local nature, since, only for straight lines sufficiently close to p, can we assert that they intersect each of the hypersurfaces X_α only at one point, as it was for the line p. The Grassmann three-webs will be studied in detail in Section 3.3.

Example 1.4 A Grassmann three-web is called *algebraic* and denoted AW if the hypersurfaces X_α, defining it, belong to the same cubic hypersurface. We note the following special cases of algebraic three-webs:

a) All three hypersurfaces X_α are hyperplanes, and

b) The hypersurface X_1 is a hyperplane and the hypersurfaces X_2 and X_3 belong to a hyperquadric.

On a three-web W a natural system of local coordinates can be introduced. As we have already mentioned, the foliations λ_α of general position that form a three-web define fibrations in a neighborhood U of a point $p \in X$, and these fibrations. This implies that the neighborhood U is diffeomorphic to the direct product of the bases of two of the web fibrations: for example, $U \sim X_1 \times X_2$. Then any point $p \in U$ has coordinates $p(x^i, y^j)$, where x^i and y^j are local coordinates in the bases X_1 and X_2 respectively (here and in the sequel $i, j, k, l, m \ldots = 1, 2, \ldots, r$). Thus, the leaves of the first and second foliations of the web W in neighborhood U are defined by the equations $x^i = \text{const}$ and $y^i = \text{const}$. The third foliation of the web W is given by the equations $f^i(x^1, \ldots, x^r, y^1, \ldots, y^r) = \text{const}$, or simply $f^i(x^j, y^k) = c^i$, where f^i are

[1] For brevity, we will use these words instead of the words "each point p of a manifold X carrying a three-web W".

smooth functions. Since the leaves of the web are in general position, at any point of neighborhood U the following conditions hold:

$$\det\left(\frac{\partial f^i}{\partial x^j}\right) \neq 0, \quad \det\left(\frac{\partial f^i}{\partial y^k}\right) \neq 0. \tag{1.11}$$

The vector-valued function $f^i(x^j, y^k)$ is called the *web function*, and the equation

$$z^i = f^i(x^j, y^k), \tag{1.12}$$

or simply $z = f(x, y)$, is called the *web equation*. The geometric meaning of the equation (1.12) is that it connects the parameters x, y and z of three leaves of the web passing through the point $p(x^i, y^j)$ of neighborhood U.

Note that if we take the foliations λ_1 and λ_3 or λ_3 and λ_2 as the base foliations, then the web equation is written in the form $y = g(x, z)$ or respectively $x = h(z, y)$, and each of these web equations is obtained by solving the equation (1.12) with respect to one of the variables.

Of course, the functions f^i, defined in different neighborhoods, are C^∞-compatible in the intersections. However, this does not imply that there exists a web function, defined globally.

Example 1.5 Let $G(\cdot)$ be a Lie group of dimension r, and let X be the direct product $G \times G$. Three leaves $x = a, y = b$ and $x \cdot y = a \cdot b$ pass through a point $(a, b) \in X$, and it is obvious that these three leaves are in general position. Therefore, the foliations $x = \text{const}$, $y = \text{const}$ and $x \cdot y = \text{const}$ form a three-web on X. This three-web is called the *group three-web*. Its equation is $z = x \cdot y$.

2. The web equation (1.12) is not uniquely defined. It is defined up to local diffeomorphisms of the form

$$\tilde{x} = J_1(x), \quad \tilde{y} = J_2(y), \quad \tilde{z} = J_3(z), \tag{1.13}$$

which give transformations of local coordinates in the bases X_α of the foliations λ_α. On the other hand, equations (1.13) have another meaning. In addition to the web $W = (X, \lambda_\alpha)$, let us consider another web $\widetilde{W} = (\widetilde{X}, \widetilde{\lambda}_\alpha)$ of the same codimension r. Denote local coordinates in the bases $\widetilde{X}_\alpha$ of the foliations $\widetilde{\lambda}_\alpha$ by $\tilde{x}^i, \tilde{y}^i, \tilde{z}^i$ and write the equation of the web $\widetilde{W}$ in the form:

$$\tilde{z} = \tilde{f}(\tilde{x}, \tilde{y}). \tag{1.14}$$

Then locally the equations(1.13) give bijective mappings of the foliations λ_α onto the foliations $\widetilde{\lambda}_\alpha$. If this mapping preserves the incidence, i.e. if any three leaves of the web W passing through a point will be mapped onto three leaves of the web $\widetilde{W}$ also passing through a point, then the functions f and $\tilde{f}$ are related by the condition:

$$\tilde{f}(J_1(x), J_2(y)) = J_3(f(x, y)), \tag{1.15}$$

("the image of the product is the product of the images").

Definition 1.6 Two webs W and $\widetilde{W}$ of the same codimension are said to be *equivalent* if there exists a triple $J = (J_1, J_2, J_3)$ of local diffeomorphism $J_\alpha : X_\alpha \to \widetilde{X}_\alpha$ such that the functions f and $\tilde{f}$ of these webs satisfy the condition (1.15). Since the mappings J_α preserve the web leaves, then the triple $J = (J_1, J_2, J_3)$ is called the *local isotopy* of the web W on the web $\widetilde{W}$.

In particular, three-webs equivalent to the parallel, Grassmann and algebraic three webs, which were considered above, are called *parallelizable, Grassmannizable* and *algebraizable* respectively. A local isotopy of the form (1.3) generates a local diffeomorphism $\varphi : X \to \widetilde{X}$, which transfers a point $p(x, y)$ into the point $\tilde{p}(J_1(x), J_2(y))$. The converse is also valid: any local diffeomorphism $\varphi : X \to \widetilde{X}$, which maps the leaves of the web W onto the corresponding leaves of the web $\widetilde{W}$ and preserves the incidence of leaves, generates a triple of mappings of the form (1.13) acting in the bases of the fibrations of the web W.

3. A submanifold $\widetilde{X}$ of dimension 2ρ of the manifold X, carrying a three-web W, is said to be a *transversal submanifold* or a *transversal surface* of the web W if any point $p \in X$ possesses a neighborhood $U_p \subset X$, in which the manifold $\widetilde{X}$ intersects each leaf of the web W, having at least one common point with $\widetilde{X}$, along a submanifold of dimension ρ.

In general, a web W does not possess transversal submanifolds. Local conditions of their existence will be discussed in Section 1.9. Here we will note only the following trivial fact: on any transversal submanifold (if the latter exists) the leaves of a web W cut a three-web $\widetilde{W}$, formed by the foliations of dimension ρ. This three-web is called a *subweb* of the web W.

For example, Grassmann webs GW have Grassmann subwebs of any dimension. In fact, let a $(\rho+1)$-dimensional submanifold $P^{\rho+1} \subset P^{r+1}$ intersect the hypersurfaces X_α generating the Grassmann web GW (see Example 1.3) along surfaces of dimension ρ. Then these surfaces define a Grassmann three-web $\widetilde{W}$, which is a subweb of GW.

Let us now define the direct product of three-webs $W_1 = (X_1, \underset{1}{\lambda}_\alpha)$ and $W_2 = (X_2, \underset{2}{\lambda}_\alpha), \dim X_1 = 2r_1, \dim X_2 = 2r_2$. Let the leaves $F_{1\alpha}$ of the web W_1 have a common point $p_1 \in X_1$ and the leaves $F_{2\alpha}$ of the web W_2 have a common point $p_2 \in X_2$. Then the three r-dimensional leaves $F_{1\alpha} \times F_{2\alpha} \stackrel{\text{def}}{=} F_\alpha$ pass through the point $p = (p_1, p_2)$ of the direct product $X_1 \times X_2 \stackrel{\text{def}}{=} X$, where $r = r_1 + r_2$. It follows from the definitions of a three-web and a direct product that the leaves F_α are in general position. Therefore, a three-web is defined in a neighborhood of a point $p \in X$. This three-web is called the *direct product* of the webs W_1 and W_2 and is denoted by $W_1 \times W_2$.

Note that submanifolds of the form (p_1, X_2) and $X_1, p_2)$ of the submanifold $X = X_1 \times X_2$ are transversal submanifolds of the direct product $W = W_1 \times W_2$.

The semi-direct product of webs and the quotient web (factor web) will be defined in Section 1.9.

1.3 Geometry of the Tangent Space of a Multidimensional Three-Web

Denote by $\underset{\alpha}{\omega}, \alpha = 1,2,3$, a linear form with its values in an r-dimensional space, which is defined in a domain $U \subset X$ and annihilated on the foliation λ_α of a web W. Denote by $\underset{\alpha}{\omega}^i$ the corresponding coordinate forms. Since each pair $\lambda_\alpha, \lambda_\beta, \alpha \neq \beta$, of the foliations of W is in general position in X, then the system of forms $\{\underset{\alpha}{\omega}, \underset{\beta}{\omega}\}, \alpha \neq \beta$, is linearly independent. Let us take the forms $\underset{1}{\omega}$ and $\underset{2}{\omega}$ as basis forms. Then the forms $\underset{3}{\omega}$ are the linear combination:

$$\underset{3}{\omega} = A\underset{1}{\omega} + B\underset{2}{\omega}.$$

Here the r by r matrices A and B depend on local coordinates in U. Since these equations are solvable with respect to $\underset{1}{\omega}$ and $\underset{2}{\omega}$, the matrices A and B are non-singular. Therefore, the substitution $A\underset{1}{\omega} \rightarrow \underset{1}{\omega}, B\underset{2}{\omega} \rightarrow \underset{2}{\omega}, \underset{3}{\omega} \rightarrow \underset{3}{\omega}$ is admissible. After this substitution the above equation becomes symmetric:

$$\underset{1}{\omega} + \underset{2}{\omega} + \underset{3}{\omega} = 0. \tag{1.16}$$

Denote the tangent space to the manifold X at the point p by T_p and the tangent spaces to the leaves $\mathcal{F}_\alpha$ of the web W at this point by $T_\alpha, \alpha = 1,2,3$. Consider three frames $\underset{\alpha}{e}_i, \alpha = 1,2,3$, at the point p, such that $\underset{\alpha}{e}_i \in T_\alpha$. Then, by definition of the forms $\underset{\alpha}{\omega}$, the following equations hold:

$$\underset{\alpha}{\omega}^i(\underset{\alpha}{e}_j) = 0. \tag{1.17}$$

In addition, we set

$$\underset{1}{\omega}^i(\underset{2}{e}_j) = \underset{2}{\omega}^i(\underset{3}{e}_j) = \underset{3}{\omega}^i(\underset{1}{e}_j) = \delta^i_j.$$

Making use of equation (1.16), we easily find that for $(\alpha, \beta) = (1,2), (2,3), (3,1)$ the relations

$$\underset{\beta}{\omega}^i(\underset{\alpha}{e}_j) = -\underset{\alpha}{\omega}^i(\underset{\beta}{e}_j) = -\delta^i_j \tag{1.18}$$

hold. It follows from (1.18) that the basis vectors $\underset{\alpha}{e}_i$ satisfy the equation

$$\underset{1}{e}_i + \underset{2}{e}_i + \underset{3}{e}_i = 0, \tag{1.19}$$

and an arbitrary vector $\xi \in T_p$ can be written in one of the following three forms:

$$\xi = \underset{1}{\omega}^i(\xi)\underset{2}{e}_i - \underset{2}{\omega}^i(\xi)\underset{1}{e}_i = \underset{2}{\omega}^i(\xi)\underset{3}{e}_i - \underset{3}{\omega}^i(\xi)\underset{2}{e}_i = \underset{3}{\omega}^i(\xi)\underset{1}{e}_i - \underset{1}{\omega}^i(\xi)\underset{3}{e}_i. \tag{1.20}$$

The basis forms $\underset{\alpha}{\omega}^i$ are defined up to transformations

$$'\underset{\alpha}{\omega}^i = A^i_j \underset{\alpha}{\omega}^j, \quad \det(A^i_j) \neq 0.$$

Thus, the group of admissible transformations of the frame $\{\underset{1}{e_i}, \underset{2}{e_i}\}$ of the space T_p consists of the matrices of the form $\begin{pmatrix} A & 0 \\ 0 & A \end{pmatrix}$ and is isomorphic to the general linear group $\mathbf{GL}(r)$. This implies that a G-structure with the structure group $\mathbf{GL}(r)$ is defined in a $(2r)$-dimensional manifold X carrying a three-web W. We call this a G_W*-structure*. We denote by $\mathcal{R}(W)$ the bundle of adapted frames connected with a three-web W.

Let us now consider vectors $\eta_\alpha = \underset{\alpha}{\eta^i} e_i$ in the space T_α. These vectors belong to a two-dimensional subspace since, by (1.19), they are related by $\eta_1 + \eta_2 + \eta_3 = 0$. We will call this subspace *transversal* to the triple of spaces T_α. It is determined by a bivector

$$H = \eta_1 \wedge \eta_2 = \eta_2 \wedge \eta_3 = \eta_3 \wedge \eta_1,$$

which we will call a *transversal bivector* of the web W. The family of transversal subspaces at a point p depends on $r - 1$ parameters.

Note that if there exists a two-dimensional transversal surface V^2 on a three-web W, then the tangent spaces of this surface are transversal subspaces of the web.

There is another way to define the transversal subspaces. Consider two leaves $\mathcal{F}_1$ and $\mathcal{F}_2$ of a web W, passing through a point $p, \mathcal{F}_1 \in \lambda_1, \mathcal{F}_2 \in \lambda_2$. The foliation λ_3 defines a bijective transformation $\varphi_{12} : \mathcal{F}_1 \longrightarrow \mathcal{F}_2$, under which points $q \in \mathcal{F}_1$ and $\varphi_{12}(q)$ belong to the same leaf of the third foliation. Since a leaf of the third foliation is given by the equation $\underset{3}{\omega} = 0$, namely this equation defines the tangent mapping $(\varphi_{12})_* = d\varphi_{12}\big|_p$. If $\underset{3}{\omega} = 0$, the formula (1.20) gives $\xi = \underset{2}{\omega^i}(\xi)\underset{3}{e_i}$. This proves that the mapping $(\varphi_{12})_*$ is a projection of the space T_1 onto the space T_2 parallel to the space T_3 (Figure 1.1). The mappings $\varphi_{\alpha\beta}$ and $(\varphi_{\alpha\beta})_*$ are defined in a similar way.

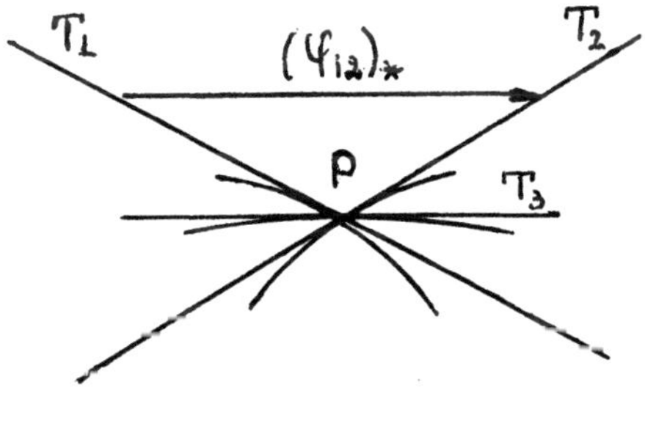

Figure 1.1

Proposition 1.7 *Transversal subspaces are invariant under transformations* $(\varphi_{\alpha\beta})_*$.

Proof. We will prove this for the mapping $(\varphi_{12})_*$. Let $\eta = \eta^i \underset{1}{e}_i$ be a vector from T_1. Set $(\varphi_{12})_*\big|_p(\eta_1) = {}'\eta_2$ and show that ${}'\eta_2 = -\eta_2 = -\eta^i \underset{2}{e}_i$. In fact, as was proved above, ${}'\eta_2 - \eta_1 \in T_3$, from which it follows that $\underset{3}{\omega}({}'\eta_2 - \eta_1) = 0$, or $\underset{3}{\omega}({}'\eta_2) = \underset{3}{\omega}(\eta_1)$. But, by (1.18), $\underset{1}{\omega}(\eta_1) = 0$. Therefore, from the equations (1.20) we obtain the equation $\eta_1 = \underset{3}{\omega}^i(\eta_1)\underset{1}{e}_i$, which implies $\eta^i = \underset{3}{\omega}^i(\eta_1)$. Now, using the second expansion ξ from (1.20) and the condition $\underset{2}{\omega}({}'\eta_2) = 0$, which is obtained from (1.18), we finally obtain:

$$ {}'\eta_2 = -\underset{3}{\omega}^i({}'\eta_2)\underset{2}{e}_i - \underset{3}{\omega}^i(\eta_1)\underset{2}{e}_i = -\eta^i \underset{2}{e}_i = -\eta_2. $$

This proves Proposition 1.7. ∎

Our proof of Proposition 1.7 gives more: namely, any two-dimensional subspace in T_p, which is invariant under the transformations $\varphi_{\alpha\beta *}$, is transversal. Therefore, the property of transversal subspaces, outlined in Proposition 1.7, characterizes these subspaces, i.e. it can be taken as a definition of these subspaces.

Proposition 1.7 implies one more important consequence. As was outlined above, tangent spaces to any two-dimensional transversal surface of a web are transversal subspaces of this web. Therefore, two-dimensional transversal surfaces of the web (if they exist) are invariant with respect to the transformations $\varphi_{\alpha\beta}$.

With a triple of r-dimensional tangent subspaces T_α of the space T_p there is also associated a one-parameter family of r-dimensional subspaces defined by the vectors

$$ \zeta_i = \zeta^1 \underset{2}{e}_i - \zeta^2 \underset{1}{e}_i = \zeta^2 \underset{3}{e}_i - \zeta^3 \underset{2}{e}_i = \zeta^3 \underset{1}{e}_i - \zeta^1 \underset{3}{e}_i, $$

where $\zeta^1 + \zeta^2 + \zeta^3 = 0$. We will call these spaces *isoclinic spaces* relative to the space T_α and the r-vectors $Z = \zeta_1 \wedge \zeta_2 \wedge \ldots \zeta_r$ *isoclinic r-vectors* of the web W.

The transversal 2-space, defined by the bivector H, and the isoclinic r-space, defined by r-vector Z, have a common straight line [1] l with the directional vector

$$ \xi = \eta^i(\zeta^1 \underset{2}{e}_i - \zeta^2 \underset{1}{e}_i), \tag{1.21} $$

since $\xi = \zeta^1\eta_2 - \zeta^2\eta_1 = \eta^i\zeta_i$. The set of all such straight lines l, passing through a point p, forms a cone with the vertex at the point p. It follows from our considerations that this cone carries an $(r-1)$-parameter family of two-dimensional flat generators (transversal 2-planes) and a one-parameter family of r-dimensional flat generators (isoclinic r-planes). Because of this, we will denote it by $C(2,r)$. The equation (1.21) is a vector equation of the cone $C(2,r)$. This equation shows that the dimension of this cone is $r+1$.

[1] For convenience we use here the "point–vector" terminology.

Let us consider the vector space T_p. Its *projectivization* PT_p is a projective space, which is obtained from T_p by factorization relative to the collinearity of vectors. The projectivization PT_p is a set of straight lines, passing through the point p. Under projectivization, the cone $C(2,r)$ becomes a manifold $S(1,r-1)$ of dimension r. The latter carries an $(r-1)$-parameter family of rectilinear generators, corresponding to transversal generators of the cone $C(2,r)$, and a one-parameter family of $(r-1)$-dimensional flat generators, corresponding to isoclinic generators of this cone. The generators of the same family of the manifold $S(1,r-1)$ do not have common points, and the generators of different families intersect each other at a point. Hence, the manifold $S(1,r-1)$ is a realization of the product of the projective spaces P^1 and P^{r-1} in the projective space PT_p of dimension $2r-1$. Such manifolds are called the *Segre varieties* [HP 47]. This is the reason why the cone $C(2,r)$, constructed in the tangent space T_p to the manifold X, is called the *Segre cone*.

If $r=2$, the projectivization PT_p is a three-dimensional projective space, and the Segre manifold $S(1,1)$ is a ruled surface of second order. One family of its rectilinear generators corresponds to transversal bivectors of the three-web W, and the second family corresponds to its isoclinic bivectors. The Segre cone $C(2,2)$ is a cone of second order, and it carries two two-parameter families of two-dimensional flat generators.

1.4 Structure Equations of a Multidimensional Three-Web

1. Let us consider a three-web W in a domain U of a manifold X. According to Section 1.1, the foliations λ_α of this web can be given by three completely integrable systems of Pfaffian forms $\underset{\alpha}{\omega}^i$, $\alpha = 1,2,3$, that are normalized by the condition (1.16). Conditions of complete integrability of each of these systems are written in the form (1.10):

$$d\underset{\alpha}{\omega}^i = \underset{\alpha}{\omega}^j \wedge \underset{\alpha}{\omega}{}^i_j, \tag{1.22}$$

where the forms $\underset{\alpha}{\omega}{}^i_j$ depend on the differentials of the secondary parameters defining the bundle of frames in the domain U. These forms are not linearly independent. In fact, exterior differentiation of equations (1.16), connecting these forms, leads to

$$\underset{1}{\omega}^j \wedge \underset{1}{\omega}{}^i_j + \underset{2}{\omega}^j \wedge \underset{2}{\omega}{}^i_j + \underset{3}{\omega}^j \wedge \underset{3}{\omega}{}^i_j = 0. \tag{1.23}$$

Using equations (1.16), we exclude from equations (1.23), for example, the forms $\underset{3}{\omega}^j$. This leads to the equations:

$$\underset{1}{\omega}^j \wedge (\underset{1}{\omega}{}^i_j - \underset{3}{\omega}{}^i_j) + \underset{2}{\omega}^j \wedge (\underset{2}{\omega}{}^i_j - \underset{3}{\omega}{}^i_j) = 0.$$

It follows from these equations that the forms $\underset{1}{\omega}{}^i_j - \underset{3}{\omega}{}^i_j$ and $\underset{2}{\omega}{}^i_j - \underset{3}{\omega}{}^i_j$, and therefore their differences, are principal forms on the manifold X, i.e. they are linear combinations of the basis forms $\underset{1}{\omega}^i$ and $\underset{2}{\omega}^i$:

$$\underset{1}{\omega}{}^i_j - \underset{2}{\omega}{}^i_j = \underset{1}{a}{}^i_{jk}\underset{1}{\omega}^k - \underset{2}{a}{}^i_{jk}\underset{2}{\omega}^k.$$

We rewrite the last equations in the form:

$$\underset{1}{\omega}{}^i_j - \underset{1}{a}{}^i_{jk}\underset{1}{\omega}^k = \underset{2}{\omega}{}^i_j - \underset{2}{a}{}^i_{jk}\underset{2}{\omega}^k \stackrel{\text{def}}{=} \omega^i_j.$$

It follows that

$$\underset{1}{\omega}{}^i_j = \omega^i_j + \underset{1}{a}{}^i_{jk}\underset{1}{\omega}^k, \quad \underset{2}{\omega}{}^i_j = \omega^i_j + \underset{2}{a}{}^i_{jk}\underset{2}{\omega}^k. \tag{1.24}$$

As a result, equations (1.23) become

$$(\underset{1}{\omega}^i + \underset{2}{\omega}^i) \wedge (\underset{3}{\omega}{}^i_j - \omega^i_j) = \underset{1}{a}{}^i_{[jk]}\underset{1}{\omega}^j \wedge \underset{1}{\omega}^k + \underset{2}{a}{}^i_{[jk]}\underset{2}{\omega}^j \wedge \underset{2}{\omega}^k. \tag{1.25}$$

These relations must be identically satisfied in the whole manifold X, in particular, they are identically satisfied on the leaves of the third foliation, which are determined by the equations $\underset{3}{\omega}^i = 0$, or by (1.25), $\underset{1}{\omega}^i + \underset{2}{\omega}^i = 0$. Substituting $\underset{1}{\omega}^i + \underset{2}{\omega}^i = 0$ into equation (1.25), we obtain: $\underset{1}{a}{}^i_{[jk]} + \underset{2}{a}{}^i_{[jk]} = 0$. Denote

$$\underset{1}{a}{}^i_{[jk]} = -\underset{2}{a}{}^i_{[jk]} \stackrel{\text{def}}{=} a^i_{jk}.$$

and note that the quantities a^i_{jk} are skew-symmetric in the lower indices. Substitute expressions of $\underset{1}{\omega}^i$ and $\underset{2}{\omega}^i$ from (1.24) into (1.22). Using the notations introduced above, we can write two series of equations (1.22), which give the conditions of complete integrability of the forms $\underset{1}{\omega}^i$ and $\underset{2}{\omega}^i$, in the form:

$$\begin{aligned} d\underset{1}{\omega}^i &= \underset{1}{\omega}^j \wedge \omega^i_j + a^i_{jk}\underset{1}{\omega}^j \wedge \underset{1}{\omega}^k, \\ d\underset{2}{\omega}^i &= \underset{2}{\omega}^j \wedge \omega^i_j + a^i_{jk}\underset{2}{\omega}^j \wedge \underset{2}{\omega}^k. \end{aligned} \tag{1.26}$$

The integrability of the third series of forms $\underset{3}{\omega}^i$ also follows from the system (1.26), since addition of equations (1.26) gives

$$d\underset{3}{\omega}^i = \underset{3}{\omega}^j \wedge \omega^i_j + a^i_{jk}\underset{3}{\omega}^j \wedge (\underset{1}{\omega}^k - \underset{2}{\omega}^k). \tag{1.27}$$

The forms $\underset{1}{\omega}^i, \underset{2}{\omega}^i$ and ω^i_j in the equations (1.26) depend on the principal parameters (local coordinates on the manifold X) and the secondary parameters, defining a location of a moving frame $\{\underset{1}{e}_i, \underset{2}{e}_i\}$. Let us fix the principal parameters, defining a position of a point p from X. For this, we set $\underset{1}{\omega}^i = \underset{2}{\omega}^i = 0$. Then the forms ω^i_j become the forms

$$\pi^i_j \stackrel{\text{def}}{=} \omega^i_j\Big|_{\underset{1}{\omega}^i = \underset{2}{\omega}^i = 0},$$

which depend only on the secondary parameters. The geometric meaning of the forms π_j^i is that they define concordant infinitesimal transformations of the moving frames $\{\underset{\alpha}{e}_i\}$, $\alpha = 1, 2, 3$, in the space T_p:

$$\delta \underset{\alpha}{e}_i = \pi_i^j \underset{\alpha}{e}_j,$$

where δ, as was noted in Section 1.1, is the symbol of differentiation with respect to secondary parameters. It follows from this that the forms π_j^i are invariant forms of the group $\mathbf{GL}(r)$, the structure group of the G_W-structure, which was defined in Section 1.3. The forms π_j^i can be expressed in terms of the quantities A_j^i (see Section 1.2) defining transformations of the moving frames associated with the point p and their differentials (see, for example, [Li 55], n° 37).

2. Let us find differential prolongations of the equations (1.26). Their exterior differentiation gives

$$\begin{aligned} &\Omega_j^i \wedge \underset{1}{\omega}^j - \nabla a_{jk}^i \wedge \underset{1}{\omega}^j \wedge \underset{1}{\omega}^k - 2a_{jk}^m a_{ml}^i \underset{1}{\omega}^j \wedge \underset{1}{\omega}^k \wedge \underset{1}{\omega}^l = 0, \\ &\Omega_j^i \wedge \underset{2}{\omega}^j - \nabla a_{jk}^i \wedge \underset{2}{\omega}^j \wedge \underset{2}{\omega}^k - 2a_{jk}^m a_{ml}^i \underset{2}{\omega}^j \wedge \underset{2}{\omega}^k \wedge \underset{2}{\omega}^l = 0, \end{aligned} \tag{1.28}$$

where we use the notation

$$\Omega_j^i = d\omega_j^i - \omega_j^k \wedge \omega_k^i,$$

and ∇ is a differential operator defined by the formula:

$$\nabla a_{jk}^i = da_{jk}^i + a_{jk}^m \omega_m^i - a_{mk}^i \omega_j^m - a_{jm}^i \omega_k^m.$$

Let us now prove that the forms ∇a_{jk}^i are linear combinations of the forms $\underset{1}{\omega}^i$ and $\underset{2}{\omega}^i$, and the quadratic forms Ω_j^i are linear combinations of the exterior products of these forms. In fact, let us write the most general expression of the forms ∇a_{jk}^i and Ω_j^i:

$$\begin{aligned} \nabla a_{jk}^i &= \underset{1}{a}_{jkl}^i \underset{1}{\omega}^l + \underset{2}{a}_{jkl}^i \underset{2}{\omega}^l + \theta_{jk}^i, \\ \Omega_j^i &= \underset{1}{b}_{jkl}^i \underset{1}{\omega}^k \wedge \underset{1}{\omega}^l + b_{jkl}^i \underset{1}{\omega}^k \wedge \underset{2}{\omega}^l + \underset{2}{b}_{jkl}^i \underset{2}{\omega}^k \wedge \underset{2}{\omega}^l \\ &\quad + \underset{1}{\theta}_{jk}^i \wedge \underset{1}{\omega}^k + \underset{2}{\theta}_{jk}^i \wedge \underset{2}{\omega}^k + \Theta_j^i, \end{aligned} \tag{1.29}$$

where the linear forms $\theta_{jk}^i, \underset{1}{\theta}_{jk}^i$ and $\underset{2}{\theta}_{jk}^i$ are expressed in terms of some independent forms θ^α complementing the co-basis $\{\underset{1}{\omega}^i, \underset{2}{\omega}^i\}$, and the quadratic forms Θ_j^i contain only the products of the form $\theta^\alpha \wedge \theta^\beta$. The coefficients $\underset{1}{a}_{jkl}^i, \underset{2}{a}_{jkl}^i$ and the forms θ_{jk}^i in the expansions (1.29) are skew-symmetric in j and k, and the coefficients $\underset{1}{b}_{jkl}^i$ and $\underset{2}{b}_{jkl}^i$ are skew-symmetric in k and l. Substituting ∇a_{jk}^i and Ω_j^i from (1.29) into (1.28), we find:

$$
\begin{aligned}
(\underset{1}{b}{}^i_{jkl} - \underset{1}{a}{}^i_{jkl} - 2a^m_{jk}a^i_{ml})\underset{1}{\omega}{}^j \wedge \underset{1}{\omega}{}^k \wedge \underset{1}{\omega}{}^l - (b^i_{jkl} - \underset{2}{a}{}^i_{jkl})\underset{1}{\omega}{}^j \wedge \underset{1}{\omega}{}^k \wedge \underset{2}{\omega}{}^l + \underset{2}{b}{}^i_{jkl}\underset{1}{\omega}{}^j \wedge \underset{2}{\omega}{}^k \wedge \underset{2}{\omega}{}^l \\
-(\underset{1}{\theta}{}^i_{jk} + \theta^i_{jk})\underset{1}{\omega}{}^j \wedge \underset{1}{\omega}{}^k - \underset{2}{\theta}{}^i_{jk}\underset{1}{\omega}{}^j \wedge \underset{2}{\omega}{}^k + \theta^i_j \wedge \underset{1}{\omega}{}^j = 0, \\
(\underset{2}{b}{}^i_{jkl} + \underset{2}{a}{}^i_{jkl} - 2a^m_{jk}a^i_{ml})\underset{2}{\omega}{}^j \wedge \underset{2}{\omega}{}^k \wedge \underset{2}{\omega}{}^l - (b^i_{jlk} - \underset{1}{a}{}^i_{jkl})\underset{2}{\omega}{}^j \wedge \underset{1}{\omega}{}^k \wedge \underset{1}{\omega}{}^l + \underset{1}{b}{}^i_{jkl}\underset{2}{\omega}{}^j \wedge \underset{1}{\omega}{}^k \wedge \underset{1}{\omega}{}^l \\
-(\underset{2}{\theta}{}^i_{jk} - \theta^i_{jk})\underset{2}{\omega}{}^j \wedge \underset{2}{\omega}{}^k - \underset{1}{\theta}{}^i_{jk}\underset{2}{\omega}{}^j \wedge \underset{1}{\omega}{}^k + \theta^i_j \wedge \underset{2}{\omega}{}^j = 0.
\end{aligned}
$$

Since all summands in the last identities are linearly independent, then each of them must vanish. It follows from this that, first of all,

$$\underset{1}{\theta}{}^i_{jk} = \underset{2}{\theta}{}^i_{jk} = \theta^i_{jk} = 0, \quad \Theta^i_j = 0.$$

Substituting these values into the equation (1.29) completes our proof on the form of the expressions of the forms ∇a^i_{jk} and Ω^i_j. Moreover, if we set to zero the terms with $\underset{1}{\omega}{}^j \wedge \underset{1}{\omega}{}^k \wedge \underset{2}{\omega}{}^l$ and $\underset{1}{\omega}{}^j \wedge \underset{2}{\omega}{}^k \wedge \underset{2}{\omega}{}^l$, we get the equations

$$
\begin{aligned}
&\underset{1}{b}{}^i_{jkl} = 0, \quad \underset{2}{b}{}^i_{jkl} = 0, \\
&\underset{2}{a}{}^i_{jkl} = b^i_{[jk]l}, \quad \underset{1}{a}{}^i_{jkl} = b^i_{[j|l|k]}.
\end{aligned}
\tag{1.30}
$$

Finally, setting to zero the terms with $\underset{1}{\omega}{}^j \wedge \underset{1}{\omega}{}^k \wedge \underset{1}{\omega}{}^l$ and $\underset{2}{\omega}{}^j \wedge \underset{2}{\omega}{}^k \wedge \underset{2}{\omega}{}^l$, we get the equations

$$\underset{1}{a}{}^i_{[jkl]} + 2a^m_{[jk}a^i_{|m|l]} = 0, \quad \underset{2}{a}{}^i_{[jkl]} - 2a^m_{[jk}a^i_{|m|l]} = 0.$$

On the other hand, if we alternate the expressions for $\underset{1}{a}{}^i_{jkl}$ and $\underset{2}{a}{}^i_{jkl}$ from (1.30) with respect all lower indices, we get the equations

$$\underset{1}{a}{}^i_{[jkl]} = b^i_{[jlk]}, \quad \underset{2}{a}{}^i_{[jkl]} = b^i_{[jkl]}.$$

Eliminating from these and preceding equations the quantities $\underset{1}{a}{}^i_{jkl}$ and $\underset{2}{a}{}^i_{jkl}$, we obtain only one series of relations

$$b^i_{[jkl]} = 2a^m_{[jk}a^i_{|m|l]}. \tag{1.31}$$

As a result, the system (1.29), the first differential prolongation of the structure equations (1.26) of a three-web, has the form

$$d\omega^i_j - \omega^k_j \wedge \omega^i_k = b^i_{jkl}\underset{1}{\omega}{}^k \wedge \underset{2}{\omega}{}^l, \tag{1.32}$$

$$\nabla a^i_{jk} = b^i_{[j|l|k]}\underset{1}{\omega}{}^l + b^i_{[jk]l}\underset{2}{\omega}{}^l, \tag{1.33}$$

where the quantities a^i_{jk} and b^i_{jkl} are related by conditions (1.31).

Equations (1.26), (1.32) and (1.33) are the fundamental equations of a three-web $W(X, \lambda_\alpha)$. They contain all the information on the web structure. We will use these equations frequently in this book. Equations (1.26) and (1.32) are called the *structure equations* of the web W.

If we fix a point p in X, i.e. put $\underset{1}{\omega}^i = \underset{2}{\omega}^i = 0$, then equations (1.32) become

$$\delta\pi_j^i = \pi_j^k \wedge \pi_k^i,$$

where the forms π_j^i and the symbol δ have the same meaning as in Section 1.1. These equations are the structure equations of the group $G = \mathbf{GL}(r)$ of admissible transformations of frames in the tangent space T_p, which is the structure group of the G_W-structure, associated with a three-web W.

It follows from equations (1.33) that, if the principal parameters are fixed, the quantities a_{jk}^i satisfy the equations

$$\nabla_\delta a_{jk}^i \overset{\text{def}}{=} \delta a_{jk}^i - a_{mk}^i \pi_j^m - a_{jm}^i \pi_k^m + a_{jk}^m \pi_m^i = 0.$$

These equations show that the quantities a_{jk}^i form a tensor relative to the group $\mathbf{GL}(r)$ of admissible transformations of adapted frames of the web W. Considering differential prolongations of equations (1.32) (see equations (1.36) in the following subsection), one can prove that the quantities b_{jkl}^i also form a tensor. These tensors are called respectively the *torsion tensor* and the *curvature tensor* of a three-web W. We call them simply a and b.

3. We will prove two theorems showing that the torsion and curvature tensor fields uniquely define a three-web $W(X, \lambda_\alpha)$.

Theorem 1.8 *If two three-webs W and $\widetilde{W}$, given on manifolds X and $\widetilde{X}$ of the same dimension $2r$, are equivalent, then in corresponding frames their tensors a_{jk}^i and $\tilde{a}_{jk}^i$, b_{jkl}^i and $\tilde{b}_{jkl}^i$ coincide.*

Proof. The structure equations of the web W have the form (1.26), (1.32). We write the structure equations of the web $\widetilde{W}$ in a similar way:

$$\begin{aligned}
d\underset{1}{\tilde{\omega}}^i &= \underset{1}{\tilde{\omega}}^j \wedge \tilde{\omega}_j^i + \tilde{a}_{jk}^i \underset{1}{\tilde{\omega}}^j \wedge \underset{1}{\tilde{\omega}}^k, \\
d\underset{2}{\tilde{\omega}}^i &= \underset{2}{\tilde{\omega}}^j \wedge \tilde{\omega}_j^i - \tilde{a}_{jk}^i \underset{2}{\tilde{\omega}}^j \wedge \underset{2}{\tilde{\omega}}^k, \\
d\tilde{\omega}_j^i - \tilde{\omega}_j^k \wedge \tilde{\omega}_k^i &= \tilde{b}_{jkl}^i \underset{1}{\tilde{\omega}}^k \wedge \underset{2}{\tilde{\omega}}^l.
\end{aligned}$$

Let a mapping $\varphi : X \to \widetilde{X}$ transfer the three-web W into an equivalent three-web $\widetilde{W}$, i.e. it transfers each of the foliations λ_α, forming the web W, into corresponding foliation $\tilde{\lambda}_\alpha$, forming the web $\widetilde{W}$. Then the basis forms $\underset{\alpha}{\omega}^i$ and $\underset{\alpha}{\tilde{\omega}}^i$ of these foliations are connected by relations of the form:

$$\underset{\alpha}{\tilde{\omega}}^i = \underset{\alpha}{A}{}_j^i \underset{\alpha}{\omega}^j, \quad \alpha = 1, 2, 3.$$

But since the forms $\underset{\alpha}{\tilde{\omega}}^i$ as well as the forms $\underset{\alpha}{\omega}^i$ satisfy equations (1.16), $\underset{1}{A}^i_j = \underset{2}{A}^i_j = \underset{3}{A}^i_j$. It follows from this that, if one takes frames in the tangent spaces $T_p(X)$ and $T_p(\widetilde{X})$ accordingly, then the above equations can be reduced to

$$\underset{\alpha}{\tilde{\omega}}^i = \underset{\alpha}{\omega}^i. \tag{1.34}$$

Exterior differentiation of the equation (1.34) by means of the structure equations of the webs W and $\widetilde{W}$ gives

$$\underset{1}{\omega}^j \wedge (\tilde{\omega}^i_j - \omega^i_j) + (\tilde{a}^i_{jk} - a^i_{jk})\underset{1}{\omega}^j \wedge \underset{1}{\omega}^k = 0,$$
$$\underset{2}{\omega}^j \wedge (\tilde{\omega}^i_j - \omega^i_j) - (\tilde{a}^i_{jk} - a^i_{jk})\underset{2}{\omega}^j \wedge \underset{2}{\omega}^k = 0.$$

It follows from this that the forms $\tilde{\omega}^i_j$ and ω^i_j can be expressed in terms of the basis forms $\underset{1}{\omega}^j$ and $\underset{2}{\omega}^j$, i.e. we can set

$$\tilde{\omega}^i_j - \omega^i_j = \lambda^i_{jk}\underset{1}{\omega}^k + \mu^i_{jk}\underset{2}{\omega}^k.$$

Substituting these expansions into the previous equations, we find that $\lambda^i_{jk} = 0$, $\mu^i_{jk} = 0$, $\tilde{\omega}^i_j = \omega^i_j$, $\tilde{a}^i_{jk} = a^i_{jk}$. Now, comparing equations (1.32) written for the webs W and $\widetilde{W}$, we obtain $b^i_{jkl} = \tilde{b}^i_{jkl}$. ■

The converse is more essential. Denote the set of adapted frames, associated with a point p of the web W by R_p and the bundle of adapted frames by $\mathcal{R}(W)$ (cf. Section 1.3).

Theorem 1.9 *Suppose that two three-webs $W = (X, \lambda_\alpha)$ and $\widetilde{W} = (\widetilde{X}, \widetilde{\lambda}_\alpha)$ are given, and there exists a local diffeomorphism $\varphi : X \rightarrow \widetilde{X}$ and a smooth field of linear mappings $\Phi_p : R_p \rightarrow R_{\varphi(p)}$, $p \in X$, such that in the corresponding frames the values of the tensors a and b coincide:*

$$a^i_{jk} = \tilde{a}^i_{jk}(\varphi(p)), \quad b^i_{jkl} = \tilde{b}^i_{jkl}(\varphi(p)).$$

Then the three-webs W and $\widetilde{W}$ are equivalent.

Proof. Suppose that the structure equations of the three-webs W and $\widetilde{W}$ are written as in the proof of Theorem 1.8., and, according to the condition of Theorem 1.9, the frames are chosen in such a way that $a = \tilde{a}, b = \tilde{b}$. On the direct product $R(W) \times R(W)$ let us consider the system of equations

$$\underset{1}{\omega}^i = \underset{1}{\tilde{\omega}}^i, \quad \underset{2}{\omega}^i = \underset{2}{\tilde{\omega}}^i, \quad \omega^i_j = \tilde{\omega}^i_j. \tag{1.35}$$

By the structure equations of the three-webs W and $\widetilde{W}$, this system is completely integrable. Therefore, it uniquely defines a mapping $\Phi_p : R_p \rightarrow R_{\varphi(p)}$ provided that a pair of corresponding adapted frames is given in a pair of corresponding points p_0 and $\varphi(p_0)$. The first two equations of system (1.35) define a mapping $\varphi : X \to \widetilde{X}$, which transfers the leaves of the web W (their equations are $\underset{1}{\omega}^i = 0, \ \underset{2}{\omega}^i = 0, \ \underset{1}{\omega}^i + \underset{2}{\omega}^i = 0$) into the corresponding leaves of the web $\widetilde{W}$. Thus, the webs W and $\widetilde{W}$ are equivalent. ■

4. Let us find now the second prolongation of the structure equations (1.26). For this we apply exterior differentiation to equations (1.32) and (1.33). This gives:

$$\begin{aligned}&(\nabla b^i_{jkl} - b^i_{jpl} a^p_{km} \underset{1}{\omega}^m + b^i_{jkp} a^p_{lm} \underset{2}{\omega}^m) \wedge \underset{1}{\omega}^k \wedge \underset{2}{\omega}^l = 0, \\ &(\nabla b^i_{[j|l|k]} - b^i_{[j|p|k]} a^p_{lm} \underset{1}{\omega}^m) \wedge \underset{1}{\omega}^l + (\nabla b^i_{[jk]l} + b^i_{[jk]p} a^p_{lm} \underset{2}{\omega}^m) \wedge \underset{2}{\omega}^l = B^i_{jklm} \underset{1}{\omega}^l \wedge \underset{2}{\omega}^m,\end{aligned} \tag{1.36}$$

where

$$B^i_{jklm} = a^i_{pj} b^p_{klm} - a^i_{pk} b^p_{jlm} + a^p_{jk} b^i_{plm}. \tag{1.37}$$

It follows from equations (1.36) that the forms ∇b^i_{jkl} are principal forms on the manifold X, and their expansions relative to the basis forms can be written in the form:

$$\nabla b^i_{jkl} = \underset{1}{c}^i_{jklm} \underset{1}{\omega}^m + \underset{2}{c}^i_{jklm} \underset{2}{\omega}^m. \tag{1.38}$$

Substitution of these expansions into equations (1.36) gives the relationship between the quantities $\underset{1}{c} = (\underset{1}{c}^i_{jklm})$, $\underset{2}{c} = (\underset{2}{c}^i_{jklm})$ and a^i_{jk} and b^i_{jkl}:

$$\underset{1}{c}^i_{j[k|l|m]} = b^i_{jpl} a^p_{km}, \quad \underset{2}{c}^i_{jk[lm]} = -b^i_{jkp} a^p_{lm}, \tag{1.39}$$

$$\underset{1}{c}^i_{[jk]lm]} - \underset{2}{c}^i_{j|l|k]m} = B^i_{jklm}. \tag{1.40}$$

There is no other relations between $a, b, \underset{1}{c}$ and $\underset{2}{c}$. In fact, such relations could be obtained as result of differentiation of (1.31). However, using (1.39) and (1.40), it is easy to check that the differentiation of (1.31) leads to the identities.

One can show by prolongation of the equations (1.38) that the quantities $\underset{1}{c}$ and $\underset{2}{c}$ as well as all subsequent quantities, which will be obtained in the further prolongations, are tensors. We will call them, as well as the tensors a and b, the *fundamental tensors* of the three-web W. They are connected by rather complicated conditions, which were found in the paper [S 87a]. We will not give these conditions here.

Note that if three-webs W and $\widetilde{W}$ are equivalent, then in the corresponding frames not only their torsion and curvature tensors a and b coincide but also the tensors $\underset{1}{c}$ and $\underset{2}{c}$ and all other fundamental tensors coincide.

5. Let us find the form of the fundamental equations of a three-web on a two-dimensional manifold, i.e. for $r = 1$. In this case all the indices will be 1, and therefore the skew-symmetric tensor a^i_{jk} vanishes, since it has just one component a^1_{11}. Setting $b^1_{111} = b, \ \underset{1}{\omega}^1 = \omega_1, \ \underset{2}{\omega}^1 = \omega_2$ anf $\omega^1_1 = \omega$, we obtain from (1.26), (1.32) and (1.36) the following equations:

$$d\omega_1 = \omega_1 \wedge \omega, \quad d\omega_2 = \omega_2 \wedge \omega,$$
$$d\omega = b\omega_1 \wedge \omega_2, \quad \nabla b \wedge \omega_1 \wedge \omega_2 = 0,$$

where $\nabla b = db - 2b\omega$. These equations coincide with the structure equations of a curvilinear three-web, which were obtained in [Bl 55], and the only component of the curvature tensor is a relative invariant and coincides with the web curvature, introduced in the same book [Bl 55].

1.5 Parallelizable and Group Three-Webs

1. In this section we consider two special classes of three-webs that are characterized by vanishing of their fundamental tensors. We recall that in Section 1.1 we called a three-web parallelizable if it is equivalent to a parallel three-web.

Theorem 1.10 *A three-web W is parallelizable if and only if its torsion and curvature tensors are equal to zero.*

Proof. Let a three-web W be equivalent to a web W_0, formed by three foliations of parallel r-planes of an $(2r)$-dimensional affine space A^{2r}. With any point $p \in A^{2r}$ we associate a moving affine frame $\{\underset{1}{e}_i, \underset{2}{e}_i\}$ in such a way that its vectors $\underset{1}{e}_i$ belong to the plane of the first foliation, the vectors $\underset{2}{e}_i$ belong to the plane of the second foliation, and the vectors $\underset{1}{e}_i + \underset{2}{e}_i$ belong to the plane of the third foliation, all planes passing through a point p. Then we can put:

$$dp = \underset{1}{\omega}^i \underset{2}{e}_i - \underset{2}{\omega}^i \underset{1}{e}_i, \tag{1.41}$$

and the plane leaves, composing the three-web W_0, are given by the systems of equations:

$$\underset{1}{\omega}^i = 0, \quad \underset{2}{\omega}^i = 0, \quad \underset{1}{\omega}^i + \underset{2}{\omega}^i = 0.$$

Since the r-planes of the first foliation, as well as the r-planes of the second foliation, are parallel to each other, the following equations hold:

$$d\underset{1}{e}_i = \underset{1}{\omega}_i^j \underset{1}{e}_j, \quad d\underset{2}{e}_i = \underset{2}{\omega}_i^j \underset{2}{e}_j. \tag{1.42}$$

They imply

$$d(\underset{1}{e}_i + \underset{2}{e}_i) = \frac{1}{2}(\underset{1}{\omega}_i^j + \underset{2}{\omega}_i^j)(\underset{1}{e}_j + \underset{2}{e}_j) + \frac{1}{2}(\underset{1}{\omega}_i^j - \underset{2}{\omega}_i^j)(\underset{1}{e}_j - \underset{2}{e}_j).$$

The planes of the third foliation are also parallel to each other and the vectors $\underset{1}{e}_j + \underset{2}{e}_j$ and $\underset{1}{e}_j - \underset{2}{e}_j$ are linearly independent. Because of this, the last equations imply

$$\underset{1}{\omega}_i^j = \underset{2}{\omega}_i^j \stackrel{\text{def}}{=} \omega_i^j.$$

This allow us to rewrite equations (1.42) in the form

$$d\underset{1}{e_i} = \omega_i^j \underset{1}{e_j}, \quad d\underset{2}{e_i} = \omega_i^j \underset{2}{e_j}. \tag{1.43}$$

Exterior differentiation of equations (1.41) and (1.43) leads to the following structure equations of a parallelizable web:

$$d\underset{1}{\omega}^i = \underset{1}{\omega}^j \wedge \omega_j^i, \quad d\underset{2}{\omega}^i = \underset{2}{\omega}^j \wedge \omega_j^i, \quad d\omega_i^j = \omega_i^k \wedge \omega_k^j. \tag{1.44}$$

Comparing these equations with equations (1.26) and (1.32), we see that for the web W_0 under consideration we have: $a^i_{jk} = 0, \;\; b^i_{jkl} = 0$.

Conversely, if the torsion and curvature tensors of a web W are equal to zero, then the structure equations (1.26) and (1.32) of this web have the form (1.44), i.e. they are the structure equations of a $(2r)$-dimensional affine space A^{2r}, where three families of parallel r-planes $\underset{1}{\omega}^i = 0$, $\underset{2}{\omega}^i = 0$, $\underset{1}{\omega}^i + \underset{2}{\omega}^i = 0$ are fixed. Therefore, the web W is parallelizable. ■

2. Let us find now structure equations of a group three-web, defined on the direct product $X = G \times G$, where $G(\cdot)$ is an r-dimensional Lie group with identity element e (see Example 1.5 in Section 1.2). Let $G_2 = (e, G) \subset X$ and $G_1 = (G, e) \subset X$ and let the operation in G_1 be the same as in G, and in G_2 we introduce the conjugate operation $(\circ)$ by the formula $x \circ y = y \cdot x$. Then the direct product $X = G \times G$ becomes a group with the operation "$\times$" defined by

$$(x, y) \times (u, v) = (x \cdot u, \; y \circ v) = (x \cdot u, \; v \cdot y).$$

Denote the co-bases of invariant forms in G_1 and G_2 by $\{\underset{1}{\omega}^i, 0\}$ and $\{0, \underset{2}{\omega}^i\}$, respectively. The natural mapping of the group G onto the groups G_1 and G_2 allows us to consider the forms $\underset{1}{\omega}^i$ and $\underset{2}{\omega}^i$ being also defined on the group G.

Consider further a leaf F_3 of the third foliation, passing through the point (e, e). According to Example 1.5, it is formed by the points (x, y) from X, for which the condition $x \cdot y = e$ holds. It follows from the definition of the operation $(\circ)$ that

$$(x \cdot y)^{-1} = y^{-1} \cdot x^{-1} = x^{-1} \circ y^{-1}. \tag{1.45}$$

This implies, first of all, that the leaf F_3 is a subgroup in X, since $(x, x^{-1}) \times (y, y^{-1}) = (x \cdot y, (y \circ x)^{-1}) = (x \cdot y, (x \cdot y)^{-1})$. Denote this subgroup by G_3.

Second, note that a leaf F_3 establishes a bijective mapping $\theta : G_1(\cdot) \to G_2(\circ)$, under which $\theta(x, e) = (e, x^{-1})$. It follows from (1.45) that the correspondence θ is an isomorphism. On the other hand, according to (1.45), this mapping is an anti-automorphism in the group G. Therefore, the invariant bases $\underset{1}{\omega}^i$ and $\underset{2}{\omega}^i$ (considered in G) can be coordinated in such a way that $\theta^*(\underset{2}{\omega}^i) = -\underset{1}{\omega}^i$, where θ^* is the anti-dragging, corresponding to the mapping θ. As a result, the structure tensors $\underset{1}{c}^i_{jk}$ and

$\underset{2}{c}{}^i_{jk}$ of the groups G_1 and G_2 are connected by relation $\underset{1}{c}{}^i_{jk} = -\underset{2}{c}{}^i_{jk}$. This implies that the structure equations (the Cartan–Maurer equations) of these groups are written in the form:

$$d\underset{1}{\omega}{}^i = \underset{1}{c}{}^i_{jk}\underset{1}{\omega}{}^j \wedge \underset{1}{\omega}{}^k, \quad d\underset{2}{\omega}{}^i = \underset{2}{c}{}^i_{jk}\underset{2}{\omega}{}^j \wedge \underset{2}{\omega}{}^k. \tag{1.46}$$

The system of forms $\{\underset{1}{\omega}{}^i, \underset{2}{\omega}{}^i\}$ is linearly independent and form an invariant co-basis in the direct product $X = G_1 \times G_2$. The equations (1.46) are structure equations of the group $X(\times)$.

One can see from equations (1.46) that the system of Pfaffian equations $\underset{1}{\omega}{}^i = 0$ is completely integrable in X and, therefore, defines a foliation λ_1 of codimension r on X. The leaf of this foliation, passing through the point (e, e), is the subgroup G_1, and other leaves of λ_1 are co-sets with respect to the subgroup G_1. Similarly, the completely integrable system of Pfaffian equations $\underset{2}{\omega}{}^i = 0$ defines a foliation λ_2 of codimension r on X, one of the leaves of this foliation is the subgroup G_2, and other leaves of λ_2 are co-sets with respect to the subgroup G_2.

Equations (1.46) imply the equations

$$d(\underset{1}{\omega}{}^i + \underset{2}{\omega}{}^i) = c^i_{jk}(\underset{1}{\omega}{}^j + \underset{2}{\omega}{}^j) \wedge (\underset{1}{\omega}{}^k - \underset{2}{\omega}{}^k),$$

which prove that the system

$$\underset{1}{\omega}{}^i + \underset{2}{\omega}{}^i = 0 \tag{1.47}$$

is completely integrable. Since equations (1.47) contain linear combinations (with constant coefficients) of invariant forms of the group $X = G_1 \times G_2$, then, according to a known fact of Lie group theory, the system (1.47) defines a subgroup in the group $X(\times)$. The corresponding foliation (denote it by λ_3) is formed by co-sets with respect to this subgroup. The subgroup, defined by the equations (1.47), is exactly the subgroup G_3, defined above on the leaf F_3. In fact, by (1.47), the first series of the equations (1.46) (the structure equations of the group G_2) is transfered to the second series of the equations (1.46) (the structure equations of the group G_1). Therefore, the relations (1.47) give the mapping θ^*, mentioned above, i.e. they single out the group G_3.

Our considerations imply that the foliations λ_α, $\alpha = 1, 2, 3$, form a three-web on the manifold X, and equations (1.46) are the structure equations of this web. Comparing them with equations (1.26), we find that $\omega^i_j = 0$, $a^i_{jk} = c^i_{jk}$. Substituting $\omega^i_j = 0$ into (1.32), we obtain $b^i_{jkl} = 0$.

Let us prove now the converse: if the curvature tensor of a three-web W vanishes, then this web is a group web. In fact, if $b^i_{jkl} = 0$, then equations (1.32) and (1.33) become

$$d\omega^j_i = \omega^k_i \wedge \omega^j_k, \tag{1.48}$$

$$\nabla a^i_{jk} = 0. \tag{1.49}$$

The equations (1.48) show that an admissible transformation of basis forms:

$$\underset{1}{\tilde{\omega}}^i = A^i_j \underset{1}{\omega}^j, \quad \underset{2}{\tilde{\omega}}^i = A^i_j \underset{2}{\omega}^j \tag{1.50}$$

with variable coefficients A^i_j can make all the forms ω^i_j vanish in the whole manifold X. In fact, exterior differentiation of equations (1.50) gives the quadratic equations

$$d\underset{1}{\tilde{\omega}}^i = \underset{1}{\tilde{\omega}}^k \wedge \tilde{\omega}^i_k + \tilde{a}^i_{jk} \underset{1}{\tilde{\omega}}^j \wedge \underset{1}{\tilde{\omega}}^k, \quad d\underset{2}{\tilde{\omega}}^i = \underset{2}{\tilde{\omega}}^k \wedge \tilde{\omega}^i_k - \tilde{a}^i_{jk} \underset{2}{\tilde{\omega}}^j \wedge \underset{2}{\tilde{\omega}}^k,$$

where we use the notation

$$\tilde{\omega}^i_k = \tilde{A}^j_k (A^i_l \omega^l_j - dA^i_j), \quad \tilde{a}^i_{jk} = A^i_l \tilde{A}^p_j \tilde{A}^q_k a^l_{pq},$$

and where $(\tilde{A}^i_j)$ is the matrix inverse of the matrix (A^i_j). Consider the equations $\tilde{\omega}^i_k = 0$, which are equivalent to the relations

$$dA^i_j - A^i_l \omega^l_j = 0.$$

Using equations (1.48), one can easily check that this system is completely integrable. Thus, it is possible to find quantities A^i_j such that in a new co-basis $\{\underset{1}{\tilde{\omega}}^i, \underset{2}{\tilde{\omega}}^i\}$, defined by formulas (1.50), the forms $\tilde{\omega}^i_j$ are equal to zero. Then equations (1.49) give $da^i_{jk} = 0$, i.e. the components of the torsion tensor become constants. After this, equations (1.26) coincide with the equations (1.46), and this means that the three-web under consideration is a group three-web. We thus have proved the following theorem:

Theorem 1.11 *A three-web W is a group three-web if and only if its curvature tensor is equal to zero.*

■

Note that for a group three-web conditions (1.31) have the form

$$a^m_{jk} a^i_{ml} + a^m_{lj} a^i_{mk} + a^m_{kl} a^i_{mj} = 0.$$

These relations coincide with the Jacobi identities for structure constants of a Lie group (see Section 1.1). Because of this, we will call relations (1.31), which are valid for any three-web, the *generalized Jacobi identities.*

Theorem 1.11 has the following consequence: *a three-web W is parallelizable if and only if it is a group three-web, generated by a commutative group G.*

1.6 Computation of the Torsion and Curvature Tensors of a Three-Web

In this section we show how to find the structure equations of a three-web W and calculate its torsion and curvature tensors, if in a domain U this web is given by a closed form equations:

$$z^i = f^i(x^j, y^k)$$

In addition, we assume that the functions f^i are of class $C^s, s \geq 3$. Recall that everywhere in the domain U conditions (1.11) are satisfied:

$$\det\left(\frac{\partial f^i}{\partial x^j}\right) \neq 0, \quad \det\left(\frac{\partial f^i}{\partial y^k}\right) \neq 0,$$

and the foliations λ_α, forming a three-web, are defined by the equations $\lambda_1 : x^i = c_1^i$, $\lambda_2 : y^i = c_2^i$, $\lambda_3 : f^i(x^j, y^k) = c_3^i$, where c_α^i are constants.

Define

$$\bar{f}_j^i = \frac{\partial f^i}{\partial x^j}, \quad \tilde{f}_j^i = \frac{\partial f^i}{\partial y^j}.$$

By (1.11), the matrices $(\bar{f}_j^i)$ and $(\tilde{f}_j^i)$ are invertible. Denote their inverse matrices by $(\bar{g}_j^i)$ and $(\tilde{g}_j^i)$ respectively. Differentiation of equations (1.12) leads to

$$dz^i = \bar{f}_j^i dx^j + \tilde{f}_j^i dy^j.$$

On the web under consideration these relations must coincide with equations (1.16). Because of this, we denote:

$$\underset{1}{\omega}^i = \bar{f}_j^i dx^j, \quad \underset{2}{\omega}^i = \tilde{f}_j^i dy^j, \quad \underset{3}{\omega}^i = -dz^i. \tag{1.51}$$

It follows from (1.51) that

$$dx^i = \bar{g}_j^i \underset{1}{\omega}^j, \quad dy^i = \tilde{g}_j^i \underset{2}{\omega}^j. \tag{1.52}$$

Applying exterior differentiation to (1.51) and using (1.52), we get

$$d\underset{1}{\omega}^i = -d\underset{2}{\omega}^i = \Gamma_{jk}^i \underset{1}{\omega}^j \wedge \underset{2}{\omega}^k, \tag{1.53}$$

where

$$\Gamma_{jk}^i = -\frac{\partial^2 f^i}{\partial x^l \partial y^m} \bar{g}_j^l \tilde{g}_k^m. \tag{1.54}$$

Comparing now equations (1.53) and (1.26), we easily get

$$\underset{1}{\omega}^j \wedge (\omega_j^i + a_{jk}^i \underset{1}{\omega}^k - \Gamma_{jk}^i \underset{2}{\omega}^k) = 0,$$
$$\underset{2}{\omega}^j \wedge (\omega_j^i - a_{jk}^i \underset{2}{\omega}^k - \Gamma_{kj}^i \underset{1}{\omega}^k) = 0.$$

Applying Cartan's lemma, we find that

$$\begin{aligned} \omega_j^i + a_{jk}^i \underset{1}{\omega}^k - \Gamma_{jk}^i \underset{2}{\omega}^k &= \underset{1}{\lambda}_{jk}^i \underset{1}{\omega}^k, \\ \omega_j^i - a_{jk}^i \underset{2}{\omega}^k - \Gamma_{kj}^i \underset{1}{\omega}^k &= \underset{2}{\lambda}_{jk}^i \underset{2}{\omega}^k, \end{aligned} \tag{1.55}$$

where the quantities $\underset{1}{\lambda}{}^i_{jk}$ and $\underset{2}{\lambda}{}^i_{jk}$ are symmetric relative to lower indices. If we exclude ω^i_j from these equations and equate the coefficients in the corresponding basis forms in both members of the obtained equations, we get

$$a^i_{jk} + \Gamma_{kj} = \underset{1}{\lambda}{}^i_{jk}, \quad a^i_{jk} - \Gamma^i_{jk} = -\underset{2}{\lambda}{}^i_{jk}.$$

Alternation of these relations in j and k gives

$$a^i_{jk} = \Gamma^i_{[jk]}. \tag{1.56}$$

If we symmetrize the same relations in j and k, we get $\underset{1}{\lambda}{}^i_{jk} = \Gamma^i_{(kj)}$, $\underset{2}{\lambda}{}^i_{jk} = \Gamma^i_{(jk)}$. Thus, equations (1.55) can be written in the form

$$\omega^i_j = \Gamma^i_{kj}\underset{1}{\omega}{}^k + \Gamma^i_{jk}\underset{2}{\omega}{}^k. \tag{1.57}$$

Equations (1.54) and (1.56) show that the torsion tensor a^i_{jk} of a three-web W and the forms ω^i_j can be expressed in terms of second order derivatives of the functions f^i defining the web.

In order to find an expression of the curvature tensor b^i_{jkl}, we must differentiate the equations (1.57). Before differentiating them, we substitute into them the expressions for Γ^i_{jk} and $\underset{1}{\omega}{}^k$, $\underset{2}{\omega}{}^k$ from (1.52) and (1.54). This gives

$$\omega^i_j = -\frac{\partial^2 f^i}{\partial x^l \partial y^k}\tilde{g}^k_j dx^l - \frac{\partial^2 f^i}{\partial x^k \partial y^l}\bar{g}^k_j dy^l.$$

Later we will need the quantities $d\bar{g}^k_j$ and $d\tilde{g}^k_j$. We can find them by differentiating the equations $\bar{f}^k_i \bar{g}^j_k = \delta^j_i$ and $\tilde{f}^k_i \tilde{g}^j_k = \delta^j_i$. After simple calculations we obtain:

$$d\bar{g}^k_j = -\bar{g}^p_j \bar{g}^k_l \left(\frac{\partial^2 f^l}{\partial x^p \partial x^m} dx^m + \frac{\partial^2 f^l}{\partial x^p \partial y^m} dy^m\right),$$
$$d\tilde{g}^k_j = -\tilde{g}^p_j \tilde{g}^k_l \left(\frac{\partial^2 f^l}{\partial y^p \partial x^m} dx^m + \frac{\partial^2 f^l}{\partial y^p \partial y^m} dy^m\right).$$

Now exterior differentiation of the forms ω^i_j by means of the last two formulas gives:

$$\begin{aligned} d\omega^i_j = &\left(\frac{\partial^3 f^i}{\partial x^l \partial y^k \partial y^m}\tilde{g}^k_j - \frac{\partial^3 f^i}{\partial x^k \partial x^l \partial y^m}\bar{g}^k_j\right) dx^l \wedge dy^m \\ &+ \frac{\partial^2 f^i}{\partial x^l \partial y^k}\tilde{g}^p_j \tilde{g}^k_q \left(\frac{\partial^2 f^q}{\partial y^p \partial x^m} dx^m + \frac{\partial^2 f^q}{\partial y^p \partial y^m} dy^m\right) \wedge dx^l \\ &+ \frac{\partial^2 f^i}{\partial x^k \partial y^l}\bar{g}^p_j \bar{g}^k_q \left(\frac{\partial^2 f^q}{\partial x^p \partial x^m} dx^m + \frac{\partial^2 f^q}{\partial x^p \partial y^m} dy^m\right) \wedge dy^l. \end{aligned}$$

Using (1.52) and (1.54), we write this expression in the form:

$$
\begin{aligned}
d\omega_j^i = \Big(&\frac{\partial^3 f^i}{\partial x^m \partial y^n \partial y^p}\tilde{g}_j^m \bar{g}_k^n \tilde{g}_l^p - \frac{\partial^3 f^i}{\partial x^m \partial x^n \partial y^p}\bar{g}_j^m \bar{g}_k^n \tilde{g}_l^p \\
&+\Gamma_{kp}^i \frac{\partial^2 f^p}{\partial y^m \partial y^n}\tilde{g}_j^m \tilde{g}_l^n - \Gamma_{pl}^i \frac{\partial^2 f^p}{\partial x^m \partial x^n}\bar{g}_j^m \bar{g}_k^n\Big)\underset{1}{\omega}^k \wedge \underset{2}{\omega}^l \\
&+\Gamma_{lm}^i \Gamma_{kj}^m \underset{1}{\omega}^k \wedge \underset{1}{\omega}^l + \Gamma_{ml}^i \Gamma_{jk}^m \underset{2}{\omega}^k \wedge \underset{2}{\omega}^l
\end{aligned}
$$

In addition, using equation (1.57), we find

$$
\begin{aligned}
\omega_j^m \wedge \omega_m^i = \Gamma_{kj}^m \Gamma_{lm}^i \underset{1}{\omega}^k \wedge \underset{1}{\omega}^l + \Gamma_{kj}^m \Gamma_{ml}^i \underset{1}{\omega}^k \wedge \underset{2}{\omega}^l \\
+\Gamma_{jk}^m \Gamma_{lm}^i \underset{2}{\omega}^k \wedge \underset{1}{\omega}^l + \Gamma_{jk}^m \Gamma_{ml}^i \underset{2}{\omega}^k \wedge \underset{2}{\omega}^l.
\end{aligned}
$$

If we subtract the last equation from the previous one and compare the obtained difference with equations (1.32), we find an expression for the tensor b_{jkl}^i:

$$
\begin{aligned}
b_{jkl}^i = &\Big(\frac{\partial^3 f^i}{\partial x^m \partial y^n \partial y^p}\tilde{g}_j^n - \frac{\partial^3 f^i}{\partial x^m \partial x^n \partial y^p}\bar{g}_j^n\Big)\bar{g}_k^m \tilde{g}_l^p \\
&+\Gamma_{kp}^i \frac{\partial^2 f^p}{\partial y^m \partial y^n}\tilde{g}_j^m \tilde{g}_l^n - \Gamma_{pl}^i \frac{\partial^2 f^p}{\partial x^m \partial x^n}\bar{g}_j^m \bar{g}_k^n + \Gamma_{jl}^m \Gamma_{km}^i - \Gamma_{kj}^m \Gamma_{ml}^i.
\end{aligned}
\tag{1.58}
$$

Thus, the curvature tensor of a web $W = (X, \lambda_\alpha)$ is expressed in terms of the third order derivatives of the functions f^i, defining this web.

Alternating the equations (1.58) in lower indices, we obtain:

$$
b_{[jkl]}^i = \Gamma_{[jl}^m \Gamma_{k]m}^i - \Gamma_{[kj}^m \Gamma_{|m|l]}^i.
$$

Using equations (1.56), one sees that the last equations are equivalent to equations (1.31), connecting the torsion and curvature tensors of a web W.

For a curvilinear three-web, given by an equation $z = f(x, y)$, formula (1.58) has the form:

$$
b = \frac{1}{f_x f_y}\Big(\frac{f_{xyy}}{f_y} - \frac{f_{xxy}}{f_x} + \frac{f_{xx} f_{xy}}{f_x^2} - \frac{f_{xy} f_{yy}}{f_y^2}\Big).
$$

1.7 The Canonical Chern Connection on a Three-Web

A three-web $W = (X, \lambda_\alpha)$ generates invariant affine onnections in the manifold X. As we will see later, the presence of these connections allows to study effectively the differential geometry of three-webs of general type and special classes of webs.

1. Let us recall the definition of an affine connection on a smooth manifold X (see, for example, [KN 63], p. 63, or [A 77b], pp. 62–67).

Let $R(X)$ be a bundle of frames over X, dim $X = n$ and let ω^u be Pfaffian forms in this bundle, defined in Section 1.1 and satisfying equations (1.2). The forms $\omega_v^u (u, v, w = 1, 2, \ldots, n)$ in the equations (1.2) are invariantly defined in the second order bundle of frames $R^2(X)$, which can be constructed in a way the bundle of frames $R(X)$ was constructed. An *affine connection* Γ on a manifold X is given in the bundle $R^2(X)$ by means of an invariant horizontal distribution Δ. The latter is defined by a system of Pfaffian forms

$$\theta_v^u = \omega_v^u - \Gamma_{vw}^u \omega^w, \tag{1.59}$$

vanishing on Δ. The distribution Δ is invariant under the group of affine transformations acting in $R(X)$. Its orbits are fibers of $R(X)$.

Using equations (1.59), we exclude the forms ω_v^u from equations (1.2). As a result, we obtain

$$d\omega^u = \omega^v \wedge \theta_v^u + R_{vw}^u \omega^v \wedge \omega^w, \tag{1.60}$$

where $R_{vw}^u = \Gamma_{[vw]}^u$. The conditions for the distribution Δ to be invariant lead to the equations

$$d\theta_v^u = \theta_v^w \wedge \theta_w^u + R_{vwz}^u \omega^w \wedge \omega^z. \tag{1.61}$$

The Pfaffian form $\theta = (\theta_v^u)$ with its values in the Lie algebra $\mathbf{gl}(n)$ of the group $\mathbf{GL}(n)$ is called the *connection form* of the connection Γ.

The quantities R_{vw}^u and R_{vwz}^u form tensors and are called the *torsion* and *curvature tensors* of the connection Γ respectively.

Conversely, if in the bundle of frames $R^2(X)$ the forms θ_v^u are given, and these forms with the forms ω^u satisfy equations (1.60) and (1.61), then they define an affine connection Γ in X, and its torsion and curvature tensors are R_{vw}^u and R_{vwz}^u (Cartan–Laptev's theorem, see [A 77b] or [ELOS 79]).

An affine connection is a G-structure in the bundle $R^2(X)$ of frames of second order.

A vector field $\xi = (\xi^u)$ is *parallel in a connection* Γ on a submanifold $\widetilde{X} \subset X$ if its coordinates ξ^u satisfy the differential equations

$$d\xi^u + \xi^v \theta_v^u = 0, \tag{1.62}$$

which hold on $\widetilde{X}$. If $\widetilde{X}$ is a curve $\gamma(t)$, then the system (1.62) becomes a system of ordinary differential equations and has a unique solution $\xi(t)$ if initial conditions $\xi_0 = \xi(0)$ are given. In this case, we will say that the vector $\xi(t)$ is obtained from the vector ξ_0 by a *parallel displacement* along the curve $\gamma(t)$.

The geodesic lines of the connection Γ are defined by the equations

$$d\xi^u + \xi^v \theta_v^u = \theta \xi^u, \tag{1.63}$$

where (ξ^u) is the tangent vector to a geodesic line and θ is a Pfaffian form.

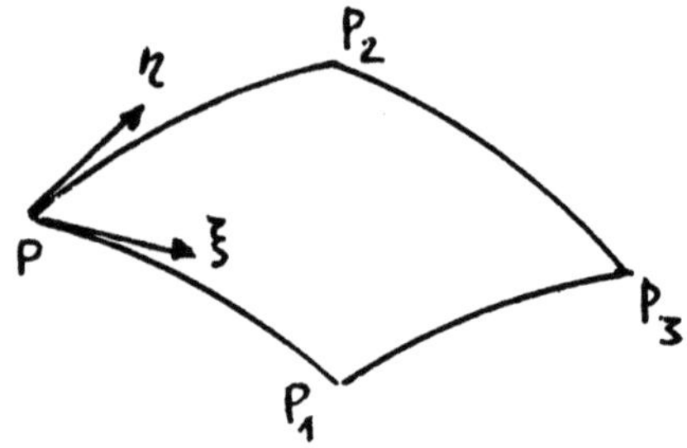

Figure 1.2

The exterior quadratic forms

$$\Omega^u = R^u_{vw}\omega^v \wedge \omega^w, \quad \Omega^u_v = R^u_{vwz}\omega^w \wedge \omega^z$$

in equations (1.60) and (1.61) are called the *torsion* and *curvature forms* of the connection Γ and have an important geometric meaning. Consider the development of a space with an affine connection Γ onto an affine space A, i.e. a family of frames $\{e_u\}$ in A, which is obtained by integration of the system of differential equations

$$dp = \omega^u e_u, \quad de_u = \theta^u_v e_v$$

along a curve $\gamma(t)$. Let V^2 be a two-dimensional surface in X, and let ξ and η be two vector fields on it. Consider on V^2 a small curvilinear parallelogram $l = pp_1p_3p_2p$ (Figure 1.2), formed by integral curves of the fields ξ and η, and suppose that the lengths pp_1 and pp_2 of this parallelogram are equal to dt and ds, respectively. The development of a closed line l in the affine space A is a line $\tilde{l}$ (not necessarily closed), and up to the infinitesimals of third order, the distance between the end points of $\tilde{l}$ is expressed by the equation:

$$\Delta p = 2R^u_{vw}\xi^v\eta^w dtdse_u. \tag{1.64}$$

In a similar way, the forms Ω^u_v allow us, up to the infinitesimals of third order, to estimate the difference Δe_u between the basis vectors e_u of the tangent space $T_p(X)$ and the vectors, which are obtained from them under a parallel displacement along a closed path l on V^2:

$$\Delta e_u = R^u_{vwz}\xi^w\eta^z dtdse_v. \tag{1.65}$$

In addition, note that if on the manifold X a G-structure with a structure group G is given, then, as we indicated in Section 1.1, the forms ω_v^u have the form (1.8). If these forms and the forms ω^u satisfy equations (1.60) and (1.61), then in X they define a connection, which is called a *G-connection.* For example, if $G = O(n)$ is the orthogonal group, then $\omega_v^u = \theta_v^u$, where $\theta_v^u = -\theta_u^v$ are skew-symmetric forms. A connection, defined by these forms, is called the *Riemannian connection.*

2. Suppose that a three-web W is given on a manifold X of dimension $2r$. Consider its structure equations (1.26) and (1.32). Comparing them with equations (1.60) and (1.61), we see that the forms

$$\omega^u = (\underset{1}{\omega^i}, \underset{2}{\omega^i}), \text{ and } \theta_v^u = \begin{pmatrix} \omega_j^i & 0 \\ 0 & \omega_j^i \end{pmatrix} \tag{1.66}$$

define an affine connection on the G_W-structure associated with a three-web W. This connection is called the *Chern connection.* We will denote it by Γ.

Equations (1.33) and (1.38) show that the differential operator ∇, which was introduced in Section 1.4, is the operator of covariant differentiation with respect to the Chern connection.

Equations (1.26) and (1.32) show that the torsion and curvature tensors of the connection Γ are:

$$\begin{aligned} R_{vw}^u &= \left\{ \begin{pmatrix} a_{jk}^i & 0 \\ 0 & 0 \end{pmatrix}, \begin{pmatrix} 0 & 0 \\ 0 & -a_{jk}^i \end{pmatrix} \right\}, \\ R_{vwz}^u &= \left\{ \begin{matrix} \begin{pmatrix} 0 & \frac{1}{2}b_{jkl}^i \\ -\frac{1}{2}b_{jkl}^i & 0 \end{pmatrix} & \begin{pmatrix} 0 & 0 \\ 0 & 0 \end{pmatrix} \\ \begin{pmatrix} 0 & 0 \\ 0 & 0 \end{pmatrix} & \begin{pmatrix} 0 & \frac{1}{2}b_{jkl}^i \\ -\frac{1}{2}b_{jkl}^i & 0 \end{pmatrix} \end{matrix} \right\}. \end{aligned} \tag{1.67}$$

Let us find geometric properties of the Chern connection.

A submanifold Y of a space X with an affine connection is said to be *autoparallel* [KN 63] (vol. 2, p. 53) if, for each vector ξ tangent to Y and for each curve l in Y, the parallel displacement of the vector ξ along l yields a vector tangent to Y.

Let Y_1 and Y_2 be two submanifolds in X, $p_1 \in Y_1, p_2 \in Y_2$ be two arbitrary points in these spaces and l be a curve joining these points. The submanifolds Y_1 and Y_2 are called *parallel* in X if any vector $\xi_1 \in T_{p_1}(Y_1)$ will be transferred to a vector $\xi_2 \in T_{p_2}(Y_2)$ by a parallel displacement along l. It is obvious that if two submanifolds are parallel, then each of them is autoparallel.

Proposition 1.12 *Each leaf of a three-web $W = (X, \lambda_\alpha)$ is an autoparallel submanifold in X relative to the Chern connection. Any two leaves of one foliation are parallel.*

Proof. To prove this proposition, we write the equations of parallel displacement of an arbitrary vector $\xi = \xi_1^i \underset{2}{e_i} - \xi_2^i \underset{1}{e_i}$, $\xi_\alpha^i = \underset{\alpha}{\omega^i}(\xi)$, in the connection Γ. Since the forms of the Chern connection have the special form (1.66), equations (1.62) of parallel displacement in this connection are written in the form:

$$d\xi_1^i + \xi_1^j \omega_j^i = 0, \quad d\xi_2^i + \xi_2^j \omega_j^i = 0. \tag{1.68}$$

Let there be at an initial point p_0 a vector ξ_0, tangent to the leaf $\mathcal{F}_1$, passing through this point. Then $\xi_{10}^i = 0$. Since the first series of equations (1.68) is homogeneous with respect to ξ_1^i, after integration of this system along a path $p(t)$ $(p(0) = p_0, 0 \leq t \leq 1)$, we obtain $(\xi_1^i)\big|_{t=1} = 0$. This means that a parallel displacement of a vector $\xi_0 \in \lambda_1$ along any path gives a vector from the foliation λ_1. Since the vector ξ_0 and the curve $p(t)$ were arbitrarily chosen, the leaves of the foliation λ_1, passing through the points $p(0)$ and $p(1)$, are parallel, and subsequently each of them is autoparallel. Using the same method, one easily prove the theorem for the leaves of the foliation λ_2.

Since the equations (1.59) imply that

$$d(\xi_1^i + \xi_2^i) + (\xi_1^j + \xi_2^j)\omega_j^i = 0,$$

and a vector ξ tangent to a leaf F_3 of the third foliation is defined by the condition $\xi_1^i + \xi_2^i = 0$, then the previous statement also valid for the third foliation λ_3. ■

In particular, Proposition 1.12 implies that the r-plane T_α is autoparallel along the leaf F_α. This means that any geodesic line, which is tangent to a web leaf, entirely belongs to this leaf. In other words, the following proposition holds:

Proposition 1.13 *The leaves of a web W are totally geodesic submanifolds of the manifold X relative to the Chern connection.*

■

The Chern connection induces affine connections on the leaves of a three-web W. On the leaves of the first foliation λ_1 the equations $\underset{1}{\omega}^i = 0$ hold. Substituting them into the structure equations (1.26) and (1.32), we find that on the leaves of the first foliation

$$d\underset{2}{\omega}^i = \underset{2}{\omega}^j \wedge \omega_j^i - a_{jk}^i \underset{2}{\omega}^j \wedge \underset{2}{\omega}^k, \quad d\omega_j^i = \omega_j^k \wedge \omega_k^i. \tag{1.69}$$

Similarly, on the leaves of the second foliation the induced connection is given by

$$d\underset{1}{\omega}^i = \underset{1}{\omega}^j \wedge \omega_j^i + a_{jk}^i \underset{1}{\omega}^j \wedge \underset{1}{\omega}^k, \quad d\omega_j^i = \omega_j^k \wedge \omega_k^i. \tag{1.70}$$

Finally, on the leaves of the third foliation λ_3, which is defined by the equations $\underset{1}{\omega}^i + \underset{2}{\omega}^i = 0$, the structure equations of the induced connection have the form:

$$\begin{aligned} d\underset{1}{\omega}^i &= \underset{1}{\omega}^j \wedge \omega_j^i + a_{jk}^i \underset{1}{\omega}^j \wedge \underset{1}{\omega}^k, \\ d\omega_j^i &= \omega_j^k \wedge \omega_k^i - b_{j[kl]}^i \underset{1}{\omega}^k \wedge \underset{1}{\omega}^l. \end{aligned} \tag{1.71}$$

Equations (1.69)–(1.71) imply

Proposition 1.14 *The leaves of the first two foliations of a web W are surfaces of absolute parallelism relative to the Chern connection. The leaves of the third foliation possess this property if and only if the curvature tensor of the web satisfies the condition $b^i_{j[kl]} = 0$, i.e. if it is symmetric in the last two lower indices.*

■

Another consequence of equations (1.69)–(1.71) is that the torsion tensors of all three induced affine connections in the leaves F_1, F_2, F_3, passing through the point $p \in X$, are equal to each other at this point and coincide with the torsion tensor of the web W.

Since the torsion and curvature tensors have a specific structure, the system (1.64) becomes

$$\Delta p = \Delta p_1 + \Delta p_2,$$

where

$$\Delta p_1 = 2a^i_{jk}\xi^j_1\eta^k_1 dtds\underset{1}{e}_i, \quad \Delta p_2 = -2a^i_{jk}\xi^j_2\eta^k_2 dtds\underset{2}{e}_i, \tag{1.72}$$

and the system (1.65) is splitted into two subsystems:

$$\begin{aligned} \Delta\underset{1}{e}_j &= b^i_{jkl}(\xi^k_1\eta^l_2 - \eta^k_1\xi^l_2)dtdse_i, \\ \Delta\underset{2}{e}_j &= b^i_{jkl}(\xi^k_1\eta^l_2 - \eta^k_1\xi^l_2)dtdse_i. \end{aligned} \tag{1.73}$$

It is assumed here that the vectors ξ and η are written in the form:

$$\xi = \xi^i_1\underset{2}{e}_i - \xi^i_2\underset{1}{e}_i, \quad \eta = \eta^i_1\underset{2}{e}_i - \eta^i_2\underset{1}{e}_i,$$

This implies the following properties of the Chern connection:

(i) Let $\xi \in T_1$, i.e. a two-dimensional area element $\xi \wedge \eta$ intersects the tangent subspace T_1 to the leaf F_1. Then $\xi^i_1 = 0$, and formula (1.72) gives $\Delta p_1 = 0$. This means that the vector Δp lies in the subspace T_1. Similarly, one can find that if a two-dimensional area element $\xi \wedge \eta$ intersects the tangent subspace T_2, then the vector Δp lies in the subspace T_2. If a two-dimensional area element $\xi \wedge \eta$ intersects the tangent subspaces T_1 and T_2, then $\Delta p = 0$.

(ii) If $\xi \in T_3$ and $\eta \in T_3$, then $\xi^i_1 + \xi^i_2 = 0$ and $\eta^i_1 + \eta^i_2 = 0$, and formulas (1.60) imply that $\Delta p_1 + \Delta p_2 = 0$, i.e. $\Delta p \in T_3$.

Properties (i) and (ii) give a complete characterization of the torsion tensor of the Chern connection. In fact, it is easy to check that these two properties imply that the torsion tensor a^u_{vw} has the form (1.67).

Let us now study the formulas (1.73).

(i$_1$) First of all note that the vectors $\Delta\underset{\alpha}{e}_i$ are expressed only in terms of the vectors $\underset{\alpha}{e}_i$, $\alpha = 1, 2, 3$, and each of these systems of vectors is transformed concordantly under parallel displacement along a closed contour l.

(ii_1)) If $\xi \in T_1$ and $\eta \in T_1$, then $\xi^i_1 = \eta^i_1 = 0$, and we obtain from formulas (1.73) that $\Delta \underset{1}{e_i} = 0$. A similar result is obtained for $\xi \in T_2, \eta \in T_2$.

Conversely, if for the connection Γ, whose structure equations have the form (1.60) and (1.61), properties (i_1) and (ii_1) hold, then the curvature tensor of this connection has the form (1.67). Therefore, properties (i_1) and (ii_1) completely characterize the structure of the curvature tensor of the Chern connection.

1.8 Other Connections Associated with a Three-Web

1. The inequality of the foliations λ_α of a three-web $W = (X, \lambda_\alpha)$ with respect to the connection $\Gamma_{\alpha\beta}$ is explained by the fact that its structure equations (1.26) and (1.32) are not symmetric relative to the systems of forms $\underset{\alpha}{\omega}^i$, defining these foliations. When we derived the structure equations, we took the forms $\underset{1}{\omega}^i$ and $\underset{2}{\omega}^i$ as basis forms. In other words, the foliations λ_1 and λ_2 played the role of the coordinate foliations on the manifold X. If we take the forms $\underset{\alpha}{\omega}^i$ and $\underset{\beta}{\omega}^i$, $\alpha \neq \beta$, as basis forms and repeat for them all the calculations performed in Section 1.4, we arrive at the equations:

$$\begin{aligned} d\underset{\alpha}{\omega}^i &= \underset{\alpha}{\omega}^j \wedge \underset{\alpha\beta}{\omega}{}^i_j + \underset{\alpha\beta}{a}{}^i_{jk}\underset{\alpha}{\omega}^j \wedge \underset{\alpha}{\omega}^k, \\ d\underset{\beta}{\omega}^i &= \underset{\beta}{\omega}^j \wedge \underset{\alpha\beta}{\omega}{}^i_j - \underset{\alpha\beta}{a}{}^i_{jk}\underset{\beta}{\omega}^j \wedge \underset{\beta}{\omega}^k, \\ d\underset{\alpha\beta}{\omega}{}^i_j &= \underset{\alpha\beta}{\omega}{}^k_j \wedge \underset{\alpha\beta}{\omega}{}^i_k + \underset{\alpha\beta}{b}{}^i_{jkl}\underset{\alpha}{\omega}^k \wedge \underset{\beta}{\omega}^l, \end{aligned} \tag{1.74}$$

which are the structure equations of a certain connection $\Gamma_{\alpha\beta}$. It is obvious that the connection Γ_{12} is the Chern connection Γ. Denote by $\overset{\alpha\beta}{\nabla}$ the operator of covariant differentiation in the connection $\Gamma_{\alpha\beta}$. Then, for the tensor $\underset{\alpha\beta}{a}{}^i_{jk}$, we obtain the equations:

$$\overset{\alpha\beta}{\nabla}\underset{\alpha\beta}{a}{}^i_{jk} = \underset{\alpha\beta}{b}{}^i_{[j|l|k]}\underset{\alpha}{\omega}^l + \underset{\alpha\beta}{b}{}^i_{[jk]l}\underset{\beta}{\omega}^l, \tag{1.75}$$

similar to equations (1.33). If we interchange the indices α and β in the equations (1.74), we obtain:

$$\underset{\alpha\beta}{\omega}{}^i_j = \underset{\beta\alpha}{\omega}{}^i_j, \quad \underset{\alpha\beta}{a}{}^i_{jk} = -\underset{\beta\alpha}{a}{}^i_{jk}, \quad \underset{\alpha\beta}{b}{}^i_{jkl} = -\underset{\beta\alpha}{b}{}^i_{jlk}. \tag{1.76}$$

This shows that the difference between the connections $\Gamma_{\alpha\beta}$ and $\Gamma_{\beta\alpha}$ is not significant. In order to find a relation among the forms $\omega^i_j = \underset{12}{\omega}{}^i_j$, $\underset{23}{\omega}{}^i_j$ and $\underset{31}{\omega}{}^i_j$, note that, using relations (1.16), we can obtain from equations (1.27) the following equations:

$$d\underset{3}{\omega}^i = \underset{3}{\omega}^j \wedge (\omega^i_j - 2a^i_{jk}\underset{2}{\omega}^k) - a^i_{jk}\underset{3}{\omega}^j \wedge \underset{3}{\omega}^k.$$

In addition, equations (1.26) can be written in the form:

$$
\begin{aligned}
d\underset{1}{\omega}^i &= \underset{1}{\omega}^j \wedge (\omega_j^i + 2a_{jk}^i \underset{1}{\omega}^k) - a_{jk}^i \underset{1}{\omega}^j \wedge \underset{1}{\omega}^k, \\
d\underset{2}{\omega}^i &= \underset{2}{\omega}^j \wedge (\omega_j^i - 2a_{jk}^i \underset{2}{\omega}^k) + a_{jk}^i \underset{2}{\omega}^j \wedge \underset{2}{\omega}^k.
\end{aligned}
$$

Thus, for the connection Γ_{23}, we have:

$$
d\underset{2}{\omega}^i = \underset{2}{\omega}^j \wedge \underset{23}{\omega}{}_j^i + a_{jk}^i \underset{2}{\omega}^j \wedge \underset{2}{\omega}^k, \quad d\underset{3}{\omega}^i = \underset{3}{\omega}^j \wedge \underset{23}{\omega}{}_j^i - a_{jk}^i \underset{3}{\omega}^j \wedge \underset{3}{\omega}^k, \tag{1.77}
$$

where

$$
\underset{23}{\omega}{}_j^i = \omega_j^i - 2a_{jk}^i \underset{2}{\omega}^k, \tag{1.78}
$$

and for the connection Γ_{31} we have:

$$
d\underset{3}{\omega}^i = \underset{3}{\omega}^j \wedge \underset{31}{\omega}{}_j^i + a_{jk}^i \underset{3}{\omega}^j \wedge \underset{3}{\omega}^k, \quad d\underset{1}{\omega}^i = \underset{1}{\omega}^j \wedge \underset{31}{\omega}{}_j^i - a_{jk}^i \underset{1}{\omega}^j \wedge \underset{1}{\omega}^k, \tag{1.79}
$$

where

$$
\underset{31}{\omega}{}_j^i = \omega_j^i + 2a_{jk}^i \underset{1}{\omega}^k. \tag{1.80}
$$

It is clear from this that the torsion tensors of all three connections Γ_{12}, Γ_{23}, and Γ_{31} are equal.

In order to find the curvature tensor of the connection Γ_{23}, we take the exterior differential of the forms $\underset{23}{\omega}{}_j^i$, defined by relations (1.78), and use equations (1.26), (1.32), and (1.33). As a result, we obtain:

$$
d\underset{23}{\omega}{}_j^i = \omega_j^k \wedge \omega_k^i - 2(a_{lk}^i \omega_j^l - a_{jk}^l \omega_l^i) \wedge \underset{2}{\omega}^k + (b_{jkl}^i - 2b_{[j|k|l]}^i) \underset{1}{\omega}^k \wedge \underset{2}{\omega}^l - 2b_{[jl]k}^i \underset{2}{\omega}^k \wedge \underset{2}{\omega}^l + 2a_{jm}^i a_{kl}^m \underset{2}{\omega}^k \wedge \underset{2}{\omega}^l.
$$

In addition, we calculate the expression $\underset{23}{\omega}{}_j^k \wedge \underset{23}{\omega}{}_k^i$:

$$
\underset{23}{\omega}{}_j^k \wedge \underset{23}{\omega}{}_k^i = \omega_j^k \wedge \omega_k^i + 2a_{jk}^l \omega_l^i \wedge \underset{2}{\omega}^k - 2a_{lk}^i \omega_j^l \wedge \underset{2}{\omega}^k + 4a_{jk}^m a_{ml}^i \underset{2}{\omega}^k \wedge \underset{2}{\omega}^l.
$$

Subtracting the last equation from the previous equation, we arrived at the equation:

$$
d\underset{23}{\omega}{}_j^i - \underset{23}{\omega}{}_j^k \wedge \underset{23}{\omega}{}_k^i = b_{lkj}^i \underset{1}{\omega}^k \wedge \underset{2}{\omega}^l - 2(b_{[jl]k}^i + 3a_{[jk}^m a_{|m|l]}^i) \underset{2}{\omega}^k \wedge \underset{2}{\omega}^l.
$$

Now, with help of (1.16), we eliminate the forms $\underset{1}{\omega}^k$ and apply (1.34). This gives:

$$
d\underset{23}{\omega}{}_j^i - \underset{23}{\omega}{}_j^k \wedge \underset{23}{\omega}{}_k^i = b_{klj}^i \underset{2}{\omega}^k \wedge \underset{3}{\omega}^l. \tag{1.81}
$$

It follows from equations (1.81) that

$$
\underset{23}{b}{}_{jkl}^i = b_{klj}^i. \tag{1.82}
$$

In a similar way, we find that

$$d\underset{31}{\omega}{}^i_j - \underset{31}{\omega}{}^k_j \wedge \underset{31}{\omega}{}^i_k = b^i_{ljk}\underset{3}{\omega}{}^k \wedge \underset{1}{\omega}{}^l \tag{1.83}$$

and

$$\underset{31}{b}{}^i_{jkl} = b^i_{ljk}. \tag{1.84}$$

Therefore, the torsion tensors of the connections Γ_{23} and Γ_{31} are obtained from the torsion tensor of the connection Γ_{12} by the cyclic permutation of indices.

We summarize the results of this subsection in Table 1.1:

Connection	Γ_{12}	Γ_{23}	Γ_{31}	Γ_{21}	Γ_{32}	Γ_{13}
Torsion tensor	a^i_{jk}	a^i_{jk}	a^i_{jk}	$-a^i_{jk}$	$-a^i_{jk}$	$-a^i_{jk}$
Curvature tensor	b^i_{jkl}	b^i_{klj}	b^i_{ljk}	$-b^i_{jlk}$	$-b^i_{lkj}$	$-b^i_{kjl}$

Table 1.1

Using formulas (1.75), (1.82) and (1.84), we find the covariant differentials of the torsion tensor in the connections Γ_{23} and Γ_{31}:

$$\overset{23}{\nabla} a^i_{jk} = b^i_{l[kj]}\underset{2}{\omega}{}^l + b^i_{[k|l|j]}\underset{3}{\omega}{}^l; \quad \overset{31}{\nabla} a^i_{jk} = b^i_{[kj]l}\underset{3}{\omega}{}^l + b^i_{l[jk]}\underset{1}{\omega}{}^l. \tag{1.85}$$

Note also that the connections $\Gamma_{\alpha\beta}$ possess the same geometric properties as the Chern connection Γ_{12}.

2. We introduce one more connection, the *middle connection* $\overset{*}{\Gamma}$, which is symmetric relative to all three foliations λ_α of a web W. We set

$$\overset{*}{\omega}{}^i_j = \frac{1}{3}(\underset{12}{\omega}{}^i_j + \underset{23}{\omega}{}^i_j + \underset{31}{\omega}{}^i_j). \tag{1.86}$$

It follows from relations (1.85) and (1.87) that

$$\begin{aligned} \underset{12}{\omega}{}^i_j &= \overset{*}{\omega}{}^i_j - \frac{2}{3}a^i_{jk}(\underset{1}{\omega}{}^k - \underset{2}{\omega}{}^k), \\ \underset{23}{\omega}{}^i_j &= \overset{*}{\omega}{}^i_j - \frac{2}{3}a^i_{jk}(\underset{2}{\omega}{}^k - \underset{3}{\omega}{}^k), \\ \underset{31}{\omega}{}^i_j &= \overset{*}{\omega}{}^i_j - \frac{2}{3}a^i_{jk}(\underset{3}{\omega}{}^k - \underset{1}{\omega}{}^k). \end{aligned} \tag{1.87}$$

As a result, in the connection $\overset{*}{\Gamma}$ the equations (1.26) and (1.27) become symmetric:

$$
\begin{aligned}
d\underset{1}{\omega}{}^i &= \underset{1}{\omega}{}^j \wedge \overset{*}{\omega}{}^i_j + \frac{1}{3} a^i_{jk} \underset{1}{\omega}{}^j \wedge (\underset{2}{\omega}{}^k - \underset{3}{\omega}{}^k),\\
d\underset{2}{\omega}{}^i &= \underset{2}{\omega}{}^j \wedge \overset{*}{\omega}{}^i_j + \frac{1}{3} a^i_{jk} \underset{2}{\omega}{}^j \wedge (\underset{3}{\omega}{}^k - \underset{1}{\omega}{}^k),\\
d\underset{3}{\omega}{}^i &= \underset{3}{\omega}{}^j \wedge \overset{*}{\omega}{}^i_j + \frac{1}{3} a^i_{jk} \underset{3}{\omega}{}^j \wedge (\underset{1}{\omega}{}^k - \underset{2}{\omega}{}^k).
\end{aligned} \tag{1.88}
$$

Let us now find the covariant differential of the torsion tensor in the connection $\overset{*}{\Gamma}$. By the definition of this connection, the covariant differential is $\overset{*}{\nabla} = \frac{1}{3}(\overset{12}{\nabla} + \overset{23}{\nabla} + \overset{31}{\nabla})$. Thus, using the formulas (1.33) and (1.92), we find that

$$
\overset{*}{\nabla} a^i_{jk} = \frac{1}{3}[(b^i_{l[jk]} - b^i_{[k|l|j]})\underset{1}{\omega}{}^l + (b^i_{[jk]l} - b^i_{l[jk]})\underset{2}{\omega}{}^l + (b^i_{[k|l|j]} - b^i_{[jk]l})\underset{3}{\omega}{}^l].
$$

If, using the equations (1.16), we consecutively exclude from the right member of the last equations the forms $\underset{3}{\omega}{}^l, \underset{1}{\omega}{}^l$ and $\underset{2}{\omega}{}^l$, we obtain respectively the following three expressions of the forms $\overset{*}{\nabla} a^i_{jk}$:

$$
\overset{*}{\nabla} a^i_{jk} = \overset{*}{b}{}^i_{[jk]l}\underset{2}{\omega}{}^l - \overset{*}{b}{}^i_{[k|l|j]}\underset{1}{\omega}{}^l = \overset{*}{b}{}^i_{[k|l|j]}\underset{3}{\omega}{}^l - \overset{*}{b}{}^i_{l[jk]}\underset{2}{\omega}{}^l = \overset{*}{b}{}^i_{l[jk]}\underset{1}{\omega}{}^l - \overset{*}{b}{}^i_{[jk]l}\underset{3}{\omega}{}^l, \tag{1.89}
$$

where we denote $\overset{*}{b}{}^i_{jkl} = b^i_{jkl} - b^i_{[jkl]}$. The tensor $\overset{*}{b}{}^i_{jkl}$ satisfies the relation

$$
\overset{*}{b}{}^i_{[jkl]} = 0. \tag{1.90}
$$

In contrast to the tensor b^i_{jkl}, the tensor $\overset{*}{b}{}^i_{jkl}$ is not connected with the torsion tensor a^i_{jk} by any relations.

We now find equations connecting the forms $\overset{*}{\omega}{}^i_j$. Exterior differentiation of equations (1.78) by means of (1.26), (1.88) and (1.90) gives

$$
d\overset{*}{\omega}{}^i_j = \frac{1}{3}(\underset{12}{\omega}{}^k_j \wedge \underset{12}{\omega}{}^i_k + \underset{23}{\omega}{}^k_j \wedge \underset{23}{\omega}{}^i_k + \underset{31}{\omega}{}^k_j \wedge \underset{31}{\omega}{}^i_k) + \frac{1}{3}(b^i_{jkl}\underset{1}{\omega}{}^k \wedge \underset{2}{\omega}{}^l + b^i_{klj}\underset{2}{\omega}{}^k \wedge \underset{3}{\omega}{}^l + b^i_{ljk}\underset{3}{\omega}{}^k \wedge \underset{1}{\omega}{}^l).
$$

Moreover, the relations (1.87) imply that

$$
\frac{1}{3}(\underset{12}{\omega}{}^k_j \wedge \underset{12}{\omega}{}^i_k + \underset{23}{\omega}{}^k_j \wedge \underset{23}{\omega}{}^i_k + \underset{31}{\omega}{}^k_j \wedge \underset{31}{\omega}{}^i_k) = \overset{*}{\omega}{}^k_j \wedge \overset{*}{\omega}{}^i_k - \frac{4}{9}(a^m_{jk}a^i_{ml} + a^m_{lj}a^i_{mk})(\underset{1}{\omega}{}^k \wedge \underset{2}{\omega}{}^l + \underset{2}{\omega}{}^k \wedge \underset{3}{\omega}{}^l + \underset{3}{\omega}{}^k \wedge \underset{1}{\omega}{}^l).
$$

In addition, using (1.31), we can write

$$
b^i_{jkl} = \overset{*}{b}{}^i_{jkl} + 2a^m_{[jk}a^i_{|m|l]}.
$$

Using these relations, we find that

$$
\begin{aligned}
d\overset{*}{\omega}{}^i_j - \overset{*}{\omega}{}^k_j \wedge \overset{*}{\omega}{}^i_k = \; & \tfrac{1}{3}(\overset{*}{b}{}^i_{jkl}\underset{1}{\omega}{}^k \wedge \underset{2}{\omega}{}^l + \overset{*}{b}{}^i_{klj}\underset{2}{\omega}{}^k \wedge \underset{3}{\omega}{}^l + \overset{*}{b}{}^i_{ljk}\underset{3}{\omega}{}^k \wedge \underset{1}{\omega}{}^l)\\
& -\tfrac{1}{3}A^i_{jkl}(\underset{1}{\omega}{}^k \wedge \underset{2}{\omega}{}^l + \underset{2}{\omega}{}^k \wedge \underset{3}{\omega}{}^l + \underset{3}{\omega}{}^k \wedge \underset{1}{\omega}{}^l),
\end{aligned} \tag{1.91}
$$

where

$$A^i_{jkl} = \frac{2}{3}(a^m_{jk}a^i_{ml} + a^m_{lj}a^i_{mk} - a^m_{kl}a^i_{mj}). \tag{1.92}$$

Equations (1.88) and (1.91) are the structure equations of the middle connection $\overset{*}{\Gamma}$.

3. Later we will use the another type of connection, generated by the web $W = (X, \lambda_\alpha)$ in the manifold X. Set

$$\underset{12}{\tilde{\omega}}{}^i_j = \omega^i_j + a^i_{jk}(\underset{1}{\omega}{}^k - \underset{2}{\omega}{}^k). \tag{1.93}$$

Then equations (1.26) and (1.27) become

$$\begin{aligned} d\underset{1}{\omega}{}^i &= \underset{1}{\omega}{}^j \wedge \underset{12}{\tilde{\omega}}{}^i_j + a^i_{jk}\underset{1}{\omega}{}^j \wedge \underset{2}{\omega}{}^k, \\ d\underset{2}{\omega}{}^i &= \underset{2}{\omega}{}^j \wedge \underset{12}{\tilde{\omega}}{}^i_j - a^i_{jk}\underset{1}{\omega}{}^j \wedge \underset{2}{\omega}{}^k, \\ d\underset{3}{\omega}{}^i &= \underset{3}{\omega}{}^j \wedge \underset{12}{\tilde{\omega}}{}^i_j. \end{aligned} \tag{1.94}$$

Denote by $\tilde{\Gamma}_{12}$ the connection defined in the manifold X by the forms $\underset{12}{\tilde{\omega}}{}^i_j$.

Along with the connection $\tilde{\Gamma}_{12}$, we also introduce the connections $\tilde{\Gamma}_{23}$ and $\tilde{\Gamma}_{12}$, defined by the forms

$$\underset{23}{\tilde{\omega}}{}^i_j = \omega^i_j + a^i_{jk}\underset{1}{\omega}{}^k, \quad \underset{31}{\tilde{\omega}}{}^i_j = \omega^i_j - a^i_{jk}\underset{2}{\omega}{}^k \tag{1.95}$$

respectively.

For the connection $\tilde{\Gamma}_{23}$, equations (1.26) and (1.27) have the form:

$$\begin{aligned} d\underset{1}{\omega}{}^i &= \underset{1}{\omega}{}^j \wedge \underset{23}{\tilde{\omega}}{}^i_j, \\ d\underset{2}{\omega}{}^i &= \underset{2}{\omega}{}^j \wedge \underset{23}{\tilde{\omega}}{}^i_j + a^i_{jk}\underset{2}{\omega}{}^j \wedge \underset{3}{\omega}{}^k, \\ d\underset{3}{\omega}{}^i &= \underset{3}{\omega}{}^j \wedge \underset{23}{\tilde{\omega}}{}^i_j - a^i_{jk}\underset{2}{\omega}{}^j \wedge \underset{3}{\omega}{}^k, \end{aligned}$$

and for the connection $\tilde{\Gamma}_{31}$ they are of the form:

$$\begin{aligned} d\underset{1}{\omega}{}^i &+ \underset{1}{\omega}{}^j \wedge \underset{31}{\tilde{\omega}}{}^i_j - a^i_{jk}\underset{3}{\omega}{}^j \wedge \underset{1}{\omega}{}^k, \\ d\underset{2}{\omega}{}^i &= \underset{2}{\omega}{}^j \wedge \underset{31}{\tilde{\omega}}{}^i_j, \\ d\underset{3}{\omega}{}^i &= \underset{3}{\omega}{}^j \wedge \underset{31}{\tilde{\omega}}{}^i_j + a^i_{jk}\underset{3}{\omega}{}^j \wedge \underset{1}{\omega}{}^k. \end{aligned}$$

Proceeding in the same way as previously for the connections $\Gamma_{\alpha\beta}$, we can construct the middle connection for the connections $\tilde{\Gamma}_{\alpha\beta}$, defined by the forms

$$\overset{*}{\tilde{\omega}}{}^i_j = \frac{1}{3}(\underset{12}{\tilde{\omega}}{}^i_j + \underset{23}{\tilde{\omega}}{}^i_j + \underset{31}{\tilde{\omega}}{}^i_j).$$

However, applying formulas (1.93) and (1.95), we see that $\overset{*}{\tilde{\omega}}{}^i_j = \overset{*}{\omega}{}^i_j$. Hence, the connection $\overset{*}{\Gamma}$ is also the middle connection relative to the connections $\tilde{\Gamma}_{\alpha\beta}$.

All the connections we constructed so far, belong to the pencil of connections defined by the forms

$$\theta^u_v = \begin{pmatrix} \theta^i_j & 0 \\ 0 & \theta^i_j \end{pmatrix}, \tag{1.96}$$

where $u, v = 1, 2, \dots, 2r$, and

$$\theta^i_j = \omega^i_j + a^i_{jk}(p\underset{1}{\omega}{}^k + q\underset{2}{\omega}{}^k). \tag{1.97}$$

We denote this pencil by $\gamma(W)$. All web leaves are totally geodesic relative to any connection of the pencil $\gamma(W)$. Let us also show that *all these connections have the same geodesic lines in the web leaves.* In fact, since the matrix of forms θ^I_J has the structure indicated above, the equations (1.63) of geodesic lines are broken up into two series:

$$d\xi^i_1 + \xi^j_1\theta^i_j = \theta\xi^i_1, \quad d\xi^i_2 + \xi^j_2\theta^i_j = \theta\xi^i_2.$$

This implies that

$$d\xi^i_3 + \xi^j_3\theta^i_j = \theta\xi^i_3$$

By (1.97) and the skew-symmetry of the tensor a^i_{jk}, these equations can be written in the form:

$$\begin{aligned} d\xi^i_1 + \xi^j_1\omega^i_j + qa^i_{jk}\xi^j_1\xi^k_2 &= \theta\xi^i_1, \\ d\xi^i_2 + \xi^j_2\omega^i_j + pa^i_{jk}\xi^j_2\xi^k_1 &= \theta\xi^i_2, \\ d\xi^i_3 + \xi^j_3\omega^i_j + (q-p)a^i_{jk}\xi^j_1\xi^k_2 &= \theta\xi^i_3. \end{aligned} \tag{1.98}$$

Let us find, for example, the equations of geodesic lines in the leaves of the first foliation. For this, we substitute $\underset{1}{\omega}{}^i = 0$ in equations (1.98). This gives

$$d\xi^i + \xi^j\omega^i_j = \theta\xi^i. \tag{1.99}$$

Equations (1.99) do not depend on the quantities p and q. Therefore, all the connections from the pencil $\gamma(W)$ have the same geodesic lines in the leaves of the foliation λ_1.

4. Later we will see that all basic types of three-webs are characterized by a special structure of the torsion tensor and especially the curvature tensor. This indicates that it is important to have a decomposition of the curvature tensor into the sum of independent components.

We introduce the following notations for the skew-symmetric parts of the curvature tensor:

$$b^i_{[jk]l} = \underset{1}{a}{}^i_{jkl}, \quad b^i_{j[l|k]} = -\underset{2}{a}{}^i_{jkl}, \quad b^i_{l[jk]} = \underset{3}{a}{}^i_{jkl}. \tag{1.100}$$

(These notations are slightly different from the notations in Section 1.4.) The tensors $\underset{\alpha}{a}{}^i_{[jk]l}$ are skew-symmetric in the indices j and k and satisfy the following conditions:

$$\underset{\alpha}{a}{}^i_{[jkl]} = b^i_{[jkl]}, \quad \underset{1}{a}{}^i_{jkl} + \underset{2}{a}{}^i_{jkl} + \underset{3}{a}{}^i_{jkl} = 3b^i_{[jkl]}. \tag{1.101}$$

Further, we introduce the tensors $\overset{*}{\underset{\alpha}{a}}{}^i_{jkl}$:

$$\overset{*}{\underset{\alpha}{a}}{}^i_{jkl} = \underset{\alpha}{a}{}^i_{jkl} - b^i_{[jkl]}. \tag{1.102}$$

These tensors are skew-symmetric in the indices j and k, as are the tensors $\underset{\alpha}{a}{}^i_{jkl}$, and by (1.101), satisfy the following conditions:

$$\overset{*}{\underset{\alpha}{a}}{}^i_{[jkl]} = 0, \quad \overset{*}{\underset{1}{a}}{}^i_{jkl} + \overset{*}{\underset{2}{a}}{}^i_{jkl} + \overset{*}{\underset{3}{a}}{}^i_{jkl} = 0. \tag{1.103}$$

The tensors $\overset{*}{\underset{\alpha}{a}}{}^i_{jkl}$ are covariant derivatives of the torsion tensor a^i_{jk} with respect to the middle connection $\overset{*}{\Gamma}$ (see Problem 19).

Consider the following identity:

$$b^i_{jkl} = b^i_{[jkl]} + \frac{2}{3}(b^i_{[jk]l} - b^i_{[jkl]}) + \frac{2}{3}(-b^i_{[l|k|j]} - b^i_{[ljk]}) + \frac{2}{3}(b^i_{j[kl]} - b^i_{[klj]}) + b^i_{(jkl)},$$

which can be easily verified. If we use formula (1.31) to substitute for the first term and the formula (1.102) to substitute for the following three terms, we obtain

$$b^i_{jkl} = 2a^m_{[jk}a^i_{|m|l]} = \frac{2}{3}(\overset{*}{\underset{1}{a}}{}^i_{jkl} + \overset{*}{\underset{2}{a}}{}^i_{jkl} + \overset{*}{\underset{3}{a}}{}^i_{jkl}) + b^i_{(jkl)}. \tag{1.104}$$

This is the desired decomposition of the curvature tensor b^i_{jkl}.

Thus, the curvature tensor b^i_{jkl} is represented as the sum of the summands of three types: the first is expressed in terms of the torsion tensor, the second is expressed in terms of the covariant derivatives of the torsion tensor with respect to the middle connection $\overset{*}{\Gamma}$, and the third does not depend on the torsion tensor. It follows from this that the differential neighborhood of third order of a three-web is defined by an assignment of the torsion tensor and the symmetric part of the curvature tensor.

The formula (1.104) allows us to make an essential addition to Theorem 1.8 on the equivalence of webs. By (1.104), for webs W and $\widetilde{W}$ to be equivalent, it is necessary and sufficient that their torsion tensors a^i_{jk} and $\tilde{a}^i_{jk}$ and the symmetric parts of their curvature tensors b^i_{jkl} and $\tilde{b}^i_{jkl}$ coincide.

1.9 Subwebs of Multidimensional Three-Webs

We recall (see Section 1.3) that a subweb of a multidimensional three-web $W = (X, \lambda_\alpha)$ is cut by its leaves on a transversal surface $\widetilde{X}$ of dimension 2ρ, $1 \leq \rho \leq r$, i.e. on each submanifold in X having a ρ-dimensional intersection with each of those web leaves which have at least one common point with $\widetilde{X}$. Not every three-web possesses such transversal submanifolds. In this section we will find conditions for existence of such submanifolds and the structure equations of a three-web containing a subweb.

In Section 1.3 we found two-dimensional transversal subspaces in the tangent space T_p at the point p of the manifold X. These transversal subspaces intersect the tangent space T_α to the web leaf, passing through the point p, along a one-dimensional subspace, generated by the vector $\underset{\alpha}{\xi} = \xi^i \underset{\alpha}{e_i}$, $\underset{\alpha}{e_i} \in T_\alpha$. In a similar way, we can define (2ρ)-dimensional transversal subspaces in T_p, generated by vectors

$$\underset{\alpha}{\xi_a} = \xi^i_a \underset{\alpha}{e_i}, \quad a = 1, \ldots, \rho,$$

and the (2ρ)-vector $H^{2\rho}$:

$$H^{2\rho} = \underset{1}{\xi_1} \wedge \underset{1}{\xi_2} \wedge \ldots \wedge \underset{1}{\xi_\rho} \wedge \underset{2}{\xi_1} \wedge \underset{2}{\xi_2} \wedge \ldots \wedge \underset{2}{\xi_\rho},$$

defined by these vectors. The transversal (2ρ)-dimensional subspace intersects the tangent spaces T_α to the leaves of the web W along ρ-dimensional subspaces spanned by the vectors $\underset{\alpha}{\xi_a}$.

The results of Section 1.3 on two-dimensional transversal subspaces are also valid for (2ρ)-dimensional subspaces: the latter are also invariant with respect to the transformations $\varphi_{\alpha\beta *}$. It follows from this that *(2ρ)-dimensional submanifolds of the web W are invariant with respect to the transformations* $\varphi_{\alpha\beta}$, since their tangent spaces are transversal subspaces of this web. The last fact implies that the equations of an embedding $i: \widetilde{X} \to X$ of a transversal submanifold $\widetilde{X}$ into the manifold X, carrying a three-web W, can be written in the form:

$$\underset{1}{\omega}^i = \xi^i_a \underset{1}{\theta}^a, \quad \underset{2}{\omega}^i = \xi^i_a \underset{2}{\theta}^a, \tag{1.105}$$

where the forms $\underset{1}{\theta}^a$ and $\underset{2}{\theta}^a$ form a co-frame in the transversal submanifold $\widetilde{X}$. Moreover, the leaves of a three-subweb $\widetilde{W}$, cut on the submanifold $\widetilde{X}$ by the leaves of the web W, are defined by the equations

$$\underset{1}{\theta}^a = 0, \quad \underset{2}{\theta}^a = 0, \quad \underset{1}{\theta}^a + \underset{2}{\theta}^a = 0. \tag{1.106}$$

Note that the structure of equations (1.106) is the same as that of the equations, defining the foliations of the web W. Therefore, the forms $\underset{1}{\theta}^a$ and $\underset{2}{\theta}^a$ satisfy the structure equations of the type (1.26) and (1.32):

$$d\underset{1}{\theta}^a = \underset{1}{\theta}^b \wedge \theta^a_b + \tilde{a}^a_{bc} \underset{1}{\theta}^b \wedge \underset{1}{\theta}^c, \quad d\underset{2}{\theta}^a = \underset{2}{\theta}^b \wedge \theta^a_b - \tilde{a}^a_{bc} \underset{2}{\theta}^b \wedge \underset{2}{\theta}^c, \tag{1.107}$$

$$d\theta^a_b = \theta^a_c \wedge \theta^c_b + \tilde{b}^a_{bcd} \underset{1}{\theta}^c \wedge \underset{1}{\theta}^d. \tag{1.108}$$

On the other hand, along with equations (1.105), their differential consequences must be satisfied. Applying exterior differentiation to equations (1.105) and using equations (1.26), (1.105) and (1.107), we arrive at the system:

$$(\nabla\xi^i_a - \xi^i_b\theta^b_a) \wedge \underset{1}{\theta}^a + (\xi^i_a\tilde{a}^a_{bc} - a^i_{jk}\xi^j_a\xi^k_b)\underset{1}{\theta}^a \wedge \underset{1}{\theta}^b = 0,$$
$$(\nabla\xi^i_a - \xi^i_b\theta^b_a) \wedge \underset{2}{\theta}^a - (\xi^i_c\tilde{a}^c_{ab} - a^i_{jk}\xi^j_a\xi^k_b)\underset{2}{\theta}^a \wedge \underset{2}{\theta}^b = 0,$$

where we denote $\nabla\xi^i_a = d\xi^i_a + \xi^j_a\omega^i_j$. It follows from these quadratic equations that

$$\nabla\xi^i_a = \xi^i_b\theta^b_a, \tag{1.109}$$

$$a^i_{jk}\xi^j_a\xi^k_b = \xi^i_c\tilde{a}^c_{ab}. \tag{1.110}$$

The web $\widetilde{W}$ induces in the manifold $\widetilde{X}$ an affine connection $\widetilde{\Gamma}$, defined by the forms θ^a_b appearing in equations (1.107). The connection $\widetilde{\Gamma}$ is consistent with the Chern connection Γ in the sense that a parallel transport of a vector in the manifold $\widetilde{X}$ relative to the connection $\widetilde{\Gamma}$ coincides with a parallel transport of the same vector in the manifold X relative to the connection Γ. In fact, a parallel vector field η with coordinates $\underset{\alpha}{\tilde{\eta}}^a$ in the manifold $\widetilde{X}$ is defined by the equations:

$$d\underset{\alpha}{\tilde{\eta}}^a + \underset{\alpha}{\tilde{\eta}}^b\theta^a_b = 0, \quad \alpha = 1, 2. \tag{1.111}$$

Coordinates of the same vector field in the frame $\{\underset{\alpha}{e}_i\}$, associated with the manifold X, by (1.105), are calculated according to the formula: $\underset{\alpha}{\eta}^i = \xi^i_a\underset{\alpha}{\tilde{\eta}}^a$. Differentiating this equation and using the relations (1.109), we obtain the equations

$$d\underset{\alpha}{\eta}^i + \underset{\alpha}{\eta}^j\omega^i_j = 0,$$

which show that the vector field η is parallel in the Chern connection Γ in X. This immediately implies that the transversal submanifold $\widetilde{X}$ of the manifold X is a totally geodesic submanifold in this connection. But then the leaves of the subweb $\widetilde{W}$ are intersections of the following totally geodesic submanifolds: the leaves of the web W and its transversal submanifold $\widetilde{X}$. Thus, the leaves of the web $\widetilde{W}$ are also totally geodesic submanifolds.

It is easy to prove that all these results are valid for any connection from the pencil $\gamma(W)$.

Therefore, we have proved the following theorem:

Theorem 1.15 *Transversal submanifolds of a web $W = (X, \lambda_\alpha)$ are totally geodesic submanifolds with respect to all the connections from the pencil $\gamma(W)$ of connections induced by this web in the manifold X. A subweb $\widetilde{W}$ of the web W, cut on a transversal submanifold $\widetilde{X}$ by the leaves of W, consists of totally geodesic submanifolds with respect to these connections.*

■

This theorem explains why transversal surfaces of a web are also called transversally geodesic surfaces.

Let us find necessary and sufficient conditions for existence of transversal surfaces of a three-web. Taking exterior differentials of equations (1.109) and using relations (1.26) and (1.108), after simple calculations we arrive at the conditions:

$$b^i_{jkl}\xi^j_b\xi^k_c\xi^l_d = \xi^i_a\tilde{b}^a_{bcd}. \qquad (1.112)$$

Relations (1.100) and (1.112) are the conditions of integrability of the system (1.105), (1.109), defining the surface $\widetilde{X}$. Therefore, these relations are necessary and sufficient for existence of a fibration in X one of the fibres of which is the submanifold $\widetilde{X}$. In particular, for $\rho = 1$, the torsion tensor is equal to zero (cf. Section 1.4), and conditions (1.112) become

$$b^i_{jkl}\xi^j\xi^k\xi^l = \xi^i\tilde{b}. \qquad (1.113)$$

A vector ξ^i, satisfying the relation (1.113), is called an *eigenvector* of the tensor b^i_{jkl}.

The equations of a subweb $\widetilde{W}$ will be considerably simpler, if we use an adapted frame, associated with this web. It follows from equations (1.20) and (1.109) that

$$\xi = \underset{1}{\theta}^a(\xi^i_a \underset{2}{e}_i) - \underset{2}{\theta}^a(\xi^i_a \underset{1}{e}_i).$$

These equations show that the vectors $\xi^i_a \underset{1}{e}_i$ and $\xi^i_a \underset{2}{e}_i$ form a frame on the manifold $\widetilde{X}$, which is dual to the co-frame $\{\underset{1}{\theta}^a, \underset{2}{\theta}^a\}$. We restrict the family of frames in X by locating the vectors $\underset{\alpha}{e}_a$ in the tangent space $\widetilde{T}_\alpha$ to the leaf $\widetilde{F}_\alpha$ of the web $\widetilde{W}$. This family of frames is adapted for the embedding $i : \widetilde{X} \rightarrow X$, given by the equations (1.105). As a result, we obtain $\xi^a_b = \delta^a_b$, $\xi^u_a = 0$, $u, v, w, \ldots = \rho + 1, \ldots, r$, and equations (1.105) become:

$$\underset{1}{\theta}^a = \underset{1}{\omega}^a, \quad \underset{2}{\theta}^a = \underset{2}{\omega}^a, \quad \underset{1}{\omega}^u = \underset{2}{\omega}^u = 0.$$

Moreover, conditions (1.110) and (1.112), connecting the torsion and curvature tensors of the webs W and $\widetilde{W}$, immediately imply the equations

$$\tilde{a}^a_{bc} = a^a_{bc}, \quad \tilde{b}^a_{bcd} = b^a_{bcd}, \quad a^u_{ab} = 0, \quad b^u_{abc} = 0, \qquad (1.114)$$

which are satisfied on the manifold $\widetilde{X}$.

Suppose that the torsion and curvature tensors of the web W are reduced to the form (1.114) in the whole manifold X. Then the structure equations (1.107) are broken up into two series:

$$d\underset{1}{\omega}^a = \underset{1}{\omega}^i \wedge \omega^a_i + a^a_{jk}\underset{1}{\omega}^j \wedge \underset{1}{\omega}^k, \quad d\underset{2}{\omega}^u = \underset{2}{\omega}^i \wedge \omega^a_i - a^a_{jk}\underset{2}{\omega}^j \wedge \underset{2}{\omega}^k, \qquad (1.115)$$

and

$$\begin{aligned} d\underset{1}{\omega}^u &= \underset{1}{\omega}^v \wedge \omega_v^u + \underset{1}{\omega}^a \wedge \omega_a^u + 2a_{av}^u \underset{1}{\omega}^a \wedge \underset{1}{\omega}^v + a_{vw}^u \underset{1}{\omega}^v \wedge \underset{1}{\omega}^w, \\ d\underset{2}{\omega}^u &= \underset{2}{\omega}^v \wedge \omega_v^u + \underset{2}{\omega}^a \wedge \omega_a^u - 2a_{av}^u \underset{2}{\omega}^a \wedge \underset{2}{\omega}^v - a_{vw}^u \underset{2}{\omega}^v \wedge \underset{2}{\omega}^w. \end{aligned} \tag{1.116}$$

As a result of our transfer to the family of adapted frames, equations (1.109) become

$$\omega_b^a \equiv \theta_b^a, \quad \omega_u^a \equiv \theta_b^a \pmod{\underset{1}{\omega}^u, \underset{2}{\omega}^u}.$$

In particular, it follows from the last relations that on the manifold X the following equations hold:

$$\omega_u^a = \lambda_{av}^a \underset{1}{\omega}^v + \mu_{av}^u \underset{2}{\omega}^v. \tag{1.117}$$

Equations (1.117) and (1.118) show that the system $\underset{1}{\omega}^u = \underset{2}{\omega}^u = 0$, defining the transversal surface $\widetilde{X}$, is completely integrable. On the manifold $\widetilde{X}$, the basis forms are $\underset{1}{\omega}^a$ and $\underset{2}{\omega}^a$. As equations (1.115) show, if $\underset{1}{\omega}^u = \underset{2}{\omega}^u = 0$, these forms satisfy the equations:

$$d\underset{1}{\omega}^a = \underset{1}{\omega}^b \wedge \omega_b^a + a_{bc}^a \underset{1}{\omega}^b \wedge \underset{1}{\omega}^c, \quad d\underset{2}{\omega}^a = \underset{2}{\omega}^b \wedge \omega_b^a - a_{bc}^a \underset{2}{\omega}^b \wedge \underset{2}{\omega}^c. \tag{1.118}$$

The forms ω_b^a in (1.118) are connected by the relations:

$$d\omega_b^a = \omega_b^c \wedge \omega_c^a + b_{bcd}^a \underset{1}{\omega}^c \wedge \underset{2}{\omega}^d, \tag{1.119}$$

which are obtained from equations (1.108) for $\underset{1}{\omega}^u = \underset{2}{\omega}^u = 0$. Equations (1.118) and (1.119) are the structure equations of a subweb in the adapted frame.

Now we can supplement the results of Section 1.1 by introducing the notion of the factor web and semi-direct product of webs. In general, equations (1.116) for the forms $\underset{1}{\omega}^u$ and $\underset{2}{\omega}^u$, are not the structure equations of a web. They define a three-web if and only if they do not contain terms with the forms $\underset{1}{\omega}^a$ and $\underset{2}{\omega}^a$, i.e. if and only if the following relations hold on a web W:

$$\omega_a^u = 0, \quad a_{ai}^u = 0. \tag{1.120}$$

(In (1.120) the conditions (1.114) and (1.117) are taken into account.) A three-web, defined by equations (1.116) under conditions (1.120), is given in the factor space $X/\widetilde{X}$. We call it a *factor web*. In this case, the web $\widetilde{W}$ is called a *normal subweb*. Equations (1.120) are necessary and sufficient for a subweb $\widetilde{X}$ to be a normal subweb of a web W and for existence of a factor web $W/\widetilde{W}$.

Differentiating relations (1.120), we obtain the conditions $b_{aij}^u = 0$, $b_{[a|i|v]}^u = 0$, $b_{[av]c}^u = 0$, which imply that

$$b_{aij}^u = b_{via}^u = b_{vai}^u = 0. \tag{1.121}$$

As a result, we see that the only non-vanishing components of the torsion tensor are b^u_{vwz}.

Similarly, one can prove that if in the manifold X the relations

$$a^a_{uv} = 0, \quad \omega^a_u = \lambda^a_{ub}\underset{1}{\omega}^b + \mu^a_{ub}\underset{2}{\omega}^b \tag{1.122}$$

hold, then this manifold is stratified into transversal submanifolds of dimension $2(r-\rho)$, carrying subwebs of a web W.

If in the manifold X the equations (1.120), (1.121) and (1.122) are satisfied, we say that a web W is a *semi-direct product* of webs $\widetilde{W}$ and $\widetilde{\widetilde{W}}$. If the conditions

$$a^a_{iv} = o, \quad \omega^a_u = 0, \tag{1.123}$$

hold on a three-web W, the equations

$$b^a_{vij} = b^a_{biv} = b^a_{bvi} = 0 \tag{1.124}$$

are satisfied on this three-web, there exists a factor web $W/\widetilde{\widetilde{W}}$, and the subweb $\widetilde{\widetilde{W}}$ is a normal subweb. If, in addition, equations (1.120), (1.121) and (1.123), (1.124) are satisfied, then the web is the *direct product* of the subwebs $\widetilde{W}$ and $\widetilde{\widetilde{W}}$, since in this case equations (1.26), (1.32), (1.33) and their prolongations can be broken up into two independent serie:s

$$\begin{aligned}
d\underset{1}{\omega}^a &= \underset{1}{\omega}^b \wedge \omega^a_b + a^a_{bc}\underset{1}{\omega}^b \wedge \underset{1}{\omega}^c, & d\underset{1}{\omega}^u &= \underset{1}{\omega}^v \wedge \omega^w_v + a^u_{vw}\underset{1}{\omega}^v \wedge \underset{1}{\omega}^w,\\
d\underset{2}{\omega}^a &= \underset{2}{\omega}^b \wedge \omega^a_b - a^a_{bc}\underset{2}{\omega}^b \wedge \underset{2}{\omega}^c, & d\underset{2}{\omega}^u &= \underset{2}{\omega}^v \wedge \omega^u_v - a^u_{vw}\underset{2}{\omega}^v \wedge \underset{2}{\omega}^w,\\
d\omega^a_b &= \omega^c_b \wedge \omega^a_c + b^a_{bcd}\underset{1}{\omega}^c \wedge \underset{2}{\omega}^d, & d\omega^u_v &= \omega^w_v \wedge \omega^u_w + b^u_{vwz}\underset{1}{\omega}^w \wedge \underset{2}{\omega}^z,
\end{aligned}$$

etc.

A three-web, not posessing normal subwebs, is called a *simple web.*

PROBLEMS

1. Find equations of a three-web formed in the plane:

(i) by three pencils of straight lines;

(ii) by a pencil of straight lines with the vertex at the origin O and with the lines tangent to a circle with the center at O;

(iii) by three elliptical pencils of circles with the vertices at the points $(A, B), (B, C)$ and (C, A);

(iv) by the Cartesian coordinate lines and a family of concentric circles;

(v) by the lines $y = \text{const}$, the pencil of straight lines through the origin and a parabolic pencil of circles with vertex at the origin O and with centers lying on the y-axis;

(vi) by a pencil of straight lines with vertex at the origin O, a family of concentric circles with the vertex at the origin O and the hyperbolic pencil of circles with vertices at the points $(1,0)$ and $(-1,0)$;

(vii) by three families of straight lines: $A_\alpha(u_\alpha)x + B_\alpha(u_\alpha)y + C_\alpha(u_\alpha)z = 0, \quad \alpha = 1,2,3,$ where $A_\alpha(u_\alpha), B_\alpha(u_\alpha), C_\alpha(u_\alpha)$ are smooth functions of a parameter u_α;

(viii) by Cartesian coordinate lines and the two elliptical pencils of circles with the vertices at the points $(1,1)$ and $(-1,-1)$.

For each of the cases (i)–(viii) find a web domain.

2. The set of all straight lines in the plane, whose tangential coordinates are connected by a homogeneous algebraic equation of third degree, is called a *curve of the third class.* Prove that the latter form a three-web.

3. Prove that the set of tangents to the three-arc hypocycloid (the deltoid and the tricuspid are other names for this curve):

$$x = a(2\cos t + \cos 2t), \quad y = a(2\sin t - \sin 2t)$$

is a curve of third class.

4. Prove that the webs listed in **1**(i)–(v) are parallelizable. (*Hint*: reduce their equations to the form $z = x + y$ using admissible transformations.)

5. Prove that the three-web, given in the plane by the equation

$$Au_1u_2u_3 + B_1u_2u_3 + B_2u_1u_3 + B_3u_1u_2 + C_1u_1 + C_2u_2 + C_3u_3 + D = 0,$$

where $A, B_1, B_2, B_3, C_1, C_2, C_3$ and D are constants, is parallelizable. (*Hint*: first transform the web equation to the form

$$(a_1u_1 + b_1)(a_2u_2 + b_2)(a_3u_3 + b_3) + (a_4u_1 + b_4)(a_5u_2 + b_5)(a_6u_3 + b_6) = 0,$$

and then reduce it to the form $z = x + y$.)

Prove the statements of problems **6**–**8**.

6. The equations of a parallelizable web can be reduced to the form $z^i = x^i + y^i$.

7. If a web W is a direct product of two webs, then its equations can be written in the form: $z = f_1(x,y), \quad w = f_2(u,v)$.

8. The three-web defined by the equations

$$\begin{aligned} z^1 &= x^1 + y^1, \\ z^i &= x^i\varphi_1(x^1,y^1) + y^i\varphi_2(x^1,y^1), \quad i = 2,3,\dots,r, \end{aligned}$$

is a semi-direct product of two parallelizable webs.

9. Find forms ω_v^u and the first structure tensor for the e-structure and the Pfaffian structure.

10. Use the structure equations to prove the following statements:

(i) If the curvature of a two-dimensional web is equal to zero, then this web is parallelizable; and

(ii) If the functions $\underset{1}{c}$ and $\underset{2}{c}$ in equations (1.38) of a web W vanish, then the web curvature is zero, and the web is parallelizable.

11. If a two-dimensional web is given by the equation $z = f(x,y)$, then its curvature b can be written in the following elegant form:

$$b = -\frac{1}{f_x f_y}\frac{\partial^2}{\partial x \partial y}\left(\ln \frac{f_x}{f_y}\right).$$

In particular, if the web curvature is zero, then the function f, defining the web, satisfies St. Robert's equation:

$$\frac{\partial^2}{\partial x \partial y}\left(\ln \frac{f_x}{f_y}\right) = 0.$$

Derive from this the result of Problem **10** (i).

12. Suppose that a two-dimensional web is given by the equation $W(u_1, u_2, u_3) = 0$. Prove that its curvature b can be expressed by the formula:

$$b = A_{23} + A_{31} + A_{12},$$

where

$$A_{\alpha\beta} = \frac{W_{\alpha\alpha\beta}}{W_\alpha^2 W_\beta} - \frac{W_{\alpha\beta\beta}}{W_\alpha W_\beta^2} + \frac{W_{\alpha\beta}}{W_\alpha W_\beta}\left(\frac{W_{\beta\beta}}{W_\beta^2} - \frac{W_{\alpha\alpha}}{W_\alpha^2}\right), \quad \alpha,\ \beta,\ \gamma = 1,2,3,$$

and W_α, $W_{\alpha\beta}$, $W_{\alpha\beta\gamma}$ are partial derivatives of the function $W(u_1, u_2, u_3)$.

Use this formula to compute the curvature of the web from Problem **5**.

13. Find the structure equations and the torsion and curvature tensors of the six-dimensional web defined by the equations:

$$z^1 = x^1 + y^1 - (x^2 + y^2)x^3y^3, \quad z^2 = x^2 + y^2, \quad z^3 = x^3 + y^3. \tag{1.125}$$

Solution. Following (1.51), we set

$$\begin{aligned} \underset{1}{\omega}^1 &= dx^1 - x^3y^3dx^2 - y^3(x^2 + y^2)dx^3, \\ \underset{2}{\omega}^1 &= dy^1 - x^3y^3dy^2 - x^3(x^2 + y^2)dy^3, \\ \underset{1}{\omega}^2 &= dx^2,\ \underset{1}{\omega}^3 = dx^3,\ \underset{2}{\omega}^2 = dy^2,\ \underset{2}{\omega}^3 = dy^3. \end{aligned} \tag{1.126}$$

Applying exterior differentiation to the forms $\underset{1}{\omega}^i$ and $\underset{1}{\omega}^i$, $i, j, k = 1, 2, 3$, and using (1.126), we find

$$\begin{aligned} d\underset{1}{\omega}^1 &= \underset{1}{\omega}^2 \wedge \omega_2^1 + \underset{1}{\omega}^3 \wedge \omega_3^1 + (x^3 - y^3)\underset{1}{\omega}^2 \wedge \underset{1}{\omega}^3, \\ d\underset{2}{\omega}^1 &= \underset{2}{\omega}^2 \wedge \omega_2^1 + \underset{2}{\omega}^3 \wedge \omega_3^1 - (x^3 - y^3)\underset{2}{\omega}^2 \wedge \underset{2}{\omega}^3, \\ d\underset{1}{\omega}^2 &= d\underset{1}{\omega}^3 = d\underset{2}{\omega}^2 = d\underset{2}{\omega}^3 = 0, \end{aligned} \tag{1.127}$$

where we denote

$$\begin{aligned} \omega_2^1 &= x^3 \underset{2}{\omega^3} + y^3 \underset{1}{\omega^3} = x^3 dy^3 + y^3 dx^3 = d(x^3 y^3), \\ \omega_3^1 &= y^3 dy^2 + x^3 dx^2 + (x^2 + y^2)(dx^3 + dy^3) \\ &= d(x^2 x^3) + d(y^2 y^3) + x^2 dy^3 + y^2 dx^3. \end{aligned} \tag{1.128}$$

Comparing the first two equations from (1.127) with equations (1.26), we find that $\omega_1^1 = 0$, $a_{12}^1 = a_{13}^1 = 0$, $a_{23}^1 = \frac{1}{2}(x^3 - y^3)$. Comparing the last four equations from (1.127) with equations (1.26), we find that $\omega_k^2 = \omega_k^3 = 0$, $a_{jk}^2 = a_{jk}^3 = 0$. Exterior differentiation of the forms ω_j^i gives the following result:

$$d\omega_2^1 = 0, \quad d\omega_3^1 = \underset{1}{\omega^2} \wedge \underset{2}{\omega^3} - \underset{1}{\omega^3} \wedge \underset{2}{\omega^2}, \quad d\omega_k^2 = d\omega_k^3 = 0.$$

Comparing these equations with equations (1.32), we find the non-vanishing components of the curvature tensor: $b_{323}^1 = 1$, $b_{332}^1 = -1$. Note that $b_{j(kl)}^i = 0$, i.e. the curvature tensor is skew-symmetric in the last two lower indices.

14. Suppose that a group G is a direct product of two isomorphic r-dimensional Lie groups G_1 and G_2. Suppose also that a subgroup G_3 of G is the graph of a certain isomorphism from G_1 to G_2. Prove that

(i) The three-web W, formed on G by co-sets with respect to the subgroups G_1, G_2 and G_3, is a group web;

(ii) Any group web can be defined in this way; and

(iii) The subgroup G_3 is normal if and only if the three-web W is parallelizable.

Prove the statements of problems **15–17**.

15. A field $\xi = (\xi_1^i, \xi_2^i)$ is parallel on a web $W = (X, \lambda_\alpha)$ relative to the connection Γ_{12} if and only if the coordinates ξ_1^i and ξ_2^i satisfy the relations $b_{jkl}^i \xi_1^j = 0$ and $b_{jkl}^i \xi_2^j = 0$. In the manifold X there exist r independent parallel vector fields if and only if the web W is a group web.

16. The isoclinic r-vectors $\zeta_1 \wedge \ldots \wedge \zeta_r$, where $\zeta_i = \zeta^1 \underset{2}{e_i} - \zeta^2 \underset{1}{e_i}, \zeta^1 : \zeta^2 = \text{const}$, are parallel relative to any connection in X, whose matrix has quasi-diagonal form (1.96).

17. The leaves of the web W are autoparallel submanifolds relative to each connection of the pencil $\gamma(W)$ (cf. (1.97)).

18. A vector field $\xi = (\xi_1^i, \xi_2^i)$, given in a manifold X carrying a web $W = (X, \lambda_\alpha)$, is called *Segrean* if at any point $p \in X$ a vector $\xi(p)$ belong to the Segre cone, defined at the point p. Prove that

(i) A field ξ is Segrean if and only if $\xi_2^i = \lambda \xi_1^i$; and

(ii) Any Segrean field ξ defines in X a unique distribution of two-dimensional transversal subspaces, containing ξ; and

(iii) If a Segrean field $\xi = (\xi_1^i, \lambda \xi_1^i)$ is parallel along a curve γ, then $\lambda|_\gamma = \text{const}$.

19. Prove that the tensors $\underset{\alpha}{\overset{*}{a}}{}_{jkl}^{i}$ are covariant derivatives of the tensor a_{jk}^i relative to the middle connection $\overset{*}{\Gamma}$.

20. Find the torsion and curvature tensors of the following four-dimensional three-webs:

(i) $z^1 = x^1 e^{-2y^2} + y^1 + \frac{1}{2}x^2 y^2 e^{-2y^2}, \quad z^2 = x^2 + y^2;$

(ii) $z^1 = x^2 \frac{x^1 x^2 - y^1 y^2}{x^1 y^1 + 2x^1 x^2 + x^2 y^2}, \quad z^2 = x^2 y^2.$

Prove that the torsion tensor of each of these two webs satisfies the Jacobi identity, and their curvature tensors are symmetric in each pair of the lower indices.

21. Prove that the commutator ζ of vector fields $\xi = (\xi^i_1, \xi^i_2)$ and $\eta = (\eta^i_1, \eta^i_2)$, given in a manifold X carrying a web $W = (X, \lambda_\alpha)$, can be calculated by means of the formulas:

$$\underset{1}{\zeta}^i = \underset{1}{\xi}^j \eta^i_j - \underset{1}{\eta}^j \xi^i_j + 2a^i_{jk} \underset{1}{\xi}^j \underset{1}{\eta}^k, \quad \underset{2}{\zeta}^i = \underset{2}{\xi}^j \eta^i_j - \underset{2}{\eta}^j \xi^i_j + 2a^i_{jk} \underset{2}{\xi}^j \underset{2}{\eta}^k.$$

(*Hint*: the commutator of the fields $\xi(\xi^a)$ and $\eta(\eta^a)$ in a manifold with an affine connection can be expressed in the form:

$$[\xi, \eta]^a = \xi^b \eta^a_b - \eta^b \xi^a_b - A^a_{bc} \xi^b \eta^c,$$

where A^a_{bc} is the torsion tensor of the connection and ξ^a_b and η^a_b are covariant derivatives of ξ^a and η^a respectively (see, for example, [KN 63], vol. 1, p. 133).)

NOTES

1.1.[1] The term "three-web" was introduced by Blaschke [BB 38], and the term "quasigroup" was introduced by Moufang.

1.2. As proved in [DJ 85], the differential-geometric equivalence of smooth three-webs follows from their topological equivalence.

1.4. Differential equations of a multidimensional three-web were first obtained by Chern [C 36a], and, in more readable and convenient form (1.26)–(1.32), they were found by Akivis [A 69b]. These equations also can be considered as the structure equations of the coordinate quasigroup q_{12} of a web. As such, they generalize the Maurer–Cartan equations of a Lie group. The terms "torsion tensor" and "curvature tensor" were introduced in [A 69b].

Theorem 1.2 was proved by the authors in [AS 81]. Its converse (Theorem 1.3) is given here for the first time.

The structure equations of a two-dimensional web were derived by Blaschke [Bl 55]. The term "the web curvature" was also introduced in [Bl 55].

1.5 Parallelizable and group three-webs were considered by Chern [C 36b] as webs on which the closure conditions T and R hold (in [C 36b] these conditions are called conditions (D) and (P) respectively). The proof of Theorems 1.10 and 1.11 can be also found in [C 36b]. However, in [C 36b] it is given in more complicated and less geometric form than in this book.

1.6. The formulas (1.56) and (1.58) for calculation of the torsion and curvature tensors of a web were found in [AS 71a], although expressions of these tensors close to that in these formulas can be also found in [C 36b]. The formula for computation of the curvature of a two-dimensional three-web, given "implicitly" by the equation $F(x, y, z) = 0$, can be found in [Bl 55].

---

[1]The numbers **1.1**,**1.2**, etc. mean the notes to Section 1.1, Section 1.2, etc.

1.7. The canonical affine connection Γ on a three-web was defined in [A 69b]. It was considered in more detail in [AS 81]. Kikkawa [Ki 85] named it the Chern connection.

1.8. The connections indicated in this section were studied in [AS 81]. The middle connection $\overset{*}{\Gamma}$ was defined in [AS 71b] in relation to the study of Moufang webs.

1.9. Subwebs and factor webs were first defined in [S 81b]. However, they were given a special consideration later, in [AS 85b] and [AS 88].

Problems. Webs with a symmetric curvature tensor were studied in [S 81b], [A 81a], [To 81] and [To 82]. The webs, discussed in Problem **20**, were considered in [To 81] and [To 82].

Chapter 2

Algebraic Structures Associated with Three-Webs

2.1 Quasigroups and Loops

1. Certain algebraic systems, first of all quasigroups and loops, are associated with three-webs.

A groupoid Q with a binary operation

$$q(x,y) \stackrel{\text{def}}{=} x \cdot y$$

is called a *quasigroup* if for any $a, b, c \in Q$ the equations

$$a \cdot y = c \quad \text{and} \quad x \cdot b = c. \tag{2.1}$$

are solvable and if each has a unique solution. If a quasigroup Q has an *identity element*, it is called a *loop*. A loop is different from a group by the fact that, in general, associativity $x \cdot (y \cdot z) = (x \cdot y) \cdot z$ does not hold in a loop. An associative loop is a *group*. An example of a non-associative loop is a set of non-zero octaves with respect to multiplication (see, for example, [Po 82], p. 307). Other division algebras, such as the fields of real and complex numbers or the division ring of quaternions, are associative loops with respect to multiplication, i.e. groups. A set of points on a straight line with an operation $x \cdot y = z$, where y is the midpoint of a segment xz, is a quasigroup but not a loop.

In what follows we will need a notion of a *three-base quasigroup*. This is a triple of sets X_1, X_2 and X_3 with an operation $q : X_1 \times X_2 \to X_3$ in which equations (2.1) are solvable for this operation. Here, as above, $q(x,y) = x \cdot y$, $x \in X_1, y \in X_2$ and $x \cdot y \in X_3$. We denote a three-base quasigroup by $q(\cdot, X_\alpha)$, $\alpha = 1,2,3$, or simply $q(\cdot)$. Note that the notion of the identity element has a meaning only in a one-base quasigroup.

Sets X_α in the definition of a three-base quasigroup are of the same cardinality, since the first of the relations (2.1), with fixed y, establishes a one-to-one correspondence between X_1 and X_3, and the second one, with fixed x, establishes a one-to-one correspondence between X_2.

In the theory of quasigroups the notion of isotopy, generalizing the notion of isomorphism, plays an important role.

Two quasigroups $q = (\cdot, X_\alpha)$ and $\widetilde{q} = (\circ, \widetilde{X}_\alpha)$ are called *isotopic* if there exists a triple $J = (J_1, J_2, J_3)$ of bijections $J_\alpha : X_\alpha \rightarrow \widetilde{X}_\alpha$, $\alpha = 1,2,3$, such that for any $x \in X_1$ and $y \in X_2$ the relation

$$J_1(x) \circ J_2(y) = J_3(x \cdot y) \tag{2.2}$$

holds. The triple $J = (J_\alpha)$ is called an *isotopy* of the quasigroup $q(\cdot)$ upon a quasigroup $\widetilde{q}(\circ)$. This definition is also valid in the cases when one or both of quasigroups $q(\cdot)$ and $\widetilde{q}(\circ)$ are one-base quasigroups.

The isotopies are the widest class of mappings, transfering a quasigroup into a quasigroup, since a groupoid, which is isotopic to a quasigroup, is itself a quasigroup.

The isotopy $J = (J_\alpha)$, $\alpha = 1,2,3$, is *regular* if one of the mappings J_α is the identity map. In particular, the isotopy of the form $(J_1, J_2,$ id is called a *principal isotopy.*

A composition of isotopies can be defined in a natural way:

$$J \cdot \widetilde{J} = (J_1 \cdot \widetilde{J}_1, \;\; J_2 \cdot \widetilde{J}_2, \;\; J_3 \cdot \widetilde{J}_3)$$

Action from the left is assumed here: $(J \cdot \widetilde{J})(x) = J(\widetilde{J}(x))$.

If $q(\cdot)$ and $\widetilde{q}(\circ)$ are one-base quasigroups, then an isotopy $J = (J_\alpha)$ of q upon $\widetilde{q}$, where $J_1 = J_2 = J_3$, is an isomorphism.

Let $q(\cdot, X_\alpha)$ be a three-base quasigroup, Q be a set and $J_\alpha : X_\alpha \rightarrow Q$ be bijections. Define an operation "$\circ$"in Q by means of the equation (2.2). A groupoid $Q(\circ)$ is a quasigroup isotopic to the initial quasigroup $q(\cdot)$. In particular, if $J_3 =$ id, then $X_3 \equiv Q$, and the quasigroup $X_3(\circ)$ is a *principal isotope* of a quasigroup $q(\cdot, X_\alpha)$.

2. We note some properties of isotopies.

Proposition 2.1 *An isotopy of a three-base quasigroup $q(\cdot, X_\alpha)$ upon a one-base quasigroup $Q(\circ)$ is a composition of a principal isotopy and isomorphism.*

Proof. Let $J = (J_\alpha)$ be an isotopy under consideration. Then $J_\alpha : X_\alpha \rightarrow Q$, and condition (2.2) holds. We define a new operation "$\times$" in X_3:

$$J_3^{-1}J_1(x) \times J_3^{-1}J_2(y) = x \cdot y, \; x \in X_1, \; y \in X_2, \; x \cdot y \in X_3.$$

By (2.2), we have the following relation:

$$J_3^{-1}J_1(x) \times J_3^{-1}J_2(y) = x \cdot y = J_3^{-1}(J_3(x \cdot y)) = J_3^{-1}(J_1(x) \circ J_2(y)),$$

which implies that the triple of mappings $\tilde{J}^{-1} = (J_3^{-1},\ J_3^{-1},\ J_3^{-1})$ of the quasigroup $Q(\circ)$ onto $X_3(\times)$ is an isomorphism. Therefore, the composition $\tilde{J}^{-1}J$ has the form

$$\tilde{J}^{-1} \cdot J = (J_3^{-1}J_1,\ J_3^{-1}J_2,\ \mathrm{id}),$$

i.e. it is a principal isotopy of the quasigroup $q(\cdot)$ upon the quasigroup $X_3(\times)$. Proposition 2.1 follows from the following trivial equation: $J = \tilde{J}(\tilde{J}^{-1}J)$. ■

Proposition 2.2 *Every pair of elements a from X_1 and b from X_2 define a principal isotopy of a three-base quasigroup $q(\cdot, X_\alpha)$ upon a loop $X_3(\circ)$, whose operation $\circ$ depends on a choice of elements a and b.*

Proof. In fact, consider the bijections $R_b : X_1 \to X_3$ and $L_a : X_2 \to X_3$, defined by the equations $R_b(x) = x \cdot b$, $L_a(y) = a \cdot y$. Let $u = R_b(x)$ and $v = L_a(y)$, where $u, v \in X_3$. We define a new operation "$\circ$" in X_3 setting

$$u \circ v = x \cdot y = R_b^{-1}(u) \cdot L_a^{-1}(v). \tag{2.3}$$

The quasigroup $X_3(\circ)$ is a principal isotope of the quasigroup $q = (\cdot, X_\alpha)$. The fact, that this quasigroup is a loop with the identity element $e = a \cdot b$, follows from the relations:

$$u \circ e = R_b^{-1}(u) \cdot L_a^{-1}(e) = x \cdot b = u,$$
$$e \circ v = R_b^{-1}(e) \cdot L_a^{-1}(v) = a \cdot y = v. ■$$

A loop $X_3(\circ)$ is called an *LP-isotope* of the quasigroup $q = (\cdot)$ and is denoted $l(a,b)$. Equation (2.3) shows that the isotopy $q(\cdot) \to X_3(\circ)$ has the form (R_b, L_a, id). The importance of the LP-isotopes is clarified by the following proposition.

Proposition 2.3 *Every principal isotope of a quasigroup $q(\cdot, X_\alpha)$, which is a loop, is an LP-isotope of this quasigroup.*

Proof. Let the principal isotopy $q(\cdot) \to Q(\circ)$ have the form (J_1, J_2, id). Then we have $Q = X_3$ and, for any $x \in X_1$ and $y \in X_2$, the relation

$$J_1(x) \circ J_2(y) = x \cdot y \tag{2.4}$$

holds. Let e be the identity element of the loop $Q(\circ)$. Denote $J_1^{-1}(e) = a$ and $J_2^{-1}(e) = b$. It follows from the relations $J_1(a) \circ J_2(y) = a \cdot y$ and $J_1(a) = e$ that $J_2(y) = a \cdot y$, i.e. $J_2 = L_a$. Similarly we can prove that $J_1 = R_b$. ■

An isotopy of a quasigroup $q(\cdot, X_\alpha)$ upon itself is called an *autotopy* of $q(\cdot, X_\alpha)$. If $A = (A_1, A_2, A_3)$ is an autotopy, then, for any pair $(x, y) \in X_1 \times X_2$, we have

$$A_1(x) \cdot A_2(y) = A_3(x \cdot y) \tag{2.5}$$

Properties of autotopies will be studied in detail in Chapter 6.

Loop manifold	Identity in a loop	Name of the identity
Commutative loops	$x \cdot y = y \cdot x$	Commutativity
Monoassociative loops	$x^2 \cdot x = x \cdot x^2$	Monoassociativity
Right-alternative loops	$x \cdot y^2 = (x \cdot y) \cdot y$	Right alternativity
Left-alternative loops	$x^2 \cdot y = x \cdot (x \cdot y)$	Left alternativity
Alternative loops	$x^2 \cdot y = x \cdot (x \cdot y),\ x \cdot y^2 = (x \cdot y) \cdot y$	
Elastic loops	$(x \cdot y) \cdot x = x \cdot (y \cdot x)$	Elasticity
Left Bol loops	$(x \cdot (y \cdot x)) \cdot z = x \cdot (y \cdot (x \cdot z))$	Left Bol identity
Right Bol loops	$x \cdot ((y \cdot z) \cdot y) = ((x \cdot y) \cdot z) \cdot y$	Right Bol identity
Moufang loops	$(x \cdot y) \cdot (z \cdot x) = x \cdot ((y \cdot z) \cdot x)$	Moufang identity

Table 2.1

3. As noted earlier, an associative loop is a group. However, there are classes of loops between groups and general loops. For loops of these classes a certain weakened associativity condition holds. These classes of loops play an important role in the theory of webs.

A set of loops, for which a certain identity S holds, is called a *manifold of loops* and is denoted by $V(S)$. Thus, the groups form a manifold of loops defined by the associativity identity $(x \cdot y) \cdot z = x \cdot (y \cdot z)$.

In Table 2.1 we have listed more most important loop manifolds and corresponding identities. The identities, from the second to the fifth in Table 2.1, generalize the associativity identity. They are connected by some relations. Some of these relations are obvious: for example, alternativity or elasticity implies mono-associativity. Other less obvious relations will be considered in the problem set and following sections.

In general, loop manifolds are not invariant with respect to isotopy. This means that if an identity S holds in a loop Q, then, in general, it does not hold in its isotopes. For example, as Problem **10** shows, the manifold of commutative loops is not invariant relative to isotopies.

On the other hand, Alberts' theorem is valid: if a quasigroup is isotopic to a group, then they are isomorphic, and the quasigroup itself is a group.[1] Alberts' theorem implies that the manifold of groups is invariant relative to isotopies. It turns out be that there are other manifolds of loops with this property. The identities defining such manifolds are said to be *universal.* A fundamental problem associated with universal identities is: find out whether a given identity S is universal, and if the answer is no, find a maximal submanifold of the manifold $V(S)$ which is invariant relative to isotopies. We will show in Section 2.3 that to any universal identity there corresponds a class of three-webs. This will allow us to prove that the Moufang and Bol identities are universal.

In what follows, we will write the product $x \cdot y$ in quasigroups and loops without the dot sign if this will not lead to ambiguities.

2.2 Configurations in Abstract Three-Webs

1. In Chapter 1 we considered three-webs formed by smooth foliations in a differentiable manifold of dimension $2r$. We will call such webs *geometric.* However, there exists a simplier structure preserving those properties of geometric webs related to the intersection of leaves, i.e. incidence properties, but not possessing the differential-geometric properties of these webs. This incidence structure is called an abstract three-web.

We give now the precise definition. Consider a set X whose elements we will call *points* and consider sets (or families) λ_α, $\alpha = 1, 2, 3$, whose elements we will call *lines.*

Definition 2.4 The sets X and λ_α form an *abstract three-web* $W(X, \lambda_\alpha)$ if their elements are connected by an incidence relation satisfying the following axioms:

A1. Every point of the set X is incident to just one line from each family λ_α.

A2. Any two lines from different families are incident to exactly one point of X.

The following property of an abstract three-web follows from these two axioms.

A3. Two lines from the same family λ_α are disjoint.

---

[1] A proof of this theorem can be found in [Bel 67], p. 17. Another, purely geometric proof will be given in Section 2.3.

We will denote elements of the sets X and λ_α by small Latin letters. If a point p is incident to a line x, then we say that p lies on x, or x passes through p. If the points p and q lie on a line of a family λ_α, we will write $p\alpha q$.

Axiom A1 always holds on a geometric web, but axiom A2 does not. Consider, for example, a three-web W, formed in a plane by the lines $x = \text{const}$, $y = \text{const}$ and the pencil of circles with vertices at the points $(-1, -1)$ and $(1, 1)$ (see Problem **1** (iii) to Chapter 1). If the web W is given in the open square $D = (-1, 1) \times (-1, 1)$, then axiom A2 holds. If we extend the web domain to an open circle bounded by the circumference S, circumscribed around the square D, then axiom A2 does not hold, since not just any two chords of the circumference S, that are parallel to different coordinate axes, have a common point inside S. If we further extend a web domain, there appear points and circles belonging to the web W that have two common points.

On the other hand, all axioms of an abstract three-web hold for a parallel three-web formed by three families of r-dimensional planes in an affine space A^{2r}.

An abstract three-web $W = (X, \lambda_\alpha)$ define six mappings $q_{\alpha\beta} : \lambda_\alpha \times \lambda_\beta \rightarrow \lambda_\gamma$, where α, β, $\gamma = 1, 2, 3$; $\alpha \neq \beta \neq \gamma \neq \alpha$ and $q_{\alpha\beta}(x_\alpha, x_\beta) = x_\gamma$, if lines x_α, x_β and $x_\gamma (x_\alpha \in \lambda_\alpha, \alpha = 1, 2, 3)$ have a common point. Since these mappings are uniquely solvable relative to x_α and x_β, they are three-base quasigroups. These quasigroups are called *coordinate quasigroups* of a three-web W.

It follows immediately from Definition 2.4 that the sets of points of two distinct lines of an abstract three-web have the same cardinality. Moreover, there is a bijection between these two lines established by the lines of the web. Since each point of the set X is the intersection of two lines, for example, x_1 and x_2, where $x_1 \in \lambda_1$ and $x_2 \in \lambda_2$, the set X is bijective to the direct product $\lambda_1 \times \lambda_2$: $X \sim \lambda_1 \times \lambda_2$. To lines of the first family then there corresponds the set of pairs of the form (a, y), where a is a fixed element from λ_1 and y is any line from λ_2. To lines of the second family there corresponds the set of points of the form (x, b). To lines of the third family there corresponds the set of pairs of the form (x, y) such that $q_{12}(x, y) = \text{const}$. Similarly, an abstract three-web can be assigned coordinates with the help of any quasigroup $q_{\alpha\beta}$.

All coordinate quasigroups of an abstract three-web W can be obtained from one of these quasigroups, for example, from q_{12}. If $q_{12}(x, y) = z$, then $q_{21}(x, y) = q_{12}(y, x)$, $q_{32}(z, y) = q_{23}(y, z) = x$ and $q_{13}(x, z) = q_{31}(z, x) = y$. The quasigroups q_{13} and q_{32} are called the *right inverse quasigroup* and the *left inverse quasigroup* for the quasigroups q_{12}, $q_{13} = q_{12}^{-1}$ and $q_{32} = {}^{-1}q_{12}$, respectively. The operation in the quasigroup q_{12}^{-1} is also denoted by "\" and the operation in the quasigroup ${}^{-1}q_{12}$ by "/". Thus, $q_{13}(x, z) = x \backslash z$ and $q_{32}(z, y) = z / y$. The transition from one coordinate quasigroup to another is called a *parastrophy*. In the following discussion we will often use the quasigroup q_{12} as the basic one, will denote it simply by q and denote the operation in it by the dot.

Conversely, a three-base quasigroup $q : X_1 \times X_2 \rightarrow X_3$ uniquely defines an abstract three-web W in the direct product $X = X_1 \times X_2$. Lines of the first family are subsets of points of the form (a, y) from X, where a is a fixed element from X_1. Lines of the second family are subsets of points of the form (x, b), where b is a fixed element from X_2. Lines of the third family are pairs of the form (x, y), for which $x \cdot y = c$, where $c \in X_3$. Since in a quasigroup q each of the equations is uniquely solvable, then axioms A1 and A2 hold for the three-web we have constructed. The initial quasigroup q is a coordinate quasigroup q_{12} of this web.

Thus, with an abstract three-web W there are associated its three coordinate quasigroups $q_{\alpha\beta}$, and conversely, a three-base quasigroup q generates an abstract three-web, for which it is a coordinate quasigroup.

	a	b	c
a	c	a	b
b	b	c	a
c	a	b	c

Figure 2.1

Consider, for example, a quasigroup q consisting of three elements, for which the table of multiplication is shown in Figure 2.1. The corresponding three-web is shown in Figure 2.2. Each of its leaves contains three points.

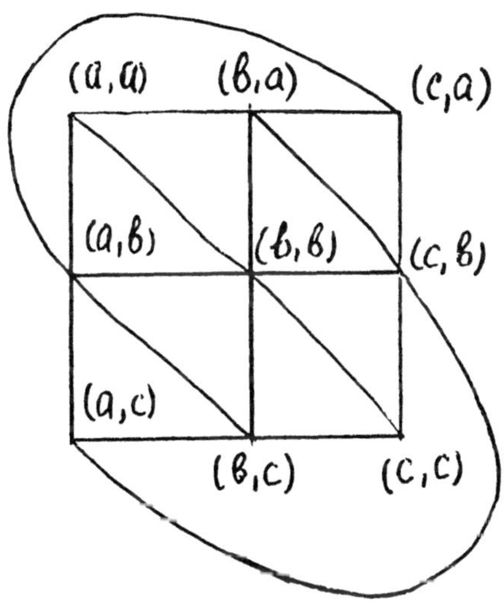

Figure 2.2

The correspondence between quasigroups and abstract three-webs indicated above shows that an isotopic transformation of a coordinate quasigroup implies the corresponding transformation of the corresponding web. We will also call the latter an *isotopy*. Equation (2.2), defining an isotopic transformation of a coordinate quasigroup q of a three web W into a coordinate quasigroup $\tilde{q}$ of a three-web $\widetilde{W}$, means that three lines of the three-web W, passing through a point, are transformed into three lines of the three-web $\widetilde{W}$, also passing through a point. *Isotopy is an equivalence relation in the set of abstract three-webs.*

Recall that an isotopy of geometric three-webs was defined in Section 1.2. However, in that definition the mappings J_α were local diffeomorphisms and not necessarily global bijections.

2. Classification of three-webs, both abstract and geometric, is closely connected with *closure conditions* of configurations, formed by points and leaves of a web.

Figure 2.3 shows a *hexagonal figure* or the *figure* H. Here and in what follows the leaves of the first, second and third families of a web are represented by vertical, horizontal and oblique lines respectively.

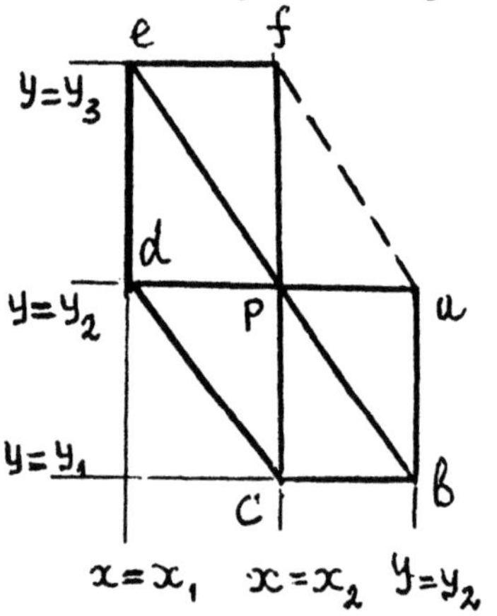

Figure 2.3

$$
\begin{aligned}
x_1y_2 &= x_2y_1, \\
x_1y_3 = x_2y_2 &= x_3y_1, \\
x_2y_3 &= x_3y_2.
\end{aligned}
\tag{2.6}
$$

The figure H can be obtained in the following way. Let p be an arbitrary point of the set X. Three lines of the web $W = (X, \lambda_\alpha)$ pass through it. Take a point a on one of these lines and construct the points b, c, d, e and f as shown in Figure 2.3. The existence of these points follows from Definition 2.4 of an abstract three-web. In general, a line of the family λ_3, passing through a point f, does not pass through a. However, if the points a and f belong to a line of the family λ_3, we say that the hexagonal figure $(abcdef)$ is closed. Hexagonal figures are denoted by the letter H. Let us write the closure condition H with help of the operation of multiplication in the coordinate quasigroup q. Denote the lines of the first two families forming the figure H by x_1, x_2, x_3 and y_1, y_2, y_3, as shown in Figure 2.3. Recall that in the coordinate quasigroup q the equation $z = x \cdot y$ holds if a line z of the third family passes through the point of intersection of lines x and y of the first and the second families. Thus, to the closed figure H shown in Figure 2.3 there corresponds the following equations held in the coordinate quasigroup q: The conditions (2.6) can be written in the form of an implication:

$$x_1y_2 = x_2y_1, \quad x_1y_3 = x_2y_2 = x_3y_1 \Longrightarrow x_2y_3 = x_3y_2,$$

which is called a *conditional identity* [Bel 66].

Other basic figures are shown in Figures 2.4–2.9. Next to these figures the corresponding closure conditions are written.

The Thomsen figure T:

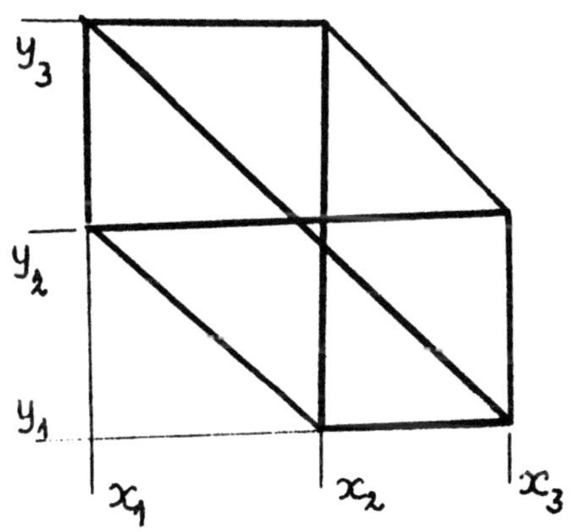

$$x_2y_1 = x_1y_2, \quad x_3y_1 = x_1y_3, \quad x_2y_3 = x_3y_2. \tag{2.7}$$

Figure 2.4

The Reidemeister figure R:

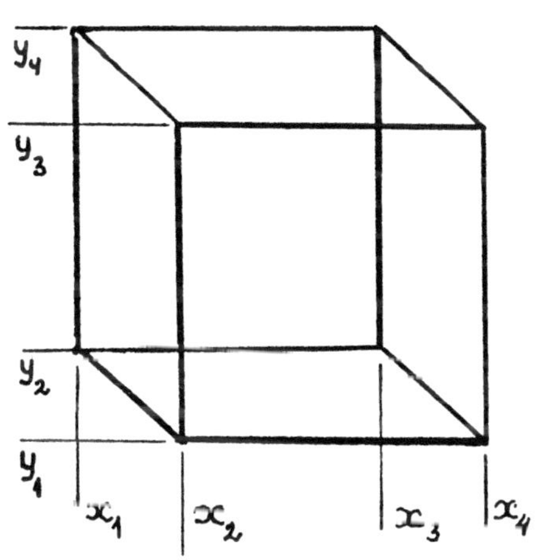

$$x_2y_1 = x_1y_2, \quad x_4y_1 = x_3y_2, \quad x_1y_4 = x_2y_3, \quad x_3y_4 = x_4y_3. \tag{2.8}$$

Figure 2.5

The left Bol figure B_l:

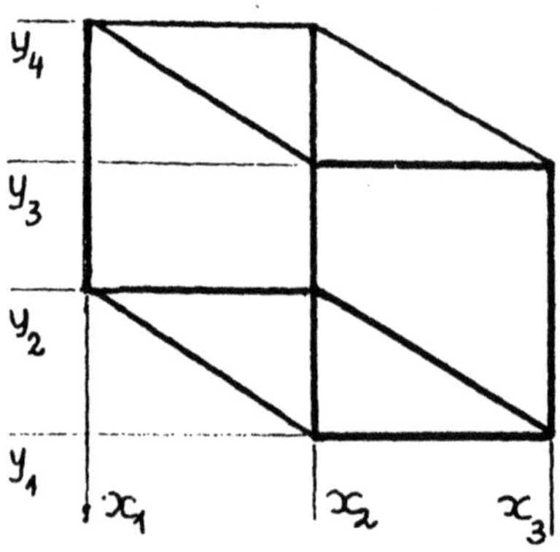

$$\begin{aligned} x_1y_2 &= x_2y_1, \\ x_1y_4 &= x_2y_3, \\ x_2y_2 &= x_3y_1, \\ x_2y_4 &= x_3y_3. \end{aligned} \tag{2.9}$$

Figure 2.6

The right Bol figure B_r:

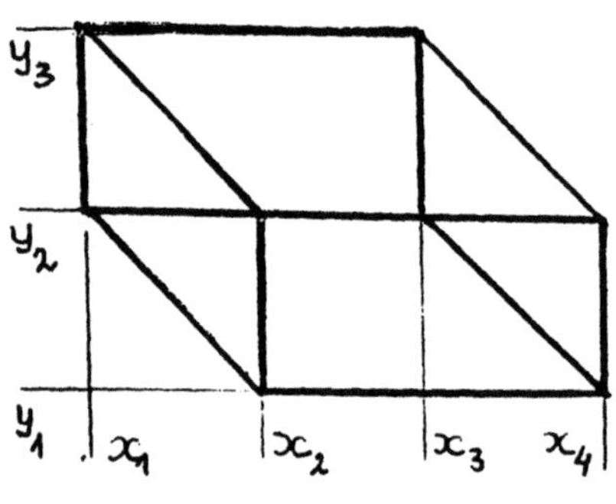

$$\begin{aligned} x_1y_2 &= x_2y_1, \\ x_1y_3 &= x_2y_2, \\ x_3y_2 &= x_4y_4, \\ x_3y_3 &= x_4y_2. \end{aligned} \tag{2.10}$$

Figure 2.7

The middle Bol figure B_m:

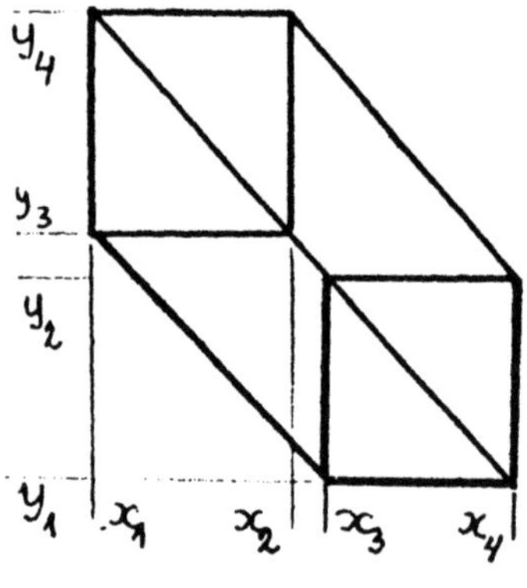

$$\begin{gathered} x_1y_2 = x_3y_4, \\ x_1y_1 = x_2y_2 = x_3y_3 = x_4y_4, \\ x_2y_1 = x_4y_3. \end{gathered} \tag{2.11}$$

Figure 2.8

The figure E:

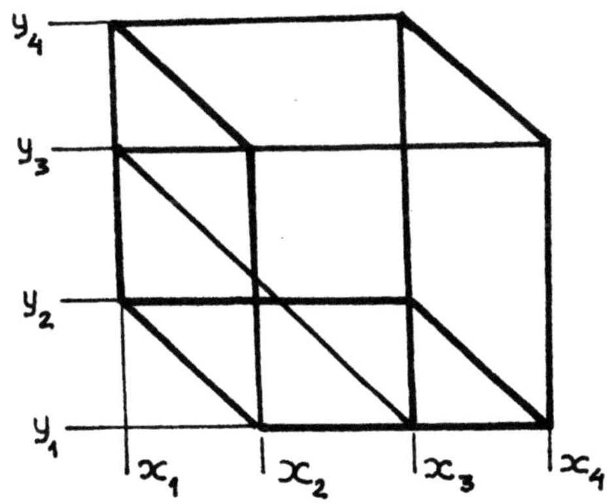

$$
\begin{aligned}
x_1y_2 &= x_2y_1,\\
x_1y_3 &= x_3y_1,\\
x_3y_2 &= x_4y_1,\\
x_2y_3 &= x_1y_4,\\
x_4y_3 &= x_3y_4.
\end{aligned}
\tag{2.12}
$$

Figure 2.9

Webs in which all figures of one of the types indicated above are closed are called the *webs* H, T, R, B_l, B_r, B_m and E, respectively. Figures 2.3–2.9 show that the figures H, B_l, B_r, B_m and E are particular cases of the figure R and can be obtained from it by imposing additional conditions. Similarly, the figure H can be obtained from the figures B_l, B_r, B_m and E. For example, the figure B_l can be obtained from the figures R shown in Figure 2.5 by making two middle vertical leaves coincide, i.e. by setting $x_2 \equiv x_3$. If, in addition, we set $y_2 \equiv y_3$, we obtain the figure H. The latter can be also obtained from the figure T if in T three interior lines have a common point.

The following theorem clarifies a relationship between classes T and R.

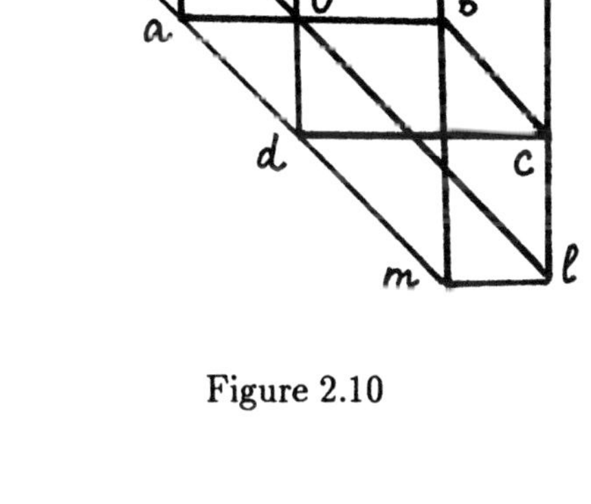

Figure 2.10

Theorem 2.5 *A three-web T is a web R.*

Proof. Let all figures T be closed in an abstract three-web W. We will prove that all figures R are also closed in this web. Consider an arbitrary figure $R = (abcdefgh)$ in W (Figure 2.10) and prove that it is closed, i.e. that the points f and g belong to a line of the third family (this condition can be written as $f3g$). Construct the points n, k, l and m as shown in Figure 2.10 and consider two figures T: $T_1 = (oanklh)$ and $T_2 = (obclmd)$. Since these two figures must be closed, the conditions $n1k$ and $l2m$ are satisfied. Next, consider the figure $T_3 = (lmnkfg)$. Since this figure must be also closed, the condition $f3g$ is satisfied. Therefore, the figure $R = (abcdefgh)$ is also closed. ■

The so-called webs M (*Moufang webs*) constitute another important class of webs. For them, the Bol figures of all three types, B_l, B_r and B_m, are closed. The following theorem "minimizes" this definition.

Theorem 2.6 *If the Bol figures of two types are closed in a three-web W, then the Bol figures of the third type are also closed.*

Proof. Let us prove, for example, that $B_l \& B_r \Rightarrow B_m$. Suppose that the figures B_l and B_r are closed in a three-web W. Consider an arbitrary figure $B_m = (abcdefgh)$ in W (Figure 2.11) and prove that it is closed, i.e. that the condition $b3f$ holds.

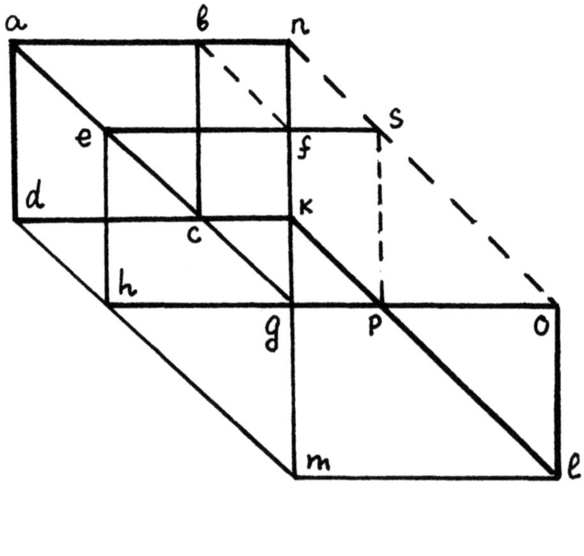

Figure 2.11

Construct consecutively the points n, k, m, l, o and p. The points d, k, l, m, a, n, o and g form the figure B_r. Since it must be closed, this implies $n3o$. Denote the intersection of the lines ef and on by s. The points h, e, g, m, p, s, o and l form the figure B_r. Since it must be closed, this implies pls. Next, the points b, n, s, f, c, k, p and g form the figure B_l. Since it must be also closed, this implies $b3f$. Therefore, the figure $B_m = (abcdefgh)$ is also closed. ■

Thus, the closure conditions considered above are related by the following implications:

$$T \Longrightarrow R \Longrightarrow M \begin{array}{c} \nearrow B_l \searrow \\ \Longrightarrow B_m \Longrightarrow \\ \searrow B_r \nearrow \end{array} H, \quad R \Longrightarrow E \Longrightarrow H$$

In conclusion, note that the Bol figures B_l, B_r and B_m (Figures 2.6–2.8) represent the same configuration but with different location relative to the families λ_α forming a three-web. It would be more precise to denote them by $\underset{12}{B_l}, \underset{12}{B_r}$ and $\underset{12}{B_m}$, relating them with the coordinate quasigroup q_{12}. If we transfer, for example, to the coordinate quasigroup q_{32}, then oblique lines must be considered as lines of the first family and vertical lines as lines of the third family. Thus, the figure $\underset{12}{B_m}$ coincides with the figure $\underset{32}{B_l}$, i.e. with the figure B_l for the coordinate quasigroup q_{32}. Similarly, one can prove that the figure $\underset{12}{B_m}$ coincides with the figure $\underset{13}{B_r}$. Therefore, we have

$$\underset{12}{B_m} \Longleftrightarrow \underset{32}{B_l} \Longleftrightarrow \underset{13}{B_r}. \tag{2.13}$$

2.3 Identities in Coordinate Loops and Closure Conditions

1. Relations (2.6)–(2.12), representing the closure conditions for different configurations in an abstract three-web $W = (X, \lambda_\alpha)$, take a simpler form if, instead of using the coordinate quasigroup $q : \lambda_1 \times \lambda_2 \to \lambda_3$, one uses LP-isotopes of W, which are called *coordinate loops* of the web W. According to Section 2.1, an LP-isotope $l(a, b)(\circ)$ is defined by formula (2.3):

$$u \circ v = R_b^{-1}(u) \circ L_a^{-1}(v) = x \cdot y,$$

where $u, v \in \lambda_3$, $u = x \cdot b$ and $v = a \cdot y$. The element $u \circ v$ can be constructed in the web W as shown in Figure 2.12.

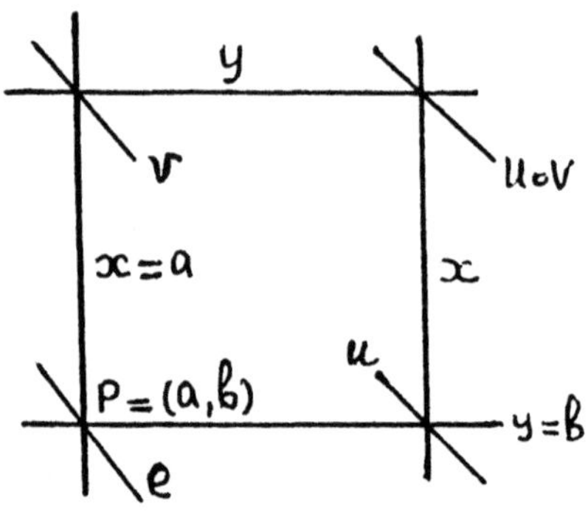

Figure 2.12

Thus, the coordinate loop $l(a,b)$ is connected with every point $p=(a,b)$. We denote this loop also by l_p. Since LP-isotopes are principal isotopes of the coordinate quasigroup $q(\cdot)$, they are principal isotopes of each other. It follows from Propositions 2.2 and 2.3 that the class of loops, isotopic to each other, corresponds to a three-web W.

Suppose that elements u and v of the coordinate loop $l(a,b)$ commute, i.e. $u \circ v = v \circ u$. Using Figure 2.12, we construct the products $u \circ v$ and $v \circ u$ (Figure 2.13).

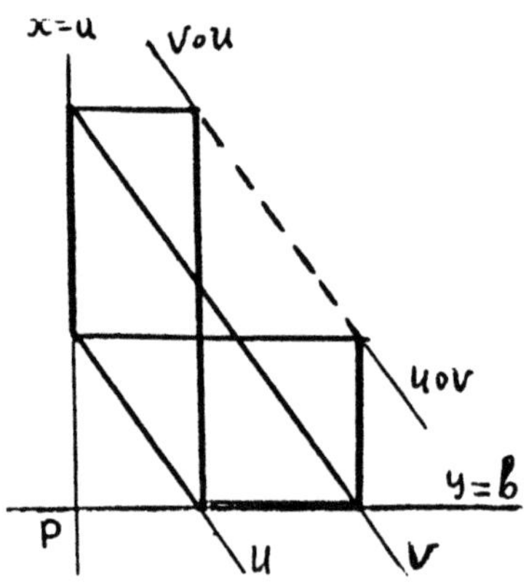

Figure 2.13

Comparing Figures 2.5 and 2.13, we see that the condition $u \circ v = v \circ u$ is equvalent to the closure of a figure T. We can obtain figures T shown in Figures 2.13–2.15 by taking different locations for lines u and v with respect to the point p.

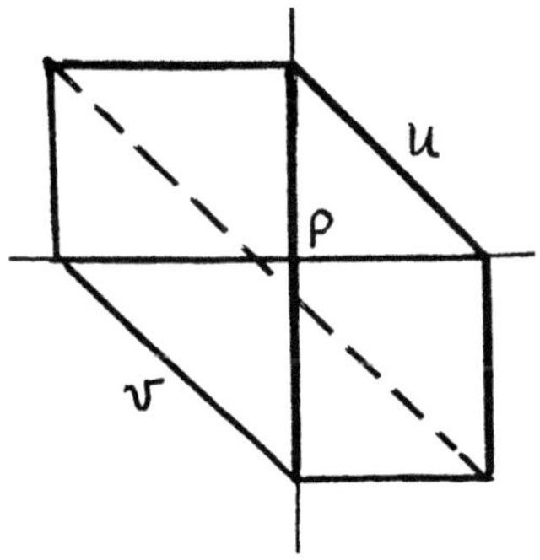

Figure 2.14

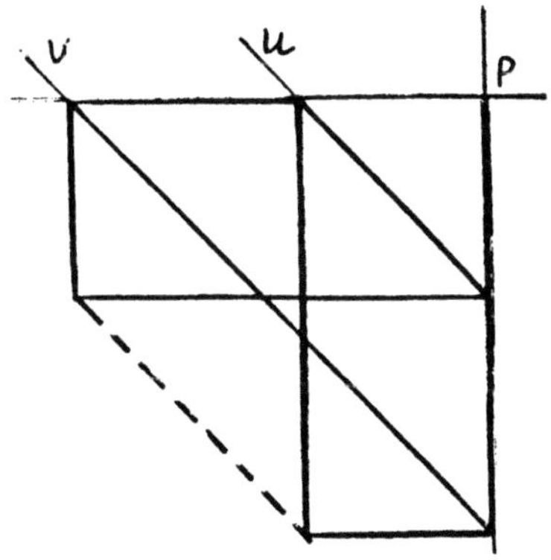

Figure 2.15

Suppose now that the coordinate loop $l(a, b)$ is commutative. Then all possible figures T, situated relative to lines $x = a$ and $y = b$ as shown in Figures 2.13–2.15, are closed in the web W under consideration. We call these figures *coordinate figures* associated with the loop $l(a, b)$. Interchanging a and b, we obtain all figures T in a three-web W. This implies that *an abstract three-web is a web T if and only if all its coordinate loops are commutative.*

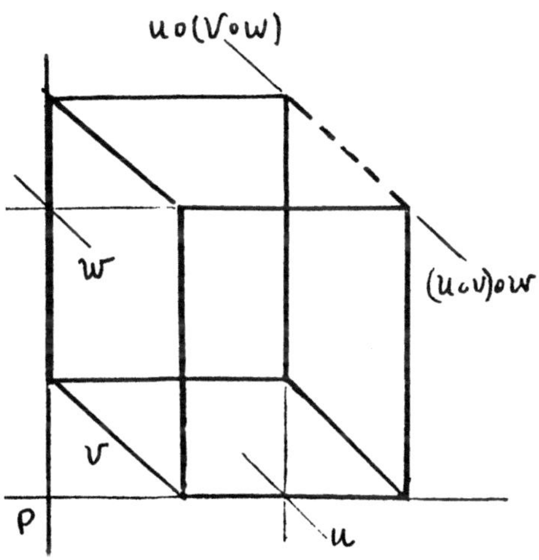

Figure 2.16

Next, let us take in the loop $l(a, b)$ three arbitrary elements u, v and w and construct the products $(u \circ v) \circ w$ and $u \circ (v \circ w)$ (Figure 2.16). Comparison of Figures 2.16 and 2.6 shows that the identity of these products leads to the closure of a figure R. If in the loop $l(a, b)$ associativity $(u \circ v) \circ w = u \circ (v \circ w)$ holds, then all possible figures R, connected with the loop $l(a, b)$, are closed in the web (see Problem 16).

Interchanging a and b, we come to the conclusion that an *abstract three-web is a web R if and only if all its coordinate loops are associative loops, i.e. groups.*

Comparing Figures 2.6, 2.7, 2.9 and 2.3 with Figure 2.16, we find that the figures B_l, B_r, E and H can be obtained from the figure R if one takes $u = v, v = w, u = w$ and $u = v = w$, respectively. The identities $u^2 \circ w = u \circ (u \circ w)$, $u \circ v^2 = (u \circ v) \circ v$, $(u \circ v) \circ u = u \circ (v \circ u)$ and $u^2 \circ u = u \circ u^2$ correspond to the figures B_l, B_r, E and H, respectively.

Further, in Section 2.2 we proved that a web T is a web R. This implies that, in addition to commutativity, associativity also holds in its coordinate loops, i.e. coordinate loops of a web T are abelian groups.

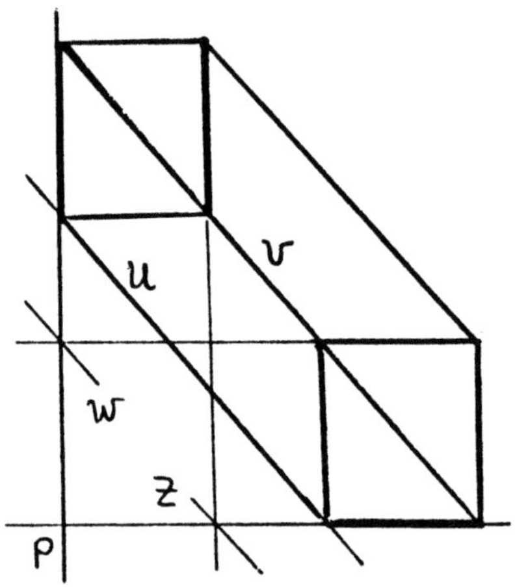

Figure 2.17

To find an identity corresponding to the figure B_m, we introduce variables u, v, w and z in the way shown in Figure 2.17. Using Figure 2.12, we obtain the following equations:

$$u \circ w = v, \quad z \circ u = v, \quad v \circ w = z \circ v.$$

The first two equations imply that $w = u \backslash v$ and $z = v/u$. Substituting these values into the last equation, we get the desired identity:

$$u \circ (u \backslash v) = (v/u) \circ v. \tag{2.14}$$

Comparing these results with Table 2.1, we arrive at the following theorem.

Theorem 2.7 *An abstract web W is a web R, T, B_l, B_r, M, E, H or B_m if and only if all its coordinate loops are groups, abelian groups, left alternative, right alternative, alternative, elastic, monoassociative loops or loops where identity* (2.14) *holds, respectively.* ■

2. In the previous subsection we considered three-webs in whose coordinate loops a certain identity holds. We call $W(S)$ a class of three-webs with coordinate loops belonging to a manifold $V(S)$ (cf. Section 2.1).

Suppose that the identity S is universal. Since all coordinate loops are isotopic to each other, we conclude that a web W belongs to a class $W(S)$ if and only if the identity S holds at least in one of the coordinate loops of W.

In contrast, if the identity S is not universal, then a web W belongs to a class $W(S)$ if and only if all its coordinate loops belong to the manifold $V(S)$.

In the last case it is important to indicate a universal identity $\tilde{S}$, which characterizes the loop manifold $V(S)$ or the web W. This universal identity can be found as follows. Since the web W belongs to the class $W(S)$, then all figures S, in a certain way connected with coordinate loops (in previous subsection we called them coordinate figures), are closed. Consider an arbitrary figure S, which is not a coordinate figure for the loop $l(a,b)$. To this figure S in the loop $l(a,b)$ there corresponds a certain identity $\tilde{S}$, which is universal. In fact, if the identity $\tilde{S}$ holds in at least one of the loops $l(a,b)$, then all possible figures S are closed in the web W. Therefore, the identity $\tilde{S}$ holds in all coordinate loops of the web.

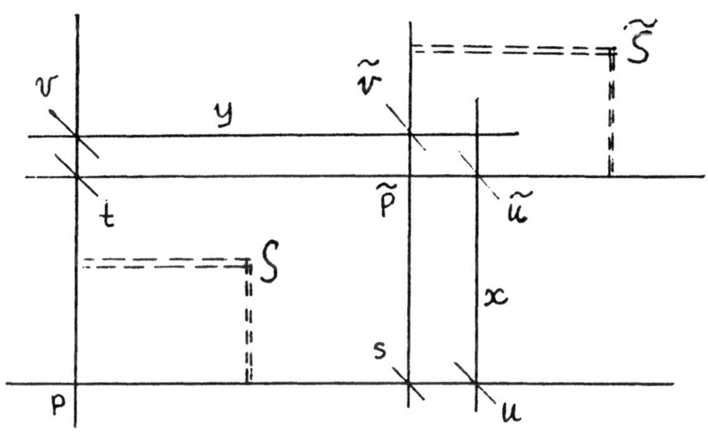

Figure 2.18

As a result, we have the following algorithm for finding a universal identity $\tilde{S}$ corresponding to an identity S. In Figure 2.18, S is a certain coordinate figure associated with a coordinate loop $l_p(\circ)$ and $\tilde{S}$ is the same figure located arbitrarily relative to a point p. The figure $\tilde{S}$ is a coordinate figure for another coordinate loop $l_{\tilde{p}}(\tilde{\circ})$. The definition of LP-isotope in Section 2.1 implies the following identity (see Figure 2.18):

$$\tilde{u}\tilde{\circ}\tilde{v} = x \cdot y = u \circ v.$$

Setting $\tilde{u} = u \circ t$, $\tilde{v} = s \circ v$, we can write the last equation in the form:

$$\tilde{u}\tilde{\circ}\tilde{v} = (\tilde{u}/t) \circ v(s\backslash\tilde{v}), \tag{2.15}$$

where $\backslash$ and $/$ are the inverse operations of the operation $(\circ)$ (cf. Section 2.2).

In the loop $l_{\tilde{p}}(\tilde{\circ})$ the identity S corresponds to the identity $\tilde{S}$. To get a corresponding identity in the loop $l_{\tilde{p}}(\tilde{\circ})$, we apply equation (2.15) to each product $\tilde{u}\tilde{\circ}\tilde{v}$ in the identity S. This gives a universal identity $\tilde{S}$, which we were looking for. The identity $\tilde{S}$ is called a *derived identity* of the identity S and contains additional variables s and t in addition to the variables in the initial identity S.

For instance, the derived identity

$$(u/t \circ s\backslash u) \circ s\backslash u = u/t \circ s\backslash(u/t \circ s\backslash u), \tag{2.16}$$

with three variables u, s and t, corresponds to monoassociativity $u^2 \circ u = u \circ u^2$.

Besides the derived identity $\tilde{S}$, other universal identities, which do not contain inverse operations and depend on a smaller number of variables than the derived identity, can exist in the loop manifold $V(S)$. Such universal identity (2.17) holdidentities are connected not with an arbitrary figure but with a "semi-coordinate" figure S (see p. 109). To prove that such identities are universal, one must prove that if a semi-coordinate figure S, connected with one of the coordinate loops of a three-web, is closed, then all other such figures S are closed in this three-web.

Theorem 2.8 *Associativity*

$$(u \circ v) \circ w = u \circ (v \circ w) \tag{2.17}$$

is a universal identity.

Proof. Let identity (2.17) hold in a loop Q and let W be a three-web, for which this loop is one of coordinate loops l_p. Since this loop is associative, our considerations in the first subsection show that all coordinate figures R, connected with the loop l_p, are closed in the web W.

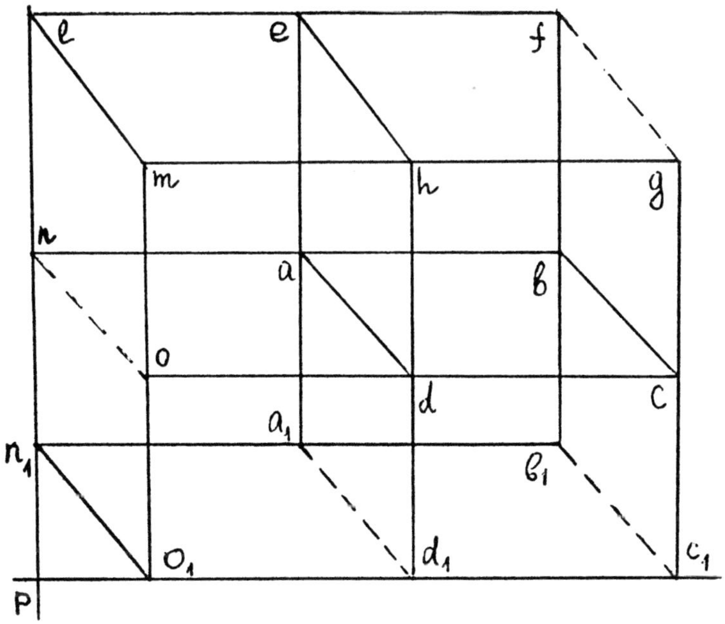

Figure 2.19

Consider now an arbitrary figure $R = (abcdefgh)$ (Figure 2.19). We will show that this figure is closed, i.e. that its points f and g belong to the same line of the third family. Construct the points $l, m, o, n, n_1, o_1, a_1, d_1, b_1$ and c_1 as shown in Figure 2.19. Since coordinate figures R are closed, the figure $R_1 = (elmha_1n_1o_1d_1)$ is also closed, i.e. the condition a_13d_1 holds. Consider further the coordinate figure $R_2 = (anoda_1n_1o_1d_1)$. Since it is also closed, we have $n3o$. Similarly, from the closure of the coordinate figure $R_3 = (bnocb_1n_1o_1c_1)$ follows the condition b_13c_1. Finally, from the closure of the coordinate figure $R_4 = (flmgb_1n_1o_1c_1)$ we find that the condition $f3g$ holds. Therefore, the figure $R = (abcdefgh)$ is closed.

Thus, all figures R are closed in the three-web W considered. Because of this, all coordinate loops of this three-web are groups. We proved earlier in Proposition 2.1 that, up to an isomorphism, the coordinate loops give all possible isotopes of Q. This implies that any isotope of an associative loop is also an associative loop. This proves that associativity is a universal identity. ■

Theorem 2.9 *A loop is a commutative group if and only if the identity T,*

$$T : u \circ (v \circ w) = v \circ (u \circ w), \tag{2.18}$$

holds in it. This identity is universal.

Proof. On one hand, the identity (2.18) follows from commutativity and associativity. Conversely, if we take $w = e$, we obtain commutativity. The latter and (2.18) give associativity. This proves the first part of Theorem 2.9.

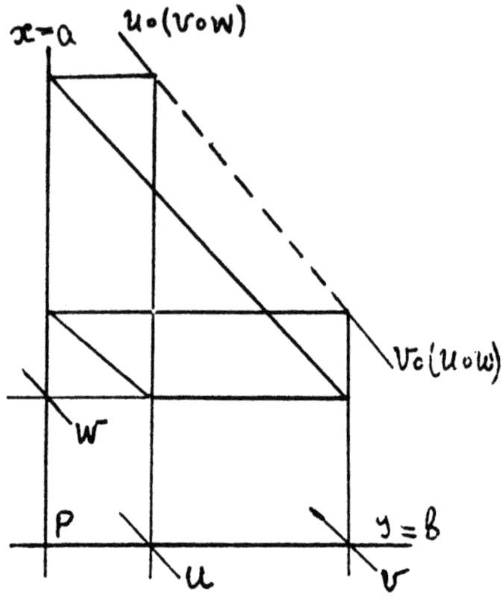

Figure 2.20

Suppose now that the identity (2.18) holds in a coordinate loop l_p of a three-web W. The figure shown in Figure 2.20 corresponds to this identity. This figure is a figure T, in which, unlike for coordinate figures T considered above, only one side belongs to the line $x = a$. We call this figure a *semi-coordinate figure* T.

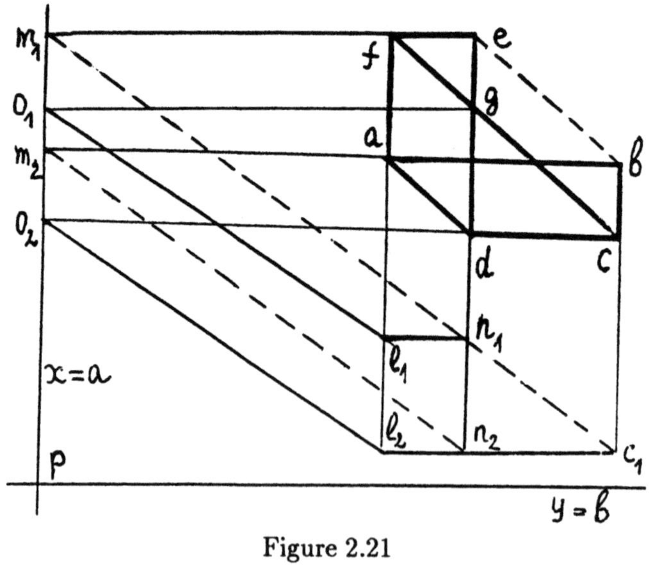

Figure 2.21

Consider further an arbitrary Thomsen figure $T = (abcdef)$, formed by the lines of the web W (Figure 2.21). We will prove that this figure is closed, i.e. its points b and e lie on the same line of the third family of the web W. Construct the points m_1, o_1, l_1 and n_1, as shown in Figure 2.21, and consider the figure $T_1 = (o_1gn_1l_1fm_1)$. It is a semi-coordinate figure, and thus the condition m_13n_1 holds. Construct further the points m_2, o_2, n_2 and l_2 and consider the semi-coordinate figure $T_2 = (o_2dn_2l_2am_2)$. Since this figure is closed, we get m_23n_2. After constructing the point c_1, we obtain another semi-coordinate figure $T_3 = (o_2cc_1l_2fm_1)$, which leads to the condition m_13c_1. Finally, consider the semi-coordinate figure $T_4 = (m_2bc_1l_2fm_1)$. Its closure gives the condition $b3f$. Therefore, the figure $T = (abcdef)$ is closed. Since this figure was taken arbitrarily, all figures T are closed in the web. Thus, all coordinate loops of the web considered are abelian groups, and identity (2.18) holds in them. But this means that this identity is universal. ■

Theorem 2.10 *The left Bol identity B_l,*

$$B_l : (u \circ (w \circ u)) \circ v = u \circ (w \circ (u \circ v)), \tag{2.19}$$

holds in the coordinate loops of the webs B_l, and only in such webs. The left Bol identity is universal.

Proof. Suppose that identity (2.19) holds in coordinate loops of a web. Denote by b_l the figure, corresponding to this identity. This figure is shown in Figure 2.22 (without the dotted line aa_1).

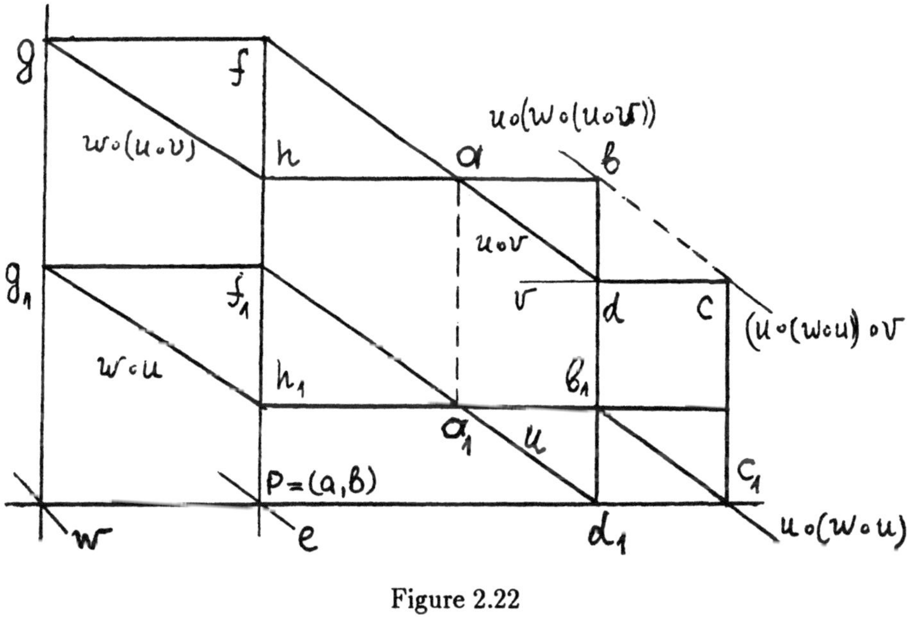

Figure 2.22

Particular cases of this figure are the left Bol figures B_l' and B_l, shown in Figures 2.23 and 2.24.

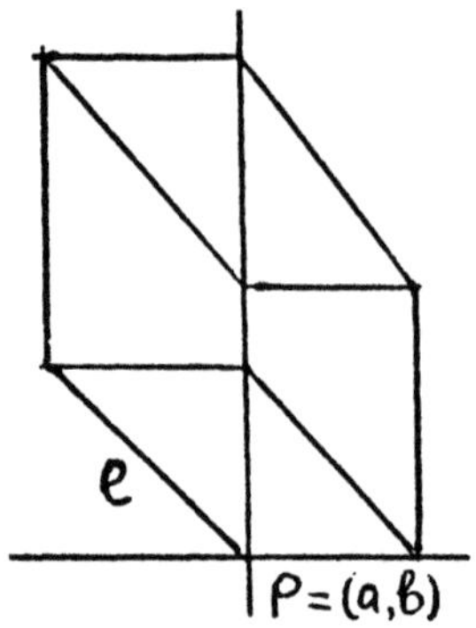

Figure 2.23

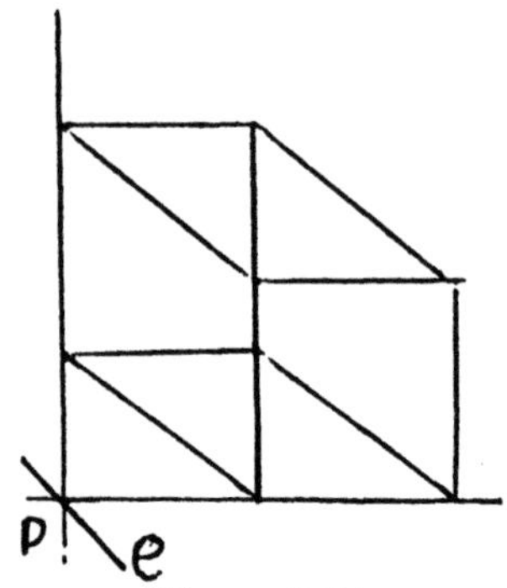

Figure 2.24

The former, B_l', can be obtained for $w \circ u = e$ and the latter for $w = e$, where e is the identity element of the loop l_p. First, we will prove that if the coordinate figures B_l' shown in Figure 2.23 are closed, then the semi-coordinate figures B_l' shown in Figure 2.25 are also closed.

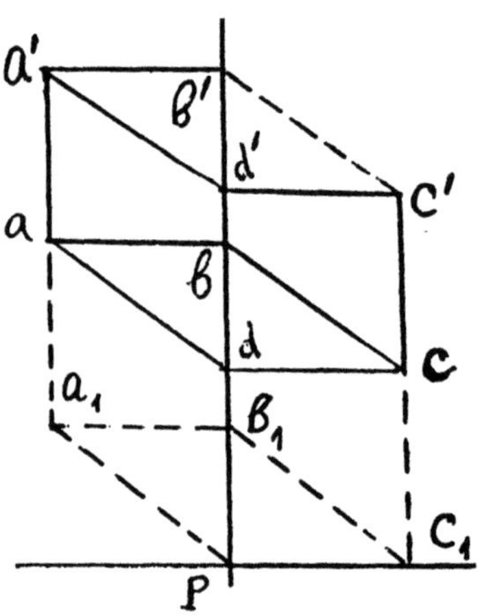

Figure 2.25

In fact, let us construct the points a_1, b_1 and c_1 as shown in Figure 2.25. We get the coordinate figure $B_l' = (abcda_1b_1c_1p)$. If the latter is closed, we get $b_1 3 c_1$. This gives another coordinate figure $B_l' = (a'b'c'd'a_1b_1c_1p)$ which implies that $b'3c'$.

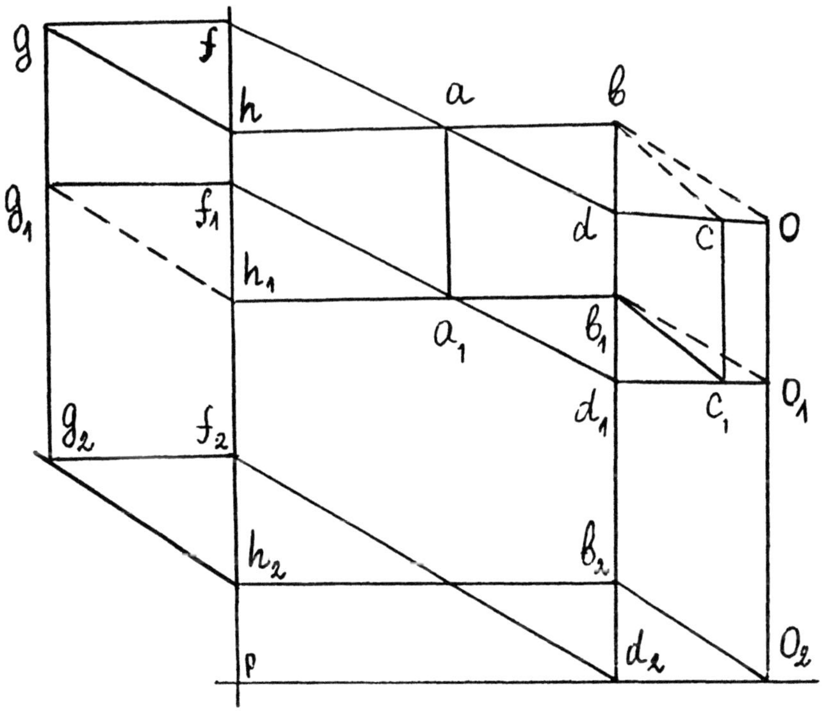

Figure 2.26

Consider now an arbitrary left Bol figure $B_l = (abcda_1b_1c_1d_1)$ (Figure 2.26) and prove that it is closed, i.e. the condition $b3c$ holds. First, we construct the points $f, h, g, d_2, f_2, g_2, h_2, o_2$ and o as shown in Figure 2.26, and consider the figure $b_l = (bodfghb_2o_2d_2f_2g_2h_2)$. Since, according to the conditions of the theorem, this figure is closed, we have $b3o$. Next, we construct the points f_1, g_1 and h_1 and consider the semi-coordinate figure $B_l' = (afgha_1f_1g_1h_1)$. As we proved above, this figure must be closed. Thus, we have g_13h_1. Finally, we construct the point o_1 and consider the figure $b_l = (b_1o_1d_1f_1g_1h_1b_2o_2d_2f_2g_2h_2)$. According to the conditions of the theorem, this figure is closed. This implies that b_13o_1. By the definition of a web, one and only one line of the third family passes through the point b_1. Therefore, $c_1 \equiv o_1$. This implies $c \equiv o$ and the figure $B_l = (abcda_1b_1c_1d_1)$ is closed. But the figure B_l was taken arbitrarily. Thus, all figures B_l are closed in the web W, i.e. this web is a web B_l.

Conversely, suppose that a web W is a left Bol web, i.e. all figures B_l are closed in this web. We will prove that all figures b_l shown in Figure 2.22 are also closed in W. In fact, each figure b_l is made up from two Bol figures B_l: $(abcda_1b_1c_1d_1)$ and

$(afgha_1f_1g_1h_1)$ that are "glued" together. Since, according to the conditions of the theorem, both these figures are closed, we have $a1a_1$ and $c3b$. This completes the proof of the first part of the theorem.

We will now prove that identity (2.19) is universal. Let this identity hold in a loop Q. Consider a three-web W, for which the loop Q is one of the coordinate loops l_p. Then, in the three-web W the coordinate figures b_l, associated with the loop l_p, are closed. These figures are shown in Figure 2.26. The first part of the theorem implies that in the web W the Bol figures B_l are closed. Consider further an arbitrary (not coordinate) figure b_l. One can prove in the same way as above, that this figure is closed since it is "glued" from two figures B_l. Thus, all figures B_l are closed in the web W. This means that the identity (2.19) is universal. ■

A similar proof can be performed to prove the following theorem.

Theorem 2.11 *The right Bol identity B_r,*

$$B_r : u \circ ((v \circ w) \circ v) = ((u \circ v) \circ w) \circ v, \tag{2.20}$$

holds in the coordinate loops of the webs B_r, and only in such webs. The right Bol identity is universal. ■

We recall that in Section 2.2 we defined a Moufang web as a web in which the Bol figures of all three kinds are closed. The last two theorems, Theorem 2.6 and Problems **8** and **9** of Chapter 1 imply

Theorem 2.12 *The Moufang identity M,*

$$M : (u \circ v) \circ (w \circ u) = u \circ ((v \circ w) \circ u), \tag{2.21}$$

holds in the coordinate loops of Moufang webs, and only in such webs. The Moufang identity is universal. ■

Theorem 2.13 *The identity B_m,*

$$B_m : w \circ ((u \circ v)\backslash w) = (w/v) \circ (u\backslash w), \tag{2.22}$$

holds in the coordinate loops of the webs B_m, and only in such webs. This identity is universal.

Proof. All figures B_m are closed in a web B_m. In particular, this is the case for the semi-coordinate figures B_m, associated with the loop l_p and shown in Figure 2.27. Identity (2.22) corresponds to these figures. If $u = e$, this identity is the same as identity (2.14). This proves the first part of the theorem.

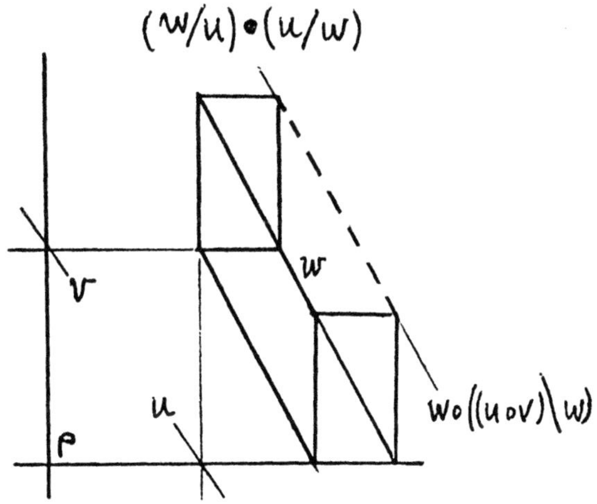

Figure 2.27

Suppose now that only semi coordinate figures B_m, associated with the loop l_p, are closed in a web W. We will prove that all such figures are closed in this web. Consider an arbitrary figure $B_m = (abcda_1b_1c_1d_1)$ shown in Figure 2.28.

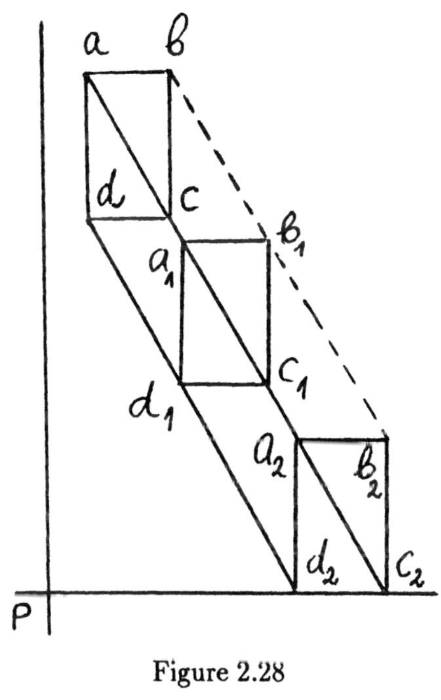

Figure 2.28

If we construct the points d_2, c_2, a_2, b_2 as shown in Figure 2.28, we obtain the semi-coordinate figure $B_m^1 = (a_1 b_1 c_1 d_1 a_2 b_2 c_2 d_2)$. The latter is closed according to the conditions of the theorem. This implies $b_1 3 b_2$. For the same reason, the semi-coordinate figure $B_m^2 = (abcda_2 b_2 c_2 d_2)$ is also closed, and this leads to $b3b_2$. By the definition of a three-web, we find that the points b, b_1 and b_2 belong to the same line of the third family, i.e. the figure $B_m = (abcda_1 b_1 c_1 d_1)$ is closed. Thus, we have proved that the closure of the semi-coordinate figures B_m implies the closure of all figures B_m. Therefore, identity (2.22) corresponding to a semi-coordinate figure B_m is universal. ■

Loops in which the identity B_m holds are called *middle Bol loops.*

Recall that for each of the classes of webs, R, T, B_l, B_r, M and B_m, in addition to the universal identities (2.17)–(2.22), we have found, there exists one more derived identity. For example, in coordinate loops of a web B_l we have a universal identity (2.19) derived from left alternativity $u^2 \circ v = u \circ (u \circ v)$. However, this derived identity contains four variables and inverse operations and consequently looks much more complicated than the Bol identity B_l. For webs H the derived identity has the form (2.16) and for webs E it can be obtained from elasticity by the method described above. For webs H and E the problem of existence of universal identities, which are simpler than the derived identities indicated above, remains unsolved.

2.4 Local Differentiable Loops and Their Tangent Algebras

1. For studying multidimensional geometric three-webs we will need the notion of a local differentiable or analytic quasigroup which differs from the algebraic notion of a quasigroup in Section 2.1 in the same way as the notion of a local Lie group differs from the notion of a group in algebra.

Consider three manifolds X, Y and Z of the same dimension r. A partial mapping $q : X \times Y \to Z$ is called a *local quasigroup* if for any $a \in X, b \in Y$ and $c \in Z$ such that $q(a, b) = c$, the following three conditions hold:

i) for any neighborhood U_c of the point c, there exist neighborhoods U_a and U_b of the points a and b, such that for any $x \in U_a$ and any $y \in U_b$ their product $q(x, y) = z$ is defined and lies in U_c;

ii) for any neighborhood U_a of the point a, there exist such neighborhoods U_b and U_c of the points b and c such that for any $y \in U_b$ and $z \in U_c$ there exists a unique $x \in U_a$ satisfying the equation $q(x, y) = z$;

iii) for any neighborhood U_b of the point b, there exist such neighbourhoods U_a and U_c of the points a and c such that for any $x \in U_a$ and $z \in U_c$ there exists a unique $y \in U_b$ satisfying the equation $q(x, y) = z$.

If the manifolds X, Y and Z and the mapping q, satisfying the conditions i–iii, are of class $C^s, s \geq 1$, we say that there is given an *local differentiable quasigroup of class s*. If $s = \omega$, we say that we are given a *local analytic quasigroup*. The mapping q is called *multiplication* and denoted by a dot or by other symbols as in the algebraic theory. A local three-base quasigroup is also denoted by the symbol $q = (\cdot, X_\alpha)$, where $\alpha = 1, 2, 3, X_1 \equiv X, X_2 \equiv Y$ and $X_1 \equiv Z$.

A *local differentiable loop* is defined as an r-dimensional differentiable manifold Q with an operation $q(u, v) = u \cdot v$, given in a certain neighborhood of a fixed point e, provided that this operation satisfies the conditions i–iii for $a = b = c = e$ and the conditions $u \cdot e = u$ and $e \cdot v = v$ for all u and v from Q sufficiently close to e. The element e is called the *identity element*. A local differentiable loop is denoted by $Q(\cdot)$. Conditions i–iii mean that the equation $w = u \cdot v$ is uniquely solvable with respect to u if v and w (sufficiently close to e) are given and with respect to v if u and w (sufficiently close to e) are given.

All notions and theorems, which we considered in Sections 2.1 and 2.2 for algebraic quasigroups and loops, are valid for local differentiable quasigroups and loops with the only difference being that all those notions and theorems should be considered locally. For example, an isotopy $J = (J_\alpha)$ of a local quasigroup $q(\cdot)$ upon a local quasigroup $\tilde{q}(\circ)$ is a triple of diffeomorphisms satisfying condition (2.2); one should consider all configurations being sufficiently small etc.

By Proposition 2.1, using local diffeomorphisms

$$u = q(x, b), \quad v = q(a, y), \quad z = z,$$

one can construct an LP-isotope $X_3(\circ)$ in the manifold X_3 for any local differentiable quasigroup $q = (\cdot, X_\alpha)$. The operation in this LP-isotope is defined by formula (2.3). The point $q(a, b) = e$ is the identity element of the local differentiable loop $X_3(\circ)$.

2. Consider a local differentiable loop $Q(\cdot)$ of class C^s, $s \geq 3$ and in a neighborhood U_e of its identity e define a coordinate system in such a way that the coordinates of the point e are equal to zero. The coordinates of points u and v from U_e we denote by u^i and v^i, respectively. Suppose that the product $u \cdot v$ also belongs to the neighborhood U_e. Then the coordinates of this product are differentiable functions of the coordinates of its factors u and v:

$$(u \cdot v)^i = f^i(u^j, v^k) \tag{2.23}$$

We will also write equations (2.23) in the form $u \cdot v = f(u, v)$ using the same symbol u for the set of coordinates (u^i) of the point $u \in Q$. This will not lead to a misunderstanding in what follows.

Using the Taylor formula in a neighborhood of the point e, we can expand the function $f(u, v)$. Since all the coordinates of the identity element e are zero, the function f satisfies the conditions $f(u, 0) = u$ and $f(0, v) = v$. Because of this the Taylor formula has the form:

$$u \cdot v = u + v + \Lambda(u, v) + \frac{1}{2} M(u, u, v) + \frac{1}{2} N(u, v, v) + o(\rho^3), \tag{2.24}$$

where $\rho = \max(\| u \|, \| v \|)$, $u = \max\limits_i u^i$ and

$$\begin{gathered} \Lambda^i(u,v) = \lambda^i_{jk} u^j v^k, \quad M^i(u,u,v) = \mu^i_{jkl} u^j u^k v^l, \quad N^i(u,v,v) = \nu^i_{jkl} u^j v^k v^l, \\ \lambda^i_{jk} = \left.\frac{\partial^2 (u,v)^i}{\partial u^j \partial v^k}\right|_{u=v=0}, \quad \mu^i_{jkl} = \left.\frac{\partial^3 (u,v)^i}{\partial u^j \partial u^k \partial v^l}\right|_{u=v=0}, \quad \nu^i_{jkl} = \left.\frac{\partial^3 (u,v)^i}{\partial u^j \partial v^k \partial v^l}\right|_{u=v=0}. \end{gathered} \tag{2.25}$$

To the polynomials $M(u,u,v)$ and $N(u,v,v)$ there correspond trilinear forms $M(u,v,w)$ and $N(u,v,w)$ of variables u^i, v^i and w^i. The first of these forms is symmetric in u and v and the second in v and w.

Local coordinates u^i in a local loop $Q(\cdot)$ are defined up to transformations of the form

$$u^{i'} = u^{i'}(u^j), \tag{2.26}$$

where

$$u^{i'}(0) = 0, \quad \left.\frac{\partial u^{i'}}{\partial u^i}\right|_0 = \gamma^{i'}_i, \quad \det(\gamma^{i'}_i) \neq 0. \tag{2.27}$$

One can easily check that under these transformations the coefficients λ^i_{jk} undergo a change according to the following formulas:

$$\lambda^{i'}_{j'k'} = \gamma^{i'}_i \gamma^j_{j'} \gamma^k_{k'} \lambda^i_{jk} + \gamma^{i'}_{jk} \gamma^j_{j'} \gamma^k_{k'},$$

where $\gamma^j_{j'}$ is the inverse matrix of the matrix $\gamma^{i'}_i$ and

$$\gamma^{i'}_{jk} = \left.\frac{\partial^2 u^{i'}}{\partial u^j u^k}\right|_0, \quad \gamma^{i'}_{jk} = \gamma^{i'}_{kj}.$$

It follows from this that the coefficients λ^i_{jk} do not form a tensor. However, using them, we can construct a tensor

$$\alpha^i_{jk} = \lambda^i_{jk} - \lambda^i_{kj}. \tag{2.28}$$

This implies that the skew-symmetric form

$$A(u,v) = \Lambda(u,v) - \Lambda(v,u) \tag{2.29}$$

of variables u and v is a bilinear function in the space $T_e \times T_e$ with its values in T_e, where T_e is the tangent space to the loop $Q(\cdot)$ at the identity element e.

Similarly, the forms M and N (see (2.25)) are not invariant under admissible coordinate transformations in Q, but they allow us to construct an invariant trilinear form

$$B(u,v,w) = M(u,v,w) - N(u,v,w) + \Lambda(\Lambda(u,v),w) - \Lambda(u,\Lambda(v,w)). \tag{2.30}$$

Alternating equation (2.30) in the variables u, v and w and using the symmetry of the forms M and N, we obtain the relation

$$\begin{aligned} & B(u,v,w) + B(v,w,u) + B(w,u,v) - B(v,u,w) - B(w,v,u) - B(u,w,v) \\ = \; & A(A(u,v),w) + A(A(v,w),u) + A(A(w,u),v), \end{aligned}$$

which can be written in the form

$$\mathrm{alt} B(u,v,w) = \frac{1}{2} \mathrm{alt} A(A(u,v),w). \tag{2.31}$$

We have thus proved the following theorem.

Theorem 2.14 *The bilinear and trilinear forms $A(u,v)$ and $B(u,v,w)$ connected by relation (2.31) are invariantly defined in the tangent space T_e of a local differentiable loop $Q(\cdot)$.* ■

The coordinates of the vector $B(u,v,w)$ can be written as

$$B^i(u,v,w) = \beta^i_{jkl} u^j v^k w^l,$$

where, by (2.29), the quantities β^i_{jkl} are expressed in terms of the coefficients of the Taylor expansion (2.24):

$$\beta^i_{jkl} = \mu^i_{jkl} - \nu^i_{jkl} + \lambda^m_{jk} \lambda^i_{ml} - \lambda^m_{kl} \lambda^i_{jm}. \tag{2.32}$$

Under admissible coordinate transformations in U_e the quantities β^i_{jkl} are transformed according to the tensor law.

3. Let us find an algebraic meaning of the forms $A(u,v)$ and $B(u,v,w)$. Denote by ${}^{-1}x$ and x^{-1} the left and right inverse elements of the element x from Q, i.e. ${}^{-1}x \cdot x = x \cdot x^{-1} = e$.

In a loop Q we further consider the *left and right commutators*

$$\alpha_l(u,v) = {}^{-1}(v \cdot u) \cdot (u \cdot v), \quad \alpha_r(u,v) = (u \cdot v) \cdot (v \cdot u)^{-1}, \tag{2.33}$$

as well as the *left and right associators*

$$\beta_l(u,v,w) = {}^{-1}(u \cdot (v \cdot w)) \cdot ((u \cdot v) \cdot w), \quad \beta_r(u,v,w) = ((u \cdot v) \cdot w) \cdot (u \cdot (v \cdot w))^{-1}. \tag{2.34}$$

Theorem 2.15 *Up to infinitesimals of higher orders, the commutators and associators of the loop $Q(\cdot)$ coincide with the forms $A(u,v)$ and $B(u,v,w)$ defined above, i.e. the following relations hold:*

$$\alpha_l(u,v) = A(u,v) + o(\rho^2), \quad \alpha_r(u,v) = A(u,v) + o(\rho^2), \tag{2.35}$$

and

$$\beta_l(u,v,w) = B(u,v,w) + o(\rho^3), \quad \beta_r(u,v,w) = B(u,v,w) + o(\rho^3). \tag{2.36}$$

Proof. First let us express the elements ${}^{-1}z$ and z^{-1} through the element z in the loop $Q(\cdot)$. Since ${}^{-1}z \cdot z = e$, formula (2.24) gives

$$ {}^{-1}z = -z - \Lambda({}^{-1}z, z) - \frac{1}{2}M({}^{-1}z, {}^{-1}z, z) - \frac{1}{2}N({}^{-1}z, z, z) + o(\rho^3). $$

If we substitute for ${}^{-1}z$ on the right hand side of this equation its values taken from the same equation, we arrive at the equation

$$ {}^{-1}z = -z + \Lambda(z, z) + \sigma(z, z, z) + o(\rho^3), \tag{2.37} $$

where

$$ \sigma(z, z, z) = -\frac{1}{2}(M(z, z, z) - N(z, z, z)) - \Lambda(\Lambda(z, z), z). \tag{2.38} $$

Proceeding in the same way as above for the left inverse element, we calculate the right inverse element:

$$ z^{-1} = -z + \Lambda(z, z) + \tau(z, z, z) + o(\rho^3), \tag{2.39} $$

where

$$ \tau(z, z, z) = \frac{1}{2}(M(z, z, z) - N(z, z, z)) - \Lambda(z, \Lambda(z, z)). $$

Next, since (2.24) implies

$$ v \cdot u = v + u + \Lambda(v, u) + o(\rho^2), $$

by (2.35), we have:

$$ \begin{aligned} {}^{-1}(v \cdot u) &= -v - u - \Lambda(v, u) + \Lambda(v + u, v + u) + o(\rho^2) \\ &= -v \cdot u + \Lambda(u + v, u + v) + o(\rho^2). \end{aligned} \tag{2.40} $$

We now calculate the commutator $\alpha_l(u, v)$. Using formulas (2.24) and (2.40), we find that

$$ \begin{aligned} \alpha_l(u, v) &= {}^{-1}(v \cdot u) + u \cdot v + \Lambda({}^{-1}(v \cdot u), u \cdot v) + o(\rho^2) \\ &= -\, v \cdot u + \Lambda(u + v, u + v) + u \cdot v + \Lambda(-v - u, u + v) + o(\rho^2) \\ &= u \cdot v - v \cdot u + o(\rho^2) = \Lambda(u, v) - \Lambda(v, u) + o(\rho^2) = A(u, v) + o(\rho^2). \end{aligned} $$

The second formula in (2.35) can be proved in the same way.

To prove relations (2.36), we first calculate the products $(u \cdot v) \cdot w$ and $u \cdot (v \cdot w)$. Applying formula (2.26) twice, we get

$$ \begin{aligned} (u \cdot v) \cdot w &= t + \widetilde{\Lambda} + \widetilde{M} + \widetilde{N} + M(u, v, w) + \Lambda(\Lambda(u, v), w) + o(\rho^3), \\ u \cdot (v \cdot w) &= t + \widetilde{\Lambda} + \widetilde{M} + \widetilde{N} + N(u, v, w) + \Lambda(u, \Lambda(v, w)) + o(\rho^3), \end{aligned} \tag{2.41} $$

where

$$\begin{aligned} t &= u+v+w, \quad \widetilde{\Lambda} = \Lambda(u,v)+\Lambda(u,w)+\Lambda(v,w), \\ \widetilde{M} &= \tfrac{1}{2}(M(u,u,v)+M(u,u,w)+M(v,v,w)), \\ \widetilde{N} &= \tfrac{1}{2}(N(u,v,v)+N(u,w,w)+N(v,w,w)). \end{aligned}$$

With the help of formula (2.37) we obtain

$$^{-1}(u\cdot(v\cdot w)) = -\, u\cdot(v\cdot w)+\Lambda(t+\widetilde{\Lambda},t+\widetilde{\Lambda})+\sigma(t,t,t)+o(\rho^3)$$

Applying equation (2.34), we arrived at

$$\begin{aligned} \beta_l(u,v,w) = &\ {}^{-1}(u\cdot(v\cdot w))\cdot((u\cdot v)\cdot w) = -u\cdot(v\cdot w)+\Lambda(t+\widetilde{\Lambda},t+\widetilde{\Lambda}) \\ &+\sigma(t,t,t)+(u\cdot v)\cdot w+\Lambda(-t-\widetilde{\Lambda}+\Lambda(t,t),t+\widetilde{\Lambda}) \\ &+\tfrac{1}{2}M(t,t,t)-\tfrac{1}{2}N(t,t,t)+o(\rho^3) \\ = &\ -u\cdot(v\cdot w)+(u\cdot v)\cdot w+\sigma(t,t,t)+\Lambda(\Lambda(t,t),t) \\ &+\tfrac{1}{2}M(t,t,t)-\tfrac{1}{2}N(t,t,t)+o(\rho^3). \end{aligned}$$

By (2.28), the last formula implies that

$$\beta_l(u,v,w) = (u\cdot v)\cdot w - u\cdot(v\cdot w)+o(\rho^3).$$

Substituting for the terms on the right hand side their values taken from equations (2.41), we arrive at the following expression:

$$\beta_l(u,v,w) = M(u,v,w)-N(u,v,w)+\Lambda(\Lambda(u,v),w)-\Lambda(u,\Lambda(v,w))+o(\rho^3).$$

By (2.30), this formula is identical with the first relation in (2.36). The second relation in (2.36) can be proved by the same method. ■

4. The commutators and associators of the loop $Q(\cdot)$ considered above allow us to define, in the tangent space T_e of this loop, binary and ternary operations which are called *commutation* and *association*, respectively.

Let $u(t)$ and $v(t)$ be two smooth curves in the loop $Q(\cdot)$ passing through the point e. We parametrize these lines in such a way that $u(0) = v(0) = e$ and denote the tangent vectors to these curves at the point e by ξ and η. In order not to make our notation complicated, we identify a vector with a set of its coordinates (cf. p. 116):

$$\xi = \lim_{t\to 0}\frac{u(t)}{t}, \quad \eta = \lim_{t\to 0}\frac{v(t)}{t},$$

Construct two more curves in the loop $Q(\cdot)$:

$$\alpha_l(t) = \alpha_l(u(t),v(t)) \ \text{ and } \ \alpha_r(t) = \alpha_r(u(t),v(t)),$$

where α_l and α_r are the commutators of the loop $Q(\cdot)$. These lines also pass through the point e for $t = 0$. Consider those parts of these curves where $t > 0$ and define a new parameter for these lines setting $t = \sqrt{s}$.

Theorem 2.16 *The curves $\alpha_l(\sqrt{s})$ and $\alpha_r(\sqrt{s})$ have the common tangent vector $\zeta = A(\xi, \eta)$ at the point e.*

Proof. In fact, by the formula (2.33), we have

$$\alpha_l(t) = A(u(t), v(t)) + o(t^2), \quad \alpha_r(t) = A(u(t), v(t)) + o(t^2).$$

It follows from this that

$$\zeta = \lim_{s\to 0} \frac{\alpha_l(\sqrt{s})}{s} = \lim_{s\to 0} \frac{\alpha_r(\sqrt{s})}{s} = \lim_{t\to 0} \frac{1}{t^2}(A(u(t), v(t)) + o(t^2)) = A(\xi, \eta). \blacksquare$$

The vector ζ is called the *commutator* of the vectors ξ and η and is denoted by $[\xi, \eta]$. Therefore, we have

$$[\xi,\ \eta] = A(\xi, \eta). \tag{2.42}$$

It follows from this that the commutation is bilinear and skew-symmetric:

$$[\xi,\ \eta] = -[\eta,\ \xi]. \tag{2.43}$$

Next, consider in the loop $Q(\cdot)$ three smooth curves $u(t), v(t)$ and $w(t)$ parametrized in such a way that $u(0) = v(0) = w(0) = e$ and denote the tangent vectors to these curves at the point e by ξ, η and ζ:

$$\xi = \lim_{t\to 0} \frac{u(t)}{t}, \quad \eta = \lim_{t\to 0} \frac{v(t)}{t}, \quad \zeta = \lim_{t\to 0} \frac{w(t)}{t}.$$

Construct two more curves in the loop $Q(\cdot)$ both also passing through the point e:

$$\begin{aligned} \beta_l(t) &= {}^{-1}(u(t) \cdot (v(t) \cdot w(t))) \cdot ((u(t) \cdot v(t)) \cdot w(t)), \\ \beta_r(t) &= ((u(t) \cdot v(t)) \cdot w(t)) \cdot (u(t)(v(t) \cdot w(t)))^{-1}. \end{aligned}$$

Define a new parameter for these lines setting $t = \sqrt[3]{s}$.

Theorem 2.17 *The curves $\beta_l(\sqrt[3]{s})$ and $\beta_r(\sqrt[3]{s})$ have the common tangent vector $\Theta = B(\xi, \eta, \zeta)$ at the point e.*

Proof. By (2.34), we have

$$\beta_l(t) = B(u(t), v(t), w(t)) + o(t^3),$$
$$\beta_r(t) = B(u(t), v(t), w(t)) + o(t^3).$$

This implies that

$$\begin{aligned}\Theta &= \lim_{s\to 0}\frac{\beta_l(\sqrt[3]{s})}{s} = \lim_{s\to 0}\frac{\beta_r(\sqrt[3]{s})}{s}\\ &= \lim_{t\to 0}\frac{1}{t^3}(B(u(t),v(t),w(t)) + o(t^3)) = B(\xi,\eta,\zeta) \blacksquare\end{aligned}$$

The vector Θ is called the *associator* of the vectors ξ,η,ζ. It follows from the previous formula that

$$(\xi,\eta,\zeta) = B(\xi,\eta,\zeta), \tag{2.44}$$

and relation (2.30) shows that the operations of commutation and association are connected by the relation

$$\operatorname{alt}(\xi,\eta,\zeta) = J(\xi,\eta,\zeta), \tag{2.45}$$

where

$$J(\zeta,\eta,\zeta) - [[\xi,\eta],\zeta] + [[\eta,\zeta],\xi] + [[\zeta,\xi],\eta]$$

is the Jacobian of the elements ξ,η and ζ. Relation (2.45) is called the *generalized Jacobi identity* for the pair of forms $[\xi,\eta]$ and (ξ,η,ζ).

Definition 2.18 A *W-algebra* is a linear space T where two multilinear operations $[\xi,\eta]$ and (ξ,η,ζ) are defined, provided that the first operation is symmetric and both operations are connected by the generalized Jacobi identity.

We have the following important result.

Theorem 2.19 *The tangent space T_e of the loop $Q(\cdot)$ is a W-algebra with respect to the operations of commutation and association.* ■

This W-algebra is called the *tangent to the loop* $Q(\cdot)$.

In particular, if a differentiable loop $Q(\cdot)$ is a Lie group, then its associators β_l and β_r defined by formulas (2.36) are identically equal to e. In this case the operation of association in the space T_e is trivial, i.e. $(\xi,\eta,\zeta) = 0$ for any vectors ξ,η and ζ from T_e. Moreover, identity (2.45) becomes the Jacobi identity $J(\xi,\eta,\zeta) = 0$, and the tangent W-algebra of the loop $Q(\cdot)$ becomes a Lie algebra. Namely this was the reason for calling relation (2.45) the generalized Jacobi identity.

2.5 Tangent Algebras of a Multidimensional Three-Web

Consider a smooth three-web $W = (X,\lambda_\alpha)$, $\dim X = 2r$, $\alpha,\beta,\gamma = 1,2,3$, and denote as above the base of the foliation λ_α by X_α. The three-web W generates partial mappings $q_{\alpha\beta} : X_\alpha \times X_\beta \to X_\gamma$ (α,β and γ all distinct), and, to a leaf $F_\alpha \in \lambda_\alpha$

and a leaf $F_\beta \in \lambda_\beta$ with a common point p the mapping $q_{\alpha\beta}$ sets in correspondence a leaf $F_\gamma \in \lambda_\gamma$ also passing through the point p. It is easy to see that the partial mapping $q_{\alpha\beta}$ is a local differential quasigroup. This quasigroup is called a *coordinate quasigroup* of the three-web W. Thus, six coordinate quasigroups are associated with a smooth three-web W. As in the case of an abstract three-web, they are called *parastrophs* of the quasigroup $q = q_{12}$.

There is a natural correspondence $\Gamma_{\alpha\beta} \leftrightarrow q_{\alpha\beta}$ between the parastrophs $q_{\alpha\beta}$ of the coordinate quasigroup q of the geometric web $W = (X, \lambda_\alpha)$ and the connections $\Gamma_{\alpha\beta}$ introduced in Section 1.8, since to each parastroph $q_{\alpha\beta}$ there corresponds the choice of forms $\underset{\alpha}{\omega}^i$ and $\underset{\beta}{\omega}^i$ as basis forms of the manifold X of the three-web. This allows us to give added meaning to Table 1.1 (see p. 32) associating the tensors of the three-web W with the parastrophs $q_{\alpha\beta}$, i.e. with a different coordinatization of this web.

In the following considerations, we will assume that the web W is coordinatized by means of the quasigroup $q = q_{12}$, i.e. the manifold X is locally diffeomorphic to the direct product of the bases X_1 and X_2 of the foliations λ_1 and λ_2.

Consider a point $p = (a, b)$ of the manifold X $(a \in X_1,\ b \in X_2)$ through which three leaves $x = a$, $y = b$ and $z = q(a, b)$ pass. A local coordinate quasigroup $l(a, b)$ is connected with this point. The operation in this quasigroup is defined, as in Section 1.3 for an abstract three-web, by formula (2.3) which can be also written in the form:

$$u \circ v = q({}^{-1}q(u, b), q^{-1}(a, v)). \tag{2.46}$$

However, unlike an abstract three-web, the loop $l(a, b)$ is defined only in a sufficiently small neighborhood of the point $c = q(a, b)$ of the manifold X_3.

Let us expand the product $u \circ v$ (in the coordinate loop $l(a, b)$) into the series (2.24). Consider the bilinear skew-symmetric form $A(u, v)$ defined by equation (2.29) and the trilinear form $B(u, v, w)$ defined by equation (2.30). They satisfy condition (2.31). Next, in the tangent space T_e to the loop $l(a, b)$ at its identity element e, we define the operations of commutation and association by relations (2.42) and (2.43) and assume that these two operations satisfy identities (2.43) and (2.45). This transforms the space T_e into a W-algebra which is called the *tangent W-algebra* of the three-web W. Thus, with a point $p = (a, b)$ of a three-web W there is associated its local W-algebra, and the fibration of the W-algebra arises in the manifold X carrying this three-web.

We will now find a relation between the torsion and curvature tensors of a three-web W, introduced in Chapter 1, and the operations of commutation and association in its local W-algebras. To find this relation, we parametrize a three-web W in a neighborhood of the point p in a special way. We define the leaves $\mathcal{F}_1$ and $\mathcal{F}_2$ of the first two foliations, passing through the point p, by the equations $x = 0$ and $y = 0$ and assign the same value of a parameter to any two leaves intersecting on the leaves

$\mathcal{F}_1$ and $\mathcal{F}_2$ (Figure 2.29).

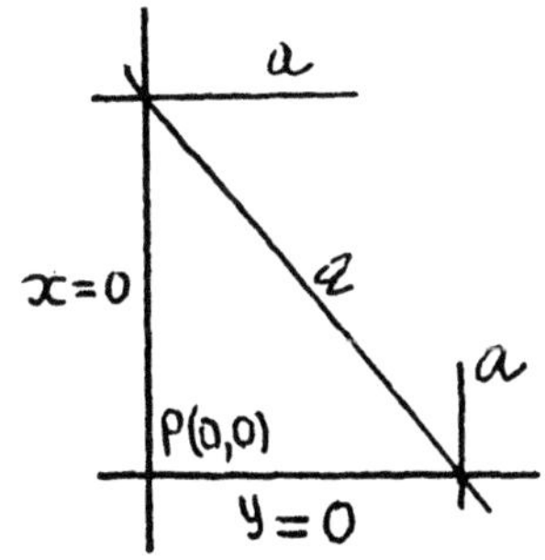

Figure 2.29

Then the web function $z = f(x,y)$, defined in a neighborhood of the point $p = (0,0)$, satisfies the conditions $f(0,y) = y$ and $f(x,0) = x$. It follows from these conditions that the coordinate quasigroup q of the web W becomes a loop with identity element $e(0,\ldots,0)$. This loop coincides with the coordinate loop $l(0,0)$. We will call this parametrization *standard.*

Given the standard parametrization of the web $W = (X, \lambda_\alpha)$, the bases X_α of the foliations λ_α are identified among each other and with the coordinate loop l_p. Thus, the standard parametrization is preserved under concordant transformations of local coordinates in the bases X_α, and the identity element e of the loop l_p preserves its coordinates being zero under these transformations.

Such transformations induce linear transformations of the form $\begin{pmatrix} A & 0 \\ 0 & A \end{pmatrix}$ in the tangent space $T_p(X)$ to the manifold X. These linear transformations leave invariant the subspaces T_α tangent to the leaves of the three-web W passing through the point p. These transformations form the group $G = \mathbf{GL}(r)$. As was indicated in Section 1.3, the latter is the structure group of the G_W-structure defined by a three-web W in a manifold X. They also induce linear transformations in the tangent space T_e to the loop l_p at its identity element e. Because of this, all tensors associated with a three-web can be considered as tensors in the r-dimensional space T_e and all tensor fields as tensor fields in the fibration whose base is the manifold X of dimension $2r$ and whose fibers are r-dimensional linear spaces tangent to the coordinate loops.

So, let a three-web W be parametrized in a neighborhood of the point $p = (0,0)$ in a standard way. Then the Taylor expansion formula for the function $z = f(x,y)$ has the form (2.24). Using this expansion, we calculate the partial derivatives of the function f up to the third order at the point p. We obtain:

$$\begin{gathered} \frac{\partial f^i}{\partial x^j} = \delta^i_j, \quad \frac{\partial f^i}{\partial y^j} = \delta^i_j, \quad \bar{g}^i_j = \delta^i_j, \quad \tilde{g}^i_j = \delta^i_j, \\ \frac{\partial^2 f^i}{\partial x^j \partial y^k} = \lambda^i_{jk}, \quad \frac{\partial^3 f^i}{\partial x^j \partial x^k \partial y^l} = \mu^i_{jkl}, \quad \frac{\partial^3 f^i}{\partial x^j \partial y^k \partial y^l} = \nu^i_{jkl}, \end{gathered} \tag{2.47}$$

and all other derivatives are equal to zero.

Substituting the derivatives from (2.47) into formulas (1.45) and (1.47), we find that at the point p

$$\Gamma^i_{jk} = -\lambda^i_{jk}, \quad a^i_{jk} = -\lambda^i_{[jk]},$$

from which, by (2.28) it follows that at this point

$$a^i_{jk} = -\frac{1}{2}\alpha^i_{jk}. \tag{2.48}$$

Therefore, the torsion tensor of a three-web W is simply a constant times 1 the tensor α^i_{jk}, defining the commutator of its local loop l_p.

Next, by (2.47), formula (1.49) gives the following expression for the curvature tensor of a web W at the point p:

$$b^i_{jkl} = \nu^i_{kjl} - \mu^i_{jkl} + \lambda^m_{jl}\lambda^i_{km} - \lambda^m_{kj}\lambda^i_{ml}. \tag{2.49}$$

Comparing equations (2.49) and (2.32), we find that

$$b^i_{jkl} = -\beta^i_{kjl}, \tag{2.50}$$

i.e. the curvature tensor of a three-web W differs from the tensor β^i_{jkl}, defining the associator of its local loop l_p, only by sign and permutation of the indices.

Denote by $a(\xi,\eta)$ and $b(\xi,\eta,\zeta)$ multilinear forms defined by the torsion and curvature tensors of a three-web. Then the previous relations can be written in the form

$$a(\xi,\eta) = -\frac{1}{2}A(\xi,\eta), \quad b(\xi,\eta,\zeta) = -B(\eta,\xi,\zeta). \tag{2.51}$$

This gives the following form for the operations in the tangent W-algebra:

$$[\xi,\eta] = -2a(\xi,\eta), \quad (\xi,\eta,\zeta) = -b(\eta,\xi,\zeta)$$

The following theorem gives necessary and sufficient conditions for a three-web W to belong to one of the basic classes connected with closure conditions.

Theorem 2.20 *If the closure condition, indicated in the first column of Table 2.2, holds in a three-web W, then the torsion and curvature tensors of this web satisfy the conditions indicated in the second column of this table.*

Closure condition	Conditions on tensors a and b
T	$a(\xi,\eta)=0,\ \ b(\xi,\eta,\zeta)=0$
R	$b(\xi,\eta,\zeta)=0$
B_l	$b(\xi,\eta,\zeta)=-b(\eta,\xi,\zeta)$
B_r	$b(\xi,\eta,\zeta)=-b(\zeta,\eta,\xi)$
B_m	$b(\xi,\eta,\zeta)=-b(\xi,\zeta,\eta)$
M	$b(\xi,\eta,\zeta)=\text{alt}\ b(\xi,\eta,\zeta)$
H	$\text{symm}\ b(\xi,\eta,\zeta)=0$

Table 2.2

Proof. For example, let us derive a condition for a web B_l. According to Theorem 2.7, the left alternativity $u\circ(u\circ v)=(u\circ u)\circ v$ holds in the coordinate loops of this web. On the other hand, formula (2.36) gives

$$^{-1}(u\circ(u\circ v))\circ((u\circ u)\circ v)=B(u,u,v)+o(\rho^3).$$

By left alternativity, the left hand side of this equation vanishes. This gives $B(u,u,v)=0$. Then from (2.51) we have $b(u,u,v)=0$, and after linearization we have $b(u,v,w)+b(v,u,w)=0$. ■

We will now prove the sufficiency of the conditions given in Table 2.2 for the classes T and R. By Theorem 1.10, the relations $a=0$ and $b=0$ characterize the parallelizable three-webs for which, obviously, all listed closure figures are closed, including the figures T. Therefore, this class of geometric webs is identical to the class of parallelizable webs and is characterized by the condition $a=b=0$.

As proved in Theorem 1.11, the condition $b=0$ characterizes the group three-webs. By the definition of a group three-web given in Section 2.2, its coordinate quasigroup is a Lie group G. Alberts' theorem (p. 51) implies that all coordinate loops of a group three-web are also groups since they are isotopic to the coordinate group G. By Theorem 2.7, we have that a group three-web is a web R. Thus, the class of group geometric three-webs is identical to the class of webs R and is characterized by the condition $b = 0$.

The sufficiency of the tensor conditions for webs B_r, B_l, B_m and M will be proved in the following chapters.

2.6 Canonical Coordinates in a Local Analytic Loop

1. The importance of the canonical coordinates in the theory of Lie groups is well-known. In these coordinates:

a) Any one-parametric subgroup is given by linear equations, and the parameters are added when multiplication is performed in this subgroup;

b) The Taylor series of the function $u \cdot v$, which is called the Campbell–Hausdorff series, is completely defined by the structure tensor of a group, i.e. all terms of the series can be expressed in terms of the commutator $[u, v]$; and

c) An automorphism of a Lie group written in canonical coordinates is a linear transformation.

As we will see, in general, an arbitrary local analytic loop $Q(\cdot)$ does not possess one-parametric subgroups and even one-dimensional subloops. In addition, it does not have non-trivial automorphisms. However, it is still possible to introduce canonical coordinates in a local analytic loop, and these coordinates have the properties listed above.

Let $Q(\cdot)$ be a local analytic loop of dimension r with identity element e. We define local coordinates in a neighborhood of e in such a way that the coordinates of e are zero. As in Section 1.4, we write multiplication in the chosen coordinates as $w = u \cdot v = f(u, v)$ and the expansion of the function f in the Taylor series in the form:

$$u \cdot v = f(u, v) = \sum_{s=0}^{\infty} \underset{s}{\Lambda}(u, v), \tag{2.52}$$

where $\underset{s}{\Lambda}(u, v)$ is a homogeneous polynomial of degree s in the local coordinates u^i and v^i. By the equations $f(u, 0) = f(0, u) = u$, the polynomials $\underset{s}{\Lambda}(u, v)$ satisy the equations

$$\underset{1}{\Lambda}(u, v) = u + v, \quad \underset{s}{\Lambda}(0, v) = \underset{s}{\Lambda}(u, 0) = 0, \quad s \geq 2, \tag{2.53}$$

from which it follows that they have the form:

$$\underset{s}{\Lambda}(u,v) = \sum_{k=1}^{s-1} \frac{1}{(s-k)!k!} \underset{s-k,k}{\Lambda}(u,\dots,u,v,\dots,v), \tag{2.54}$$

where $\underset{s-k,k}{\Lambda}(u,\dots,u,v,\dots,v)$ is a homogeneous polynomial of degree $s-k$ in u and degree k in v. Note that the notations in the expansions (2.52) are slightly different from that in formula (2.24) which has terms only up to third order.

Let us consider another system of analytic coordinates $\tilde{u}^i$ in the loop Q, and suppose that multiplication in these coordinates is written in the form $\tilde{w} = \tilde{f}(\tilde{u},\tilde{v})$. The old and new variables are connected by the relations

$$u = \gamma(\tilde{u}), \quad v = \gamma(\tilde{v}), \quad w = \gamma(\tilde{w}), \tag{2.55}$$

where γ is a local diffeomorphism satisfying the condition $\gamma(0) = 0$. This condition means that the identity element e preserves its zero coordinates, and this implies that the Taylor series of the function $\tilde{f}$ also has the properties (2.53), and the expansion for γ, up to non-singular linear transformations, has the form:

$$\gamma(\tilde{u}) = \tilde{u} + \sum_{s=2}^{\infty} P_s(\tilde{u}), \tag{2.56}$$

where $P_s(\tilde{u})$ is a homogeneous polynomial of degree s.

Definition 2.21 Local coordinates $\tilde{u}$ in the loop Q, defined in a neighborhood U of the identity element $e(0)$, are said to be *canonical* if for any $\tilde{u} \in U$, such that $\tilde{f}(\tilde{u},\tilde{u}) \in U$, the following equation holds:

$$\tilde{f}(\tilde{u},\tilde{u}) = 2\tilde{u}. \tag{2.57}$$

Since equation (2.57) holds for the canonical coordinates in a Lie group, the notion introduced above generalizes these coordinates.

Theorem 2.22 *Canonical coordinates can be defined in a neighborhood of the identity element of a local analytic loop. These coordinates are unique up to a linear transformation.*

Proof. By (2.55), the functions f and $\tilde{f}$ are connected by the relation:

$$\gamma(\tilde{f}(\tilde{u},\tilde{v})) = f(\gamma(\tilde{u}),\gamma(\tilde{v})).$$

From this, by (2.57), we obtain

$$\gamma(2\tilde{u}) = f(\gamma(\tilde{u}),\gamma(\tilde{u})) = f_d(\gamma(\tilde{u})), \tag{2.58}$$

where f_d denotes the restriction of f on the diagonal: $f_d(u) \stackrel{\text{def}}{=} f(u,u)$. From (2.52) we get

$$f_d(u) = 2u + \sum_{s=2}^{\infty} \underset{s}{\Lambda}_d(u), \tag{2.59}$$

where $\underset{s}{\Lambda}_\alpha(u)$ is a notation for $\underset{s}{\Lambda}(u,v)\Big|_{u=v}$. Substituting (2.56) and (2.59) into equations (2.58) and suppressing the tilde over u, we obtain

$$\begin{aligned} &2u + P_2(2u) + P_3(2u) + \ldots = 2[u + P_2(u) + P_3(u) + \ldots] \\ &+\underset{2}{\Lambda}_d(u + P_2(u) + P_3(u) + \ldots) + \underset{3}{\Lambda}_d(u + P_2(u) + P_3(u) + \ldots) + \ldots. \end{aligned}$$

Equating to zero polynomials of the same degree and taking into account their homogeneity, we find that

$$\begin{aligned} &4P_2(u) = 2P_2(u) + \underset{2}{\Lambda}_d(u), \\ &8P_3(u) = 2P_3(u) + 2\underset{2}{\Lambda}(u, P_2(u)) + \underset{3}{\Lambda}_d(u), \\ &\ldots\ldots\ldots\ldots\ldots\ldots\ldots\ldots\ldots\ldots\ldots\ldots\ldots\ldots \\ &2^k P_k(u) = 2P_k(u) + \textstyle\sum_{i_1+i_2=k} \underset{2}{\Lambda}(P_{i_1}(u), P_{i_2}(u)) \\ &+ \textstyle\sum_{i_1+i_2+i_3=k} \underset{3}{\Lambda}(P_{i_1}(u), P_{i_2}(u), P_{i_3}(u)) + \ldots + \underset{k}{\Lambda}(P_1(u), \ldots, P_1(u)) \end{aligned}$$

This gives the following recursive formula for the polynomials $P_k(u), k \geq 2$:

$$P_k(u) = A_k \Big[\sum_{s=2}^{k} \sum_{i_1+\ldots+i_s=k} \underset{s}{\Lambda}(P_{i_1}(u), P_{i_2}(u), \ldots, P_{i_s}(u))\Big], \tag{2.60}$$

where $A_k = (2^k - 2)^{-1}$ and $i_1, i_2, \ldots, i_k$ are natural numbers. Applying this formula, we have, for example,

$$P_2(u) = \frac{1}{2}\underset{2}{\Lambda}_d(u), \quad P_3(u) = \frac{1}{6}[\underset{2}{\Lambda}(u, P_2(u)) + \underset{3}{\Lambda}_d(u)],$$

etc. Thus, if the series in equation (2.56) converges, then the uniquiness of the function $\gamma(u)$ is obvious. Let us prove the convergence of this series.

We recall that the terms $\underset{k}{\Lambda}$ of the Taylor expansion of any real analytic function $f(u)$ satisfy the Cauchy inequalities:

$$||\underset{k}{\Lambda}|| < MR^{-k}, \tag{2.61}$$

where the norm $\underset{k}{\Lambda}$ is defined as $\sup \underset{k}{\Lambda}$ on the unit sphere $||u|| = 1$ and M and R are constants. Moreover, if $f(u)$ is a vector function of a vector variable $u = (u^i)$, then $\underset{k}{\Lambda}(u)$ are vector-valued forms in the variables u^i.

Conversely, if the conditions (2.61) hold, then the series composed of the norms $\| \Lambda_k \|$ is convergent. Therefore, the series $\Lambda_0 + \Lambda_1(u) + \Lambda_2(u) + \ldots$ converges to the function $f(u)$ (see for, example, [Sch 70]).

To prove the convergence of the series (2.56), we must show the existence of constants N and ρ such that the inequalities

$$|\rho_k| < N\rho^{-k}. \tag{2.62}$$

hold.

Since the series in equation (2.59) converges to the function f_d, there exist constants M and R such that inequalities (2.61) hold for the polynomials $\underset{k}{\Lambda}$ in the expansion (2.59). We will show that condition (2.62) will be satisfied if we set

$$\rho < N < 2R^2M^{-1}. \tag{2.63}$$

The proof will be performed by induction with respect to k. If $k = 1$, then $P_1(u) = u$, and by (2.61), $\| P_1 \| = 1$ and $\| P_1 \| \leq N\rho^{-1}$. Suppose next that the inequalities (2.62) hold for all natural numbers less than k and prove them for k. From (2.60) we have

$$\begin{aligned} \| P_k \| \leq A_k \Big(& \textstyle\sum_{i_1+i_2=k} \| \underset{2}{\Lambda} \| \cdot \| P_{i_1} \| \cdot \| P_{i_2} \| \\ & + \textstyle\sum_{i_1+i_2+i_3=k} \| \underset{3}{\Lambda} \| \cdot \| P_{i_1} \| \cdot \| P_{i_2} \| \cdot \| P_{i_3} \| + \ldots + \| \underset{k}{\Lambda} \| \cdot \| P_1 \|^k \Big). \end{aligned} \tag{2.64}$$

The number of terms in the first sum is equal to the number of representations of the number k as the sum of two natural numbers with permutations counted, i.e. $k-1$. The number of terms in the second sum is equal to the number of decompositions of the number k into three summands, i.e. it is $\binom{k-1}{2}$. The following sums contain $\binom{k-1}{3}, \ldots, \binom{k-1}{k-1}$ terms. Since each of the indices $i_1, i_2, \ldots$ less than k, according to the induction assumption, $\| P_{i_s} \| < N\rho^{-i_s}$. Because of this, inequality (2.64) implies another inequality:

$$\| P_k \| < \frac{MA_k}{\rho^k}\left(\binom{k-1}{1}\frac{N^2}{R^2} + \binom{k-1}{2}\frac{N^3}{R^3} + \ldots + \binom{k-1}{k-1}\frac{N^k}{R^k}\right)$$

(we used here relations $i_1 + i_2 = k,\ i_1 + i_2 + i_3 = k, \ldots$). The right-hand side of this inequality can be transformed to the form:

$$A_k\rho^{-k}MNR^{-1}\Big[(1+\frac{N}{R})^{k-1} - 1\Big] = \rho^{-k}MNR^{-1}\Big[(1+\frac{N}{R})^{k-1} - 1\Big](2^k-2)^{-1}.$$

It follows from (2.63) that $MN < 2R^2$, which leads to

$$\| P_k \| < R\rho^{-k}[(1+NR^{-1})^{k-1} - 1](2^{k-1}-1)^{-1}.$$

As equation (2.59) shows, $\| \underset{2}{\Lambda} \| = 2$, and inequality (2.61) for $k = 1$ gives $2 < MR^{-1}$. Using inequality (2.63), we find that

$$NR^{-1} < 2RM^{-1} < 1,$$

which leads to

$$\frac{(1+NR^{-1})^{k-1}-1}{2^{k-1}-1} = \frac{N}{R}\frac{1+(1+NR^{-1})+(1+NR^{-1})^2+\ldots++(1+NR^{-1})^{k-2}}{1+2+2^2+\ldots+2^{k-2}} <$$

and $\| P_k \| < R\rho^{-k} \cdot NR^{-1} = N\rho^{-k}$. ■

The expansion (2.52) written in canonical coordinates is called the *canonical expansion* of the loop $Q(\cdot)$. The polynomials $\underset{s}{\Lambda}$ in the canonical expansion satisfy the relations

$$\underset{s}{\Lambda}(u,u) = 0, \tag{2.65}$$

which can be obtained from (2.52) by setting $u = v$ and applying the definition of canonical coordinates. Using relations (2.54), we write equations (2.65) in the expanded form:

$$\underset{1,1}{\Lambda}(u,u) = 0, \quad \underset{2,1}{\Lambda}(u,u) + \underset{1,2}{\Lambda}(u,u) = 0, \quad \ldots, \quad \sum_{\substack{p=1\\p+q=s}}^{s-1} \binom{s}{p} \underset{p,q}{\Lambda}(u,u) = 0. \tag{2.66}$$

The first of equations (2.66) means that the form $\underset{2}{\Lambda} = \underset{1,1}{\Lambda}$ is skew-symmetric. In the coordinate form these relations can be written as follows:

$$\sum_{\substack{p=1\\p+q=s}}^{s-1} \binom{s}{p} \underset{p,q}{\Lambda}{}^i_{(j_1 j_2 \ldots j_p k_{p+1} \ldots k_{p+q})} = 0. \tag{2.67}$$

2. As we noted earlier, the canonical coordinates in the loop Q are defined up to a non-singular linear transformation. Under such transformations, the forms $\underset{k,l}{\Lambda}$, defining the canonical expansion, are changed according to the tensor law. Therefore, they are invariantly defined in the tangent space T_e to the loop Q. Using these forms, we introduce two linear operations in T_e, one binary and two ternary, by defining the products of vectors $\xi_1, \xi_2, \ldots$ as values of the forms

$$\underset{1,1}{\Lambda}(\xi_1,\xi_2), \quad \underset{2,1}{\Lambda}(\xi_1,\xi_2,\xi_3), \quad \underset{1,2}{\Lambda}(\xi_1,\xi_2,\xi_3), \quad \ldots, \underset{p,q}{\Lambda}(\xi_1,\xi_2,\ldots,\xi_{p+q}). \tag{2.68}$$

These operations are not independent: they are connected by relations (2.66). In addition, they satisfy the symmetry conditions: as follows from the definition of the forms $\underset{p,q}{\Lambda}$, they are symmetric in the first p indices and the last q indices.

Definition 2.23 The set of operations $\underset{p,q}{\Lambda}$ for $p+q \leq k$ is called the *tangent Λ_k-algebra* of the loop Q, and the set of all such operations, i.e. for any k, is called the *tangent Λ-algebra* of this loop.

The binary operation in the tangent Λ-algebra of the loop Q coincides with one-half of the commutator defined in Section 2.4, and the associator of this loop can be expressed through the operations in its Λ-algebra as follows:

$$(\xi_1, \xi_2, \xi_3) = \underset{2,1}{\Lambda}(\xi_1, \xi_2, \xi_3) - \underset{1,2}{\Lambda}(\xi_1, \xi_2, \xi_3) + \underset{1,1}{\Lambda}(\underset{1,1}{\Lambda}(\xi_1, \xi_2), \xi_3) - \underset{1,1}{\Lambda}(\xi_1, \underset{1,1}{\Lambda}(\xi_2, \xi_3)).$$

The canonical coordinates allow us to define naturally an exponential mapping of a neighborhood of the identity element of the tangent space T_e to an analytic loop Q onto this loop: $\exp u = (u^i)$, where u^i are the canonical coordinates in Q.

In the following theorems the properties of the canonical coordinates of Lie groups, which were listed in the beginning of this section, are transfered to analytic loops.

Theorem 2.24 *In canonical coordinates every one-parametric subloop of a local analytic loop is defined by a linear equation.*

Proof. Let, as above, multiplication in the loop Q be written in the form $w = f(u,v)$, and suppose that the identity element e of Q has zero coordinates. Let $u(t)$, $t \in I \subset \mathbf{R}$ be a subloop and the product of points $u(t)$ and $v(t)$ of this subloop have a parameter $\tau = \tau(t,s)$, $\tau \in C^\omega$. Then, for any admissible values t and s from the interval I, we have the condition:

$$f(u(t), u(s)) = u(\tau(t,s)). \tag{2.69}$$

Let $u(0) = e$. Setting first $t = 0$ and next $s = 0$ in (2.69), we obtain $\tau(0,s) = s$ and $\tau(t,0) = t$, respectively. Thus, the Taylor seies of the function $\tau(t,s)$ has the form:

$$\tau(t,s) = t + s + \mu_2(t,s) + \mu_3(t,s) + \dots, \tag{2.70}$$

where $\mu_l(t,s)$ is a homogeneous polynomial of degree l in t and s. Setting $t = s$ in (2.69), we obtain:

$$f(u(t), u(t)) = u(\tau(t,t)).$$

Suppose now that the coordinates u^i of the point u are the canonical coordinates. Then, by their definition, the last equation gives

$$2u^i(t) = u^i(\tau(t,t)). \tag{2.71}$$

Since $u^i(0) = 0$, the Taylor series of the function $u^i(t)$ has the form:

$$u^i(t) = \xi_1^i t + \xi_2^i t^2 + \ldots. \tag{2.72}$$

Moreover, it follows from (2.70) that

$$\tau(t,t) = 2t + m_2 t^2 + m_3 t^3 + \ldots.$$

Substituting these expansions into (2.71), we get

$$2\sum_{k=1}^{\infty} \xi_k^i t^k = \sum_{k=1}^{\infty} (t + \sum_{l=2}^{\infty} \mu_l t^l)\xi_k^i.$$

Comparing the coefficients for like powers of t, we arrive at the relations:

$$\begin{aligned}
2\xi_2^i &= \mu_2 \xi_1^i + 4\xi_2^i, \\
2\xi_3^i &= \mu_3 \xi_1^i + 4\mu_2 \xi_2^i + 8\xi_3^i, \\
&\ldots\ldots\ldots\ldots\ldots\ldots\ldots\ldots\ldots\ldots \\
2\xi_k^i &= \mu_k \xi_1^i + (\ldots)\xi_2^i + \ldots + (\ldots)\xi_{k-1}^i + 2^k \xi_k^i, \ldots
\end{aligned}$$

From this we subsequently find that the vectors ξ_2, $\xi_3, \ldots$ are collinear to the vector ξ_1. ■

Theorem 2.25 *Let Q be a local analytic loop given in canonical coordinates u^i and $u^i(t) = \xi^i t$ be a curve in Q passing through the identity element e. The line $u(t)$ is a one-parameter subloop in Q if and only if its tangent space at the point e, defined by the vector ξ, is a subalgebra of the tangent Λ-algebra of this loop.*

Proof. Since the coordinates u^i are canonical, the functions $u^i(t)$ satisfy condition (2.69). By (2.52), this condition can be written in the form:

$$(t+s)\xi + \frac{1}{2}\underset{2,1}{\Lambda}_d(\xi)t^2 s + \frac{1}{2}\underset{1,2}{\Lambda}_d(\xi)ts^2 + \ldots = \tau(s,t)\xi,$$

where ξ is the vector with the coordinates ξ^i and by (2.66), $|\underset{1,1}{\Lambda}(\xi,\xi)| = 0$. The last relation must be satisfied identically with respect to the variables s and t. This is possible if and only if each of the vectors $\underset{p,q}{\Lambda}_d(\xi)$ is collinear to the vector ξ, i.e. if the relations

$$\underset{p,q}{\Lambda}(\xi,\xi,\ldots,\xi) = \underset{p,q}{\lambda}\xi, \quad \underset{p,q}{\lambda} \in \mathbf{R} \tag{2.73}$$

hold. This condition means precisely that the one-dimensional subspace defined by the vector ξ is a subalgebra with respect to the operation $\underset{p,q}{\Lambda}$ in the tangent Λ-algebra of the loop Q. ■

A vector ξ satisfying condition (2.73) is called an *eigenvector* of the tensor $\underset{p,q}{\Lambda}$. Thus, Theorem 2.25 can be reformulated as follows: *a line $u(t)$ is a subloop if and only if its tangent vector at the point e is an eigenvector for any of the tensors $\underset{p,q}{\Lambda}$.*

Suppose that all quantities $\underset{p,q}{\Lambda}$ in the equations (2.73) vanish. Then the relation (2.69) tells us $f^i(u(t), u(s)) = (t+s)\xi^i$, i.e. the subloop $u(t)$ becomes a subgroup and the parameter t becomes the canonical parameter of this group. Conversely, suppose that $u(t)$ is a subgroup and define in it a canonical parameter by the condition $u^i(t) = t\xi^i$. Then, in the expansion of the function $f(u(t), u(s))$ all terms except the first one vanish: $\underset{p,q}{\Lambda}_d(\xi) = 0$. This gives that the product of any two vectors from the one-dimensional space defined by the vector ξ relative to any of the operations $\underset{p,q}{\Lambda}$ is equal to zero. According to conventional terminology, all products of elements of an algebra A form the so-called derived algebra. Thus, these considerations imply the following theorem.

Theorem 2.26 *A one-parametric subloop of a local analytic loop Q is a subgroup if and only if its tangent vector at the point e is an eigenvector of each of the tensors $\underset{p,q}{\Lambda}$ corresponding to the eigenvalue zero.* ■

In other words, this vector generates a one-dimensional subalgebra in the tangent Λ-algebra of the loop Q, and the derived Λ-algebra of this subalgebra is zero.

Note that any one-dimensional subalgebra of a Lie group possesses this property.

Theorem 2.25 implies that an analytic loop Q possesses the maximal possible number of one-parametric subloops if and only if each vector ξ of its tangent Λ-algebra defines a one-dimensional subalgebra. If, in addition, the derived subalgebra of each of these subalgebras is trivial, and only in this case, a loop Q possesses the maximal possible number of one-parametric subgroups.

The following two theorems can be also proved by the same kind of uncomplicated considerations.

Theorem 2.27 *A local analytic loop Q possesses the maximal possible number of one-parametric subgroups if and only if the associativity of the powers with natural exponents holds in it, i.e. for any $m, n \in \mathrm{N}$ and any $x \in Q$ the equation*

$$x^m \cdot x^n = x^{m+n}$$

holds. ■

Theorem 2.28 *Let Q be a local analytic loop to which we have assigned canonical coordinates. If $\varphi \in Aut\, Q$, $\varphi \in C^\omega$, then φ is a linear transformation.* ■

Theorem 2.29 *A diffeomorphism $\varphi : Q \rightarrow Q$ of a local analytic loop Q is an automorphism of this loop if and only if $\varphi(e) = e$, and the derived mapping $d\varphi|_e$ is an automorphism of the tangent Λ-algebra of the loop Q.* ■

3. In conclusion note that the canonical coordinates can be also introduced in C^k-smooth loops, $k < \infty$. In this case the relations (2.65)–(2.67) preserve their meaning for $s \leq k$. They can be obtained by successive simplifications of the series (2.52) by means of admissible transformations preserving conditions (2.53). The details can be found in the paper [A 69a].

2.7 Algebraic Properties of the Chern Connection

The correspondence between three-webs and quasigroups allows us to interpret properties of the Chern connection by means of the coordinate quasigroups and loops of a three-web. In this vein, in Section 2.5 we established that the torsion and curvature tensors of a three-web W are connected with the commutators and associators of its coordinate loops. In this section we will describe a parallel transport in the Chern connection by means of the coordinate loops and establish a correspondence between subwebs and subquasigroups.

Let, as in Section 1.2, a three-web $W = (X, \lambda_\alpha)$, $\alpha = 1,2,3$, be defined in a neighborhood U of a manifold X by the equation $z = f(x, y)$. An arbitrary vector field ξ can be written in the form (see Section 1.2):

$$\xi = \xi^i_1 \underset{2}{e}_i - \xi^i_2 \underset{1}{e}_i = \xi^i_2 \underset{3}{e}_i - \xi^i_3 \underset{2}{e}_i = \xi^i_3 \underset{1}{e}^i - \xi^i_1 \underset{3}{e}_i,$$

where the vectors $\underset{\alpha}{e}_i$ are tangent to the leaf of the foliation λ_α and the coordinates ξ^i_α are connected by the relation

$$\xi^i_1 + \xi^i_2 + \xi^i_3 = 0.$$

Lemma 2.30 *A vector field ξ is parallel in the Chern connection if and only if its coordinates ξ^i_α satisfy the condition*

$$\tilde{f}^i_k \frac{\partial}{\partial x^m}(\tilde{g}^k_j \xi^j_\alpha) dx^m + \bar{f}^i_k \frac{\partial}{\partial y^m}(\bar{g}^k_j \xi^j_\alpha) dy^m = 0, \tag{2.74}$$

where $\bar{f}^i_j = \frac{\partial f}{\partial x^j}$, $\tilde{f}^i_j = \frac{\partial f}{\partial y^j}$ and the matrices $\bar{g}^i_j$ and $\tilde{g}^i_j$ are the inverse matrices of the matrices $\bar{f}^i_j$ and $\tilde{f}^i_j$, respectively.

Proof. Since the connection forms of the Chern connection have a special structure (see Section 1.7), the coordinates ξ^i_α of the vector field ξ, for different α, satisfy the same system of equations:

$$d\xi^i_\alpha + \omega^i_j \xi^j_\alpha = 0. \tag{2.75}$$

By formulas from Section 1.6 (see p. 23), this equation becomes

$$d\xi^i_\alpha - \Big(\frac{\partial^2 f^i}{\partial x^m \partial y^k}\tilde{g}^k_j dx^m + \frac{\partial^2 f^i}{\partial x^k \partial y^m}\bar{g}^k_j dy^m\Big)\xi^j_\alpha = 0,$$

or

$$d\xi^i_\alpha - \Big(\frac{\partial \bar{f}^i_k}{\partial x^m}\tilde{g}^k_j dx^m + \frac{\partial \tilde{f}^i_k}{\partial y^m}\bar{g}^k_j dy^m\Big)\xi^j_\alpha = 0. \tag{2.76}$$

Consider relations:

$$\bar{f}^i_k \bar{g}^k_j = \delta^i_j, \quad \tilde{f}^i_k \tilde{g}^k_j = \delta^i_j, \tag{2.77}$$

connecting the elements of inverse matrices. Differentiating the first of equations (2.77) with respect to x^m and the second with respect to y^m, we obtain

$$\frac{\partial \bar{f}^i_k}{\partial y^m}\bar{g}^k_j = -\frac{\partial \bar{g}^k_j}{\partial y^m}\bar{f}^i_k, \quad \frac{\partial \tilde{f}^i_k}{\partial x^m}\tilde{g}^k_j = -\frac{\partial \tilde{g}^k_j}{\partial x^m}\tilde{f}^i_k.$$

Applying these relations, we can write equations (2.76) in the form:

$$d\xi^i_\alpha + \Big(\tilde{f}^i_k \frac{\partial \tilde{g}^k_j}{\partial x^m} dx^m + \bar{f}^i_k \frac{\partial \bar{g}^k_j}{\partial y^m} dy^m\Big)\xi^j_\alpha = 0,$$

or

$$\Big(\frac{\partial \xi^i_\alpha}{\partial x^m} + \tilde{f}^i_k \frac{\partial \tilde{g}^k_j}{\partial x^m}\xi^j_\alpha\Big)dx^m + \Big(\frac{\partial \xi^i_\alpha}{\partial y^m} + \bar{f}^i_k \frac{\partial \bar{g}^k_j}{\partial y^m}\xi^j_\alpha\Big)dy^m = 0.$$

Taking into account relations (2.77), the last equations can be written in the form:

$$\tilde{f}^i_k\Big(\tilde{g}^k_j \frac{\partial \xi^j_\alpha}{\partial x^m} + \frac{\partial \tilde{g}^k_j}{\partial x^m}\xi^j_\alpha\Big)dx^m + \bar{f}^i_k\Big(\bar{g}^k_j \frac{\partial \xi^j_\alpha}{\partial y^m} + \frac{\partial \bar{g}^k_j}{\partial y^m}\xi^j_\alpha\Big)dy^m = 0,$$

and this leads to equation (2.74). ■

We will call a vector field ξ tangent to the leaves of the first foliation of a web *vertical* and a vector field ζ tangent to the leaves of the second foliation *horizontal.* Lemma 2.30 implies the following corollary.

Corollary 2.31 *A vertical vector field* $\underset{1}{\xi} = -\xi^i_2 \underset{1}{e_i}$ *is parallel in the Chern connection along a vertical leaf if and only if its coordinates* ξ^i_2 *satisfy the equation:*

$$\xi^i_2 = \frac{\partial f^i}{\partial x^j} c^j_2, \tag{2.78}$$

and a horizontal vector field $\underset{2}{\xi} = \xi_1^i \underset{2}{e}_i$ *is parallel in the Chern connection along a horizontal leaf if and only if its coordinates* ξ_1^i *satisfy the equation:*

$$\xi_1^i = \frac{\partial f^i}{\partial y^j} c_1^j, \tag{2.79}$$

where c_1^i *and* c_2^i *are constants.*

Proof. In fact, in a leaf of the first foliation we have $x^i = \text{const}$. Thus, in it the condition (2.74) of parallelism gives for $\alpha = 2$:

$$\bar{f}_k^i d(\bar{g}_j^k \xi_2^j) = 0.$$

According to the definition of a three-web, in a domain of its regularity the condition $\det(\bar{f}_k^i) \neq 0$ holds. Therefore, using equations (2.77), we obtain $\bar{g}_j^k \xi_1^j = c_1^k$. Applying equations (2.77), we obtain from this equations (2.78). Equations (2.78) can be proved in a similar way. ■

Equations (2.78) and (2.79) yield a visual geometric interpretation if we use the standard parametrization (see Section 2.5). Let p be an arbitrary point of the domain U of the manifold X carrying the three-web W. We define the leaves of the first and the second foliations of the web, passing through the point p, by the equations $x^i = 0$ and $y^i = 0$. On these leaves equations (2.78) and (2.79) have the form:

$$\xi_2^i(0, y) = \frac{\partial f^i}{\partial x^j}(0, y) c_2^j, \quad \xi_1^i(x, 0) = \frac{\partial f^i}{\partial y^j}(x, 0) c_1^j. \tag{2.80}$$

Suppose that our parametrization is standard. Then $f(x, y) = x \cdot y$ is multiplication in the coordinate loop l_p and $L_x(y) = x \cdot y$, $R_y(x) = x \cdot y$ are shifts in l_p. Thus, we have

$$\frac{\partial f}{\partial x}(0, y) = (R_y)_*\Big|_{x=0}, \quad \frac{\partial f}{\partial y}(x, 0) = (L_x)_*\Big|_{y=0}.$$

The first of these equations is written under assumption that the loop l_p acts in the leaf $x = 0$ and the tangent space T_e to this loop at the identity element e is identified with the tangent space T_1 to this leaf at the point $p(0, 0)$. Since $R_y(e) = y$, we have $(R_y)_*\Big|_e : T_e \to T_y$, and the first equation of (2.80) gives

$$\underset{1}{\xi}(0, y) = (R_y)_*\Big|_e (\underset{1}{\xi}(0, 0)), \quad \underset{1}{\xi}(0, 0) \in T_1 \equiv T_e. \tag{2.81}$$

Similarly, we find that the following relation holds in the leaf $y = 0$:

$$\underset{2}{\xi}(x, 0) = (L_x)_*\Big|_e (\underset{2}{\xi}(0, 0)), \quad \underset{2}{\xi}(0, 0) \in T_2 \equiv T_e. \tag{2.82}$$

Thus, the following theorem is valid.

Theorem 2.32 *Any vertical vector field $\underset{1}{\xi}$, which is parallel along a vertical leaf of a web W in the Chern connection, can be written in the form (2.81), and any horizontal vector field $\underset{2}{\xi}$, which is parallel along a horizontal leaf of a web W in the Chern connection, can be written in the form (2.82), where L_x and R_y are shifts in the coordinate loop l_p of the web W.* ■

Since the leaves $x = 0$ and $y = 0$ of the web W carry the structure of its coordinate loop $l(0,0)$, the restriction of the Chern connection on these leaves can be considered as an affine connection in the loop $l(0,0)$. This is also valid for any affine connection associated with a three-web (see Sections 1.7 and 1.8). Without loss of generality, we can assume that the loop $l(0,0)$ is an arbitrary local analytic loop and thus all affine connections mentioned above are defined in any such loop. For us the most interesting connections are the connections induced by the Chern connection in the leaves of the first and second foliations of the web. Denote them by Γ_1 and Γ_2, respectively. The structure equations of the connections Γ_1 and Γ_2 can be obtained from the structure equations (1.26) and (1.33) of the Chern connection by setting $\underset{1}{\omega}^i = 0$ or $\underset{2}{\omega}^i = 0$:

$$\begin{aligned} \Gamma_1: \ & d\underset{2}{\omega}^i + \underset{2}{\omega}^i \wedge \omega_j^i = -a^i_{jk}\underset{2}{\omega}^j \wedge \underset{2}{\omega}^k, \quad d\omega_j^i = \omega_j^k \wedge \omega_k^i; \\ \Gamma_2: \ & d\underset{1}{\omega}^i + \underset{1}{\omega}^i \wedge \omega_j^i = a^i_{jk}\underset{1}{\omega}^j \wedge \underset{1}{\omega}^k, \quad d\omega_j^i = \omega_j^k \wedge \omega_k^i. \end{aligned} \tag{2.83}$$

The connections Γ_1 and Γ_2 are curvature-free and, therefore, possess absolute parallelism. If a web W is a group web, then the forms ω_j^i can be reduced to zero (see Section 1.5), the quantities a^i_{jk} become constants, and equations (2.83) are the Maurer–Cartan equations of a Lie group. These equations are also the structure equations of one of the canonical connections of Cartan associated with a Lie group. Therefore, the connections Γ_1 and Γ_2 are a generalization of Cartan's group connections. We will call them the *canonical connections* in the loop Q. It follows from Theorem 2.32 that parallel vector fields in these connections can be obtained from vectors defined in the identity element e of the loop Q by means of shifts defined in this loop, i.e. in the same way as invariant fields of a Lie group. However, if a loop Q is not a group, then, because of the lack of associativity, the indicated fields are not invariant relative to shifts.

It is known that geodesic lines of Cartan's group connections are one-parametric subgroups. However, the geodesic lines of the canonical connections Γ_1 and Γ_2 do not have this property since, as shown in Section 2.6, in general, there are no subgroups (or even subloops) in a loop. The following problem arises from this fact: if there is one-parametric subloop in a loop Q, is this loop geodesic in the connections Γ_1 and Γ_2? The following theorem solves this problem:

Theorem 2.33 *Every one-parametric subloop of a local analytic loop Q is geodesic in the canonical connections Γ_1 and Γ_2.*

Proof. We will prove the theorem, for example, for the connection Γ_1. Denote by W a three-web for which the loop Q is a coordinate loop. Suppose that this web has the standard parametrization. Then the condition of parallelism of a vector field $\xi = \underset{1}{\xi}$ in the loop Q can be written in the form (2.81) or (2.80, 1). Further, suppose that $y(t)$ is a subloop in Q. Define the canonical coordinates in Q. By Theorem 2.24, the equations of the subloop $y(t)$ have the form $y^i(t) = a^i t$. We prove that this subloop is geodesic, i.e. there exists a parallel vector field of the form $\underset{2}{\xi}^i = \lambda(t)\frac{dy^i}{dt} = \lambda(t)a^i$ in this subloop. Substituting $\underset{2}{\xi}^i$ into equations (2.80), we get

$$\lambda(t)a^i = \frac{\partial f^i}{\partial x^j}(0, at)\underset{2}{c}^j. \tag{2.84}$$

We can find the function $\lambda(t)$ from equation (2.84). To prove this, it is sufficient to show that the right hand side of this equation is proportional to the vector $a = (a^i)$. In fact, the canonical expansion of the function $z = f(x, y)$ has the form

$$f(x, y) = x + y + \dots,$$

from which it follows that

$$\frac{\partial f}{\partial x}(0, 0) = \text{id}.$$

Thus, setting $t = 0$ in (2.84), we obtain $\lambda(0)a^i = \underset{2}{c}^i$, and the equations (2.84) can be written in the form:

$$\lambda(t)a^i = \lambda(0)\frac{\partial f^i}{\partial x^j}(0, at)a^j.$$

To calculate the right hand side of this equation, we note that if we use the canonical expansion of the function f, we obtain

$$\begin{aligned}\tfrac{\partial f}{\partial x}(0, at)(a) &= a + \underset{1,1}{\Lambda}(a, at) + \frac{1}{2}\underset{1,2}{\Lambda}(a, at, at) + \frac{1}{6}\underset{1,3}{\Lambda}(a, at, at, at) + \dots \\ &= a + \tfrac{1}{2}t^2\underset{1,2}{\Lambda}(a, a, a) + \frac{1}{6}t^3\underset{1,3}{\Lambda}(a, a, a, a) + \dots .\end{aligned}$$

We used here the property $\underset{1,1}{\Lambda}(a, a) = 0$ since the form $\underset{2}{\Lambda}$ is skew-symmetric (see (2.66)). By Theorem 2.25, the vector a is an eigenvector of each of the forms $\underset{p,q}{\Lambda}$. Thus, if we use the notations (2.73), we finally obtain

$$\frac{\partial f}{\partial x}(0, at)(a) = a + \frac{1}{2}t^2\underset{1,2}{\lambda}a + \frac{1}{6}t^3\underset{1,3}{\lambda}a + \dots .$$

Therefore,

$$\lambda(t) = \lambda(0)\Big(1 + \sum_{s=2}^{\infty}\frac{1}{s!}t^s\underset{1,s}{\lambda}\Big).$$

This completes our proof. ■

The geodesic property of one-parametric subloops also follows from other, more general considerations. The correspondence between subwebs and subquasigroups will play the main role in these considerations. As in Section 2.5, we will use for the local coordinate quasigroup of the web $W = (X, \lambda_\alpha)$ the notation q, $q : X_1 \times X_2 \to X_3$, where X_α is the base of the foliation λ_α, $\alpha = 1, 2, 3$, dim $X_\alpha = r$.

Theorem 2.34 *To 2ρ-dimensional subweb $\widetilde{W}$ of a three-web W there corresponds a ρ-dimensional subquasigroup $\widetilde{q}$ of the local coordinate quasigroup q. Conversely, if there exists a subquasigroup $\widetilde{q}$ of dimension ρ in the quasigroup $q, \rho < r =$ dim q, then this subquasigroup defines a 2ρ-dimensional subweb in the web W.*

Proof. By Section 1.2, a subweb $\widetilde{W}$ of dimension 2ρ is cut by the leaves of the three-web $W = (X, \lambda_\alpha)$ in 2ρ-dimensional transversal submanifold $\widetilde{X}$ of the manifold X. The leaves of the foliation λ_α that cut the subweb $\widetilde{W}$ in a neighborhood $U \subset \widetilde{X}$, form a ρ-dimensional subfibration $\widetilde{\lambda}_\alpha$ in λ_α. Denote the base of the foliation $\widetilde{\lambda}_\alpha$ by $\widetilde{X}_\alpha$.

Consider two leaves of the web $\widetilde{W}$, one from λ_1 and another one from λ_2. They have a common point on the manifold $\widetilde{X}$. The leaf of the third foliation, passing through this common point, belongs to the manifold $\widetilde{\lambda}_3$. Therefore, the restriction $\widetilde{q}$ of the coordinate quasigroup q to the set $\widetilde{X}_1 \times \widetilde{X}_2$, acts from $\widetilde{X}_1 \times \widetilde{X}_2$ to $\widetilde{X}_3$ and is thus a local subquasigroup of q.

Conversely, let the local coordinate quasigroup $q(\cdot) : X_1 \times X_2 \to X_3$ of a three-web $W = (X, \lambda_\alpha)$ have a subquasigroup $\widetilde{q}$ of dimension ρ, $\widetilde{q} : \widetilde{X}_1 \times \widetilde{X}_2 \to \widetilde{X}_3$, $\widetilde{X}_\alpha \subset X_\alpha$. To the submanifold $\widetilde{X}_\alpha$ there corresponds a ρ-dimensional subfoliation in the foliation λ_α. We will denote this subfibration by $\widetilde{\lambda}_\alpha$. On the manifold X we consider a submanifold $\widetilde{X}$ formed by the points of intersections of all leaves of the foliations λ_1 and λ_2. Since dim $\widetilde{\lambda}_\alpha = \rho$, we have dim $\widetilde{X}_\alpha = 2\rho$. Next, since the equation $\widetilde{x}_1 \cdot \widetilde{x}_2 = \widetilde{x}_3$, $\widetilde{x}_\alpha \in \widetilde{\lambda}_\alpha$, is uniquely solvable in the quasigroup $\widetilde{q}$, the manifold $\widetilde{X}$ is cut only by the leaves of $\widetilde{\lambda}_\alpha$ but not by any leaf of λ_α. Further, since dim $\widetilde{X}_\alpha = 2\rho$ and dim $\widetilde{\lambda}_\alpha = \rho$, each leaf of $\widetilde{\lambda}_\alpha$ intersects X along a manifold of dimension ρ. Therefore, $\widetilde{X}$ is a transversal submanifold in X, and the leaves of the web W cut a subweb in this submanifold. ■

We will now establish a correspondence between subwebs and subloops. Let $\widetilde{X}$ be a 2ρ-dimensional transversal manifold in X and p be an arbitrary point in $\widetilde{X}$. A coordinate loop $l_p(\circ)$ of the web W is associated with the point p. The product $u \circ v$ in this loop is defined as shown in Figure 2.11. If the leaves u and v from λ_3 intersect the transversal surface $\widetilde{X}$, then their product also satisfies this property. Thus, we get a ρ-dimensional coordinate loop $\widetilde{l}_p$ of the subweb $\widetilde{W}$ cut in the manifold $\widetilde{X}$ by the leaves of the web W. It follows from our previous considerations that $\widetilde{l}_p$ is a subloop of l_p.

The following theorem generalizes Theorem 2.33.

Theorem 2.35 *A ρ-dimensional subloop of a local analytic loop Q is a totally geodesic submanifold in the connections Γ_1 and Γ_2 in Q.*

Proof. Denote by W a three-web for which the loop Q is the coordinate loop $l(0,0)$. As was shown earlier (see Theorem 1.12, p. 38), the leaves of a web W, its transversal surfaces as well as their intersections with the leaves of the web W are totally geodesic submanifolds in the Chern connection. This implies that the same intersections are totally geodesic submanifolds in the connections Γ_1 and Γ_2 induced in W by the Chern connection. Suppose that in a neighborhood of the point $p(0,0)$ the three-web the standard parametrization is introduced (see Section 2.5). Then its horizontal and vertical leaves, passing through the point p, carry the structure of the coordinate loop $l(0,0) = Q$. Moreover, the connections Γ_1 and Γ_2 become the canonical connections in Q, and the totally geodesic submanifolds mentioned above, which are cut in the horizontal and vertical leaves of the three-web, become ρ-dimensional subloops of the loop Q. This completes the proof of Theorem 2.35. Theorem 2.33 follows from this if we take $\rho = 1$. ■

PROBLEMS

1. Prove that if an isotopy $J = (J_\alpha)$ of a loop $Q(\cdot)$ upon a loop $\widetilde{Q}(\circ)$ satisfies the condition $J_1(e) = J_2(e) = \widetilde{e}$, where e and $\widetilde{e}$ are the identity elements in Q and $\widetilde{Q}$, then J is an isomorphism.

2. Prove that ${}^{-1}(x^{-1}) = ({}^{-1}x)^{-1}$ in any loop.

A loop $Q(\cdot)$ is said to be *IP-loop* if it has the left and right inverse properties:

$$(x \cdot y) \cdot y^{-1} = x, \quad {}^{-1}x \cdot (x \cdot y) = y. \tag{2.85}$$

3. Prove the following properties of IP-loops:

i) The solution of the equation $a \cdot x = b$ has the form $x = {}^{-1}a \cdot b$, and the solution of the equation $y \cdot b = c$ has the form $y = c \cdot b^{-1}$;

ii) $x^{-1} = {}^{-1}x, \quad (x^{-1})^{-1} = {}^{-1}({}^{-1}x) = x$;

iii) $(x \cdot y)^{-1} = y^{-1} \cdot x^{-1}$;

iv) $L_a^{-1} = L_{a^{-1}}, \quad R_b^{-1} = R_{b^{-1}}$;

Define a mapping $I : Q \to Q$ in a loop Q by the equation $I(x) = x^{-1}$. Then, for IP-loops prove the following properties:

v) $IR_aI = L_a^{-1}, \quad IL_aI = R_a^{-1}$;

vi) If $A = (A_\alpha)$ is an autotopy in Q, then (IA_1I, A_3, A_2), (A_3, IA_2I, A_1), (A_2, IA_3I, IA_1I), (IA_3, A_1, IA_2I) and (IA_2I, IA_1I, IA_3I) are also autotopies.

Solution. vi) We prove, for example, that (IA_1I, A_3, A_2) is an autotopy. From the definition (2.5) of autotopy, we have: $A_1(x^{-1}) \cdot A_2(x \cdot y) = A_3(y)$, from which it follows that

$(A_1(x^{-1}))^{-1} = A_2(x \cdot y) \cdot A_3^{-1}(y)$. On the other hand, applying (2.85) and the property iii), we get

$$(IA_1I)(x) \cdot A_3(y) = (A_1(x^{-1}))^{-1} \cdot A_3(y) == (A_2(x \cdot y) \cdot A_3^{-1}(y)) \cdot A_3(y) = A_2(x \cdot y) \blacksquare$$

4. Prove that the left (right) Bol loop is a) a left (right) alternative, and b) a left (right) inverse loop. (*Hint*: In the identity B_l set a) $y = e$ and b) $y = {}^{-1}$.)

5. Prove that the Moufang loop is a) elastic; b) a left inverse loop and a right inverse loop, i.e. an IP-loop; and c) left and right alternative.

(*Hint*: b) In the Moufang identity (see Table 2.1, p. 49) set first $y = z^{-1}$, next $y = {}^{-1}$ and apply the elasticity identity.); c) In the Moufang identity set $z = x$ and subsequently $y = x$, and apply the elasticity identity.)

6. Prove that a Moufang loop possesses autotopies of the form a) $L_y, R_y, L_yR_y)$; b) $(L_y, IL_y^{-1}I, L_yR_y)$, $(IR_y^{-1}I, R_y, L_yR_y)$, and c)$(L_yR_y, L_y^{-1}, L_y)$, (R_y^{-1}, L_yR_y, R_y), where, as above $I(x) = x^{-1}$. (*Hint*: a) Apply the Moufang identity; b) Apply the results of Problems 5 and 3v); c) Apply the result of Problem 3vi).)

7. Prove that in a Moufang loop the following identities hold:

$$M_1 : x \cdot (y \cdot (x \cdot z)) = ((x \cdot y) \cdot x) \cdot z; \quad M_2 : ((z \cdot x) \cdot y) \cdot x = z \cdot (x \cdot (y \cdot x)).$$

Prove that any two of three identities M (see Table 2.1, p. 49), M_1 and M_2 are equivalent.

(*Hint*: Apply 6c).)

8. Check that the table of multiplication given in Table 2.3 defines the LP-isotope $l(6,3)(\circ)$ of the commutative loop $Q(\cdot)$ with the table given in Table 2.4. Is the loop $Q(\cdot)$ a group?

	1	2	3	4	5	6
1	1	2	3	4	5	6
2	2	3	4	1	6	5
3	3	4	6	5	1	2
4	4	1	5	6	2	3
5	5	6	1	2	3	4
6	6	5	2	3	4	1

Table 2.3

	6	5	2	3	4	1
3	1	2	3	4	5	6
4	2	3	4	1	6	5
6	3	4	6	5	1	2
5	4	1	5	6	2	3
1	5	6	1	2	3	4
2	6	5	2	3	4	1

Table 2.4

9. Prove that a Moufang loop is a left and right Bol loop. (*Hint*: The left Bol identity follows from the identity M_1 and elasticity.)

10. Prove that a left alternative right (right alternative left) Bol loop is a Moufang loop. If a loop Q is both a left Bol loop and a right Bol loop, then it is a Moufang loop.

11. Prove that if one family λ_α of an abstract three-web $W = (X, \lambda_\alpha)$ contains n lines, then the set X contains n^2 points.

12. Prove that if e is the identity element of a loop Q, then e is the left unit in the quasigroup Q^{-1} and the right unit in the quasigroup ${}^{-1}Q$.

13. Prove that $B_l \& B_m \Rightarrow B_r$ and that $B_r \& B_m \Rightarrow B_l$.

14. Write the derived identity of the elasticity identity $(x \cdot y) \cdot x = x \cdot (y \cdot x)$ and construct the corresponding coordinate figure in a three-web.

15. Construct the coordinate figures corresponding to the associativity identity and the left Bol identity for different locations of lines u, v and w relative to the line e as this was done for the figure T in Figures 2.13–2.15. What identities coresspond to these figures ?

16. Use Figure 2.12 to construct the elements $x^{-1}, {}^{-1}x, (x^{-1})^{-1}, {}^{-1}({}^{-1}x), x/y$ and $y\backslash x$. Find the coordinate figures corresponding to the following identities: $x^{-1} = {}^{-1}x$, $(x^{-1})^{-1} = x$ and ${}^{-1}({}^{-1}x) = x$.

17. Prove that the webs B_m are also characterized by the following universal identity: $(w/v) \circ (u\backslash w) = (w/(u \circ v)) \circ w$.

18. Construct the coordinate figure corresponding to the identity $(y \circ z) \circ x = (y \circ x) \circ z$. Is this identity universal?

19. Prove that the webs B_l and B_r are characterized by the universal identities $(v \circ (u\backslash v)) \circ w = v \circ (u\backslash (v \circ w))$ and $((u \circ v)/w) \circ v = u \circ ((v/w) \circ v)$, respectively. (*Hint*: See the proof of Theorem 2. 13.)

20. Prove that the identities $x^{-1} = {}^{-1}x$ and ${}^{-1}({}^{-1}x) = x$ hold in the coordinate loop of webs H and only in such quasigroups.

21. Construct the coordinate figure corresponding to the identities $(x \cdot y) \cdot y^{-1} = x$ and ${}^{-1}x \cdot (x \cdot y) = y$. Prove that the corresponding closure conditions characterize the webs B_r and B_l, respectively.

22. Prove that if in the coordinate loops the identity $(x \cdot y)^{-1} = {}^{-1}y \cdot {}^{-1}x$ holds, then this web is a Moufang web.

23. Prove geometrically that in the coordinate loops of the webs B_l, B_m and E the identities $x^m \cdot (y \cdot x^n) = (x^m \cdot y) \cdot x^n$; $(x \cdot y^m) \cdot y^n = x \cdot y^{m+n}$ and $x^m \cdot (y \cdot x^n) = (x^m \cdot y) \cdot x^n$ hold respectively (see Section 7.5).

24. Prove that an identity S is universal if and only if it is equivalent to its own derived identity.

25. Prove Theorem 2.7 analytically deriving it directly from the closure conditions (2.6)–(2.12). (*Hint*: Apply formula (2.3).)

26. Prove that in the three-dimensional loop given by the equations

$$z^1 = x^1 + y^1 - (x^2 + y^2)x^3y^3, \quad z^2 = x^2 + y^2, \quad z^3 = x^3 + y^3, \tag{2.86}$$

the elasticity identity holds and the identities B_l and B_r do not. (*Hint*: Compare the first coordinates of the products $(x \cdot y) \cdot x, x \cdot (y \cdot x), x \cdot (y \cdot y), (x \cdot y) \cdot y, x \cdot (x \cdot y)$ and $(x \cdot x) \cdot y$.)

27. Prove that the three-web defined by equations (2.86) is a web E, i.e. all figures E are closed on this web (see Figure 2.9).

Solution. Equations (2.86) define multiplication $(\cdot)$ in the coordinate quasigroup q of the web considered. Thus, equations (2.12), implying the closure of an arbitrary figure E, have the form: (2.87)&(2.88)&(2.89) $\Rightarrow$ (2.90), where

$$\begin{cases} x_2^3 + y_1^3 = x_1^3 + y_2^3, \\ x_3^3 + y_1^3 = x_1^3 + y_3^3, \\ x_2^3 + y_3^3 = x_1^3 + y_4^3, \\ x_3^3 + y_2^3 = x_4^3 + y_1^3; \end{cases} \tag{2.87}$$

$$\begin{cases} x_2^2 + y_1^2 = x_1^2 + y_2^2, \\ x_3^2 + y_1^2 = x_1^2 + y_3^2, \\ x_2^2 + y_3^2 = x_1^2 + y_4^2, \\ x_3^2 + y_2^2 = x_4^2 + y_1^2; \end{cases} \tag{2.88}$$

$$\begin{cases} x_2^1 + y_1^1 - (x_2^2 + y_1^2)x_2^3y_1^3 = x_1^1 + y_2^1 - (x_1^2 + y_2^2)x_1^3y_2^3, \\ x_3^1 + y_1^1 - (x_3^2 + y_1^2)x_3^3y_1^3 = x_1^1 + y_3^1 - (x_1^2 + y_3^2)x_1^3y_3^3, \\ x_2^1 + y_3^1 - (x_2^2 + y_3^2)x_2^3y_3^3 = x_1^1 + y_4^1 - (x_1^2 + y_4^2)x_1^3y_4^3, \\ x_3^1 + y_2^1 - (x_3^2 + y_2^2)x_3^3y_2^3 = x_4^1 + y_1^1 - (x_4^2 + y_1^2)x_4^3y_1^3; \end{cases} \tag{2.89}$$

$$\begin{cases} x_4^3 + y_3^3 = x_3^3 + y_4^3, \\ x_4^2 + y_3^2 = x_3^2 + y_4^2, \\ x_4^1 + y_3^1 - (x_4^2 + y_3^2)x_4^3y_3^3 = x_3^1 + y_4^1 - (x_3^2 + y_4^2)x_3^3y_4^3, \end{cases} \tag{2.90}$$

The first equation of (2.90) is obtained from the system (2.87) if one adds the first two equations and subtracts the third. In the same way, the second equation of (2.90) is obtained from the system (2.88). If we construct the same combination from equations (2.89), we arrive at:

$$\begin{aligned} &x_4^1 + y_3^1 - (x_1^2 + y_2^2)x_1^3y_2^3 + (x_1^2 + y_4^2)x_1^3y_4^3 - (x_4^2 + y_1^2)x_4^3y_1^3 \\ &= x_3^1 + y_4^1 - (x_2^2 + y_1^2)x_2^3y_1^3 + (x_2^2 + y_3^2)x_2^3y_3^3 - (x_3^2 + y_2^2)x_3^3y_2^3. \end{aligned}$$

Let us prove that the last equation is identical to the third equation of (2.90), or equivalently, that the difference of these two equations is zero. Denote this difference by P. We have:

$$\begin{aligned} P = \ & (x_3^2 + y_4^2)x_3^3y_4^3 + (x_1^2 + y_2^2)x_1^3y_2^3 - (x_1^2 + y_4^2)x_1^3y_4^3 + (x_4^2 + y_1^2)x_4^3y_1^3 \\ & -(x_4^2 + y_3^2)x_4^3y_3^3 - (x_2^2 + y_1^2)x_2^3y_1^3 + (x_2^2 + y_3^2)x_2^3y_3^3 - (x_3^2 + y_2^2)x_3^3y_2^3. \end{aligned}$$

Rearranging the terms of this equation, we find:

$$\begin{aligned} P = \ & x_4^2x_4^3(y_1^3 - y_3^3) + y_3^2y_3^3(x_2^3 - x_4^3) + x_2^2x_2^3(y_3^3 - y_1^3) + y_1^2y_1^3(x_4^3 - x_2^3) \\ & +x_3^2x_3^3(y_4^3 - y_2^3) + y_2^2y_2^3(x_1^3 - x_3^3) + y_4^2y_4^3(x_3^3 - x_1^3) + x_1^2x_1^3(y_2^3 - y_4^3). \end{aligned}$$

Next, we find from (2.87) and (2.88) that

$$\begin{aligned} &y_1^3 - y_3^3 = x_1^3 - x_3^3, \\ &x_2^3 - x_4^3 = x_4^3 + y_3^3 - y_1^3 - (x_3^3 + y_2^3 - y_1^3) = x_1^3 - x_3^3, \\ &y_2^3 - y_4^3 = x_2^3 + y_1^3 - x_1^3 - (x_2^3 + y_3^3 - x_1^3) = y_1^3 - y_3^3, = x_1^3 - x_3^3. \end{aligned}$$

As a result, the sum P has the form $(x_1^3 - x_3^3)R$ where

$$R = x_4^2x_4^3 + y_3^2y_3^3 - x_2^2x_2^3 - y_1^2y_1^3 - x_3^2x_3^3 + y_2^2y_2^3 - y_4^2y_4^3 + x_1^2x_1^3.$$

Using equations (2.87) and (2.88), we calculate the terms of the last equation:

$$\begin{aligned}
x_4^2 x_4^3 = &\ (x_3^2 + y_2^2 - y_1^2)(x_3^3 + y_2^3 - y_1^3) = (x_3^2 + x_2^2 - x_1^2)(x_3^3 + x_2^3 - x_1^3) \\
= &\ -(x_3^2 + x_2^2)x_1^3 - x_1^2(x_3^3 + x_2^3) + (x_3^2 + x_2^2)(x_3^3 + x_2^3) + x_1^2 x_1^3; \\
y_3^2 y_3^3 = &\ (x_3^2 + y_1^2 - x_1^2)(x_3^3 + y_1^3 - x_1^3) \\
= &\ x_3^2 x_3^3 + x_3^2(y_1^3 - x_1^3) + x_3^3(y_1^2 - x_1^2) + (y_1^2 - x_1^2)(y_1^3 - x_1^3); \\
y_2^2 y_2^3 = &\ (x_2^2 + y_1^2 - x_1^2)(x_2^3 + y_1^3 - x_1^3) \\
= &\ x_2^2 x_2^3 + x_2^2(y_1^3 - x_1^3) + x_2^3(y_1^2 - x_1^2) + (y_1^2 - x_1^2)(y_1^3 - x_1^3); \\
y_4^2 y_4^3 = &\ (x_2^2 + y_3^2 - x_1^2)(x_2^3 + y_3^3 - x_1^3) \\
= &\ (x_2^2 + x_3^2 + y_1^2 - 2x_1^2)(x_2^3 + x_3^3 + y_1^3 - 2x_1^3) \\
= &\ [(x_2^2 + x_3^2) + (y_1^2 - x_1^2) - x_1^2][(x_2^3 + x_3^2) + (y_1^3 - x_1^3) - x_1^3] \\
= &\ (x_2^2 + x_3^2)(x_2^3 + x_3^3) + (y_1^2 - x_1^2)(x_2^3 + x_3^3) - x_1^2(x_2^3 + x_3^3) \\
&\ +(x_2^2 + x_3^2)(y_1^3 - x_1^3) + (y_1^2 - x_1^2)(y_1^3 - x_1^3) - x_1^2(y_1^3 - x_1^3) \\
&\ -(x_2^2 + x_3^2)x_1^3 - (y_1^2 - x_1^2)x_1^3 + x_1^2 x_1^3.
\end{aligned}$$

Substituting the results of this calculation into R, after some simple manipulations we obtain $R = 0$. This implies $P = 0$. ■

Denote the three-web E defined by equations (2.86) by E_1. As we have shown, this web is neither a left Bol web nor a right Bol web, i.e. it is not a Moufang web.

28. Let $x(t) = ta$ be a one-parameter subloop of a local analytic loop Q and let $\underset{p,q}{\lambda}$ be the eigenvalues of the vector a with respect to the form $\underset{p,q}{\Lambda}$ (see (2.73)). Prove that the numbers $\underset{p,q}{\lambda}$ are related by the condition:

$$\sum_{\substack{p=1 \\ p+q=s}}^{s-1} \binom{s}{p} \underset{p,q}{\lambda} = 0.$$

(*Hint*: apply equation (2.67).)

29. Prove that if a loop Q possesses the maximal number of two-dimensional subloops, then its equations can be written in the form: $z = \lambda(x,y)x + \mu(x,y)y$, where $\lambda(x,y)$, $\mu(x,y) \in \mathbf{R}$ and $\lambda(x,x) + \mu(x,x) = 2$, $x, y \in \mathbf{R}^r$. Prove that the converse is also true. (*Hint*: As a preliminary exercise, prove that the tangent Λ-algebra of the loop Q is characterized by the relations:

$$\underset{p,q}{\Lambda}{}^{i}_{j_1 \dots j_p k_1 \dots k_q} = \delta^i_{(j_1} \lambda_{j_2 \dots j_p) k_1 \dots k_q} + \mu^i_{j_1 \dots j_p (k_1 \dots k_{q-1}} \delta^i_{k_q)}.$$

30. Prove that a loop Q is monoassociative if and only if the coefficients of its canonical expansion are related by the condition:

$$\sum_{\substack{p=1 \\ p+q=s}}^{s-1} \frac{2^q - 2^p}{p!q!} \underset{p,q}{\Lambda}|_d = 0.$$

31. Prove that if the the right (left) alternativity identity holds in a local analytic loop Q, then the associativity of powers also holds in this loop.

32. Find an expression for the forms $\underset{p,q}{\Lambda}$ in the canonical expansion of a local analytic loop Q possessing the maximal number of one-parametric subloops (in particular, subgroups).

Solution. Suppose that local coordinates in the loop Q are canonical. The condition of Theorem 2.25 leads to relations (2.73) which must be satisfied for any value of a. Therefore, the quantities $\underset{p,q}{\lambda}$ are symmetric forms of degree $p+q-1$ in a: $\underset{p,q}{\lambda} = \underset{p,q}{\lambda}(a, a, \dots, a)$. Substituting this into (2.73), we get the relation

$$\underset{p,q}{\Lambda}(a, a, \dots a) = \underset{p,q}{\lambda}(a, a, \dots a) \cdot a,$$

which implies the following conditions on the coordinates of the forms $\underset{p,q}{\Lambda}$:

$$\underset{p,q}{\Lambda}{}^i_{(j_1 \dots j_{p+q})} = \underset{p,q}{\lambda}{}^i_{(j_1 \dots j_{p+q-1}} \delta^i_{j_{p+q})}.$$

Conversely, the last equations yield that any vector a of the one-dimensional tangent Λ-algebra of the loop Q defines a one-dimensional subalgebra.

If all one-parameter subloops are groups, all eigenvalues $\underset{p,q}{\lambda}$ are zero, and we arrive at the conditions:

$$\underset{p,q}{\Lambda}{}^i_{(j_1 \dots j_{p+q})} = 0. \blacksquare$$

33. Prove that there exist canonical coordinates in a local analytic loop Q satisfying the condition:

$$f(x, \sigma x) = (1+\sigma)x, \quad \sigma \neq 0, -1,$$

where $x \in Q$ and $f(x, y) = x \cdot y$ is the multiplication in Q.

34. Prove that all canonical coordinates introduced in Problem **33** (i.e. considered for different σ) coincide if and only if a loop Q is monoassociative.

35. Prove that a vector field ξ is parallel on the leaves of the third foliation of a web W if and only if its coordinates satisfy the equation:

$$Q^k_n \frac{\partial}{\partial x^m}(\tilde{g}^n_j \xi^j) = P^n_m \frac{\partial}{\partial y^n}(\bar{g}^k_j \xi^j),$$

where $P^n_m = \tilde{g}^n_k \bar{f}^k_m$, $Q^k_n = \bar{g}^k_m \tilde{f}^m_n$, $P^n_m Q^k_n = \delta^k_m$.

36. Let canonical coordinates are assigned to a loop Q. A straight line $z^i = a^i t$ is geodesic with respect to canonical connections in Q if and only if the vector a^i is an eigenvector of the tensors $\underset{1,2}{\Lambda}, \underset{1,3}{\Lambda}, \dots, \underset{1,s}{\Lambda}, \dots$ or the tensors $\underset{2,1}{\Lambda}, \underset{3,1}{\Lambda}, \dots$.

37. Prove that the coefficients μ^i_{jkl} and ν^i_{jkl} of the canonical expansion of the coordinate loop l_p of a three-web W can be expressed in terms of the curvature tensor of this three-web as follows:

$$\begin{aligned} \mu^i_{jkl} &= -\tfrac{1}{2} b^i_{(jkl)} + \tfrac{4}{3}(b^i_{[l|k|j]} + b^i_{[l|j|k]}) + \tfrac{2}{3}(a^m_{kl} a^i_{jm} + a^m_{jl} a^i_{km}), \\ \nu^i_{jkl} &= \tfrac{1}{2} b^i_{(jkl)} + \tfrac{4}{3}(b^i_{[kj]l} + b^i_{[lj]k}) + \tfrac{2}{3}(a^m_{lj} a^i_{km} + a^m_{kj} a^i_{lm}). \end{aligned} \tag{2.91}$$

(We assume that the canonical expansion has the form (2.24.)

Solution. Conditions (2.67) taken for $s = 2, 3$ imply that

$$\lambda^i_{(jk)} = 0, \quad \nu^i_{(jkl)} + \mu^i_{(jkl)} = 0.$$

Next, using the skew-symmetry of the quantities λ^i_{jk} and applying formula (2.49), we obtain

$$\begin{aligned} b^i_{[jk]l} &= \nu^i_{[kj]l} + \lambda^m_{[j|l|}\lambda^i_{k]m} - \lambda^m_{kj}\lambda^i_{ml}, \\ b^i_{[j|k|l]} &= -\mu^i_{k[jl]} + \lambda^m_{jl}\lambda^i_{km}, -\lambda^m_{k[j}\lambda^i_{|m|l]}. \end{aligned}$$

The last two equations yield

$$\nu^i_{(jkl)} = -\mu^i_{(jkl)} = \frac{1}{2} b^i_{(jkl)}.$$

Since the quantities μ^i_{jkl} are symmetric in j and k and the quantities ν^i_{jkl} are symmetric in k and l, the following identities hold:

$$\begin{aligned} \mu^i_{jkl} &= \mu^i_{(jkl)} + \tfrac{2}{3}(\mu^i_{[j|k|l]} + \mu^i_{[k|j|l]}), \\ \nu^i_{jkl} &= \nu^i_{(jkl)} + \tfrac{2}{3}(\nu^i_{[jk]l} + \nu^i_{[jl]k}). \end{aligned}$$

Substituting for the terms in the right hand sides the expressions found above for these terms, we arrive at formulas (2.91). ■

38. Prove that if a three-web $\widetilde{W}$ is a subweb of a web W, prove that the tangent $\widetilde{W}_k$-subalgebra of the web $\widetilde{W}$ is a subalgebra of the corresponding W_k-algebra of the web W. If $\widetilde{W}$ is a normal subweb (see Section 1.9), then its $\widetilde{W}_k$-algebras are ideals of W_k-algebras.

39. Let V be an r-dimensional smooth submanifold of a manifold X. Prove that a three-web $W = (X, \lambda_\alpha)$ induces a binary operation $x \circ y$ in V defining a local idempotent quasigroup in V which is isotopic to one of the coordinate quasigroups of the web W.

NOTES

2.1. In this section we give only basic facts from the algebraic theory of loops. A more detailed exposition can be found in the books [Bel 67], [Br 71] [Pf 90], in the surveys [Br 51], [Ga 88] and in the paper [Ac 65]. The terms "loop" and "principal isotope" were introduced by A.A. Albert [Al 43a]. Our exposition is specific in the sense that all our formulations are given in the most general setting, i.e. for the three-base quasigroups. The topological and analytical loops are investigated in [Ho 58a], [Ho 58b], [Ho 59] and [HS 90].

The term "universal identity" was introduced in [Br 51].

2.2. Abstract three webs are also called three-nets. A precise definition of an abstract three-web was given by G. Pickert [Pi 75] although the notion of a k-net was used by G. Thomsen as far back as 1929. In [Ac 65], J. Aczel used the term "algebraic web". The authors of this book also used this term earlier. However, this term has also another meaning: it relates to three-webs generated by cubic hypersurfaces (see Chapter 3). This is the reason the authors stopped using it and switched to the term "abstract web".

The terms "coordinate quasigroup" and "coordinate loop", probably first appeared in [Bel 67].

The relationship between configurations in a three-web and algebraic properties of quasigroups was investigated starting in the late 1920's by the founders of the web theory W. Blaschke, G. Bol and K. Reidemeister. A survey of papers of this period can be found in the monograph [BB 38].

The systematic exposition of the material of Section 2.2 (excluding the configuration E) was first given by J. Aczel in [Ac 65]. The corresponding closure conditions (2.6)–(2.11) are given in [Ac 65] slightly differently, in the form of conditional identities (see p. 55). We write them in a symmetric form since it does not matter for us on which side the figures are closed.

The figure E first appeared in [SS 85].

2.3. The notion of the derived identity arose in [Go 71].

Theorem 2.8 follows from Albert's theorem stated on p. 51. Its geometrical proof is due to M.A. Shestakova.

The identity (2.18) was obtained by J. Aczel [Ac 81] and the identity (2.22) by V.D. Belousov [Bel 67]. Note that the proof that the identity (2.22) is universal was algebraic and rather complicated. The simple geometric approach which we gave was substantiated in [SS 84] where the identities (2.18)–(2.20) and (2.22) were obtained by this geometric method.

2.4. The tangent W-algebra to a local C^3-smooth loop was defined by M.A. Akivis in [A 76a]. K. Strambach and K. H. Hofmann suggested calling it the "Akivis algebra" [HS 86b] (see also [HS 90]).

2.6. A.I. Mal'cev proved in [Ma 55] that the canonical coordinates in a local analytic alternative loop are introduced in the same way as in a Lie group since in such loops there exists a one-parameter subgroup in any direction emanating from the identity element (in [Ma 55] a loop is called alternative if any pair of its elements generates an associative subloop). E.N. Kuz'min proved this fact for a wider class of loops, so-called power-asociative loops [Ku 71]. Even earlier, in [A 69a], the canonical coordinates were defined by M.A. Akivis for an arbitrary local analytic loop. According to [A 69a], the canonical coordinates are introduced by the normalization $(x \cdot x)^i = (1+\sigma)x^i$, where σ is a real number different from 0 and -1. For the power-associative loops and only for them, all canonical coordinates (i.e. for different σ's) coincide since in these loops all points with the coordinates $(1+\sigma)x^i$, for fixed σ and any x, belong to one one-parameter subgroup. The existence of the canonical coordinates defined in [A 69a] was proved in [AS 86]; a close result is in [DJ 85].

2.7. Algebraic properties of the Chern connection were studied by P. Nagy in [N 85] and [N 89]. In particular, Lemma 2.30 has been proved by Nagy in a more general setting.

Problems. Idempotent quasigroups defined by a three-web (see Problem **39**) were studied by G.A. Tolstikhina in [To 87] and [To 88].

Chapter 3

Transversally Geodesic and Isoclinic Three-Webs

3.1 Transversally Geodesic and Hexagonal Three-Webs

1. Let a three-web $W = (X, \lambda_\alpha)$ be given on a manifold X. Then, as shown in Section 1.3, there exists an r-parameter family of transversal bivectors H at every point $p \in X$, such that they contain the vectors $\xi_\alpha = \xi^i \underset{\alpha}{e}_i$ tangent at the point p to the lines of leaves $\mathcal{F}_\alpha$ of the web W and corresponding to each other in the correspondence $\varphi_{\alpha\beta}$. A two-dimensional surface V^2 on X, tangent to a transversal bivector at each of its points, is a transversal surface of the web W. Since $H = \xi_1 \wedge \xi_2$, where $\xi_1 = \xi^i \underset{1}{e}_i$ and $\xi_2 = \xi^i \underset{2}{e}_i$, a transversal surface V^2 (if it exists) is defined on the manifold X by equations (1.105) which for $a = 1$, can be written in the form:

$$\underset{1}{\omega}^i = \xi^i \theta_1, \quad \underset{2}{\omega}^i = \xi^i \theta_2. \tag{3.1}$$

Here, θ_1 and θ_2 are linearly independent Pfaffian forms defining a displacement of the point p along the two-dimensional surface V^2.

Leaves of the first foliation λ_1 of the web W intersect the transversal surface V^2 along the lines defined on it by the equations $\theta_1 = 0$ and, by (3.1), tangent to the vectors $\xi_1 = \xi^i \underset{1}{e}_i$. Leaves of the foliation λ_2 intersect the surface V^2 along the lines $\theta_2 = 0$ tangent to the vectors $\xi_2 = \xi^i \underset{2}{e}_i$, and leaves of the foliation λ_3 intersect the surface V^2 along the lines $\theta_1 + \theta_2 = 0$ tangent to the vectors $\xi_3 = \xi^i \underset{3}{e}_i$. These three families of lines form a three-web $W^2 = (V^2, \lambda'_\alpha)$ on V^2 where $\lambda'_\alpha = \lambda_\alpha|_{V^2}$.

2. As shown in Section 2.3, a transversal surface V^2 is invariant with respect to the mappings $\varphi_{\alpha\beta}$.

Equations (3.1) as well as their differential consequences hold on the surface V^2. Exterior differentiation of equations (3.1) leads to equations (1.107)–(1.109) which, for $\rho = 1$, have the form:

$$d\xi^i + \xi^j \omega_j^i = \xi^i \theta, \tag{3.2}$$

$$d\theta_1 = \theta_1 \wedge \theta, \quad d\theta_2 = \theta_2 \wedge \theta, \tag{3.3}$$

where θ is a new Pfaffian form.

Equations (3.2) exactly coincide with equations (1.99) and geometrically express the fact that a transversal surface V^2 intersects leaves of the web W along geodesic lines. Moreover, this surface is a totally geodesic surface of the manifold X in any of the connections defined by the forms (1.97).

Exterior differentiation of the system (3.3) gives the equations:

$$d\theta = b\theta_1 \wedge \theta_2, \tag{3.4}$$

obtained from (1.108) by setting $\rho = 1$, and the equations (1.112):

$$b^i_{jkl}\xi^j\xi^k\xi^l = b\xi^i, \tag{3.5}$$

expressing the fact that the vector ξ^i is an eigenvector of the curvature tensor.

Next, we differentiate equations (3.5) by means of (1.36), (3.1), (3.3) and the equation

$$db - 2b\theta = c_1\theta_1 + c_2\theta_2,$$

obtained if one makes the differential prolongation of the equation (3.4). As a result, we obtain

$$\underset{1}{c}{}^i_{jklm}\xi^j\xi^k\xi^l\xi^m = c_1\xi^i, \quad \underset{2}{c}{}^i_{jklm}\xi^j\xi^k\xi^l\xi^m = c_2\xi^i \tag{3.6}$$

Thus, the vector ξ^i is also an eigenvector of the covariant derivatives of the curvature tensor. Differentiating (3.6), we get a series of similar relations for covariant derivatives of higher orders. This proves the following theorem:

Theorem 3.1 *The tangent vectors to the lines of the web $W^2 = (V^2, \lambda'_\alpha)$, cut on a transversal surface V^2 by the leaves of the three-web $W = (X, \lambda_\alpha)$, are eigenvectors of the curvature tensor of the web and all its covariant derivatives.* ∎

Note also that equations (3.3) and (3.4) are the structure equations of the web W^2, and the quantity b is the curvature of this web.

3. We will now give the definition of a transversally geodesic web.

Definition 3.2 A three-web $W = (X, \lambda_\alpha)$ is said to be a *transversally geodesic* web if each of its transversal bivector is tangent to a transversal surface V^2.

Definition 3.2 implies that a three-web W is transversally geodesic if and only if equations (3.5) and (3.6) and equations similar to them are satisfied identically. For this, it is sufficient to have equations (3.5) being satisfied since the other equations mentioned above are differential consequences of equations (3.5). Thus, equations (3.5) are necessary and sufficient conditions for a three-web W to be a transversally geodesic web.

But these conditions are satisfied if the factor b in equations (3.5) is a homogeneous polynomial of second degree in variables ξ^i, i.e. if we have

$$b = b_{ij}\xi^i\xi^j, \quad b_{ij} = b_{ji}. \tag{3.7}$$

Then equations (3.5) have the form

$$b^i_{jkl}\xi^j\xi^k\xi^l = \xi^i b_{kl}\xi^k\xi^l.$$

Since they are satisfied for any vector ξ^i, we obtain the equations

$$b^i_{(jkl)} = \delta^i_{(j}b_{kl)}. \tag{3.8}$$

Contracting equations (3.8) in the indices i and j, we express the quantities b_{kl} in terms of the curvature tensor:

$$b_{kl} = \frac{3}{r+2}b^i_{(ikl)}. \tag{3.9}$$

Thus, we have proved the following theorem:

Theorem 3.3 *A three-web $W = (X, \lambda_\alpha)$ is transversally geodesic if and only if the symmetric part of its curvature tensor is expressed as in* (3.8). ■

Let us now recall that the tensors a^i_{jk} and b^i_{jkl} define a binary and ternary operation in the tangent W-algebras of the three-web W (see Section 2.5):

$$[\xi, \eta] = -2a(\xi, \eta), \quad (\xi, \eta, \zeta) = -b(\eta, \xi, \zeta).$$

Thus, relations (3.8) can be written in the form:

$$(\xi, \xi, \xi) = -\xi b(\xi, \xi).$$

Since, in addition, we have $[\xi, \xi] = 0$, Theorem 3.3 can be reformulated in terms of the W-algebras as follows: a three-web $W = (X, \lambda_\alpha)$ is transversally geodesic if and only if each vector ξ of a local W-algebra of this web generates a one-dimensional subalgebra relative to the ternary operation.

Another characteristic property of a transversally geodesic three-web follows from the results of Section 2.7. In this section we established that to transversal surfaces of a three-web there correspond subloops of its coordinate loops. Since a transversally geodesic three-web carries the maximal possible number of transversal two-dimensional surfaces, we have the following theorem.

Theorem 3.4 *A three-web $W = (X, \lambda_\alpha)$ is transversally geodesic if and only all its coordinate loops possess the maximal number of one-parameter subloops.* ■

4. Consider a two-dimensional three-web $W^2 = (V^2, \lambda'_\alpha)$ cut by the leaves of the web $W = (X, \lambda_\alpha)$ on a transversal surface V^2. The structure equations of the web W^2 have the form (3.3) and (3.4) and its curvature b is determined by formula (3.7). It follows from this formula that the curvature b vanishes only on those surfaces V^2, for which the vector ξ defining such surfaces, satisfies the equation

$$b_{ij}\xi^i\xi^j = 0.$$

Suppose now that a two-dimensional three-web W^2 on each of these transversal surfaces is hexagonal. Then the last equation is satisfied for any vector ξ, i.e. we have $b_{ij} = 0$. Because of this, relations (3.8) give

$$b^i_{(jkl)} = 0. \tag{3.10}$$

The converse is trivial: equations (3.10) lead to the hexagonality of three-subwebs on all transversal surfaces V^2.

As proved in Chapter 2 (see Table 2.2, p. 71), the condition (3.10) is necessary and sufficient for a web W to be hexagonal. Using the results of this section, we can now prove the converse: if the symmetric part of the curvature tensor of a web W vanishes, then the web W is hexagonal.

In fact, if relations (3.10) are satisfied, then, as shown above, the web W is transversally geodesic and all its transversal surfaces V^2 carry hexagonal webs composed of geodesic lines.

Consider an arbitrary sufficiently small hexagonal figure $p_1p_2\ldots p_7$ with the center p formed by the leaves of the web W (Figure 3.1).

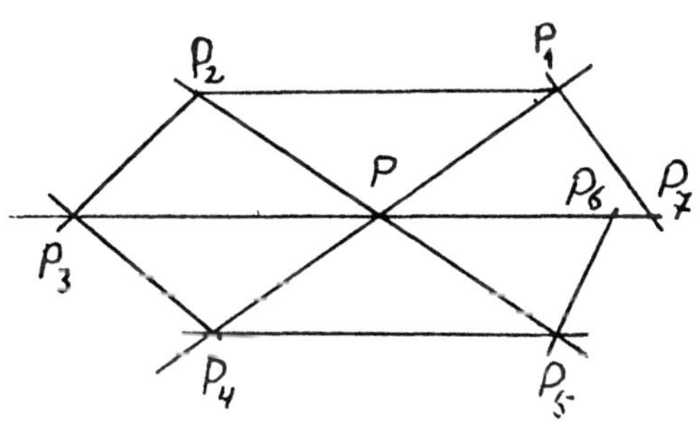

Figure 3.1

Construct the geodesic rays $pp_1, pp_2, \ldots, pp_7$ through the point p. Let ξ_1 be the tangent vector to the geodesic ray pp_1. This vector defines the transversal bivector

$H = \xi_1 \wedge \xi_2$ in the tangent space T_p. In turn, the bivector H defines a transversal surface V^2 in X. Moreover, the geodesic ray pp_1 lies on the surface V^2 since this surface is totally geodesic in the manifold X (see Section 2.7).

The surface V^2 is invariant relative to the mapping φ_{12}, established by the leaves of the third foliation, and $p_2 = \varphi_{12}(p_1)$. Thus, the point p_2 also lies on the surface V^2, and the whole geodesic ray pp_2 belongs to the surface V^2.

In a similar way one can prove that the points $p_3, p_4, \ldots, p_7$, and the geodesic rays $pp_3, pp_4, \ldots, pp_7$, defined by these points, belong to the surface V^2. Moreover, the rays pp_1 and pp_4 belong to the same geodesic line. The same is true for the rays pp_2 and pp_4 as well as for the rays pp_3, pp_6 and pp_7. This implies that the points $p_1, p_2, \ldots, p_7$ are the vertices of a hexagonal figure on the surface V^2.

Thus, we proved that the vertices of a hexagonal figure, formed by the leaves of the web W under consideration, lie on a surface V^2 and are the vertices of a hexagonal figure of the web W^2 on the surface V^2. However, as shown above, the web W^2 is hexagonal, i.e. all hexagonal figures are closed on it. Therefore, the three-web W is also hexagonal.

Note also that relation (3.10) implies the identity $(\xi, \xi, \xi) = 0$, i.e. the monoassociativity of all local W-algebras of the three-web W.

The following theorem combines the results obtained in this subsection.

Theorem 3.5 *For a multidimensional three-web $W = (X, \lambda_\alpha)$, the following statements are equivalent:*

i) A three-web W is hexagonal.

ii) A three-web W is transversally geodesic and all its two-dimensional subwebs are hexagonal.

iii) The curvature tensor of a three-web W satisfies condition (3.10).

iv) All local W-algebras of a three-web W are monoassociative.

v) All coordinate loops of a three-web W are monoassociative.

vi) All coordinate loops of a three-web W possess the maximal number of one-parameter subgroups.

■

3.2 Isoclinic Three-Webs

1. In Section 3.1 we showed that a one-parameter family of isoclinic r-vectors

$$Z = \zeta_1 \wedge \zeta_2 \ldots \wedge \zeta_r,$$

where

$$\zeta_i = \zeta^1 \underset{2}{e}_i - \zeta^2 \underset{1}{e}_i, \tag{3.11}$$

is connected with each point of the manifold X carrying a three-web $W = (X, \lambda_\alpha)$.

Definition 3.6 A submanifold V^r of dimension r of the manifold X is said to be an *isoclinic surface* of the three-web $W = (X, \lambda_\alpha)$ if it is tangent to an isoclinic r-vector at each of its points.

Let us find conditions under which a manifold X carrying a three-web W possesses an isoclinic surface V^r.

A tangent vector to an isoclinic surface is a linear combination of the vectors ζ_i written in the form (3.11):

$$\xi = \theta^i \zeta_i = \theta^i (\zeta^1 \underset{2}{e}_i - \zeta^2 \underset{1}{e}_i).$$

On the other hand, a tangent vector to the manifold X can be written in the form (1.20). Comparing these two equations, we find that the equations

$$\underset{1}{\omega}^i = \zeta^1 \theta^i, \quad \underset{2}{\omega}^i = \zeta^2 \theta^i.$$

hold on a surface V^r. Eliminating θ^i from these equations, we obtain the equations of a surface V^r in the form

$$\underset{2}{\omega}^i + \lambda \underset{1}{\omega}^i = 0, \tag{3.12}$$

where $\lambda = -\zeta^2 / \zeta^1$.

Exterior differentiation of equations (3.13) gives the equations:

$$d\lambda \wedge \underset{1}{\omega}^i - (\lambda^2 - \lambda) a^i_{jk} \underset{1}{\omega}^j \wedge \underset{1}{\omega}^i = 0.$$

It follows from these equations that on a surface V^r the differential $d\lambda$ is a linear combination of the basis forms $\underset{1}{\omega}^i$. Next, suppose that the surface V^r is not a leaf of the second or the third foliation of the web W. Then $\lambda \neq 0, \infty, 1$, and we can set

$$\frac{d\lambda}{\lambda^2 - \lambda} = a_k \underset{1}{\omega}^k. \tag{3.13}$$

Using equation (3.13), we can write the previous quadratic equation in the form:

$$(a^i_{jk} - \delta^i_k a_j) \underset{1}{\omega}^j \wedge \underset{1}{\omega}^i = 0.$$

On a surface V^r the left-hand side of this equation must be identically equal to zero. This implies that at all points of a surface V^r the relations

$$a^i_{jk} = a_{[j} \delta^i_{k]}. \tag{3.14}$$

are satisfied. Thus, *if on the manifold X of a three-web W there is an isoclinic surface V^r, then at each of its points the torsion tensor has the structure* (3.14).

Note that if $r = 2$, the torsion tensor can always be written in the form (3.14). Therefore, in what follows, we will distinguish the cases $r = 2$ and $r > 2$.

2. Consider now a three-web W with a torsion tensor of the form (3.14). The structure equations of such a web can be written in the form:

$$\begin{aligned} d\underset{1}{\omega}^i &= \underset{1}{\omega}^j \wedge \omega_j^i + a_j \underset{1}{\omega}^j \wedge \underset{1}{\omega}^i, \\ d\underset{2}{\omega}^i &= \underset{2}{\omega}^j \wedge \omega_j^i - a_j \underset{2}{\omega}^j \wedge \underset{2}{\omega}^i. \end{aligned} \tag{3.15}$$

Let us find the differential prolongations of the system (3.15). Exterior differentiation of this system gives the following system of exterior cubic equations:

$$\begin{aligned} \Omega_j^i \wedge \underset{1}{\omega}^j - \nabla a_j \wedge \underset{1}{\omega}^j \wedge \underset{1}{\omega}^i &= 0, \\ \Omega_j^i \wedge \underset{2}{\omega}^j + \nabla a_j \wedge \underset{2}{\omega}^j \wedge \underset{2}{\omega}^i &= 0, \end{aligned} \tag{3.16}$$

where

$$\Omega_j^i = d\omega_j^i - \omega_j^k \wedge \omega_k^i, \quad \nabla a_j = da_j - a_k \omega_j^k. \tag{3.17}$$

It follows from equations (3.16) that the forms Ω_j^i and ∇a_j can be expressed in terms of the basis forms:

$$\nabla a_j = p_{jk} \underset{1}{\omega}^k + q_{jk} \underset{2}{\omega}^k, \tag{3.18}$$

and

$$\Omega_j^i = b_{jkl}^i \underset{1}{\omega}^k \wedge \underset{2}{\omega}^l. \tag{3.19}$$

Substituting these expansions into equations (3.16), we obtain the relations:

$$b_{[jk]l}^i = q_{l[j}\delta_{k]}^i, \quad b_{[j|l|k]}^i = p_{l[j}\delta_{k]}^i, \tag{3.20}$$

and for $r > 2$ we get the additional relations:

$$p_{jk} = p_{kj}, \quad q_{jk} = q_{kj}. \tag{3.21}$$

Definition 3.7 A three-web W is called an *isoclinic three-web* if a one-parameter family of isoclinic surfaces passes through any point of this web.

Theorem 3.8 *A three-web $W = (X, \lambda_\alpha)$, $\dim X = 2r$, $r > 2$, is isoclinic if and only if its torsion tensor has the special structure (3.14).*

Proof. In fact, if there exists at least one isoclinic surface passing through a point p, then, as was shown above, the torsion tensor of the web has the structure (3.14) at this point.

Conversely, let the torsion tensor of the web W have the form (3.14). Then, if $r > 2$, in general, relations (3.21) do not hold on the web. Consider the system of equations (3.12) and (3.13) on X. By equations (3.14) and (3.21), this system is completely integrable. Therefore, it defines an $(r+1)$-parameter family of isoclinic surfaces. Moreover, a one-parameter family of isoclinic surfaces, defined by different values of the parameter λ, passes through any point of the manifold X. Thus, the web $W = (X, \lambda_\alpha)$ is isoclinic. ■

Consider now the case $r = 2$. As was said above, in this case the torsion tensor of an arbitrary web W can always be written in the form (3.14). However, in general, relations (3.21) do not hold. Thus, for $r = 2$ the corresponding theorem can be stated as follows.

Theorem 3.9 *Let a three-web W be given in a four-dimensional manifold X and let the torsion tensor of this web be written in the form (3.14). The web W is isoclinic if and only if the covariant derivatives of the co-vector a_i are symmetric.*
■

3. Let us find the most general form of the curvature tensor of an isoclinic three-web. To establish this form, we will use the following identity:

$$b^i_{jkl} = b^i_{(jkl)} + \frac{1}{3} b^i_{[jk]l} + \frac{1}{3} b^i_{[jl]k} + b^i_{[lk]j} + \frac{4}{3} b^i_{[j|k|l]} + \frac{2}{3} b^i_{[k|l|j]}. \tag{3.22}$$

which can be checked by inspection. Substituting the expressions for the alternated parts of the tensor b^i_{jkl} from (3.20) into (3.22), we obtain

$$b^i_{jkl} = b^i_{(jkl)} + \frac{2}{3} p_{jk}\delta^i_l - \frac{1}{3} p_{kl}\delta^i_j - \frac{1}{3} p_{lj}\delta^i_k - \frac{1}{3} q_{jk}\delta^i_l - \frac{1}{3} q_{kl}\delta^i_j + \frac{2}{3} q_{lj}\delta^i_k.$$

Further, we set:

$$p_{jk} = b^1_{jk} - b^3_{jk}, \quad q_{jk} = b^2_{jk} - b^3_{jk}, \tag{3.23}$$

where b^3_{jk}, at the moment, is an arbitrary symmetric tensor. As a result, the expression for b^i_{jkl} has the form:

$$b^i_{jkl} = b^i_{(jkl)} - (b^1_{(jk} + b^2_{(jk} + b^3_{(jk})\delta^i_{l)} + b^1_{jk}\delta^i_l + b^2_{lj}\delta^i_k + b^3_{kl}\delta^i_j.$$

Next, introduce a symmetric tensor:

$$a^i_{jkl} = b^i_{(jkl)} - (b^1_{(jk} + b^2_{(jk} + b^3_{(jk})\delta^i_{l)}. \tag{3.24}$$

Then, finally we obtain

$$b^i_{jkl} = a^i_{jkl} + b^1_{jk}\delta^i_l + b^2_{lj}\delta^i_k + b^3_{kl}\delta^i_j. \tag{3.25}$$

The tensor b^3_{kl} can be found by setting the additional condition for the tensor a^i_{jkl}:

$$a^i_{ikl} = 0. \tag{3.26}$$

In fact, contracting relation (3.24) in the indices i and j and using (2.26), we obtain

$$b^i_{ikl} = \frac{1}{3}(r+2)(b^1_{kl} + b^2_{kl} + b^3_{kl}).$$

From this, by means of (3.23), we obtain the tensor b^3_{kl}:

$$b^3_{kl} = \frac{1}{r+2} b^i_{(ikl)} - \frac{1}{3}(p_{kl} + q_{kl}).$$

Conversely, one can easily check that the curvature tensor of an isoclinic three-web can be represented in the form (3.25) where a^i_{jkl} is a symmetric tensor satisfying condition (3.26).

Relations (3.14) and (3.25) can be also interpreted in terms of the W-algebras. The first of these relations is equivalent to the equation

$$[\xi, \eta] = a(\eta)\xi - a(\xi)\eta \tag{3.27}$$

where $a(\xi) = a_i \xi^i$ is a scalar linear form. Relation (3.27) means that the operation of commutation in the W-algebras satisfies the so-called axiom of planes: any two vectors of a W-algebra define a two-dimensional subalgebra with respect to the operation of commutation.

By equation (3.25), the operation of association in the W-algebras of an isoclinic three-web can be written in the form:

$$(\xi, \eta, \zeta) = -a(\xi, \eta, \zeta) - b^1(\xi, \eta)\zeta - b^2(\eta, \zeta)\xi - b^3(\zeta, \xi)\eta, \tag{3.28}$$

where

$$a^i(\xi, \eta, \zeta) = a^i_{jkl}\xi^j \eta^k \zeta^l$$

is a symmetric trace-free trilinear form with its values in a W-algebra, and

$$b^\alpha(\xi, \eta) = b^\alpha_{ij}\xi^i \eta^j, \quad \alpha = 1, 2, 3,$$

are symmetric scalar bilinear forms.

It follows from the theorems proved in the last two subsections that the specific form (3.27) and (3.28) of the operations of commutation and association is the necessary and sufficient condition of isoclinity of a multidimensional three-web. This proves the following theorem.

Theorem 3.10 *A three-web $W = (X, \lambda_\alpha)$ is isoclinic if and only if the the operations of commutation and association in all its W-algebras have the form (3.27) and (3.28).*
■

4. Consider now a three-web which is simultaneously isoclinic and transversally geodesic.

Theorem 3.11 *For an isoclinic three-web to be transversally geodesic it is necessary and sufficient that its curvature tensor can be written in the form:*

$$b^i_{(jkl)} = b^1_{jk}\delta^i_l + b^2_{lj}\delta^i_k + b^3_{kl}\delta^i_j. \tag{3.29}$$

Proof. In fact, according to formula (3.9), a transversally geodesic three-web is characterized by the following structure of its curvature tensor:

$$b^i_{(jkl)} = b_{(jk}\delta^i_{l)}.$$

On the other hand, formula (3.25), defining the form of the curvature tensor of an isoclinic three-web, implies that

$$b^i_{(jkl)} = a^i_{jkl} + (b^1_{(jk} + b^2_{(jk} + b^3_{(jk})\delta^i_{l)}.$$

It follows from these two conditions that the tensor a^i_{jkl} has the following special form:

$$a^i_{jkl} = a_{(jk}\delta^i_{l)}.$$

Using condition (3.26), we find that

$$a^i_{ikl} = (r+2)a_{kl} = 0,$$

i.e. $a_{kl} = 0$, and consequently $a^i_{jkl} = 0$. Thus, the torsion tensor of the three-web under consideration has the form (3.29). ■

It follows from formulas (3.27). (3.28) and (3.29) that the operations of commutation and association in the W-algebras of a three-web W, which is simultaneously isoclinic and transversally geodesic, can be written as follows:

$$\begin{aligned} &[\xi,\eta] = a(\eta)\xi - a(\xi)\eta, \\ &(\xi,\eta,\zeta) = -b^1(\xi,\eta)\zeta - b^2(\eta,\zeta)\xi - b^3(\zeta,\xi)\eta, \end{aligned} \tag{3.30}$$

where $a(\xi)$ is a linear form and $b^\alpha(\xi,\eta)$ are symmetric bilinear scalar forms. Thus, the operation of commutation in the W-algebras of such webs satisfies the axiom of 2-planes and the operation of association satisfies the axiom of 3-planes.

In what follows, we will need the differential prolongations of equations (3.18) and (3.19) under the assumption that a three-web is simultaneously isoclinic and transversally geodesic. Exterior differentiation of these equations and application of conditions (3.21) and (3.29) gives the following exterior quadratic equations:

$$\begin{aligned} &(\nabla p_{jk} + p_{jk}a_l\underset{1}{\omega}^l) \wedge \underset{1}{\omega}^k + (\nabla q_{jk} - q_{jk}a_l\underset{2}{\omega}^l) \wedge \underset{2}{\omega}^k, \\ &+(b^1_{jk}a_l + b^2_{lj}a_k + b^3_{kl}a_j)\underset{1}{\omega}^k \wedge \underset{2}{\omega}^l = 0, \\ &[\delta^i_l \nabla b^1_{jk} + \delta^i_k \nabla b^2_{lj} + \delta^i_j \nabla b^3_{kl} - (b^1_{jk}\delta^i_l + b^2_{lj}\delta^i_k + b^3_{kl}\delta^i_j)a_m(\underset{1}{\omega}^m - \underset{2}{\omega}^m)] \wedge \underset{1}{\omega}^k \wedge \underset{2}{\omega}^l = 0, \end{aligned}$$

where ∇ is the operator of covariant differentiation in the connection Γ_{12}.

Let us exclude the quantities p_{ij} and q_{ij} from these equations by means of relations (3.23) and set

$$\overset{\circ}{\nabla} b^\alpha_{ij} = \nabla b^\alpha_{ij} + b^\alpha_{ij} a_l (\underset{1}{\omega}^l - \underset{2}{\omega}^l).$$

Then we obtain the equations

$$\overset{\circ}{\nabla} b^1_{jk} \wedge \underset{1}{\omega}^k + \overset{\circ}{\nabla} b^2_{jk} \wedge \underset{2}{\omega}^k - \overset{\circ}{\nabla} b^3_{jk} \wedge (\underset{1}{\omega}^k + \underset{2}{\omega}^k) + 3a_{(j} b^3_{kl)} \underset{1}{\omega}^k \wedge \underset{2}{\omega}^l = 0 \tag{3.31}$$

and

$$(\delta^i_l \overset{\circ}{\nabla} b^1_{jk} + \delta^i_k \overset{\circ}{\nabla} b^2_{lj} + \delta^i_j \overset{\circ}{\nabla} b^3_{kl}) \underset{1}{\omega}^k \wedge \underset{2}{\omega}^l = 0. \tag{3.32}$$

Contracting equation (3.32) in the indices i and j, we find that

$$(\overset{\circ}{\nabla} b^1_{kl} + \overset{\circ}{\nabla} b^2_{kl} + \overset{\circ}{\nabla} b^3_{kl}) \underset{1}{\omega}^k \wedge \underset{2}{\omega}^l = 0. \tag{3.33}$$

It follows from equations (3.32) and (3.33) that the forms $\overset{\circ}{\nabla} b^\alpha_{kl}$ are prinicipal forms on the manifold X. Thus, we set

$$\overset{\circ}{\nabla} b^\alpha_{ij} = \underset{2}{b}^\alpha_{ijk} \underset{1}{\omega}^k - \underset{1}{b}^\alpha_{ijk} \underset{2}{\omega}^k, \quad \alpha = 1, 2, 3. \tag{3.34}$$

Next, using relations (3.31) and (3.33), one can show that the tensors on the right-hand sides of expansions (3.34) are symmetric in all lower indices. Next, substituting expansions (3.34) into equations (3.32), we find that $\underset{2}{b}^2_{ijk} = 0$ and $\underset{1}{b}^1_{ijk} = 0$. Therefore, for $\alpha = 1, 2$, the expansions (3.34) are equivalent to the following quadratic equations:

$$\overset{\circ}{\nabla} b^1_{ij} \wedge \underset{1}{\omega}^j = 0, \quad \overset{\circ}{\nabla} b^2_{ij} \wedge \underset{2}{\omega}^j = 0, \tag{3.35}$$

and for $\alpha = 3$ equations (3.34) imply the equation

$$\overset{\circ}{\nabla} b^3_{ij} \underset{1}{\omega}^i \wedge \underset{2}{\omega}^j = 0. \tag{3.36}$$

3.3 Grassmann Three-Webs

1. In this section we will study in detail the Grassmann three-webs defined in Section 1.2. They are interesting first due to the fact that they can be constructed geometrically very simply. At the same time, this class of webs is sufficiently large since, in general, smooth hypersurfaces X_α defining a Grassmann three-web W are arbitrary hypersurfaces. As we will see in what follows, the latter reason allows us to find diverse examples in the class of Grassmann webs. These examples will illustrate practically all notions and results of web theory.

In Chapter 1 we denoted by $G(1, r+1)$ the Grassmannian of straight lines of a projective space P^{r+1}. It admits a bijection into a compact algebraic manifold Ω of dimension $2r$ of a projective space P^N , where $N = \binom{r+2}{2} - 1$. This bijection can be defined in the following way.

Let $x(x^u)$ and $y(y^u)$ be two points of the space P^{r+1}, $u, v, w, \ldots = 0, 1, \ldots, r+1$. The minors of second order of the matrix

$$\begin{pmatrix} x^0 & x^1 & \ldots & x^{r+1} \\ y^0 & y^1 & \ldots & y^{r+1} \end{pmatrix}$$

are called the *Plucker coordinates* of the straight line $p = [x, y]$. There are $\binom{r+2}{2}$ such minors. Thus, they define a point in a projective space P^N of dimension $N = \binom{r+2}{2} - 1$. This gives a mapping of the Grassmannian $G(1, r+1)$ into P^N. This mapping is called the *Grassmann imbedding*.

The Plucker coordinates of a straight line are not independent. They are connected by a series of quadratic relations. The latter can be obtained from equations of the form

$$\begin{vmatrix} x^u & x^v & x^w & x^z \\ y^u & y^v & y^w & y^z \\ x^u & x^v & x^w & x^z \\ y^u & y^v & y^w & y^z \end{vmatrix} = 0,$$

(all indices u, v, w and z are distinct) if one expands the determinant using the Laplace expansion formula. The equations obtained as a result of such expansions define the manifold Ω in P^N. The details can be found in [HP 52].

Let p and q be two straight lines in the space P^{r+1} intersecting at the point x. They define a plane pencil. A rectilinear generator of the manifold Ω corresponds to this pencil. Similarly, an r-dimensional flat generator of the manifold Ω corresponds to the bundle of straight lines with the vertex at the point x in P^{r+1}, and a 2-dimensional flat generator of the manifold Ω corresponds to a plane field of straight lines in P^{r+1}.

Consider in P^{r+1} a set of straight lines intersecting a fixed straight line p. This set (we will denote it by $\mathcal{K}$) is a manifold of dimension $r+1$. It contains a one-parameter family of bundles of straight lines with vertices on the line p and carries an $(r-1)$-parameter family of plane fields whose carriers are 2-planes passing through the point p. Because of this, to the set $\mathcal{K}$ there corresponds a cone with the vertex p on the manifold Ω, and this cone carries a one-parameter family of r-dimensional flat generators and an $(r-1)$-parameter family of 2-dimensional flat generators. This cone is called the *Segre cone* and is denoted by $C_p(2, r)$ (see Section 1.2). It is the intersection of the manifold Ω and its tangent space $T_p(\Omega)$.

Next, consider in P^{r+1} a smooth hypersurface X. In a domain D it defines a fibration of the Grassmannian $G(1, r+1)$ into pencils of straight lines whose vertices belong to X. To each fiber there corresponds a flat r-dimensional generator of the manifold Ω. Three hypersurfaces X_α, $\alpha = 1, 2, 3$ of the space P^{r+1} define in $G(1, r+1)$ a Grassmann three-web which is denoted by GW. The corresponding point image on Ω is formed by three r-parameter families of flat r-dimensional generators.

A coordinate quasigroup of a three-web GW can be obtained in the following way. Let p_0 be a straight line intersecting the hypersurfaces X_α at the points a_α. Then each straight line p intersecting X_1 and X_2 at the points x_1 and x_2 sufficiently close to a_1 and a_2, also intersects the hypersurface X_3 at a point x_3. This defines the local mapping $q : X_1 \times X_2 \to X_3$. The mapping q is differentiable and locally invertible. Thus, this mapping is a local differentiable quasigroup, a coordinate quasigroup of the three-web GW.

2. Let us find the differential equations of a three-web GW. We associate the space P^{r+1} with a moving frame $\{A_u\}$, $u, v, w, z, \ldots = 0, 1, \ldots, r+1$. The equations of infinitesimal displacement of this frame have the form:

$$dA_u = \theta_u^v A_v, \tag{3.37}$$

where the Pfaffian forms θ_u^v satisfy the structure equations of the projective space P^{r+1} (see, for example, [A 77b]):

$$d\theta_u^v = \theta_u^w \wedge \theta_w^v. \tag{3.38}$$

We consider now such moving frames connected with points $x_1 \in X_1$ and $x_2 \in X_2$, for which $x_1 \equiv A_0$, $x_2 \equiv A_{r+1}$ and the points A_i, $i = 1, \ldots, r$, are in the $(r-1)$-dimensional intersection of the planes $T_{A_0}(X_1)$ and $T_{A_{r+1}}(X_2)$ tangent to X_1 and X_2 at A_0 and A_{r+1} respectively.

In this frame, the equations of the hypersurfaces X_1 and X_2 have the form:

$$\theta_0^{r+1} = 0, \quad \theta_{r+1}^0 = 0, \tag{3.39}$$

and the forms θ_0^i and θ_{r+1}^i are the basis forms of the Grassmannian $G(1, r+1)$.

Let us find the equations of the hypersurface X_3 generated by the point x_3. Since this point lies on the straight line A_0A_{r+1} and does not coincide with either A_0 or A_{r+1}, we can specialize the frames in such a way that $x_3 = A_0 + A_{r+1}$. Differentiating this relation and using (3.37), we obtain

$$dx_3 = \theta_0^0 A_0 + \theta_{r+1}^{r+1} A_{r+1} + (\theta_0^i + \theta_{r+1}^i) A_i.$$

Eliminating the point A_{r+1} from the last equation, we get the equation

$$dx_3 = \theta_{r+1}^{r+1} x_3 + (\theta_0^0 - \theta_{r+1}^{r+1}) A_0 + (\theta_0^i + \theta_{r+1}^i) A_i.$$

Since the point x_3 generates an r-dimensional surface and the forms $\theta_0^i + \theta_{r+1}^i$ are linearly independent, the form $\theta_0^0 - \theta_{r+1}^{r+1}$ is their linear combination:

$$\theta_0^0 - \theta_{r+1}^{r+1} = a_i(\theta_0^i + \theta_{r+1}^i). \tag{3.40}$$

Equation (3.40) is the differential equation of the hypersurface X_3 in the adapted moving frame.

In this adapted frame, the bundles of straight lines, which form the web on the Grassmannian $G(1, r+1)$, are defined by the following systems of equations:

$$\theta_0^i = 0, \quad \theta_{r+1}^i = 0, \quad \theta_0^i + \theta_{r+1}^i = 0.$$

Thus, the forms $\theta_0^i = \underset{1}{\omega}^i$ and $\theta_{r+1}^i = \underset{2}{\omega}^i$ are the basis forms of the Grassmann web GW.

3. We will now find the structure equations of a web GW. From equations (3.38) we find that

$$d\theta_0^i = \theta_0^0 \wedge \theta_0^i + \theta_0^j \wedge \theta_j^i, \quad d\theta_{r+1}^i = \theta_{r+1}^{r+1} \wedge \theta_{r+1}^i + \theta_{r+1}^j \wedge \theta_j^i. \tag{3.41}$$

Equations (3.40) show that the forms θ_0^0 and θ_{r+1}^{r+1} can be represented in the form:

$$\theta_0^0 = \theta + a_i\theta_0^i, \quad \theta_{r+1}^{r+1} = \theta - a_i\theta_{r+1}^i, \tag{3.42}$$

where θ is an 1-form. Substituting the expressions for the forms θ_0^0 and θ_{r+1}^{r+1} from (3.42) into (3.41), we obtain the equations:

$$\begin{aligned} d\theta_0^i &= \theta_0^j \wedge \omega_j^i + a_j\delta_k^i\theta_0^j \wedge \theta_0^k, \\ d\theta_{r+1}^i &= \theta_{r+1}^j \wedge \omega_j^i - a_j\delta_k^i\theta_{r+1}^j \wedge \theta_{r+1}^k, \end{aligned} \tag{3.43}$$

where we use the notation

$$\omega_j^i = \theta_j^i - \delta_j^i\theta. \tag{3.44}$$

Comparing equations (3.43) with the structure equations (1.26) of a general three-web, we see that the forms ω_j^i are the forms of the canonical connection of the web GW, and its torsion tensor has the form

$$a_{jk}^i = a_{[j}\delta_{k]}^i. \tag{3.45}$$

4. In order to find the curvature tensor of the three-web GW, we calculate the second fundamental forms of the hypersurfaces X_α. Exterior differentiation of equations (3.39) and application of Cartan's lemma give the equations

$$\theta_i^{r+1} - b_{ij}^1\theta_0^j, \quad \theta_i^0 = -b_{ij}^2\theta_{r+1}^j, \tag{3.46}$$

where the quantities b_{ij}^1 and b_{ij}^2 satisfy the conditions: $b_{ij}^1 = b_{ji}^1$ and $b_{ij}^2 = b_{ji}^2$. Since

$$\begin{aligned} d^2A_0 &= (\ldots)A_0 + (\ldots)^iA_i + \theta^i\theta_i^{r+1}A_{r+1}, \\ d^2A_{r+1} &= (\ldots)A_{r+1} + (\ldots)^iA_i + \theta_{r+1}^i\theta_i^0A_0, \end{aligned}$$

the forms

$$b^1 = b^1_{ij}\theta^i_0\theta^j_0, \quad b^2 = b^2_{ij}\theta^i_{r+1}\theta^j_{r+1}$$

are the second fundamental forms of the hypersurfaces X_1 and X_2.

Applying exterior differentiation to equations (3.40), we find

$$(\nabla a_i + \theta^0_i - \theta^{r+1}_i) \wedge (\theta^i_0 + \theta^i_{r+1}) = 0,$$

where $\nabla a_i = da_i - a_j\omega^j_i$. It follows from this that

$$\nabla a_i + \theta^0_i - \theta^{r+1}_i = -b^3_{ij}(\theta^j_0 + \theta^j_{r+1}), \tag{3.47}$$

where the quantities b^3_{ij} are symmetric, i.e. $b^3_{ij} = b^3_{ji}$. Taking into account equations (3.46), we can write equations (3.47) in the form:

$$\nabla a_i = (b^1_{ij} - b^3_{ij})\theta^j_0 + (b^2_{ij} - b^3_{ij})\theta^j_{r+1}, \tag{3.48}$$

The quantities b^3_{ij} are the coefficients of the second fundamental form of the hypersurface X_3 (see Problem **1**).

If we apply exterior differentiation to the form ω^i_j, defined by equation (3.44), and use equations (3.38), (3.42), (3.44) and (3.47), we arrive at the equations:

$$d\omega^i_j = \theta^0_j \wedge \theta^i_0 + \theta^k_j \wedge \theta^i_k + \theta^{r+1}_j \wedge \theta^i_{r+1} - \delta^i_j d\theta,$$
$$d\theta = -b^3_{ij}\theta^i_0 \wedge \theta^j_{r+1}.$$

By (3.44) and (3.46), from these equations we find the second group of the structure equations of the web GW:

$$d\omega^i_j - \omega^k_j \wedge \omega^i_k = (b^1_{jk}\delta^i_l + b^2_{lj}\delta^i_k + b^3_{kl}\delta^i_j)\theta^k_0 \wedge \theta^l_{r+1}.$$

Comparing these equations with the similar equations (1.30) for an arbitrary three-web, we find the following expression for the curvature tensor of the web GW:

$$b^i_{jkl} = b^1_{jk}\delta^i_l + b^2_{lj}\delta^i_k + b^3_{kl}\delta^i_j. \tag{3.49}$$

As shown in Section 3.2, the special form (3.45) of the torsion tensor in the case $r > 2$ characterizes isoclinic three-webs. Thus, a web GW is isoclinic if $r > 2$. Moreover, by Theorem 3.9 it is also isoclinic if $r = 2$ since the covariant derivatives of the tensor a_i defined by formula (3.48) are symmetric in the lower indices.

Further, expressions (3.49) and (3.29) for the curvature tensor coincide. According to Theorem 3.11, it follows from this that the web GW under consideration is not only isoclinic but also transversally geodesic.

This proves the following theorem.

Theorem 3.12 *A Grassmann three-web is both isoclinic and transversally geodesic.* ■

Since equivalent webs have the same torsion and curvature tensors, Theorem 3.12 is also valid for Grassmannizable three-webs.

5. Let us now find transversal and isoclinic surfaces of a Grassmann three-web GW. For a Grassmann web GW, equations (3.1) defining two-dimensional transversal surfaces V^2, become

$$\theta_0^i = \xi^i \theta_1, \quad \theta_{r+1}^i = \xi^i \theta_2.$$

The vector ξ in these equations satisfies equation (3.3):

$$d\xi^i + \xi^j \omega_j^i = \xi^i \theta.$$

By means of these equations, we have in P^{r+1}:

$$\begin{aligned} dA_0 &= \theta_0^0 A_0 + \theta_1(\xi^i A_i), \\ dA_{r+1} &= \theta_{r+1}^{r+1} A_{r+1} + \theta_2(\xi^i A_i), \\ d(\xi^i A_i) &= \theta(\xi^i A_i) + \xi^i(\theta_i^0 A_0 + \theta_i^{r+1} A_{r+1}). \end{aligned}$$

It follows from this, that the two-dimensional plane π defined by the points A_0, A_{r+1} and $\xi^i A_i$ is fixed under transformations of frames. Thus, a transversal surface of a Grassmann three-web GW is a plane field of straight lines in the plane π passing through the straight line $A_0 A_{r+1}$.

The plane π intersects the surfaces X_α defining a Grassmann web GW along lines l_α. The pencils of straight lines whose vertices lie on the lines l_α form a two-dimensional Grassmann subweb W^2 in the plane π. This subweb represents an alignment chart, one of the simplest nomograms studied in nomography (see Preface).

An r-parameter family of planes π (transversal surfaces of a Grassmann web GW) passes through the line $A_0 A_{r+1}$.

Isoclinic surfaces V^r of a three-web are defined by equations (3.12) and (3.13). For a Grassmann web GW, we have: $\underset{1}{\omega}^i = \theta_0^i$ and $\underset{2}{\omega}^i = \theta_{r+1}^i$. Therefore, by means of (3.40), we can write equations (3.12) and (3.13) in the form:

$$\theta_{r+1}^i + \lambda \theta_0^i = 0, \tag{3.50}$$

and

$$d\lambda + \lambda(\theta_0^0 - \theta_{r+1}^{r+1}) = 0. \tag{3.51}$$

Consider the point $A_{r+1} + \lambda A_0$ on the line $A_0 A_{r+1}$ and calculate its differential using equations (3.50) and (3.51):

$$d(A_{r+1} + \lambda A_0) = \theta_{r+1}^{r+1}(A_{r+1} + \lambda A_0).$$

This equation implies that the point $A_{r+1} + \lambda A_0$ is fixed. Thus, on the Grassmannian $G(1, r+1)$, equations (3.50) and (3.51) define a bundle of straight lines with vertex $A_{r+1} + \lambda A_0$. Therefore, the isoclinic surfaces of a Grassmann web GW passing through the line $A_0 A_{r+1}$ are the bundles of straight lines with vertices located on this line.

6. Suppose that a Grassmann three-web GW generated by the hypersurfaces X_α of a projective space P^{r+1} is hexagonal. Then its curvature tensor satisfies the condition $b^i_{(jkl)} = 0$ (Theorem 3.5). Using this condition and equations (3.42), we arrive at the relations

$$b^1_{kl} + b^2_{kl} + b^3_{kl} = 0. \tag{3.52}$$

This gives the following relation among the second fundamental forms of the hypersurfaces X_α:

$$b^1(\xi,\eta) + b^2(\xi,\eta) + b^3(\xi,\eta) = 0. \tag{3.53}$$

If a web GW is hexagonal, then all its subwebs W^2 cut by transversal two-dimensional planes π are also hexagonal. Then the Graf–Sauer theorem (see Preface) tells us that the lines l_α of intersections of the plane π and the hypersurfaces X_α belong to one cubic curve. Since the plane π was taken arbitrarily, we conclude that each two-dimensional plane π passing through the line A_0A_{r+1} or through any of straight lines sufficiently close to the line A_0A_{r+1} possesses the same property, i.e. it intersects the hypersurfaces X_α along the lines belonging to one cubic curve. This implies that the hypersurfaces X_α themselves belong to one cubic hypersurface (hypercubic).

The Graf–Sauer theorem also implies the converse: if hypersurfaces X_α in P^{r+1} belong to one cubic hypersurface, then the Grassmann three-web GW generated by these hypersurfaces is hexagonal. We arrive at the following theorem.

Theorem 3.13 *A Grassmann three-web GW is hexagonal if and only if the hypersurfaces X_α generating the web GW belong to one hypercubic.* ■

We recall that a three-web of the type described in Theorem 3.13 is called algebraic and a three-web which is equivalent to an algebraic three-web is called algebraizable and denoted by AW.

The algebraic three-webs, for which a generating hypercubic is decomposed, will be studied in Section 4.2.

3.4 An Almost Grassmann Structure Associated with a Three-Web. Problems of Grassmannization and Algebraization

1. As we indicated in the beginning of Section 3.4, the tangent space to the Grassmannian Ω intersects it along the Segre cone $C_p(2,r)$. This allows us to introduce the following generalization.

Consider a manifold X of dimension $2r, r \geq 2$ and assign at each point of the tangent space $T_p(X)$ of any point $p \in X$ the Segre cone $C_p(2,r)$ with the vertex at p. We will assume that the field of the Segre cones on X is differentiable. The differential geometric structure on X defined by the field of Segre cones is called an *almost Grassmann structure* and is denoted by $AG(1, r+1)$.

Since a Segre cone carries a one-parameter family of r-dimensional flat generators and an $(r-1)$-parameter family of two-dimensional flat generators, its parametric equations in the tangent space $T_p(X)$ can be written as follows:

$$x_i^\sigma = t_i s^\sigma, \quad i = 1, 2, \ldots, r; \ \sigma = 1, 2.$$

On the Segre cone, the r-dimensional generators are defined by the equations $s^\sigma = c^\sigma s$ and the two-dimensional generators are defined by the equations $t_i = c_i t$ (here c^σ and c_i are constants). Thus, the Segre cone remains invariant under transformations of the group $\mathbf{GL}(r) \times \mathbf{SL}(2)$ which is a subgroup of the general linear group $\mathbf{GL}(2r)$ of transformations of the $(2r)$-dimensional space $T_p(X)$. The group $\mathbf{GL}(r) \times \mathbf{SL}(2)$ is the structure group of the almost Grassmann structure $AG(1, r+1)$.

On X, the r-dimensional generators of the Segre cones form a $(2r+1)$-parameter family, and the two-dimensional generators form a $(3r-1)$-parameter family. An almost Grassmann structure $AG(1, r+1)$ is called *r-semi-integrable* if on X there is an $(r+1)$-parameter family of r-dimensional subvarieties V^r which are tangent to the r-dimensional generators mentioned above at each of their points and each such generator is tangent to one and only one subvariety V^r. A *2-semi-integrable* almost Grassmann structure and the corresponding subvarieties V^2 can be defined in a similar way.

An almost Grassmann structure which is both r- and 2-semi-integrable is called *integrable*. The following theorem can be proved (the proof of this theorem can be found for example in [M 78]).

Theorem 3.14 *An integrable almost Grassmann structure $AG(1, r+1)$ is locally Grassmann, i.e. a neighborhood of any point p of a manifold X carrying such a structure admits a differentiable mapping into the Grassmannian Ω which maps the subvarieties V^r into the r-dimensional flat generators of Ω and maps the subvarieties V^2 into the two-dimensional flat generators of Ω.* ■

2. Consider now a three-web W on a differentiable manifold X of dimension $2r$, $r \geq 2$. As shown in Section 1.3, the Segre cone $C_p(2, r)$, invariantly connected with the web W, is defined in every tangent space $T_p(X)$ of this manifold. The family of these cones defines an almost Grassmann structure in X. This proves the following theorem.

Theorem 3.15 *A three-web $W(X, \lambda_\alpha)$ given in a manifold X of dimension $2r$, $r \geq 2$ defines an almost Grassmann structure $AG(1, r+1)$ on it.* ■

On the other hand, with an arbitrary three-web W, there is associated the G_W-structure whose structure group $\mathbf{GL}(r)$ leaves invariant the subspaces $T_p(\mathcal{F}_\alpha)$ tangent to the leaves $\mathcal{F}_\alpha$ passing through the point p. This G_W-structure is a substructure of the almost Grassmann structure $AG(1, r+1)$ whose structure group is the group $\mathbf{GL}(r) \times \mathbf{SL}(2)$.

The isoclinity and the transversal geodesicity of a three-web W are connected with the semi-integrability of the almost Grassmann structure $AG(1, r+1)$ associated with the web. As follows from the definitions given above, *the isoclinity of a three-web W is equivalent to the r-semi-integrability of the corresponding almost Grassmann structure $AG(1, r+1)$, and the transversal geodesicity of a three-web W is equivalent to the 2-semi-integrability of this structure.*

The following theorem holds:

Theorem 3.16 *For a three-web W to be a Grassmann web, it is necessary and sufficient that this web be both isoclinic and transversally geodesic.*

Proof. In fact the necessity of this theorem has been proved in Section 3.3. Its sufficiency follows from Theorem 3.14.

However, the sufficiency can be also proved directly. To do this, we construct a mapping of an isoclinic transversally geodesic three-web $W = (X, \lambda_\alpha)$ into the Grassmannian $G(1, r+1)$ of a projective space P^{r+1}. As in Section 3.3, let the space P^{r+1} be associated with a projective frame A_u, $u, v = 0, 1, \ldots, r+1$. Then, equations (3.37) hold, and the forms from these equations satisfy the structure equations (3.38). Let a moving line in P^{r+1} be defined by the points A_0 and A_{r+1}. We define a mapping mentioned above by the equations:

$$\theta_0^i = \underset{1}{\omega}{}^i, \quad \theta_{r+1}^i = \underset{2}{\omega}{}^i, \tag{3.54}$$

where the forms $\underset{1}{\omega}{}^i$ and $\underset{1}{\omega}{}^i$ are basis forms of the manifold X carrying the web W. As shown in Section 3.2, these forms satisfy equations (3.15) and all their differential consequences which were given in Section 3.2.

Exterior differentiation of equations (3.54) and application of the structure equations (3.15) and (3.38) lead to the equations:

$$\begin{aligned} \theta_0^j \wedge (\theta_j^i - \delta_j^i \theta_0^0 - \omega_j^i + \delta_j^i a_k \theta_0^k) + \theta_0^{r+1} \wedge \theta_{r+1}^i = 0, \\ \theta_{r+1}^j \wedge (\theta_j^i - \delta_j^i \theta_0^0 - \omega_j^i + \delta_j^i a_k \theta_{r+1}^k) + \theta_0^{r+1} \wedge \theta_0^i = 0. \end{aligned} \tag{3.55}$$

Equations (3.55) imply that the forms θ_0^{r+1} and θ_{r+1}^0 can be expressed in terms of the basis forms θ_0^j and θ_{r+1}^j as follows:

$$\theta_0^{r+1} = \lambda_j \theta_0^j, \quad \theta_{r+1}^0 = \mu_j \theta_{r+1}^j. \tag{3.56}$$

Using equations (3.56) and (3.37), we find that

$$dA_0 = \theta_0^0 A_0 + \theta_0^j (A_j + \lambda_j A_{r+1}), \quad dA_{r+1} = \theta_{r+1}^{r+1} A_{r+1} + \theta_{r+1}^j (A_j + \mu_j A_0).$$

These equations show that the points A_0 and A_{r+j} generate hypersurfaces in P^{r+1}. The tangent hyperplanes to these hypersurfaces are defined by the points A_0, $A_j + \lambda_j A_{r+1}$ and $A_{r+1}, A_j + \mu_j A_{r+1}$, respectively.

Let us locate the points A_i of the frame in the intersection of these hyperplanes. Then we have $\lambda_i = \mu_i = 0$, and equations (3.56) become:

$$\theta_0^{r+1} = 0, \quad \theta_{r+1}^0 = 0. \tag{3.57}$$

Using equations (3.57) and Cartan's lemma, we get from equations (3.55):

$$\begin{aligned} \theta_j^i - \omega_j^i - \delta_j^i(\theta_0^0 - a_k\theta_0^k) &= \rho_{jk}^i \theta_0^k, \\ \theta_j^i - \omega_j^i - \delta_j^i(\theta_{r+1}^{r+1} + a_k\theta_{r+1}^k) &= \sigma_{jk}^i \theta_{r+1}^k, \end{aligned} \tag{3.58}$$

where ρ_{jk}^i and σ_{jk}^i are symmetric in the lower indices. Subtracting the second of equations (3.58) from the first of equations (3.58), we find that

$$\delta_j^i(\theta_{r+1}^{r+1} - \theta_0^0 + a_k(\theta_0^k + \theta_{r+1}^k)) = \rho_{jk}^i\theta_0^k - \sigma_{jk}^i\theta_{r+1}^k.$$

It follows from this that the Pfaffian form on the left-hand side is principal, i.e. it can be expressed in terms of the basis forms θ_0^k and θ_{r+1}^k. We write this expression in the form:

$$\theta_{r+1}^{r+1} - \theta_0^0 + a_k(\theta_0^k + \theta_{r+1}^k) = \rho_k\theta_0^k - \sigma_k\theta_{r+1}^k.$$

Substituting this expression into the previous equation and equating to zero the coefficients in independent forms, we obtain

$$\rho_{jk}^i = \delta_j^i \rho_k, \quad \sigma_{jk}^i = \delta_j^i \sigma_k.$$

Alternating these equations in the indices j and k and using the symmetry of the quantities ρ_{jk}^i and σ_{jk}^i, we deduce that if $r \geq 2$, the equations $\rho_k = \sigma_k = 0$, $\rho_{jk}^i = \sigma_{jk}^i = 0$ hold. Thus, we get the equations:

$$\theta_0^0 - \theta_{r+1}^{r+1} = a_k(\theta_0^k + \theta_{r+1}^k). \tag{3.59}$$

Using equations (3.59) and (3.30), we calculate the differential of the point $A_0 + A_{r+1}$:

$$d(A_0 + A_{r+1}) = \theta_{r+1}^{r+1}(A_0 + A_{r+1}) + (\theta_0^k + \theta_{r+1}^k)(A_k + a_k A_0).$$

Thus, the point $A_0 + A_{r+1}$ describes a hypersurface whose tangent hyperplane in the variable point $A_0 + A_{r+1}$ is defined by the points $A_0 + A_{r+1}$ and $A_k + a_k A_0$.

It follows from equations (3.59) and (3.54) that we can set

$$\theta_0^0 = \theta + a_k \underset{1}{\omega}^k, \quad \theta_{r+1}^{r+1} = \theta - a_k \underset{2}{\omega}^k, \tag{3.60}$$

where θ is a Pfaffian form. As a result, equations (3.58) are reduced to the equations:

$$\theta_j^i = \omega_j^i + \delta_j^i \theta. \tag{3.61}$$

Along with equations (3.57), (3.60) and (3.61), all their differential consequences also hold. Differentiating equations (3.57) and using Cartan's lemma, we find that the following equations are valid:

$$\theta_i^{r+1} = \widetilde{b}_{ij}^1 \underset{1}{\omega^j}, \quad \theta_i^0 = -\widetilde{b}_{ij}^2 \underset{2}{\omega^j}, \tag{3.62}$$

where $\widetilde{b}_{ij}^1$ and $\widetilde{b}_{ij}^2$ are symmetric tensors.

Further, differentiating equations (3.60) and using equations (3.38) and (3.18), we arrive at the equations:

$$d\theta = (q_{ij} - \widetilde{b}_{ij}^2) \underset{1}{\omega^i} \wedge \underset{2}{\omega^j}, \quad d\theta = (p_{ij} - \widetilde{b}_{ij}^1) \underset{1}{\omega^i} \wedge \underset{2}{\omega^j},$$

which lead to the equations

$$p_{ij} - \widetilde{b}_{ij}^1 = q_{ij} - \widetilde{b}_{ij}^2 \stackrel{\text{def}}{=} -\widetilde{b}_{ij}^3,$$

where $\widetilde{b}_{ij}^3$ is a new symmetric tensor. Thus, the form $d\theta$ can be written as

$$d\theta = -\widetilde{b}_{ij}^3 \underset{1}{\omega^i} \wedge \underset{2}{\omega^j}. \tag{3.63}$$

Differentiate now equations (3.61). Using relations (3.38), (3.19), (3.62) and (3.63) and the fact that the curvature tensor of an isoclinic transversally geodesic three-web has the form (3.29), we arrive at the relations:

$$(b_{jk}^1 - \widetilde{b}_{jk}^1)\delta_l^i + (b_{lj}^2 - \widetilde{b}_{lj}^2)\delta_k^i + (b_{kl}^3 - \widetilde{b}_{kl}^3)\delta_j^i = 0.$$

If $r \geq 2$, we derive from this that $b_{jk}^\alpha = \widetilde{b}_{jk}^\alpha$. As a result, equations (3.62) and (3.63) become

$$\theta_i^{r+1} = b_{ij}^1 \underset{1}{\omega^j}, \quad \theta_i^0 = -b_{ij}^2 \underset{2}{\omega^j}, \quad d\theta = -b_{ij}^3 \underset{1}{\omega^i} \wedge \underset{2}{\omega^j}. \tag{3.64}$$

Consider the system of equations (3.54), (3.57), (3.60), (3.61) and (3.64). Using the structure equations (3.15) and (3.38) and relations (3.35), (3.36), one can establish that this system is closed with respect to the operation of exterior differentiation. Therefore, the system is completely integrable, and the forms θ_v^u defined by this system, satisfy the structure equations of a projective space P^{r+1}. Because of this, the equations listed above define a mapping of a manifold X carrying an isoclinic transversally geodesic three-web W into the Grassmannian $G(1, r+1)$ of straight lines of a projective space P^{r+1}. As was indicated above, the points A_0, A_{r+1} and $A_0 + A_{r+1}$ of a space P^{r+1} describe the hypersurfaces. Denote these hypersurfaces by X_1, X_2 and X_3, respectively. The systems of equations $\theta_0^i = 0$, $\theta_{r+1}^i = 0$ and $\theta_0^i + \theta_{r+1}^i = 0$ are completely integrable in P^{r+1} and define the bundles of straight lines with their vertices on the hypersurfaces X_1, X_2 and X_3, respectively. By equations (3.47), these bundles are the images of the leaves of an isoclinic transversally geodesic three-web. ■

Since a hexagonal three-web $W = (X, \lambda_\alpha)$ is transversally geodesic, Theorems 3.16 and 3.13 imply the following theorem.

Theorem 3.17 *For a three-web W to be algebraizable, it is necessary and sufficient that it be both isoclinic and hexagonal.* ■

Theorems 3.16 and 3.17 solve the problems of Grassmannization and algebraization for three-webs.

3.5 Isoclinicly Geodesic Three-Webs. Three-Webs over Algebras

1. Let us return to isoclinic three-webs and show that, in general, their r-dimensional isoclinic surfaces are not totally geodesic in the Chern connection Γ. In fact, the differential equations of geodesic lines in this connection, found in Section 1.6, have the form:

$$d\xi_1^i + \xi_1^j \omega_j^i = \theta \xi_1^i, \quad d\xi_2^i + \xi_2^j \omega_j^i = \theta \xi_2^i, \tag{3.65}$$

where (ξ_1^i, ξ_2^i) is the tangent vector to a geodesic line.

The isoclinic surfaces of a web are defined by equations (3.12) and (3.13):

$$\underset{2}{\omega}^i + \lambda \underset{1}{\omega}^i = 0, \tag{3.66}$$

$$\frac{d\lambda}{\lambda^2 - \lambda} = a_i \underset{1}{\omega}^i. \tag{3.67}$$

Consider a line l on an isoclinic surface V^r. By the first equation of (3.66), the tangent vector to l has the coordinates $(\xi^i, -\lambda\xi^i)$. Suppose that the line l is geodesic. Substituting the coordinates of its tangent vector into equations (3.65), we obtain the equations:

$$d\xi^i + \xi^j \omega_j^i = \theta \xi^i, \quad d(\lambda \xi^i) + \lambda \xi^j \omega_j^i = \theta \lambda \xi^i.$$

It follows from these equations that $d\lambda = 0$, i.e. the quantity λ must be constant on the isoclinic surface V^r. Thus, the isoclinic surfaces of a three-web are totally geodesic in the connection Γ if and only if they are defined by equations (3.66) where λ is constant. Three-webs with this property are called *isoclinicly geodesic*.

Note that the parameter λ has a simple geometric meaning: it is equal to the anharmonic ratio of four r-vectors three of which are tangent to the leaves of the web W passing through the point p, and the fourth one is tangent to the isoclinic surface V^r, defined by equations (3.66) and also passing through the point p. On an isoclinicly geodesic web, this anharmonic ratio is constant along an isoclinic surface V^r.

Relations (3.67) give the following test for isoclinic geodesicity of a three-web W.

Theorem 3.18 *A three-web W is isoclinicly geodesic if and only if its torsion tensor vanishes.*

Proof. In fact, if a three-web W is isoclinicly geodesic, then on all its isoclinic surfaces the equation $d\lambda = 0$ holds. By (3.67), this implies that $a_i = 0$, and then equation (3.14) gives $a^i_{jk} = 0$.

Conversely, if the torsion tensor of a web W vanishes, then, by (1.26), equations (3.66) are completely integrable for any constant value of λ. Therefore, the web W is isoclinicly geodesic. ■

Note also that if $a^i_{jk} = 0$, equations (1.25) imply the relations:

$$b^i_{[j|l|k]} = 0, \quad b^i_{[jk]l} = 0. \tag{3.68}$$

Equations (3.68) mean that the curvature tensor of an isoclinicly geodesic three-web is symmetric in all three lower indices.

2. We turn now to the existence problem for the isoclinicly geodesic three-webs. Their structure equations can be obtained from equations (1.26) and (1.32) by setting $a^i_{jk} = 0$ in them:

$$d\underset{1}{\omega}^i = \underset{1}{\omega}^j \wedge \omega^i_j, \quad d\underset{2}{\omega}^i = \underset{2}{\omega}^j \wedge \omega^i_j, \tag{3.69}$$

$$d\omega^i_j - \omega^k_j \wedge \omega^i_k = b^i_{jkl}\underset{1}{\omega}^k \wedge \underset{2}{\omega}^l, \tag{3.70}$$

where the tensor b^i_{jkl} is symmetric in all three lower indices.

Exterior differentiation of the last equations gives the exterior cubic equations:

$$\nabla b^i_{jkl} \wedge \underset{1}{\omega}^k \wedge \underset{2}{\omega}^l = 0. \tag{3.71}$$

The system of equations (3.69)–(3.71), defining the isoclinicly geodesic three-webs, is closed with respect to the operation of exterior differentiation. Let us apply the Cartan test [V 83] to this system.

Equations (3.69) and (3.70) enable us to find the exterior differential forms $d\underset{1}{\omega}^i$, $d\underset{2}{\omega}^i$ and $d\omega^i_j$ and have no effect on the compatibility of the system. Thus, we should consider only the subsystem (3.71). The latter contains r^2 equations for $q = \frac{1}{6}r^2(r+1)(r+2)$ characteristic forms ∇b^i_{jkl}. Since this subsystem does not give conditions for one- and two-dimensional integral elements, its characters s_0 and s_1 are equal to zero: $s_0 = s_1 = 0$. The rank of the polar system of equations which can be obtained from the system (3.71) and is used for finding a three-dimensional integral element, is equal to r^2: $s_2 = r^2$.

On the other hand, it follows from system (3.71) that the forms ∇b^i_{jkl} are linear combinations of the basis forms $\underset{1}{\omega}^i$ and $\underset{2}{\omega}^i$:

$$\nabla b^i_{jkl} = \underset{1}{c}^i_{jklm}\underset{1}{\omega}^m + \underset{2}{c}^i_{jklm}\underset{2}{\omega}^m,$$

and the coefficients of these expansions are symmetric in all four lower indices. Therefore, the number of these coefficients is $N = \frac{1}{12}r^2(r+1)(r+2)(r+3)$.

We will continue the proof of existence for the dimension $r = 2$. In this case, system (3.71) contains only four equations and $q = 8$ characteristic forms. Thus, its characters are: $s_0 = s_1 = 0, s_2 = s_3 = 4, s_4 = q - s_0 - s_1 - s_2 - s_3 = 0$. The Cartan number Q is: $Q = s_1 + 2s_2 + 3s_3 + 4s_4 = 20$. The number of parameters N, on which the most general integral element depends and which we found above, is also equal to 20 if $r = 2$. Since $N = Q$, then, according to the Cartan test, the system (3.71) is in involution, i.e. it has a solution. Moreover, the arbitrariness of an integral manifold is determined by the highest non-zero character s_3 and hence equal to four functions of three variables.

3. Many interesting examples of isoclinicly geodesic three-webs arise when one investigates three-webs over commutative and associative algebras. Let A be a commutative and associative algebra of dimension r over the field $\mathbf{R}$ of real numbers. Denote its basis vectors by e_i, $i = 1, 2, \ldots, r$. We define multiplication in the algebra A as usual by defining the products of its basis vectors as follows:

$$e_i e_j = \gamma^k_{ij} e_k. \tag{3.72}$$

The real numbers γ^i_{jk} in formulas (3.72) are called the *structure constants* of the algebra A. Under a change of basis, they are transformed according to the tensor law.

The conditions of commutativity and associativity in the algebra A are equivalent to the relations:

$$\gamma^k_{ij} = \gamma^k_{ji}, \tag{3.73}$$

and

$$\gamma^m_{ij}\gamma^l_{mk} = \gamma^l_{im}\gamma^m_{jk}. \tag{3.74}$$

In addition, we assume that the algebra A is unitary, i.e. it has an identity element σ. Using equations (3.72), from the condition $\sigma \cdot e_i = e_i$ we obtain the equations which are satisfied by the coordinates σ^i of the identity element σ:

$$\gamma^i_{jk}\sigma^k = \gamma^i_{kj}\sigma^k = \delta^i_j. \tag{3.75}$$

Non-zero elements a and b of the algebra A are called *zero divisors* if we have $a \cdot b = 0$. The last equation implies the equations:

$$\gamma^i_{jk}a^j b^k = 0,$$

relating the coordinates of the elements a and b. Since this system has a non-trivial solution, the rank of the matrix $(\gamma^i_{jk}a^j)$ must be less than r. Conversely, if the element $a = a^i e_i$ is not a zero divisor, the matrix indicated above is non-singular.

A partial mapping $f : A \to A$ for which an increment $\Delta f = f(x + \Delta x) - f(x)$ can be represented in the form:

$$\Delta f = g \cdot \Delta x + o(\Delta x),$$

is said to be a *differentiable function over the algebra A*. The principal part $g \cdot \Delta x$ of this increment is called the *differential* of the function f and is denoted in the usual manner as

$$df = g \cdot dx. \tag{3.76}$$

The function g is called the *derivative* of the function f with respect to the variable x from A. The usual notation is used: $g = f'(x) = \frac{df}{dx}$.

Let us show that *the condition for differentiability of a function f is equivalent to the following equations for its coordinates f^i*:

$$\frac{\partial f^i}{\partial x^j} = \gamma^i_{jk}\frac{\partial f^k}{\partial x^l}\sigma^l. \tag{3.77}$$

To prove this, we consider relation (3.76). In the coordinate form this equation can be written as:

$$\frac{\partial f^i}{\partial x^j}dx^j = \gamma^i_{kj}g^k dx^j,$$

or

$$\frac{\partial f^i}{\partial x^j}\gamma^i_{kj}g^k. \tag{3.78}$$

Contracting equations (3.78) with the coordinates σ^i of the identity element and using equations (3.75), we find the quantities g^i:

$$g^i = \frac{\partial f^i}{\partial x^j}\sigma^j.$$

Substituting these values into equations (3.78), we arrive at relations (3.77).

The sufficiency of conditions (3.77) for differentiability of a function f can be also easily proved.

The equations (3.77) generalize the Cauchy–Riemann conditions for functions of a complex variable.

The notion of a function of two or a few variables over an algebra A can be given in a similar way.

4. The presence of functions differentiable over an algebra A allows us to define a two-dimensional differentiable manifold AX^2 over this algebra. The local coordinates of a point of the manifold AX^2 are pairs (x, y) of elements from the algebra A.

A three-web can be also naturally defined in the manifold AX^2. We denote this three-web by AW^2. The foliations of this web can be defined by the equations:

$$x = a, \quad y = b, \quad f(x, y) = c, \quad a, b \in A,$$

where f is a differentiable function defined in a domain D of the manifold AX^2 with its values in the algebra A.

The differential of a function $f(x, y)$ can be written in the usual form:

$$dz = \frac{\partial f}{\partial x}dx + \frac{\partial f}{\partial y}dy.$$

Since the web equation $z = f(x, y)$ is uniquely solvable with respect to x and y, the derivatives $\frac{\partial f}{\partial x}$ and $\frac{\partial f}{\partial y}$ are not equal to zero and are not zero divisors in the algebra A. Thus, the matrices $(\frac{\partial f^i}{\partial x^j})$ and $(\frac{\partial f^i}{\partial y^j})$ are non-singular.

Consider the differential forms:

$$\underset{1}{\omega} = \rho\frac{\partial f}{\partial x}dx, \quad \underset{2}{\omega} = \rho\frac{\partial f}{\partial y}dy, \quad \underset{3}{\omega} = -\rho dz, \tag{3.79}$$

where ρ is a function in the manifold AX^2 with non-zero values in the algebra A and is not a zero divisor in the domain D. Values of the forms $\underset{\alpha}{\omega}$ also belong to the algebra A. These forms are connected by the equation:

$$\underset{1}{\omega} + \underset{2}{\omega} + \underset{3}{\omega} = 0,$$

and any two of them form a basis in the AX^2.

The operations of exterior multiplication and exterior differentiation for differential forms over commutative and associative algebras can be defined in the same way as these operations were defined for differential forms over the field of real numbers. All properties of the exterior operations and the corresponding theorems are preserved for the forms over algebras. Because of this, the structure equations of a three-web AW^2 have the same form as the structure equations of a two-dimensional three-web in a real plane (cf. p. 18):

$$\underset{\alpha}{\omega} = \underset{\alpha}{\omega} \wedge \omega, \quad d\omega = b\underset{1}{\omega} \wedge \underset{2}{\omega}. \tag{3.80}$$

Here ω is a differential form over the algebra A and b is a differentiable function in the manifold AX^2. This function is called the *curvature* of the web AW^2.

A two-dimensional manifold AX^2 over an algebra A admits a realization in the form of a real $(2r)$-dimensional manifold X whose admissible coordinate transformations are determined by functions satisfying generalized Cauchy–Riemann equations

(3.77). Moreover, a three-web AW^2 is realized as a three-web $W = (X, \lambda_\alpha)$ whose r-dimensional foliations are real realizations of a one-dimensional foliations of the web AW^2.

Theorem 3.19 *A three-web $W = (X, \lambda_\alpha)$ that is a real realization of a two-dimensional three-web AW^2 over commutative, associative and unitary algebra A, is an isoclinicly geodesic three-web.*

Proof. To prove this theorem, we find the structure equations of the web under consideration in a real manifold X. We decompose all the quantities from equations (3.80) with respect to the basis of the algebra A:

$$\underset{\alpha}{\omega} = \underset{\alpha}{\omega}^i e_i, \quad \omega = \omega^i e_i, \quad b = b^i e_i.$$

Substituting these values into equations (3.80), we obtain:

$$d\underset{\alpha}{\omega}^i e_i = \underset{\alpha}{\omega}^j \wedge \omega^k (e_j \cdot e_k) = \underset{\alpha}{\omega}^j \wedge \omega^k \gamma^i_{jk} e_i, \quad d\omega^i e_i = b^j e_j \underset{1}{\omega}^k \wedge \underset{2}{\omega}^l (e_k \cdot e_l) = b^j \gamma^i_{jm} \gamma^m_{kl} \underset{1}{\omega}^k \wedge \underset{2}{\omega}^l.$$

When we derived the last equations, we assumed that the basis vectors e_i of the algebra A are constant in the manifold AX^2, i.e. $de_i = 0$. These equations imply that

$$d\underset{\alpha}{\omega}^i = \underset{\alpha}{\omega}^j \wedge \gamma^i_{jk} \omega^k, \quad d\omega^i = b^j \gamma^i_{jm} \gamma^m_{kl} \underset{1}{\omega}^k \wedge \underset{2}{\omega}^l.$$

To give the latter equations the form of the structure equations of a web, we introduce the forms:

$$\omega^i_j = \gamma^i_{jk} \omega^k.$$

Then, we obtain the system:

$$d\underset{\alpha}{\omega}^i = \underset{\alpha}{\omega}^j \wedge \omega^i_j, \quad d\omega^i_j = \gamma^i_{jp} \gamma^p_{qs} \gamma^q_{kl} b^s \underset{1}{\omega}^k \wedge \underset{2}{\omega}^l. \tag{3.81}$$

By associativity of the algebra A, we have

$$\omega^k_j \wedge \omega^i_k = \gamma^k_{jl} \omega^l \wedge \gamma^i_{km} \omega^m = \frac{1}{2}(\gamma^k_{jl} \gamma^i_{km} - \gamma^k_{jm} \gamma^i_{kl}) \omega^l \wedge \omega^m = 0.$$

Thus, the form of equations (3.81) coincides with that of equations (1.26) and (1.32), i.e. these equations are the structure equations of the web $W = (X, \lambda_\alpha)$ in the manifold X. One can see from these structure equations that the torsion tensor of the web W vanishes: $a^i_{jk} = 0$. Therefore, according to Theorem 3.18, this web is isoclinicly geodesic. Its curvature tensor is expressed by the formula:

$$b^i_{jkl} = \gamma^i_{jp} \gamma^p_{qs} \gamma^q_{kl} b^s,$$

and by (3.73) and (3.74), this tensor is symmetric in the lower indices.

PROBLEMS

1. Prove that the quantities b^3_{ij} (see (3.47)) form the second fundamental tensor of the hypersurface X^3.

2. Describe the geodesic lines in the Chern connection for a Grassmann three-web.

3. Prove that the three-web

$$z^1 = e^{x^1 y^1} + x^2 y^2, \quad z^2 = x^2 + y^2$$

is neither isoclinic nor transversally geodesic but possesses two-parameter families of isoclinic and transversally geodesic surfaces. The equations of these surfaces have the form:

$$\begin{cases} x^1 = c^1 y^1, \\ x^2 = c^2 + y^2; \end{cases} \qquad \begin{cases} x^2 = p^2, \\ y^2 = q^2, \end{cases}$$

where c^1, c^2, p^2 and q^2 are constants.

4. Prove that isoclinic surfaces of a four-dimensional three-web

$$z^1 = x^1 + y^1, \quad z^2 = -x^1 y^1 + x^2 y^2, \quad x^2 \neq 0, \quad y^2 \neq 0$$

can be given by the equations:

$$x^1 = y^1, \quad x^2 = k y^2,$$

and its transversally geodesic surfaces can be given by the equations:

$$\begin{cases} x^1 = c x^2, \\ y^1 = \frac{1}{c} y^2; \end{cases} \qquad \begin{cases} x^2 = \frac{1}{c}(x^1 + c'), \\ y^2 = c(y^1 + c'), \end{cases}$$

where c, c' and k are constants. Is this web isoclinic or transversally geodesic?

5. Prove that the three-web

$$z^1 = x^2 e^{x^1 y^1}, \quad z^2 = x^2 + y^2$$

is transversally geodesic but not isoclinic. This web does not possess isoclinic surfaces that differ from the leaves of the web, and its transversal surfaces are defined by the equations:

$$x^1 = e^{cx^2 + c^1}, \quad y^1 = e^{cy^2 + c^2},$$

where c, c^1 and c^2 are constants.

6. Prove that the four-dimensional three-web

$$z^1 = (x^1 + y^1)(x^2 - y^2), \quad z^2 = x^2 + y^2$$

is isoclinic but not transversally geodesic, and it carries the following transversal surfaces:

$$\begin{cases} x^1 = c, \\ y^1 = -c; \end{cases} \qquad \begin{cases} x^2 = c', \\ y^2 = c'', \end{cases}$$

where c, c' and c'' are constants. Find all isoclinic surfaces of this web.

7. Prove that a Grassmann three-web is isoclinicly geodesic if and only if the hypersurfaces X_α defining this web are hyperplanes of a pencil. Clarify the geometric meaning of the parameter λ which is constant on isoclinic surfaces.

8. Find the structure equations and the curvature tensors of real realizations of three-webs over the following algebras:

i) The algebra of complex numbers with the basis $e_1 = 1$, $e_2 = i$, $i^2 = -1$;

ii) The algebra of split complex numbers with the basis $e_1 = 1$, $e_2 = e$, $e^2 = 1$;

iii) The algebra of plural numbers with the basis $e_1 = 1$, $e_2 = \varepsilon$, $e_2 = \varepsilon^2$, ..., $e_r = \varepsilon^{r-1}$, $\varepsilon^r = 0$.

In Problems **9 – 11** the algebra A is associative, commutative and unitary.

9. Prove that an ideal of dimension m of the algebra A generates the fibration of the $(2r)$-dimensional real realization of a three-web AW^2 over this algebra into an $(r-m)$-parametric family of $(2m)$-dimensional subwebs. Find conditions of parallelizability of these subwebs.

10. Prove that for a reducible algebra $A = J_1 \oplus J_2$, there is the double fibration of the real realization of a three-web AW^2 into $(2m)$-dimensional and $2(r-m)$-dimensional subwebs, and that in turn, these subwebs are real realizations of some three-webs over the algebras J_1 and J_2.

11. Applying the result of Problem **10**, find the structure of a real realization of a three-web AW^2 over a semi-simple algebra A.

NOTES

3.1. Transversal geodesic and isoclinic three-webs were introduced in [A 69b] and [A 74]. In the same papers the relations (3.8) and (3.14) were derived. The case $r = 2$ was omitted in [A 74]; This was noticed by Goldberg (see [AS 81] and [G 88]). The necessary and sufficient condition (3.10) of hexagonality was first obtained by Chern in [C 36a].

3.3. The Grassmann three-webs in a projective space P^{r+1} were introduced by Akivis in [A 73a]. Different necessary and sufficient conditions of Grassmannizability of a three-web were also found in [A 73a].

If $r = 1$, the Grassmann three-webs are rectilinear webs in a plane. The so-called alignment charts, one of the simplest nomograms, can be characterized in terms of these webs. Thus, for $r = 1$, the problem of Grassmannizability is connected with the problem of representation of a function of two variables by means of an alignment chart ("the problem of anamorphosis") [Gl 61]. It is surprising that in spite of a long history, until now this problem is not completely solved although in [G 89] an essential advance has been achieved.

For $r > 1$, the condition of algebraizability of a three-web was obtained by Akivis in [A 73a], and for $r = 1$ this condition is given in the Graf–Sauer theorem (see Preface).

3.4. An almost Grassmann structure on a three-web was defined in [A 80] (see also the survey [A 83a]). If $r = 2$, an almost Grassmann structure becomes a pseudo-conformal structure since the corresponding Segre cones are cones of the second order. G. Klekovkin considered four-dimensional three-webs from this point of view in [Kl 81a] and [Kl 81b]. The relationship of isoclinicity and transversal geodesicity of an $(n+1)$-webs, $n \geq 2$, and semi-integrability of almost Grassmann structures associated with them was established in [G 75d] (see also [G 88]).

3.5. Isoclinicly geodesic three-webs appeared first under the name of paratactical three-webs in [A 69b]. Three-webs over algebras, the most important subclass of isoclinicly geodesic three-webs, were studied by V. V. Timoshenko (see [Ti 75a], [Ti 75b], [Ti 76], [Ti 77a], [Ti 77b], [Ti 78] and [Ti 79]).

The four-dimensional web from Problem **6** was introduced by Bol [B 37], and the four-dimensional webs from Problems **3–5** were introduced by V. V. Goldberg.

Chapter 4

The Bol Three-Webs and the Moufang Three-Webs

4.1 The Bol Three-Webs

1. In Section 2.2 we defined three classes of the Bol three-webs on which the figures given in Figures 5–7 are closed, respectively (see p. 54–55). These classes can be transferred into one another if we renumber the foliations according to a parastrophy transformation of the coordinate quasigroup of a web. This allows us to study only one of these three classes, for example, the middle Bol webs B_m.

In Section 2.5 we proved that the curvature tensor of the webs of this class satisfies the condition:

$$b^i_{j(kl)} = 0 \tag{4.1}$$

(see Table 2.2 (p. 83)). It follows from this that *the Bol webs are transversally geodesic webs, and moreover, they are hexagonal webs.*

For studying the webs B_m, instead of using the Chern connection Γ_{12}, we will use another connection from the bundle of connections $\gamma(W)$ associated with a three-web W (see Section 1.8). Namely, we will use the connection $\widetilde{\Gamma}_{12}$ defined by the forms:

$$\widetilde{\omega}^i_j = \underset{12}{\widetilde{\omega}^i_j} = \omega^i_j + a^i_{jk}(\underset{1}{\omega^k} - \underset{2}{\omega^k}). \tag{4.2}$$

If we replace the connection forms ω^i_j for the forms $\widetilde{\omega}^i_j$, the structure equations (1.26)–(1.27) have the form (1.94):

$$\begin{aligned} d\underset{1}{\omega^i} &= \underset{1}{\omega^j} \wedge \widetilde{\omega}^i_j + a^i_{jk}\underset{1}{\omega^j} \wedge \underset{2}{\omega^k}, \\ d\underset{2}{\omega^i} &= \underset{2}{\omega^j} \wedge \widetilde{\omega}^i_j - a^i_{jk}\underset{1}{\omega^j} \wedge \underset{2}{\omega^k}, \\ d\underset{3}{\omega^i} &= \underset{3}{\omega^j} \wedge \widetilde{\omega}^i_j. \end{aligned} \tag{4.3}$$

Let us find the curvature tensor of the connection $\widetilde{\Gamma}_{12}$. First, we apply exterior differentiation to equation (4.2) and next transform the right-hand side, using relations (1.32), (1.33), (4.2) and (4.3). As a result, we obtain equations

$$d\widetilde{\omega}_j^i - \widetilde{\omega}_j^k \wedge \widetilde{\omega}_k^i = -b^i_{k(lj)} \underset{1}{\omega}^k \wedge \underset{3}{\omega}^l + (b^i_{[jk]l} + a^m_{jk} a^i_{lm} - a^i_{jm} a^m_{lk}) \underset{3}{\omega}^k \wedge \underset{3}{\omega}^l, \qquad (4.4)$$

which imply that the curvature form of the connection $\widetilde{\Gamma}_{12}$ can be expressed in terms of the forms $\underset{3}{\omega}^i$ if and only if the curvature tensor of a web W satisfies condition (4.1). This and the third equation of (4.3) lead us to the following theorem:

Theorem 4.1 *The forms $\widetilde{\omega}_j^i$ define an affine connection in the base X_3 of the third foliation of a three-web W if and only if conditions* (4.1) *hold.* ■

Denote by $\widetilde{\gamma}$ the connection defined in X_3 by the forms $\widetilde{\omega}_j^i$. Equations (4.1) and (1.31) imply that the structure equations of this connection can be reduced to the form:

$$d\underset{3}{\omega}^i = \underset{3}{\omega}^j \wedge \widetilde{\omega}_j^i, \quad d\widetilde{\omega}_j^i - \widetilde{\omega}_j^k \wedge \widetilde{\omega}_k^i = R^i_{jkl} \underset{3}{\omega}^k \wedge \underset{3}{\omega}^l, \qquad (4.5)$$

where

$$R^i_{jkl} = \frac{1}{4}(b^i_{jkl} - 2a^i_{mj} a^m_{kl}). \qquad (4.6)$$

The quantities $R = (R^i_{jkl})$ form the curvature tensor of the connection $\widetilde{\gamma}$. The relations $R^i_{jkl} = -R^i_{jlk}$, which the tensor R must satisfy, follow from equations (4.1) and the skew-symmetry of the torsion tensor a^i_{jk} of a web.

The curvature tensor b of the web B_m under consideration and its covariant derivatives satisfy other conditions in addition to relations (4.1). From equations (1.38):

$$\nabla b^i_{jkl} = \underset{1}{c}^i_{jklm} \underset{1}{\omega}^m + \underset{2}{c}^i_{jklm} \underset{2}{\omega}^m, \qquad (4.7)$$

by relations (4.1), we get

$$\underset{1}{c}^i_{j(kl)m} = 0, \quad \underset{2}{c}^i_{j(kl)m} = 0. \qquad (4.8)$$

In addition, consider relations (1.39):

$$\underset{1}{c}^i_{j[k|l|m]} = b^i_{jpl} a^p_{km}, \quad \underset{2}{c}^i_{jk[lm]} = -b^i_{jkp} a^p_{lm}, \qquad (4.9)$$

connecting the fundamental tensors of an arbitrary three-web. Making a cyclic permutation of the indices k, l and m, we obtain two additional series of equations:

$$\underset{1}{c}^i_{j[m|k|l]} = b^i_{jpk} a^p_{ml}, \quad \underset{2}{c}^i_{jl[mk]} = -b^i_{jlp} a^p_{mk},$$
$$\underset{1}{c}^i_{j[l|m|k]} = b^i_{jpm} a^p_{lk}, \quad \underset{2}{c}^i_{jm[kl]} = -b^i_{jmp} a^p_{kl}.$$

Using equation (4.8), from these equations and equation (4.9) we obtain:

$$\underset{1}{c}{}^i_{jklm} = -\underset{2}{c}{}^i_{jklm} = -b^i_{jpk}a^p_{ml} + b^i_{jpm}a^p_{kl} + b^i_{jpl}a^p_{km} \stackrel{\text{def}}{=} c^i_{jklm}. \tag{4.10}$$

Thus, the covariant derivatives $\underset{1}{c}$ and $\underset{2}{c}$ of the curvature tensor of the Bol web B_m can be expressed in terms of the torsion and curvature tensors of this web. We will discuss this result in more detail in Chapter 5.

According to Section 1.4, the tensors $\underset{1}{c}$, $\underset{2}{c}$, b and a are also related by the series of relations (1.40). Substituting the values of $\underset{1}{c}$ and $\underset{2}{c}$ into these relations, after uncomplicated calculations we arrive at:

$$b^i_{jpk}a^p_{lm} - b^i_{kpj}a^p_{lm} = b^i_{plm}a^p_{jk} + b^p_{klm}a^i_{pj} - b^p_{jlm}a^i_{pk}. \tag{4.11}$$

There is one more series of equations connecting the tensors a and b. We will discover them after proving the following theorem.

Theorem 4.2 *The base X_3 of the third foliation of the three-web B_m is a locally symmetric space.*

Proof. We recall one of the definitions of a locally symmetric space. A manifold X with an affine connection is called a *locally symmetric space* if its torsion tensor is equal to zero and its curvature tensor is covariantly constant.

Let us calculate the covariant differential $\widetilde{\nabla} R^i_{jkl}$ where $\widetilde{\nabla}$ is the operator of covariant differentiation in the connection $\widetilde{\Gamma}$. From (4.6) we find:

$$\widetilde{\nabla} R^i_{jkl} = \frac{1}{4}\widetilde{\nabla} b^i_{jkl} - \frac{1}{2}(a^m_{kl}\widetilde{\nabla} a^i_{mj} + a^i_{mj}\widetilde{\nabla} a^m_{kl}). \tag{4.12}$$

We will calculate the covariant differential $\widetilde{\nabla} b$ from (4.7) by substituting there the forms $\widetilde{\omega}^i_j$ taken from (4.2) for ω^i_j. Applying equation (4.10), we obtain:

$$\widetilde{\nabla} b^i_{jkl} = (-b^i_{jpk}a^p_{ml} + b^i_{jpm}a^p_{kl} + b^p_{jkl}a^i_{pm} - b^i_{pkl}a^p_{jm} - b^i_{jkp}a^p_{lm})(\underset{1}{\omega}{}^m - \underset{2}{\omega}{}^m). \tag{4.13}$$

Similarly, we can reduce expression (1.33) for the tensor ∇a^i_{jk}. By (4.1), it has the form:

$$\nabla a^i_{jk} = -b^i_{[jk]l}(\underset{1}{\omega}{}^l - \underset{2}{\omega}{}^l). \tag{4.14}$$

Substituting the forms $\widetilde{\omega}^i_j$ for ω^i_j in (4.1), we get

$$\widetilde{\nabla} a^i_{jk} = (-b^i_{[jk]l} + a^m_{jk}a^i_{ml} - a^i_{mk}a^m_{jl} - a^i_{jm}a^m_{kl})(\underset{1}{\omega}{}^l - \underset{2}{\omega}{}^l).$$

If in equations (1.12) we substitute for $\widetilde{\nabla} a$ and $\widetilde{\nabla} b$ its values from the above equation and the equation (4.13), after uncomplicated calculations we arrive at the equations:

$$\widetilde{\nabla} R^i_{jkl} = (b^i_{jmp} a^m_{kl} - b^i_{pmj} a^m_{kl} + b^i_{mlk} a^m_{jp} + b^m_{pkl} a^i_{jm} + b^m_{jkl} a^i_{mp})(\underset{1}{\omega}^p - \underset{2}{\omega}^p).$$

By (4.11) these equations take the form:

$$\widetilde{\nabla} R^i_{jkl} = 0. \tag{4.15}$$

■

We are now able to find all relations connecting the torsion and curvature tensors of the three-web B_m. Applying exterior differentiation to equations (4.15), we arrive at the Ricci identities:

$$R^i_{[jkl]} = 0 \tag{4.16}$$

and the Bianci identities:

$$-R^i_{slm} R^s_{pjk} + R^s_{plm} R^i_{sjk} + R^s_{jlm} R^i_{psk} + R^s_{klm} R^i_{pjs} = 0 \tag{4.17}$$

for the connection $\tilde{\gamma}$. Equation (4.16) can be also derived from equations (4.6) and (1.31).

By (4.15), subsequent differentiation does not give new relations. One can also show that the differentiation of (4.11) also does not give new relations. Thus, the following theorem holds:

Theorem 4.3 *The structure equations of the middle Bol web B_m can be written in the form* (4.3) *and* (4.5), *where the tensor R is expressed in terms of the tensors a and b of the web by formulas* (4.6). *The tensors a, b and R satisfy equations* (4.13)–(4.15) *and are connected to each other by relations* (4.11), (4.16) *and* (4.17). ■

2. Let us show that relations (4.1) for the curvature tensor of the web B_m characterize this class of webs.

Theorem 4.4 *A three-web W whose curvature tensor satisfies relation* (4.1) *is a middle Bol web.*

Proof. *Step* 1. Consider some properties of a web W for which relations (4.1) hold. First of all, in the base X_3 of the foliation λ_3 of this web, geodesic lines of the connection $\tilde{\gamma}$ taken with respect to an affine parameter are defined by the equations:

$$d\xi^i + \xi^j \tilde{\omega}^i_j = 0,$$

where ξ^i are coordinates of a vector tangent to a geodesic line. Let $x_3(t)$ be a solution of this equation where t is an affine parameter. On a manifold X carrying a three-web W, to the solution $x_3(t)$ there corresponds a one-parameter family $\mathcal{F}_3(t)$ of leaves of the foliation λ_3. The intersection of these leaves with a leaf of the first foliation is defined by the system of equations:

$$\underset{1}{\omega}^i = 0, \quad d\xi^i + \xi^j \tilde{\omega}_j^i = 0.$$

However, according to Section 1.7, namely such a system determines geodesic lines on the leaves of the foliation λ_1 in the Chern connection. Similarly, one can prove that the leaves of the geodesic family $\mathcal{F}_3(t)$ intersect the leaves of the foliation λ_2 also at geodesic lines.

Note further that conditions (4.1) imply equations $b^i_{(jkl)} = 0$ which mean that the web B_m under consideration is hexagonal (Theorem 3.5). Thus, this web is transversally geodesic (the same theorem), and the vertices of any hexagonal figure H composed of its leaves lie on a transversally geodesic two-dimensional surface V^2. The latter intersects the leaves forming the figure H at geodesic lines which, in turn, form a hexagonal figure on the surface V^2.

Using relations (4.1) and equations (3.1) of a transversally geodesic surface V^2, we find that on this surface the structure equations (4.3) of a web W have the form:

$$d\underset{1}{\omega}^i = \underset{1}{\omega}^j \wedge \tilde{\omega}_i^j, \quad d\underset{2}{\omega}^i = \underset{2}{\omega}^j \wedge \tilde{\omega}_i^j, \quad d\tilde{\omega}_j^i = \tilde{\omega}_j^k \wedge \tilde{\omega}_k^i.$$

Hence, the system of differential equations

$$dp = \underset{1}{\omega}^i \underset{2}{e}_i - \underset{2}{\omega}^i \underset{1}{e}_i, \quad d\underset{1}{e}_i = \tilde{\omega}_i^j \underset{1}{e}_j, \quad d\underset{2}{e}_i = \tilde{\omega}_i^j \underset{2}{e}_j$$

is completely integrable on V^2 and defines the development of this surface along with a frame associated with it in an affine space A^{2r}. The form of the equations obtained is the same as that of a parallel three-web (Section 1.5). Thus, the development of the surface V^2 in A^{2r} is a two-dimensional plane π, and the development of the hexagonal three-web which is cut on V^2 is a parallel three-web in π. Moreover, the development of a hexagonal figure is a hexagon with parallel sides in the plane π. Since the center of the diagonals of this hexagon is the midpoint of the diagonals, on the surface V^2 the point corresponding to this center is the midpoint of the geodesic segments, the diagonals of the hexagonal figure H.

Step 2. Consider the foliation λ_3 of a web W. By Theorem 4.1, the forms $\tilde{\omega}_j^i$ define in the base X_3 of this foliation an affine connection $\tilde{\gamma}$. Let $\mathcal{F}_3^1$ and $\mathcal{F}_3^2$ be two arbitrary leaves from λ_3 and let x_3^1 and x_3^2 be the points in the base X_3 corresponding to these leaves. There is a unique geodesic line through the points x_3^1 and x_3^2. A family of leaves $\mathcal{F}_3(t)$ (t is an affine parameter) of the foliation λ_3 corresponds to this geodesic line. Let $\mathcal{F}_3^1 = \mathcal{F}_3(t_1)$ and $\mathcal{F}_3^2 = \mathcal{F}_3(t_2)$. We set $\mathcal{F}_3^3 = \mathcal{F}_3(t_3)$ where $t_3 = 2t_2 - t_1$. Then, the leaf $\mathcal{F}_3^2$ corresponds to the midpoint of the geodesic segment $x_3^1 x_3^2$.

Let a be an arbitrary point of the leaf $\mathcal{F}_3^1$ (see Figure 4.1). Construct the leaves

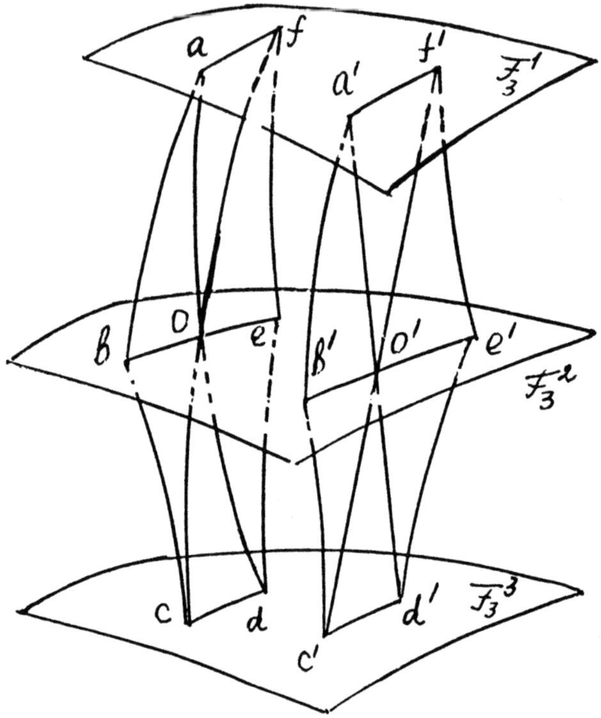

Figure 4.1

$\mathcal{F}_1$ and $\mathcal{F}_2$ through this point and denote by b and o their intersection points with the leaf $\mathcal{F}_3^2$. The points a, b and o lie on a transversal surface V^2 of the web under consideration since the points a and b correspond to each other in the mapping φ_{12} given in a neighborhood of the point o (Section 1.3). Thus, the points a, b and o determine a hexagonal figure $(abcdef)$ with the center o on the surface V^2. Moreover, the geodesic lines ab and ao are formed by the intersection points of leaves of the family $\mathcal{F}_3(t)$ with leaves $\mathcal{F}_1$ and $\mathcal{F}_1^2$. As we indicated in *Step* 1, the geodesic segments ao and od are equal. This implies that the point d and, therefore, the point c belongs to the leaf $\mathcal{F}_3^3$.

Let us take now a point $a' \in \mathcal{F}_3^1$ which is different from the point a. Repeating all our previous considerations we arrive at the points b' and o' lying in the leaf $\mathcal{F}_3^2$ and at the point c' lying in the leaf $\mathcal{F}_3^3$. But these points and the points a, b, o, c form the middle Bol figure B_m (see Figure 2.8, p. 55) which is closed since the points c and c' lie in the same leaf $\mathcal{F}_3^3$. Since the points a and a' were chosen arbitrarily, any figure

B_m defined by the leaves $\mathcal{F}_3^1$ and $\mathcal{F}_3^2$ is closed. Since these leaves were also chosen arbitrarily, we find that any figure B_m is closed on the three-web under consideration. ■

It follows from results of Section 2.2 that under parastrophy $q \to q_{32} = {}^{-1}q$ of the coordinate quasigroup q of a web W, the figures B_m are transferred into the Bol figures B_l and the curvature tensor b^i_{jkl} into the tensor $-\underset{32}{b}{}^i_{lkj}$. Similarly, under parastrophy $q \to q_{13} = q^{-1}$, the figures B_m are transferred into the Bol figures B_r and the curvature tensor b^i_{jkl} into the tensor $-\underset{13}{b}{}^i_{kjl}$. Thus, Theorem 4.4 implies the following theorem:

Theorem 4.5 *A three-web W is a left or right Bol web if and only if its curvature tensor satisfies the relation $b^i_{(jk)l} = 0$ or the relation $b^i_{(j|k|l)} = 0$, respectively.* ■

3. A symmetric structure defined in the base X_3 of the foliation λ_3 of a Bol web B_m can be naturally transferred on the leaves of its foliations λ_1 and λ_2.

Consider a leaf $\mathcal{F}$ of the first or the second foliation of web B_m. There is a unique leaf of the foliation λ_3 through any of its points. This gives a local diffeomorphism $X_3 \to \mathcal{F}$ which induces a structure of a symmetric space in $\mathcal{F}$. The corresponding connection is defined in $\mathcal{F}$ by the same equations (4.5) which define the connection $\tilde{\gamma}$ in X_3.

Next, it was shown in Section 2.5 that a leaf $\mathcal{F}$ of a three-web W is locally isomorphic to its coordinate loop Q. Since the coordinate loops of a web B_m are the middle Bol loops, the structure of a symmetric space described above is induced in an arbitrary Bol loop B_m.

The analogous conclusions can be obtained for webs B_l and B_r.

4. Let us show that the components R^i_{jkl} of the curvature tensor of the connection γ can be reduced to constant values in the whole manifold X of a three-web B_m. As in Section 1.4, we denote by $\mathcal{R}(W)$ the fibration of adapted frames of the G_W-structure in X. The forms $\underset{1}{\omega}{}^i$, $\underset{2}{\omega}{}^i$ and ω^i_j are basis forms in $\mathcal{R}(W)$. Consider in $\mathcal{R}(W)$ a distribution $\tilde{S}$ given by the equations:

$$\tilde{\omega}^i_j = R^i_{jkl}\theta^{kl} \tag{4.18}$$

where $\theta^{kl} = -\theta^{lk}$.

Proposition 4.6 *The components R^i_{jkl} of the curvature tensor of the connection γ are constant in the distribution $\tilde{S}$.*

Proof. In fact, consider equations (4.15):

$$\tilde{\nabla} R^i_{jkl} = dR^i_{jkl} + R^m_{jkl}\tilde{\omega}^i_m - R^i_{mkl}\tilde{\omega}^m_j - R^i_{jml}\tilde{\omega}^m_k - R^i_{jkm}\tilde{\omega}^m_l = 0.$$

Substituting the values of the forms $\tilde{\omega}^i_j$ from (4.18) into these equations and using relations (4.17), we obtain $dR^i_{jkl} = 0$. ■

Proposition 4.7 *The distribution $\widetilde{S}$ is involutive.*

Proof. Eliminating the parameters θ^{kl} from equations (4.18), we get an equivalent system of equations:

$$B^i_{\rho j}\widetilde{\omega}^j_i = 0, \quad \rho = 1, 2, \ldots, a. \tag{4.19}$$

In (4.19) we wrote only independent equations. If we substitute the values of the forms $\widetilde{\omega}^i_j$ from (4.18) into equations (4.19), we get identities. This gives the relations:

$$B^i_{\rho j}R^j_{ikl} = 0. \tag{4.20}$$

Since in the distribution $\widetilde{S}$ the components R^i_{jkl} are constant, the quantities $B^i_{\rho j}$, which can be expressed in terms of R^i_{jkl}, are also constant in it. Thus, differentiating relations (4.19) and using equations (4.19) and (4.5), we arrive at the relations:

$$B^i_{\rho j}(\widetilde{\omega}^k_i \wedge \widetilde{\omega}^j_k + R^j_{ikl}\underset{3}{\omega}^k \wedge \underset{3}{\omega}^l) = 0.$$

By (4.20), the second term in parentheses is equal to zero, and by (4.18) the first term can be transformed to the form:

$$B^i_{\rho j}(R^k_{ipq}R^j_{kls} - R^k_{ils}R^j_{kpq})\theta^{pq} \wedge \theta^{ls} = 0.$$

Using equations (4.17), we can write these relations in the form:

$$B^i_{\rho j}(R^j_{iks}R^k_{lpq} + R^j_{ilk}R^k_{spq})\theta^{pq} \wedge \theta^{ls} = 0.$$

By (4.20), the left-hand side of this equation vanishes. Therefore, according to the Frobenius theorem (see [CaH 67]), the system (4.58) is completely integrable, and the distribution $\widetilde{S}$ defined by this system is involutive. ■

The fact that the distribution $\widetilde{S}$ is involutive means that the manifold $\mathcal{R}(W)$ is foliated into submanifolds of dimension $2r + r^2 - a$ where a is the number of equations in system (4.19). Let $\mathcal{R}'(W)$ be one of these submanifolds. The forms $\underset{1}{\omega}^i$, $\underset{2}{\omega}^i$ and $\widetilde{\theta}^i_j = \widetilde{\omega}^i_j\big|_{\widetilde{S}}$ are independent forms in $\mathcal{R}'(W)$. These forms satisfy the structure equations (4.3) and (4.5) where the quantities R^i_{jkl} are constants.

Let us fix a point p of the manifold X, i.e. set $\underset{1}{\omega}^i = \underset{2}{\omega}^i = 0$, and denote

$$\widetilde{\theta}^i_j\Big|_{\underset{1}{\omega}^i = \underset{2}{\omega}^i = 0} = \pi^i_j.$$

The forms π^i_j satisfy the structure equations of the group $\mathbf{GL}(r)$, the stationary group of the point p. The latter equations can be obtained from equations (4.5):

$$d\pi^i_j = \pi^k_j \wedge \pi^i_k.$$

Moreover, the forms π^i_j, like the forms $\widetilde{\omega}^i_j$, are connected by relations (4.19) and the coefficients $B^i_{\rho j}$ in relations (4.19) are constants in $\mathcal{R}'(W)$. Applying a well-known theorem from Lie group theory, we find that equations (4.19) define a subgroup of the group $\mathbf{GL}(r)$. Denote this subgroup by H. Thus, the manifold $\mathcal{R}'(W)$ is the reduction of the fibration $\mathcal{R}(W)$ of adapted frames of a three-web B_m to the group H.

Let us establish a geometric meaning for the subgroup H. For this, we first recall the definition of the holonomy group of an affine connection. Let p be an arbitrary point of a manifold X with an affine connection Γ. Consider all closed smooth paths emanating from and ending in the point p. The parallel displacement of a frame $\mathcal{R}$ along a loop $l, l(0) = l(1) = p$, generates an isomorphism $\mathcal{R}(0) \to \mathcal{R}(1)$ in the tangent space $T_p(X)$. The set of all such diffeomorphisms is a group which is called the *holonomy group of the affine connection* Γ at the point p. The holonomy groups taken at different points of the connected manifold X are isomorphic. The Lie algebra of the holonomy group of an affine connection Γ is called the *holonomy algebra* of this connection.

Proposition 4.8 *The group H is the holonomy group of the connection $\widetilde{\gamma}$.*

Proof. Let X be an analytic manifold. Then the holonomy algebra h of a connection Γ given in X is generated by the curvature tensor R^i_{jkl} of this connection and its covariant derivatives of all orders, i.e. by linear transformations of the form:

$$R^i_{jkl}x^k y^l, \quad \nabla_m R^i_{jkl}x^k y^l z^m, \quad \nabla_n \nabla_m R^i_{jkl}x^k y^l z^m w^n, \ldots .$$

(see [KN 63]). For the connection $\widetilde{\gamma}$ we have: $\widetilde{\nabla} R = 0$. Thus, its holonomy algebra h is generated only by the operators $R^i_j = R^i_{jkl}x^k y^l$. Using relation (4.17), one can prove that the algebra h generated by the operators R^i_j coincides with its linear span (see Problem **2**). Because of this, the holonomy algebra of the connection $\widetilde{\gamma}$ coincides with the Lie algebra of the group H defined by relations (4.18). Therefore, the group H is the holonomy algebra of the connection $\widetilde{\gamma}$. ■

It follows from this that equations (4.18) define the reduction of the fibration of adapted frames of a three-web B_m to the holonomy group of the connection $\widetilde{\gamma}$. Propositions 4.6–4.8 constitute the so-called reduction theorem, and the fibers $\mathcal{R}'(W)$, the integral manifolds of the distribution $\widetilde{S}$, are called the the *holonomy bundles* (see [KN 63], vol. 1, p. 85).

5. As a result of reduction to the group H, the structure equations (4.5) of the connection $\widetilde{\gamma}$ take the form:

$$d\underset{3}{\omega}^i = \underset{3}{\omega}^j \wedge \widetilde{\theta}^i_j, \quad d\widetilde{\theta}^i_j = \widetilde{\theta}^k_j \wedge \widetilde{\theta}^i_k + R^j_{ikl}\underset{3}{\omega}^k \wedge \underset{3}{\omega}^l, \tag{4.21}$$

where as above $\widetilde{\theta}^i_j = \widetilde{\omega}^i_j\big|_{\widetilde{S}}$ and R^i_{jkl} are constants on $\widetilde{S}$. For definiteness, we suppose that the connection $\widetilde{\gamma}$ is realized on a leaf $\mathcal{F}$ of the first foliation. Then, the additional equations $\underset{1}{\omega}^i = 0$ should be added to equations (4.21). The forms $\underset{3}{\omega}^i$ are principal forms in $\mathcal{F}$.

By identities (4.16) and (4.17), the system (4.21), which contains only constants in addition to forms, is closed with respect to exterior differentiation. Thus, it represents the Cartan–Maurer equations of a Lie group G. The holonomy group H is a subgroup of G and is defined by the equations $\underset{3}{\omega}^i = 0$, fixing a point in the leaf $\mathcal{F}$. Thus, the fiber $\mathcal{F}$, in which the canonical connection $\tilde{\gamma}$ of a symmetric space is realized, is a homogeneous space G/H.

Equations (2.21) show that multiplication in the tangent Lie algebra of the group G is defined by means of the tensor R^i_{jkl}. This algebra can be obtained in the following way.

Let T be a vector space of dimension r. Define the ternary operation $< \xi, \eta, \zeta >$ in T by equations:

$$< \xi, \eta, \zeta >^i = R^i_{jkl} \xi^j \eta^k \zeta^l, \quad \xi, \eta, \zeta \in T. \tag{4.22}$$

The space T with the operation $< \ , \ , \ >$ is called the *triple Lie system*. It possesses the following properties:

$$\begin{aligned} &\text{a) } < \xi, \eta, \eta > = 0; \\ &\text{b) } < \xi, \eta, \zeta > + < \eta, \zeta, \xi > + < \zeta, \xi, \eta > = 0; \\ &\text{c) } << \zeta, \theta, \chi > \xi, \eta > = << \zeta, \xi, \eta >, \theta, \chi > \\ &+ < \zeta, < \theta, \xi, \eta >, \chi > + < \zeta, \theta, < \chi, \xi, \eta >> . \end{aligned} \tag{4.23}$$

Property a) follows from the skew-symmetry of the tensor R^i_{jkl} in the last two indices, and properties b) and c) are equivalent to the Bianci identities (4.16) and the Ricci identities (4.17), respectively.

Denote by $d_{\xi,\eta}$ the linear transformation in T defined by the formula:

$$d_{\xi,\eta}(\zeta) = < \zeta, \xi, \eta > . \tag{4.24}$$

Then, the third relation in (4.23) can be written in the form:

$$d_{\xi,\eta} < \zeta, \theta, \chi > = < d_{\xi,\eta}(\zeta), \theta, \chi > + < \zeta, d_{\xi,\eta}(\theta), \chi > + < \zeta, \theta, d_{\xi,\eta}(\chi) >$$

which means that the linear operator $d_{\xi,\eta}$ is a differentiation of the triple Lie system T. Differentiations of type (4.24) are called *interior*. Since $(d_{\xi,\eta})^i_j = R^i_{jkl}\xi^k \eta^l$, the space generated by all interior differentiations coincides with the holonomy algebra h of the connection $\tilde{\gamma}$ defined above. Define by g the direct sum $T \oplus h$ of the spaces T and h and let $x, y \in h$ and $\xi, \eta \in T$. Then, the following theorem holds:

Theorem 4.9 *The space g is a Lie algebra with respect to the operation $(\ , \ )$ where:*

$$
\begin{aligned}
(x,y) &= x \circ y - y \circ x \in h,\\
(x,\xi) &= -(\xi,x) = x(\xi) \in T,\\
(\xi,\eta) &= d_{\xi,\eta} \in h.
\end{aligned}
$$

The algebra g is the tangent algebra of the group G. ■

The proof of Theorem 4.9 is similar to that of more complicated Lemma 4.21 (see Section 4.4). Thus, we omit it and leave it to the reader as an exercise. Different proofs of Theorem 4.9 can be found in the books [Lo 69], [Tr 89] and other books on symmetric spaces.

The algebra g is said to be the *universal span of a triple Lie system.*

The algebra g and its subalgebra h form a so-called *symmetric pair.* There exists an involutive automorphism $\sigma : g \rightarrow g$ such that $\sigma(x) = y$ and $\sigma(\xi) = \xi$ for any $x \in h$ and any $\xi \in T$. The spaces h and T are invariant subspaces of the involutive operator σ and correspond to its eigenvalues $+1$ and -1.

6. Consider the W-algebra of a Bol three-web (see Section 2.5). Let, as was in the general case, a binary operation in the W-algebra be defined by the tensor a^i_{jk}. As to a ternary operation, we define it by means of the tensor R^i_{jkl} instead of the tensor b^i_{jkl}. Note that these two tensors are connected by relation (4.6).

As we found in Subsection **5**, this ternary operation generates a triple Lie system and satisfies relations (4.23). The binary and ternary operations are connected by relations (4.11). Substituting b^i_{jkl} for R^i_{jkl} in (4.11) and using (4.6), one can write relations (4.11) in the form:

$$R^i_{pjk}a^p_{lm} - R^i_{plm}a^p_{jk} - R^p_{klm}a^i_{pj} + R^p_{jlm}a^i_{pk} + a^i_{pq}a^p_{jk}a^q_{lm} = 0,$$

or

$$< [\zeta,\theta],\xi,\eta > - < [\xi,\eta],\zeta,\theta > -[< \eta,\zeta,\theta >,\xi] + [< \xi,\zeta,\theta >,\eta] + [[\xi,\eta],[\zeta,\theta]] = 0. \tag{4.25}$$

Definition 4.10 A linear space T with given binary and ternary operations, $[\ ,\ ]$ and $<\ ,\ ,\ >$, satisfying identities (4.23) and (4.25), is called a *Bol algebra.*

Thus, the tangent W-algebras of a Bol three-web B_m are the Bol algebras.

The system of equations (4.3), (4.5) and (4.14) defining a Bol three-web B_m is closed with respect to exterior differentiation if conditions (4.23) and (4.25), characterizing the corresponding Bol W-algebras, hold. Hence, *a Bol three-web B_m is completely defined by assigning the corresponding Bol W-algebras, i.e. the tensor fields a^i_{jk} and R^i_{jkl} satisfying conditions* (4.23) *and* (4.25).

7. We conclude this section by indicating two problems arising in the theory of Bol three-webs:

1. Can any symmetric space be obtained from a Bol three-web by the method described above? If no, find a class of spaces for which the answer is positive.

2. How many non-equivalent Bol three-webs can be associated with a given symmetric space?

Let us show that both problems can be reduced to a purely algebraic problem. Let the connection $\tilde{\gamma}$ be given by equations (4.5) where the tensor R satisfies the Ricci identities (4.16) and the Bianci identities (4.17). A three-web B_m can be reconstructed from this by means of equations (4.3) containing the tensor a^i_{jk}. This tensor satisfies only relations (4.25). Therefore, the problem of finding a Bol three-web, provided that a symmetric connection is given, is reduced to the finding the torsion tensor of this connection from equations (4.25). The curvature tensor can be calculated next from relations (4.6).

Thus, both problems indicated above lead to the structural theory of Bol algebras which is now in its initial phase of development. These problems have been completely solved only for isoclinic and hexagonal Bol three-webs and for a special class of Bol webs, the Moufang three-webs, which will be considered in the following sections of this chapter.

4.2 The Isoclinic Bol Three-Webs

1. The isoclinic Bol three-webs allow us to make a series of nice and simple geometric constructions. The latter will be used in the following discussion to model different properties of webs.

First of all, note that since any Bol web is hexagonal (see Section 4.1), by Theorem 3.17, an isoclinic Bol web is algebraizable. More precisely, the following theorem holds:

Theorem 4.11 *The class of isoclinic Bol three-webs coincides with the class of Grassmann three-webs generated by a hyperplane and a hyperquadric.*

Proof. Let B_m be an isoclinic Bol three-web. Since it is algebraizable and, therefore, Grassmannizable, its torsion and curvature tensors can be written in the form (3.45) and (3.49):

$$a^i_{jk} = a_{[j}\delta^i_{k]}, \tag{4.26}$$

and

$$b^i_{jkl} = b^1_{jk}\delta^i_l + b^2_{lj}\delta^i_k + b^3_{kl}\delta^i_j, \tag{4.27}$$

where $b^\alpha_{ij} = b^\alpha_{ji}$, $\alpha = 1,2,3$, $i,j = 1,\dots,r$, and the web B_m itself is generated in a projective space P^{r+1} by three-hypersurfaces X_α, belonging to a hypercubic.

Symmetrize expression (4.27) in the indices k and l and equate the result to zero (cf. (4.1)):

$$(b^1_{jk} + b^2_{jk})\delta^i_l + (b^1_{jl} + b^2_{jl})\delta^i_k + b^3_{kl}\delta^i_j = 0.$$

Contracting this equation first in the indices i and j and next in the indices i and l, we find that

$$b^1_{kl} + b^2_{kl} + b^3_{kl} = 0, \quad (r+1)(b^1_{jk} + b^2_{jk}) + b^3_{jk} = 0.$$

Since $r = \dim X_\alpha \geq 2$, these equations imply that

$$b^1_{kl} + b^2_{kl} = 0, \quad b^3_{kl} = 0. \tag{4.28}$$

The second equation in (4.28) tells that the asymptotic quadratic form of the hypersurface X_3 is equal to zero. Therefore, this hypersurface is a hyperplane. On the other hand, since the web under consideration is algebraizable, all three hypersurfaces, X_1, X_2 and X_3, belong to a hypercubic. It follows from this that the hypersurfaces X_1 and X_2 lie on a hyperquadric Q of the space P^{r+1}.

Conversely, suppose that the hypersurface X_3 is a hyperplane, and that the hypersurfaces X_1 and X_2 are domains of a hyperquadric Q. First we derive that $b^3_{kl} = 0$. Second, since the triple of hypersurfaces X_α belongs to a degenerate hypercubic, a Grassmann three-web generated by this hypercubic is hexagonal, and its curvature tensor satisfies the condition $b^i_{(jkl)} = 0$ (Theorem 3.5). Using the this condition, relation $b^3_{kl} = 0$ and condition (4.24), one can derive the following relations:

$$b^1_{(jk}\delta^i_{l)} + b^2_{(jk}\delta^i_{l)} = 0.$$

Contracting them in the indices i and j, we obtain

$$(r+2)(b^1_{kl} + b^2_{kl}) = 0,$$

and subsequently we find that

$$b^1_{kl} = -b^2_{kl} \stackrel{\text{def}}{=} b_{kl}.$$

Thus, the curvature tensor of the three-web under consideration can be written in the form:

$$b^i_{jkl} = b_{jk}\delta^i_l - b_{jl}\delta^i_k, \tag{4.29}$$

and satisfies the condition $b^i_{(jkl)} = 0$ characterizing the middle Bol webs. ■

2. Let us establish the structure of the configuration in a projective space P^{r+1} which corresponds to the middle Bol figure on the isoclinic Bol web under consideration. The middle Bol figure is represented in Figure 4.2 where the corresponding points of the configuration are marked by the same letters as the leaves of the figure B_m.

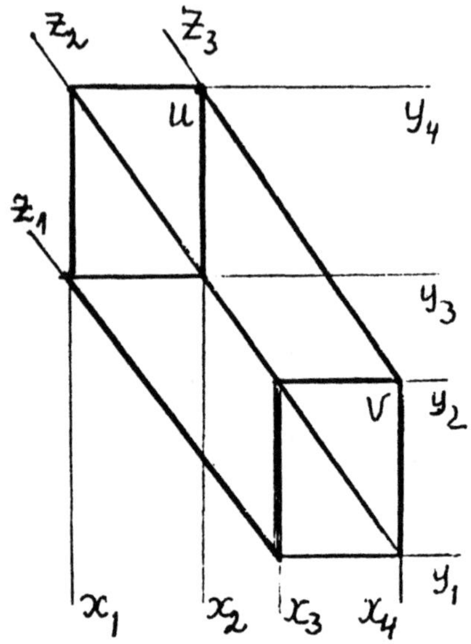

Figure 4.2

The points of the hyperquadric Q, to which the hypersurfaces X_1 and X_2 belong, represent the leaves of the first and the second foliations of the web B_m, and the points of the hypersurface X_3 represent the leaves of the third foliation. Since to three leaves of the web passing through a point, in P^{r+1}, there correspond three points of the hypersurfaces X_α located on one straight line, to the figure B_m there corresponds the configuration B'_m represented in Figure 4.3.

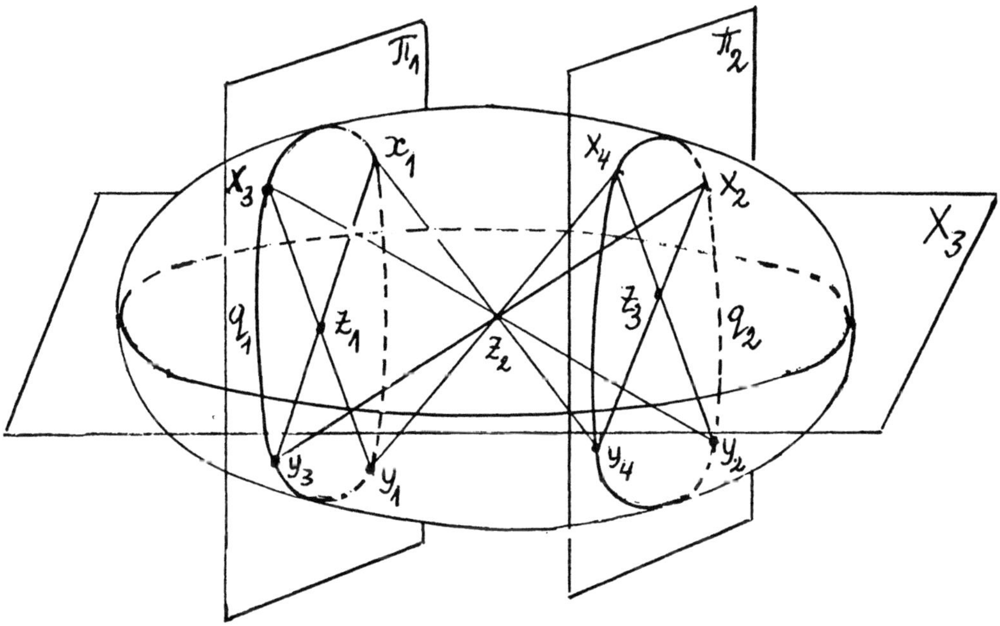

Figure 4.3

The meaning of the figure B_m represented in Figure 4.2 being closed is that the points u and v belong to one leaf z_3. In the configuration B'_m, this leads to the coincidence of the intersection points of the hyperplane X_3 and the straight lines x_2y_4 and x_4y_2.

The configuration B'_m is completely defined by its points x_1, x_3, z_1 and z_2 which are in general position. This configuration entirely belongs to a three-dimensional space P^3 spanned by these points. This allows us to make all further constructions in P^3. Denote the intersections $X_\alpha \bigcap P^3$ by the same letters X_α. This allows us to consider Figure 4.3, representing the configuration, as being located in P^3.

Denote by π_1 a two-dimensional plane, in which the points x_1, x_3, z_1, y_3 and y_1 lie, and by q_1 the second degree curve, at which this plane intersects the hyperquadric Q. Consider the second degree cone with vertex at the point z_2 which passes through the curve q_1. The intersection of this cone with the hyperquadric Q is a fourth degree curve which decomposes into two second degree curves q_1 and q_2. The points x_2, x_1, y_2 and y_4 lie on the curve q_2. Therefore, they lie in the same plane π_2.

Let us now prove that the point z_3 of the intersection of the straight lines x_2y_4 and x_4y_2 lies in the hyperplane X_3. In fact, the straight line x_2y_4 lies in the plane π_3

passing through the points x_1, y_3 and z_2, and the straight line x_4y_2 lies in the plane π_4 passing through the points x_4, y_1 and z_2. The planes π_3 and π_4 intersect one another in the straight line z_1z_2. The point z_3 is the intersection point of the planes π_2, π_3 and π_4. Hence, it belongs to the straight line z_1z_2 and subsequently to the hyperplane X_3.

Therefore, we not only described the configuration B'_m in a space P^{r+1} corresponding to the middle Bol figure B_m, but also we obtained one more, purely constructive proof of the closure of these figures on isoclinic Bol webs.

3. Consider the local W-algebras of an isoclinic Bol web. Since this web is Grassmannizable, the operation of commutation in its W-algebras can be written in the form (3.27):

$$[\xi, \eta] = a(\eta)\xi - a(\xi)\eta.$$

By (4.29), the operation of association in W-algebras can be represented in the following form:

$$(\xi, \zeta, \eta) = b(\zeta, \eta)\xi - b(\zeta, \xi)\eta,$$

where the expressions for $a(\xi)$ and $b(\xi, \eta)$ are a linear scalar form and a symmetric bilinear scalar form, respectively. It follows from this that for any vectors ξ, η and ζ from a W-algebra, the vectors $[\xi, \eta]$ and (ξ, η, ζ) can be expressed as linear combinations of only two vectors ξ and η. This means that both operations satisfy the axiom of two-dimensional planes (see Section 2.3).

It is clear that W-algebras of isoclinic Bol webs B_r and B_l have a similar property.

If on an isoclinic three-web W, all three Bol conditions, B_r, B_l and B_m, hold, then $b^\alpha_{ij} = 0$ for $\alpha = 1, 2, 3$. Hence, its curvature tensor is equal to zero. However, the simultaneous realization of all three Bol conditions characterizes the Moufang web, and the vanishing of the curvature tensor characterizes the group webs. In addition, the conditions $b^\alpha_{ij} = 0$ imply that all three hypersurfaces X_α generating the three-web under consideration are hyperplanes. Thus, the following theorem holds:

Theorem 4.12 *The following three statements are equivalent:*

a) A three-web W is an isoclinic Moufang web.

b) A three-web W is an isoclinic group web.

c) A three-web W is Grassmannizable and the three hypersurfaces, generating this web in a space P^{r+1}, are hyperplanes. ■

Note also that if the torsion tensor of an isoclinic Moufang web is non-zero, then the three hyperplanes, generating this web, are in general position. If this tensor vanishes, these hyperplanes belong to a pencil of hyperplanes. In the latter case, a web is parallelizable, its W-algebras are trivial, and the coordinate loops are abelian groups.

4. Let us find the equations of the hyperplane X_3 and the hyperquadric Q in the moving frame associated with these surfaces in the way indicated in Section 3.3. Now the points A_0 and A_{r+1} lie on the hyperquadric Q, and the point $A_0 + A_{r+1}$ is located on the hyperplane X_3 (see Figure 3.3). The points A_i of the moving frame are located in the intersection of the tangent hyperplanes to the hyperquadric Q at the points A_0 and A_{r+1}. The derivational equations of the hypersurfaces, generated by the points A_0, A_{r+1} and $A_0 + A_{r+1}$, have the form (3.39) and (3.40), and by (4.28), equations (3.46) and (3.47) can be written as follows:

$$\begin{aligned} &\theta_i^{r+1} = b_{ij}\theta_0^j, \quad \theta_i^0 = b_{ij}\theta_{r+1}^j, \\ &\nabla a_i + \theta_i^0 - \theta_i^{r+1} = 0. \end{aligned} \tag{4.30}$$

In addition, by formulas (4.26) and (4.29), equation (4.7) for the covariant differential of the curvature tensor has the form:

$$\nabla b_{jk} = -b_{jk}a_l(\underset{1}{\omega}^l - \underset{2}{\omega}^l). \tag{4.31}$$

Using the equations of infinitesimal displacements of a moving frame which were obtained in Section 3.3, we find that

$$d(A_0 + A_{r+1}) = \theta_{r+1}^{r+1}(A_0 + A_{r+1}) + (\theta_0^i + \theta_{r+1}^i)(A_i + a_i A_0).$$

Thus, the hyperplane X_3 is determined by the points $A_0 + A_{r+1}$ and $A_i + a_i A_0$, and its equation in the moving frame $\{A_0, A_i, A_{r+1}\}$ has the form:

$$a_i x^i - x^0 + x^{r+1} = 0.$$

Let us prove that the hyperquadric Q which is generated by the points A_0 and A_{r+1}, is defined by the equations:

$$b_{ij}x^i x^j - 2x^0 x^{r+1} = 0.$$

In fact, it is easy to check that the points A_0 and A_{r+1} lie in the hyperquadric Q, and the latter is tangent to the hyperplanes $x^{r+1} = 0$ and $x^0 = 0$ at these points. We need to prove only that the hyperquadric Q is fixed under the displacements of the moving frame defined by equations (3.37). Denote by $b_{\alpha\beta}$ the coefficients of the equation of the hyperquadric Q which is written in homogeneous coordinates:

$$(b_{\alpha\beta}) = \begin{pmatrix} 0 & 0 & -1 \\ 0 & b_{ij} & 0 \\ -1 & 0 & 0 \end{pmatrix}$$

Then, the conditions of immobility of this hyperquadric can be written in the form:

$$db_{\alpha\beta} - b_{\alpha\gamma}\theta_\beta^\gamma - b_{\gamma\beta}\theta_\alpha^\gamma = \rho b_{\alpha\beta}.$$

By direct calculations, using relations (4.30), (4.31), (3.40), (3.42) and (3.44), one can check that these conditions are identically satisfied.

5. We now calculate for an isoclinic web B_m the curvature tensor R^i_{jkl} of the canonical connection $\tilde{\gamma}$ which this web defines in the base of the third foliation. Substituting expressions (4.26) and (4.29) into formula (4.6), we find:

$$R^i_{jkl} = \frac{1}{4}((b_{jk} + \frac{1}{2}a_j a_k)\delta^i_l - (b_{jl} + \frac{1}{2}a_j a_l)\delta^i_k).$$

Since this tensor is covariantly constant in the connection $\tilde{\gamma}$, then the symmetric tensor

$$g_{jk} = b_{jk} + \frac{1}{2}a_j a_k$$

is also covariantly constant in this connection.

Suppose that the tensor g_{jk} is non-degenerate. Then, it defines a Riemannian or pseudo-Riemannian metric in the hyperplane X_3, and the connection $\tilde{\gamma}$ is the Levi-Civita connection of this metric. Setting $R_{ijkl} = g_{im}R^m_{jkl}$, we get

$$R_{ijkl} = \frac{1}{4}(g_{il}g_{jk} - g_{ik}g_{jl}).$$

This form of the curvature tensor characterizes Riemannian spaces of constant curvature (see, for example [KN 63]). One can show that the metric g_{ij} is induced on the hyperplane X_3 by the absolute which is the intersection $\tilde{Q}$ of the hyperplane X_3 and the hyperquadric Q. The non-singularity of the tensor g_{ij} corresponds to the non-degeneracy of the hyperquadric $\tilde{Q}$ (see Problem **5**).

As has been proved in Section 4.1, the base X_3 of the third foliation of a middle Bol web B_m is a symmetric space. Let us show how a symmetric structure in X_3 can be realized in the case when a Bol web W is isoclinic.

Consider the hyperplane X_3 and the hyperquadric Q generating this web and suppose that the quadric $\tilde{Q} = Q \bigcap X_3$ is non-degenerate. Denote by S the pole of the hyperplane X_3 relative to Q. To a point P of the three-web B_m, in the space P^{r+1}, there corresponds a straight line intersecting the hyperquadric Q at the points x_1 and x_2 and intersecting the hyperplane X_3 at the point x_3. These three points represent the leaves of the three-web B_m passing through one point. Let us now project the points x_1 and x_2 from the pole S onto the hyperplane X_3. As a result, we obtain the points $y_1 = S(x_1) \bigcap X_3$ and $y_2 = S(x_2) \bigcap X_3$ which are symmetric relative to the point x_3 in the Riemannian (or pseudo-Riemannian) metric defined in X_3 by the absolute Q. This statement follows from Problem **4**.

6. There is another reason why the class of isoclinic Bol webs is interesting: it contains all four-dimensional Bol webs. In other words, the following theorem holds:

Theorem 4.13 *Any four-dimensional Bol web is isoclinic.*

Proof. In fact, consider a four-dimensional Bol web B_m. Since the indices i, j and k now take on only two values, 1 and 2, we can represent them (using the skew-symmetry of the tensors a^i_{kl} and b^i_{jkl} in the indices k and l), in the form:

$$a^i_{kl} = a_{[k}\delta^i_{l]}, \quad b^i_{jkl} = b_{jk}\delta^i_l - b_{jl}\delta^i_k. \tag{4.32}$$

Thus, the structure equations of a four-dimensional Bol web can be written in the form:

$$\begin{aligned} d\underset{1}{\omega}^i &= \underset{1}{\omega}^j \wedge \omega^i_j + a_j \underset{1}{\omega}^j \wedge \underset{1}{\omega}^i, \\ d\underset{2}{\omega}^i &= \underset{2}{\omega}^j \wedge \omega^i_j - a_j \underset{2}{\omega}^j \wedge \underset{2}{\omega}^i, \\ d\omega^i_j &= \omega^k_j \wedge \omega^i_k + b_{jk}(\underset{1}{\omega}^k \wedge \underset{2}{\omega}^i - \underset{1}{\omega}^i \wedge \underset{2}{\omega}^k), \end{aligned} \tag{4.33}$$

and the differential equation (4.14) takes the form:

$$\delta^i_k \nabla a_j - \delta^i_j \nabla a_k = ((b_{kj} - b_{jk})\delta^i_l + (b_{jl}\delta^i_k - b_{kl}\delta^i_j))(\underset{1}{\omega}^l - \underset{2}{\omega}^l).$$

Contracting the latter equation in the indices i and k and taking into account that $\delta^i_i = 2$, we find:

$$\nabla a_j = b_{kj}(\underset{1}{\omega}^k - \underset{2}{\omega}^k). \tag{4.34}$$

Next, we consider relations (4.11) which connect the torsion and curvature tensors of a Bol web. Substituting expressions (4.32) into (4.11), after simple calculations, we arrive at the equations:

$$b_{[jk]}a_l = 0.$$

If $a_l = 0$, then, it follows from (4.34) that $b_{kj} = 0$, and in this case the web B_m is parallelizable. If $a_l \neq 0$, then $b_{[kj]} = 0$, i.e. the tensor b_{jk} is symmetric. However, this tensor is the covariant derivative of the co-vector a_j (see (4.34)). This and Theorem 3.9 imply that the four-dimensional Bol web under consideration is isoclinic. ■

It follows from Theorems 4.10 and 4.11 that any four-dimensional Bol web is equivalent to a Grassmann three-web defined in a three-dimensional projective space by a plane π and a quadric Q. In particular, this fact allows us to give a classification of the four-dimensional Bol webs, based on the type of the quadric Q and its location relative to the plane π, and to find closed form equations of the four-dimensional Bol webs.

4.3 The Six-Dimensional Bol Three-Webs

1. It follows from the results of Section 4.2 that non-algebraizable Bol three-webs may be found only if $r \geq 3$. However, for an arbitrary r, a classification of the webs B_m is rather difficult since this problem is reduced to a classification of the torsion and curvature tensors whose valences are three and four, respectively. Nevertheless, we will now show that the six-dimensional Bol three-webs, can be described relatively easily.

We will need the so-called *discriminant* tensors ϵ_{ijk} and ϵ^{ijk}. If $r = 3$, these tensors can be defined in the following way:

$$\epsilon_{ijk} = \epsilon^{ijk} = \begin{cases} 1 & \text{if the substitution } \begin{pmatrix} 1 & 2 & 3 \\ i & j & k \end{pmatrix} \text{ is even;} \\ -1 & \text{if the substitution } \begin{pmatrix} 1 & 2 & 3 \\ i & j & k \end{pmatrix} \text{ is odd;} \\ 0 & \text{if at least two indices are equal.} \end{cases}$$

(In this section all Latin indices take the values 1, 2 and 3.) The quantities ϵ_{ijk} and ϵ^{ijk} are connected by the relations:

$$\epsilon_{ikl}\epsilon^{jkl} = 2\delta_i^j, \quad \epsilon_{ijm}\epsilon^{klm} = 2\delta_{[i}^k\delta_{j]}^l. \tag{4.35}$$

Let us show that the torsion and curvature tensors of a six-dimensional Bol three-web can be written in the following form:

$$a_{jk}^i = \epsilon_{jkp}a^{ip}, \quad b_{jkl}^i = \epsilon_{klp}b_j^{ip}. \tag{4.36}$$

In fact, the indices now take on only three values and the tensor a_{jk}^i is skew-symmetric. Thus, this tensor has only nine essential components which is equal to the number of essential components of the tensor a^{ij}. Relations (4.36, 1) allow us to find the unique values of a^{pi}. To find them, one should contract equations (4.26, 1) with the tensor ϵ^{qjk} and apply equation (4.35, 1). Similarly, equations (4.36, 2) allow us to express the quantities b_j^{ip} in terms of b_{jkl}^i.

The structure equations of a six-dimensional Bol three-web can be obtained from the structure equations (1.26) and (1.32) if one substitutes expressions (4.36) into them:

$$\begin{aligned} d\underset{1}{\omega}^i &= \underset{1}{\omega}^j \wedge \omega_j^i + \epsilon_{jkl}a^{il}\underset{1}{\omega}^j \wedge \underset{1}{\omega}^k, \\ d\underset{2}{\omega}^i &= \underset{2}{\omega}^j \wedge \omega_j^i - \epsilon_{jkl}a^{il}\underset{2}{\omega}^j \wedge \underset{2}{\omega}^k, \end{aligned} \tag{4.37}$$

$$d\omega_j^i = \omega_j^k \wedge \omega_k^i + \epsilon_{klm}b_j^{im}\underset{1}{\omega}^k \wedge \underset{2}{\omega}^l. \tag{4.38}$$

The tensors a^{ij} and b_k^{ij} in equations (4.37) and (4.38) satisfy the following differential equations:

$$\nabla a^{ij} = a^{ij}\omega_p^p + \frac{1}{2}(b_m^{ij} - b_p^{ip}\delta_m^j)(\underset{1}{\omega}^m - \underset{2}{\omega}^m), \tag{4.39}$$

$$\nabla b_k^{ij} = b_k^{ij}\omega_p^p + b_k^{il}\epsilon_{pql}(2a^{pj}(\underset{1}{\omega}^q - \underset{2}{\omega}^q) - a^{pq}(\underset{1}{\omega}^j - \underset{2}{\omega}^j)), \tag{4.40}$$

and are related by the equations:

$$\begin{aligned} b_k^{ik} &= 2\epsilon_{jkl}a^{il}a^{jk}, \\ b_p^{ij}a^{pk} - b_p^{jk}a^{ip} - b_p^{ik}a^{pj} &= b_p^{ip}a^{jk} - b_p^{pk}a^{ij} \end{aligned} \tag{4.41}$$

which are obtained from equations (4.14), (4.7), (4.10), (1.31), (4.11) and (4.36). Note that when one derives equations (4.39) and (4.40), he should use the relations

$$\nabla\epsilon_{ijk} = -\epsilon_{ijk}\omega_l^l, \quad \nabla\epsilon^{ijk} = \epsilon^{ijk}\omega_l^l,$$

which can be immediately verified.

The tensors a^{ij} and b_k^{ij} are connected by an additional relation which can be obtained from (4.17) by means of (4.36). However, this relation is rather complicated and, in addition, we will not need it. Thus, we will not write it. But the readers can find it independently.

Hence, we can reduce the valency of the tensors a_{jk}^i and b_{jkl}^i by one, replacing them by the tensors a^{ij} and b_k^{ij}. The latter tensors are also called the *torsion and curvature tensors.* This allows us to classify six-dimensional Bol three-webs using the following lemma.

Lemma 4.14 *On a six-dimensional Bol three-web, it is possible to find a family of adapted frames in which the components of the tensor a^{ij} are constants.*

Proof. In fact, consider equations (4.39). Their number is nine, and they contain the same number of the forms ω_j^i. Setting $da^{ij} = 0$, we get nine equations for nine forms ω_j^i. If these equations are independent, then the forms ω_j^i can be uniquely found from them, and there is a unique field of frames satisfying the lemma's conclusion. If the rank ρ of the system of these nine equations in the forms ω_j^i is less than nine, then, at any point, there exists a $(9-\rho)$-parameter family of such frames. ■

It follows from Lemma 4.13 that classification of hexagonal Bol three-webs is connected with classification of (0,2)-tensors. Different classes of three-webs correspond to different (i.e. non-equivalent) tensors. First of all, the cases,when the matrix (a^{ij}) is symmetric, skew-symmetric or an arbitrary matrix without the symmetry or the skew-symmetry, should be considered separately. Next, each of these classes can be subdivided according to the rank of the matrix.

If the tensor a^{ij} is given, then the components of the tensor b_k^{ij} can be found from relations (4.41). Since the number of these equations is greater than the number of components of the tensor b_k^{ij} (there are 30 equations and 27 components), then the Bol web can be found not for every tensor a^{ij}.

Suppose that, for some tensor a^{ij}, the rank of the system (4.41) is equal to $27 - \rho$. Then ρ components of the tensor b_k^{ij} can be chosen arbitrarily, and the other components can be found from the system. Let the family of frames, in which the quantities a^{ij} are constants, depend on σ parameters. If $\sigma > 0$, we take a subfamily of frames in which the maximal number of components b_k^{ij} are constants. The non-constant components of this tensor are absolute invariants. Using them, we can find a final classification of Bol webs under consideration.

We will not give the complete classification of six-dimensional Bol three-webs, restricting ourselves to only most importand cases.

2. First, we will prove the following theorem:

Theorem 4.15 *A six-dimensional Bol three-web is isoclinic (and therefore algebraizable) if and only if the tensor a^{ij} is skew-symmetric.*

Proof. In fact, if the tensor a^{ij} is skew-symmetric, then it can be written as follows:

$$a^{ij} = \epsilon^{ijk} a_k,$$

where the quantities a_k can be found by contraction of the last equation with the tensor ϵ_{ijl}. Then, by (4.36) and (4.35), the torsion tensor takes the form:

$$a^i_{jk} = \epsilon_{pjk} a^{pj} = \epsilon_{pjk} \epsilon^{pil} a_l = 2\delta^i_{[j} \delta^l_{k]} a_l = 2\delta^i_{[j} a_{k]}.$$

Since we consider the case $r = 3 > 2$, it follows from Theorem 3.8 that the web under consideration is isoclinic.

Conversely, suppose that a six-dimensional Bol three-web is isoclinic. Then, by the same Theorem 3.8, its torsion tensor can be written in the form:

$$a^i_{jk} = a_{[j} \delta^i_{k]}.$$

Comparing this expression with (4.36), we find

$$\epsilon_{jkp} a^{ip} = a_{[j} \delta^i_{k]}.$$

Contracting the last equation with ϵ^{jkq} and applying relations (4.35), we find that $a^{ij} = \frac{1}{2} a_k \epsilon^{kij}$, i.e. the tensor a^{ij} is skew-symmetric. ■

Theorem 4.16 *If the tensor a^{ij} of a six-dimensional Bol three-web is symmetric and its rank is equal to zero or three, then this web is a group three-web. Moreover, in the former case, the web is parallelizable.*

Proof. In fact, if $a^{ij} = 0$, then the torsion tensor a^i_{jk} is also equal to zero. It follows from this that the curvature tensor is zero (Problem **1**). Therefore, the web under consideration is parallelizable.

Suppose now that the tensor a^{ij} is symmetric and its rank is three. Then, by the skew-symmetry of the tensor ϵ_{ijk}, relations (4.41, 1) become

$$b_k^{ik} = 0. \tag{4.42}$$

In addition, if we alternate relations (4.39) and apply (4.42), we get

$$b_k^{[ij]} = 0. \tag{4.43}$$

By means of (4.42) and (4.43), equations (4.41, 2) are transferred into the following equations:

$$b_p^{ij} a^{pk} - b_p^{jk} a^{ip} - b_p^{ik} a^{pj} = 0.$$

Symmetrizing them in the indices j and k and applying (4.43), we obtain

$$b_p^{jk} a^{ip} = 0. \tag{4.44}$$

By equations (4.44), equations (4.41, 2) are identically satisfied.

Since det $(a^{ip}) \neq 0$, equations (4.44) imply that $b_p^{ik} = 0$. Hence, the curvature tensor of the web under consideration vanishes. According to Theorem 4.1, the web is a group web. ■

3. Thus, non-trivial classes of six-dimensional Bol three-webs with a symmetric tensor a^{ij} can arise only in the case when the rank of this tensor is one or two.

Theorem 4.17 *Up to an isotopy, there exists only one six-dimensional Bol three-web with a symmetric tensor a^{ij} of rank one. In local coordinates, the equations of this web can be written in the following form:*

$$\begin{aligned} z^1 &= x^1 + y^1 - (x^2 + y^2)x^3 y^3, \\ z^2 &= x^2 + y^2, \\ z^3 &= x^3 + y^3, \end{aligned} \tag{4.45}$$

i.e. this web is the web E_1 introduced in Problems 8 *and* 9 *of Chapter* 2.

Proof. Let us take a family of frames in the six-dimensional manifold in such a way that, at each point, the matrix (a^{ij}) has the following components: $a^{11} = 1$ and $a^{ij} = 0$ for other values of the indices. Substitute these values into relations (4.42)–(4.44) connecting the fundamental tensors of the Bol web with a symmetric tensor a^{ij}. As a result, we find that

$$b_k^{ik} = 0, \quad b_k^{ij} = b_k^{ji}, \quad b_1^{ij} = 0. \tag{4.46}$$

Next, since we now have $\epsilon_{jkl} a^{jk} = \epsilon_{11l} a^{11} = 0$, equations (4.10) take the form:

$$\nabla b_k^{ij} = b_k^{ij} \omega_p^p + 2 b_k^{il} \epsilon_{1ql} a^{1j} (\underset{1}{\omega}^q - \underset{2}{\omega}^q). \tag{4.47}$$

Alternating equations (4.47) in the indices i and j and applying equations (4.46, 1), we arrive at the equations:

$$b_k^{il} \epsilon_{1ql} a^{1j} = b_k^{jl} \epsilon_{1ql} a^{1i}.$$

Setting $i = 1$ and $j = 2, 3$ in these equations, we get $b_k^{jl} = 0$, where $l = 2, 3$. As a result, the components of the tensor b_k^{ij} can be written in the form:

$$(b_1^{ij}) = \begin{pmatrix} 0 & 0 & 0 \\ 0 & 0 & 0 \\ 0 & 0 & 0 \end{pmatrix}, (b_2^{ij}) = \begin{pmatrix} b_2^{11} & b_2^{12} & b_2^{13} \\ b_2^{12} & 0 & 0 \\ b_2^{13} & 0 & 0 \end{pmatrix}, (b_3^{ij}) = \begin{pmatrix} b_3^{11} & b_3^{12} & -b_2^{12} \\ b_3^{12} & 0 & 0 \\ -b_2^{12} & 0 & 0 \end{pmatrix}.$$

Substituting the values of the tensors a^{ij} and b_k^{ij}, which we just obtained, into equations (4.39) and (4.47), we arrive at the equations:

$$\omega_1^k = \frac{1}{2} b_l^{1k} (\underset{1}{\omega}^l - \underset{2}{\omega}^l), \quad b_k^{1m} \omega_1^l = 0, \quad k, l = 2, 3,$$

which, by the independence of the basis forms $\underset{1}{\omega}^l - \underset{2}{\omega}^l$, imply that $\omega_1^k = 0$ and $b_l^{1k} = 0$. Thus, the tensor b_k^{ij} has only two non-zero components. For these components, the corresponding equations (4.47) take the form:

$$\begin{aligned} db_2^{11} + b_2^{11}(\omega_1^1 - 2\omega_2^2 - \omega_3^3) - b_3^{11}\omega_2^3 = 0, \\ db_3^{11} + b_3^{11}(\omega_1^1 - \omega_2^2 - 2\omega_3^3) - b_2^{11}\omega_3^2 = 0. \end{aligned} \tag{4.48}$$

In addition, from equations (4.37) we find:

$$d\omega_2^3 = \omega_2^2 \wedge \omega_2^3, \quad d\omega_3^2 = \omega_3^2 \wedge \omega_2^2.$$

These equations show that on the web under consideration, it is possible to choose a subfamily of frames in which $b_2^{11} = 0$ and $b_3^{11} = 1$. As a result, equations (4.48) lead to the equations

$$\omega_2^3 = 0, \quad \omega_1^1 - \omega_2^2 - 2\omega_3^3 = 0.$$

The last equations and equation (4.39) which can be written now in the form:

$$\omega_1^1 - \omega_2^2 - \omega_3^3 = \frac{1}{2}(\underset{1}{\omega}^3 - \underset{2}{\omega}^3),$$

give

$$\omega_2^3 = 0, \quad \omega_1^1 - \omega_2^2 = \underset{1}{\omega}^3 - \underset{2}{\omega}^3, \quad \omega_3^3 = \frac{1}{2}(\underset{1}{\omega}^3 - \underset{2}{\omega}^3).$$

The equations, which we have obtained, can be further simplified. Since equations (4.37) imply $d\omega_1^1 = 0$, the form ω_1^1 can be reduced to zero by reducing the family of frames. Finally, we obtain the following system of linear equations in the forms ω_j^i:

$$\omega_1^1 = \omega_1^2 = \omega_1^3 = 0, \quad \omega_2^3 = 0, \quad \omega_2^2 = -\underset{1}{\omega}^3 + \underset{2}{\omega}^3, \quad \omega_3^3 = \frac{1}{2}(\underset{1}{\omega}^3 - \underset{2}{\omega}^3).$$

In this case, the structure equations (4.37) and (4.38) have the form:

$$
\begin{aligned}
d\underset{1}{\omega}^1 &= \underset{1}{\omega}^2 \wedge \omega_2^1 + \underset{1}{\omega}^3 \wedge \omega_3^1 + 2\underset{1}{\omega}^2 \wedge \underset{1}{\omega}^3, \\
d\underset{1}{\omega}^2 &= \underset{1}{\omega}^3 \wedge \omega_3^2 - \underset{1}{\omega}^2 \wedge \Omega, \\
d\underset{1}{\omega}^3 &= \frac{1}{2}\underset{1}{\omega}^3 \wedge \Omega, \\
d\underset{2}{\omega}^1 &= \underset{2}{\omega}^2 \wedge \omega_2^1 + \underset{2}{\omega}^3 \wedge \omega_3^1 - 2\underset{2}{\omega}^2 \wedge \underset{2}{\omega}^3, \\
d\underset{2}{\omega}^2 &= \underset{2}{\omega}^3 \wedge \omega_3^2 - \underset{2}{\omega}^2 \wedge \Omega, \\
d\underset{2}{\omega}^3 &= \frac{1}{2}\underset{2}{\omega}^3 \wedge \Omega, \\
d\omega_2^1 &= \omega_2^1 \wedge \Omega, \\
d\omega_3^2 &= \tfrac{3}{2}\Omega \wedge \omega_3^2, \\
d\omega_3^1 &= \omega_3^2 \wedge \omega_2^1 + \tfrac{1}{2}\Omega \wedge \omega_3^1 + \underset{1}{\omega}^2 \wedge \underset{2}{\omega}^3 - \underset{1}{\omega}^3 \wedge \underset{2}{\omega}^2,
\end{aligned}
\tag{4.49}
$$

where we denote $\Omega = \underset{1}{\omega}^3 - \underset{2}{\omega}^3$.

It follows from equations (4.49) that $d\Omega = 0$, i.e. the form Ω is a total differential. Set $\Omega = d\ln\varphi$. Successively integrating system (4.49), we get:

$$
\begin{aligned}
\omega_2^1 &= \varphi^{-1}du, \quad \omega_3^2 = \varphi^{\frac{3}{2}}dv, \\
\underset{1}{\omega}^3 &= \varphi^{-\frac{1}{2}}da^3, \quad \underset{2}{\omega}^3 = \varphi^{-\frac{1}{2}}db^3,
\end{aligned}
$$

where u, v, a^3 and b^3 are parameters and $a^3 - b^3 = 2\varphi^{\frac{1}{2}}$. Substituting these forms in other equations of system (4.49) and continuing to integrate, we find:

$$
\begin{aligned}
\underset{1}{\omega}^2 &= \varphi(-v\,da^3 + da^2), \\
\underset{2}{\omega}^2 &= \varphi(-v\,db^3 + db^2), \\
\omega_3^1 &= \varphi^{\frac{1}{2}}(a^2db^3 + b^2da^3 + v\,du + dw), \\
\underset{1}{\omega}^1 &= -u\,da^2 - (w + a^2b^3)da^3 + a^2a^3da^3 + da^1, \\
\underset{2}{\omega}^1 &= -u\,db^2 - (w + b^2a^3)db^3 + b^2b^3db^3 + db^1,
\end{aligned}
$$

where w, a^2, b^2, a^1 and b^1 are parameters.

The first foliation of the web under consideration is defined by the system $\underset{1}{\omega}^i = 0$, whose first integrals can be written in the form $a^i = x^i$. Similarly, we find equations of the second foliation: $b^i = y^i$. The third foliation is defined by the system $\underset{1}{\omega}^i + \underset{2}{\omega}^i = 0$. Integration of this system leads to:

$$
a^3 + b^3 = z^3, \quad a^2 + b^2 = z^2, \quad a^1 + b^1 - z^2a^3b^3 = z^1.
$$

Eliminating the variables a^i and b^i from the equations of three foliations, we arrive at equations (4.45). ■

Using the same method, we can prove another theorem:

Theorem 4.18 *Closed form equations of a six-dimensional Bol three-web with a symmetric tensor a^{ij} of rank two can be written in the form:*

$$\begin{aligned} z^1 &= x^1 e^{2\epsilon(x^3+y^3)} + y^1 + 2\epsilon y^3(x^2 + y^2 e^{2\epsilon(x^3+y^3)}), \\ z^2 &= x^2 + y^2 e^{2\epsilon(x^3+y^3)}, \\ z^3 &= x^3 + y^3, \quad \epsilon = \pm 1. \blacksquare \end{aligned} \tag{4.50}$$

4. Classification of six-dimensional Bol three-web with a non-symmetric tensor a^{ij} is rather tedious, and we will not give it in this book. We refer the readers who would like to see the details, to the papers of V. I. Fedorova. We consider here only two webs of this kind: one in this section (with a matrix (a^{ij}) of rank three) and another one in Chapter 7 (with a matrix (a^{ij}) of rank one).

Let a matrix (a^{ij}) be of the form:

$$(a^{ij}) = \begin{pmatrix} -1 & 0 & 0 \\ 0 & 0 & 1 \\ 0 & -1 & 1 \end{pmatrix}. \tag{4.51}$$

First, denote the corresponding Bol web by W_6 and then study its structure. We note that since the tensor a^{ij} is not skew-symmetric, by Theorem 4.14, the web W_6 is not isoclinic. Let us find the components of its curvature tensor. Substituting values of a^{ij} from (4.51) into relations (4.41), we find:

$$(b_1^{ij}) = \begin{pmatrix} 0 & 0 & 0 \\ 0 & 0 & 0 \\ 0 & 0 & -4 \end{pmatrix}, \quad (b_2^{ij}) = \begin{pmatrix} 0 & 0 & p \\ -4 & 0 & 0 \\ -p & 0 & 0 \end{pmatrix}, \quad (b_3^{ij}) = \begin{pmatrix} 0 & 0 & -4 \\ 0 & 0 & 0 \\ 0 & 0 & 0 \end{pmatrix},$$

where p is an arbitrary function. Substituting values of a^{ij} and b_k^{ij} into differential equations (4.39) and (4.40), we arrive at the following system:

$$\begin{array}{lll} \omega_1^1 = 0, & \omega_2^1 = p(\underset{1}{\omega}^2 - \underset{2}{\omega}^2), & \omega_3^1 = -2(\underset{1}{\omega}^2 - \underset{2}{\omega}^2), \\ \omega_1^2 = 0, & \omega_2^2 = 2(\underset{1}{\omega}^1 - \underset{2}{\omega}^1), & \omega_3^2 = 0, \\ \omega_1^3 = -2(\underset{1}{\omega}^2 - \underset{2}{\omega}^2), & \omega_2^3 = -\frac{1}{4}dp + \frac{1}{2}p(\underset{1}{\omega}^1 - \underset{2}{\omega}^1), & \omega_3^3 = 0. \end{array}$$

The family of adapted frames of the web W_6 can be reduced by setting $p = 0$. Then, the system written above can be reduced to the following simpler system:

$$\begin{array}{lll} \omega_1^1 = 0, & \omega_2^1 = 0, & \omega_3^1 = -2(\underset{1}{\omega}^2 - \underset{2}{\omega}^2), \\ \omega_1^2 = 0, & \omega_2^2 = 2(\underset{1}{\omega}^1 - \underset{2}{\omega}^1), & \omega_3^2 = 0, \\ \omega_1^3 = -2(\underset{1}{\omega}^2 - \underset{2}{\omega}^2), & \omega_2^3 = 0, & \omega_3^3 = 0. \end{array}$$

Substituting these values of the forms ω_j^i into the structure equations (4.37) and (4.38), we establish that the latter are identically satisfied, and the former take the form:

$$
\begin{aligned}
&d\underset{1}{\omega}^1 = 2\underset{1}{\omega}^3 \wedge \underset{2}{\omega}^2, && d\underset{2}{\omega}^1 = -2\underset{2}{\omega}^3 \wedge \underset{1}{\omega}^3,\\
&d\underset{1}{\omega}^2 = -2\underset{1}{\omega}^2 \wedge \underset{2}{\omega}^1, && d\underset{2}{\omega}^2 = 2\underset{2}{\omega}^2 \wedge \underset{1}{\omega}^1,\\
&d\underset{1}{\omega}^3 = 2\underset{1}{\omega}^1 \wedge (\underset{2}{\omega}^2 + \underset{1}{\omega}^3), && d\underset{2}{\omega}^3 = -2\underset{2}{\omega}^1 \wedge (\underset{1}{\omega}^2 + \underset{2}{\omega}^3),
\end{aligned}
\tag{4.52}
$$

In addition to the basis forms of the three-web, system (4.52) contains only constants, and its exterior differentiation leads to the identities. Thus, this system represents the structure equations of a certain six-dimensional Lie group G. Therefore, the web under consideration can be realized on a group manifold. This web belongs to a wider class of webs, defined in homogeneous spaces, and is called G-webs. The G-webs will be studied in Chapter 6.

Let us construct a mapping of the group G into the group of transformations of a three-dimensional projective space P^3. Let $\{A_u\}, u, v, w = 0, 1, 2, 3$, be a moving frame in P^3. The infinitesimal displacements of this moving frame can be written in their usual form:

$$
dA_u = \sigma_u^v A_v. \tag{4.53}
$$

The Pfaffian forms σ_v^u satisfy the structure equations of a projective space:

$$
d\sigma_u^v = \sigma_u^w \wedge \sigma_w^v. \tag{4.54}
$$

Consider a subgroup of the group of projective transformations that leave invariant a ruled quadric Q. Denote this subgroup by $O_{2,2}$. It is the group of motions of the non-Euclidean space ${}^2S^3$ with the absolute Q. This subgroup is called the *pseudo-orthogonal group.* We write the equations of the quadric Q in the form:

$$
(x, x) \equiv g_{uv} x^u x^v = 0,
$$

and take in P^3 a subfamily of frames $\{A_u\}$ for which the straight lines A_0A_1, A_0A_2, A_3A_1 and A_3A_2 are rectilinear generators of the quadric Q. In addition, we normalize the vertices of frames by the conditions: $(A_0A_3) = (A_1A_2) = 1$. Then, the equation of the quadric Q becomes

$$
x^0x^3 + x^1x^2 = 0,
$$

and the points A_u are also connected by the following relations:

$$
(A_0A_0) = (A_1A_1) = (A_2A_2) = (A_3A_3) = 0, \;\; (A_0A_1) = (A_0A_2) = (A_3A_1) = (A_3A_2) = 0.
$$

Differentiating the last equations and applying equations (4.53), we obtain the relations connecting the forms σ_v^u:

$$
\sigma_u^v + \sigma_{3-v}^{3-u} = 0. \tag{4.55}
$$

Thus, the group $\mathbf{O}_{2,2}$ depends on six parameters, and the invariant forms of this group – the forms $\sigma_0^0, \sigma_1^1, \sigma_0^1, \sigma_1^0, \sigma_0^2$ and σ_2^0 are linear combinations of the differentials of these six parameters.

The structure equations of the group $\mathbf{O}_{2,2}$ (or the space ${}^2S^3$) can be obtained from the structure equations (4.54) by applying relations (4.55):

$$\begin{aligned} d\sigma_0^0 &= \sigma_0^1 \wedge \sigma_1^0 + \sigma_0^2 \wedge \sigma_2^0, & d\sigma_1^1 &= \sigma_1^0 \wedge \sigma_0^1 + \sigma_0^2 \wedge \sigma_2^0, \\ d\sigma_0^1 &= (\sigma_0^0 - \sigma_1^1) \wedge \sigma_0^1, & d\sigma_1^0 &= \sigma_1^0 \wedge (\sigma_0^0 - \sigma_1^1), \\ d\sigma_0^2 &= (\sigma_0^0 + \sigma_1^1) \wedge \sigma_0^2, & d\sigma_2^0 &= \sigma_2^0 \wedge (\sigma_0^0 + \sigma_1^1). \end{aligned} \tag{4.56}$$

One can directly check that equations (4.52) become equations (4.56) if we set

$$\begin{aligned} \underset{1}{\omega}^1 &= \frac{1}{2}(\sigma_1^1 - \sigma_0^0), & \underset{1}{\omega}^2 &= \sigma_0^2, & \underset{1}{\omega}^3 &= \frac{1}{2}(\sigma_1^0 - \sigma_0^1), \\ \underset{2}{\omega}^1 &= \frac{1}{2}(\sigma_0^0 + \sigma_1^1), & \underset{2}{\omega}^2 &= \sigma_0^1, & \underset{2}{\omega}^3 &= \frac{1}{2}(\sigma_2^0 - \sigma_0^2). \end{aligned} \tag{4.57}$$

In equations (4.57), the invariant forms of the group G are expressed in terms of the invariant forms of the group $\mathbf{O}_{2,2}$ with constant coefficients. Therefore, they give an isomorphism of the groups G and $\mathbf{O}_{2,2}$. Hence, we obtain that the group G, in which the web W_6 is realized, is isomorphic to the group of motions on the non-Euclidean space ${}^2S^3$. As a result, the web W_6 can be interpreted in terms of this space.

The equations

$$\underset{1}{\omega}^i = \underset{2}{\omega}^i = 0, \quad i = 1, 2, 3,$$

fix a point p in the group G. In the group $O_{2,2}$, the corresponding equations have the form:

$$\sigma_0^0 = \sigma_1^1 = \sigma_0^1 = \sigma_1^0 = \sigma_0^2 = \sigma_2^0 = 0. \tag{4.58}$$

Let us take three points M, N and L on the absolute Q in such a way that no pair of these points does not belong to one generator of Q and choose a frame $\{A_u\}$ in which $A_0 = M, A_1 + A_3 = N$ and $A_2 + A_3 = L$. Then, by (4.55) and (4.58), the points M, N and L are fixed, and conversely, if they are fixed and relation (4.55) holds, we have equations (4.58). Thus, the triple of points, M, N and L, on the quadric Q bijectively corresponds to the point p of the group G.

Let us find the images of leaves of the web W_6 under the mapping which we are considering. The leaf $\mathcal{F}_1$ of the first foliation, passing through the point p, is defined by the system $\underset{1}{\omega}^i = 0$. In the group $O_{2,2}$, this system becomes

$$\sigma_0^0 - \sigma_1^1 = 0, \quad \sigma_0^2 = 0, \quad \sigma_0^1 - \sigma_1^0 = 0. \tag{4.59}$$

In addition, from equations (4.54) we obtain

$$dA_0 = \sigma_0^0 A_0 + \sigma_0^1 A_1, \quad d(A_0 + A_1) = (\sigma_0^0 + \sigma_0^1)(A_0 + A_1),$$
$$dA_1 = \sigma_0^1 A_0 + \sigma_0^0 A_1, \quad d(A_0 - A_1) = (\sigma_0^0 - \sigma_0^1)(A_0 - A_1),$$

i.e. to the leaf $\mathcal{F}_1$, in the space P^3 there corresponds the fixed generator A_0A_1 of the absolute Q, and two points of this generator, $A_0 + A_1$ and $A_0 - A_1$ are also fixed.

In the same manner, we can prove that to the leaf $\mathcal{F}_2$ of the second foliation of the web W_6, passing through the point p, in the space P^3 there corresponds the fixed generator A_0A_2 of the absolute Q, and two points of this generator, $A_0 + A_2$ and $A_0 - A_2$ are also fixed.

The leaf $\mathcal{F}_3$ of the third foliation of the web is defined by the system $\underset{1}{\omega}^i + \underset{2}{\omega}^i = 0$, and on the group $O_{2,2}$, by the system:

$$\sigma_1^1 = 0, \quad \sigma_0^1 + \sigma_0^2 = 0, \quad \sigma_1^0 + \sigma_2^0 = 0.$$

In this case, equations (4.53) give:

$$dA_0 = \sigma_0^0 A_0 + \sigma_0^1 (A_1 - A_2),$$
$$d(A_1 - A_2) = 2\sigma_1^0 A_0 + 2\sigma_0^1 A_3,$$
$$dA_3 = -\sigma_0^0 A_3 + \sigma_1^0 (A_1 - A_2).$$

We can see from these equations that the plane $\pi = [A_0,\ A_1 - A_2,\ A_3]$ remains fixed under displacements of the moving frame. Thus, this plane corresponds to the leaf $\mathcal{F}_3$. This plane can be also defined by the points A_0,

$$T = A_0 + (A_1 - A_2) + A_3 = (A_0 + A_1) - (A_2 - A_3)$$

and

$$S = A_0 - (A_1 - A_2) + A_3 = (A_0 - A_1) + (A_2 + A_3).$$

The images of the point p and three leaves $\mathcal{F}_\alpha$, passing through this point, lead to the configuration consisting of eight generators of the quadric Q. This configuration is represented in Figure 4.4.

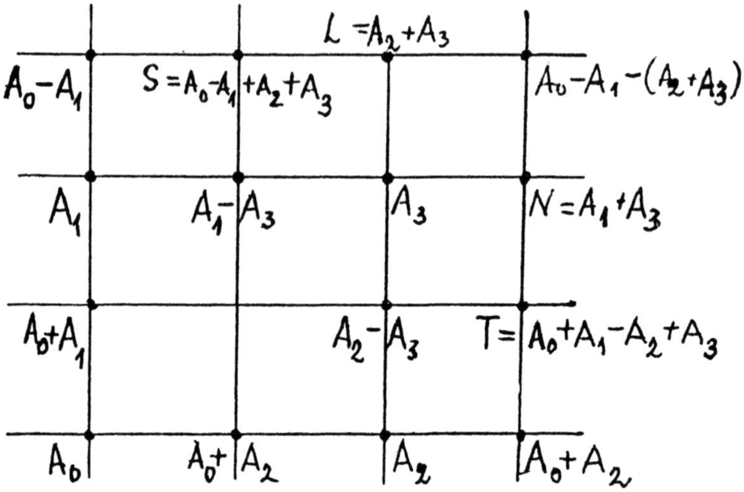

Figure 4.4

As follows from Figure 4.4, this configuration is completely defined if three points $A_0 = M, L$ and N are given. In fact, if these points are given, then six generators of the surface Q, passing through these points, will be known. Two other generators can be found as the fourth line harmonics to two triples of generators which are already determined. Hence, if the points M, L and N are given, then one can uniquely reconstruct the geometric images, corresponding to the leaves $\mathcal{F}_\alpha$ of the web W_6, passing through the point p. This corresponds to the fact that through a point of a manifold carrying a three-web, there passes one and only one leaf of each foliation of the web.

Let us find a configuration in the quadric Q corresponding to the middle Bol figure presented in Figure 4.5.

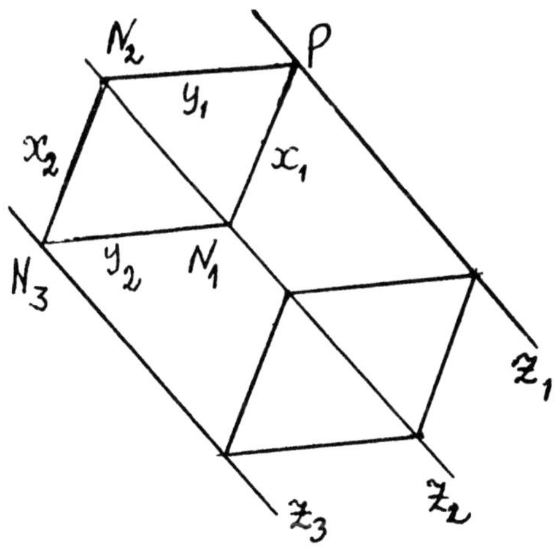

Figure 4.5

To the leaves z_1 and z_2 of the third foliation, in our interpretation there correspond two planes. These planes intersect the quadric Q in second degree curves.

Denote these curves by z_1 and z_2 (Figure 4.6).

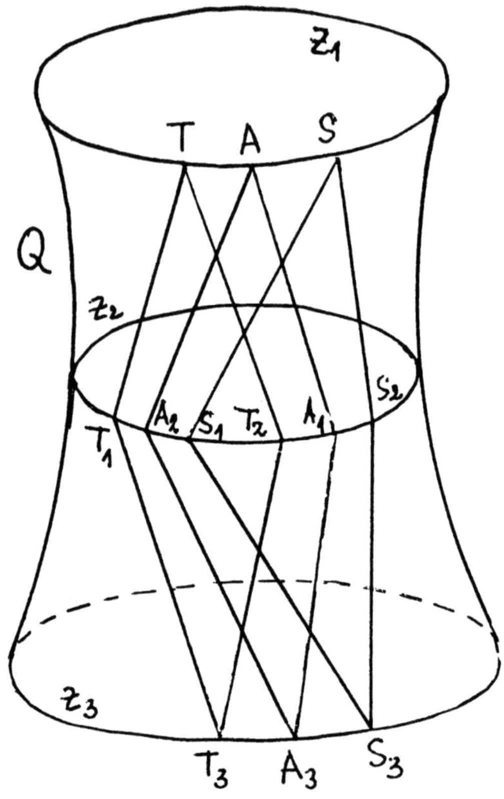

Figure 4.6

In Figure 4.6 the triple of points (A, S, T) corresponds to the point p lying in the leaf z_1. Two series of generators, AA_1, SS_2, TT_2 and AA_2, SS_1, TT_1 pass through these three points. Moreover, the triple of generators AA_1, SS_1, TT_1 corresponds to the leaf x_1 of the first foliation, and the triple of generators AA_2, SS_2, TT_2 corresponds to the leaf y_1 of the first foliation. The triple of points (A_1, S_1, T_1) represents the intersection point N_1 of the leaves z_2 and x_1, and the triple of points (A_2, S_2, T_2) represents the point $N_2 = z_2 \bigcap y_1$. Finally, the conic on the quadric Q, defined by the points A_3, S_3 and T_3, corresponds to the leaf z_3.

Thus, to the quadrangle, formed by the leaves x_1, y_1, x_2 and y_2 of the web W_6, there corresponds a configuration on the surface Q formed by three conics and three quadrangles, composed of generators.

The closure condition B_m is that the quadrangle x_1, y_1, x_2, y_2 can be moved in such a way that its vertices will be sliding along the leaves z_1, z_2 and z_3. In the interpretation we have constructed, this has the following meaning: the location of the curve z_3 in Figure 4.6 does not depend on the choice of the original triple of points A, S and T on curve z_1. This property of the ruled quadric Q is connected with the fact that its plane sections z_1 and z_3 correspond one to another in the homology H, defined by the plane π containing the curve z_2 and the pole P of this plane with respect to the quadric Q. This geometrically proves the closure of the Bol figures B_m on the web W_6.

5. In the paper [C 82] Chern formulates the Ph. Griffiths conjecture that every hexagonal three-web is algebraizable. The results of this section show that, in general, this conjecture cannot be confirmed. In fact, we proved the existence of the Bol webs whose tensor a^{ij} is not skew-symmetric. However, by Theorem 4.15, such webs are not algebraizable. Since every Bol web is hexagonal, this proves the existence of non-algebraizable hexagonal Bol webs.

On the other hand, in Section 5.2 we will show that every four-dimensional hexagonal three-web is algebraizable. This implies that the Griffiths conjecture is confirmed for four-dimensional three-webs and is not confirmed for webs of dimension higher than four.

4.4 The Moufang Three-Webs

1. We remind the reader that the Moufang three-web has been defined as a three-web, in the coordinate loops of which the Moufang identity:

$$(uv)(wu) = (u(vw))u, \tag{4.60}$$

or one of the identities, equivalent to identity (4.60) and indicated in Problem 7 of Chapter 2, holds. The curvature tensor of the Moufang three-web satisfies the relation $b = \operatorname{Alt} b$ given in the Table 2.2 (p. 83). Comparing this relation with equation (1.31) connecting the torsion and curvature tensors of an arbitrary three-web, we find that the curvature tensor of the Moufang three-web can be expressed in terms of its torsion tensor as follows:

$$b^i_{jkl} = -2a^m_{[jk}a^i_{l]m}. \tag{4.61}$$

Since the class of Moufang webs coincides with the class of webs for which all three closure conditions, B_l, B_r and B_m, hold, it follows from Theorems 4.4 and 4.5 that *relations* (4.61) *are not only necessary but also sufficient for a three web W to be a Moufang web.*

By (4.61), equations (1.33) take the form:

$$\nabla a^i_{jk} = 2a^m_{[jk}a^i_{l]m}(\underset{1}{\omega}^l - \underset{2}{\omega}^l) \tag{4.62}$$

i.e. the covariant derivatives of the torsion tensor can be expressed in terms of this tensor alone. It follows from equations (4.61) and (4.62) that *the covariant derivatives of any order of both the torsion tensor and the curvature tensor of a Moufang web, can be expressed in terms of the torsion tensor alone.* We will return to this fact in Chapter 5.

Let us find the structure equations of a Moufang three-web. Since this web is simultaneously a web B_l, B_r and B_m, then, for its description, we will use the connection $\overset{*}{\Gamma}$ – the middle connection of the connections $\widetilde{\Gamma}_{12}, \widetilde{\Gamma}_{31}$ and $\widetilde{\Gamma}_{23}$ (see Section 1.8). This connection can be defined by the forms:

$$\overset{*}{\omega}{}^i_j = \omega^i_j + \frac{2}{3} a^i_{jk} (\underset{1}{\omega}{}^k - \underset{2}{\omega}{}^k) \tag{4.63}$$

which satisfy the structure equations (1.91). The tensor

$$\overset{*}{b}{}^i_{jkl} = b^i_{jkl} + 2a^m_{[jk} a^i_{l]m},$$

in equations (1.91) vanishes by (4.61). As a result, the structure equations of a Moufang three-web take the form:

$$\begin{aligned} d\underset{1}{\omega}{}^i &= \underset{1}{\omega}{}^j \wedge \overset{*}{\omega}{}^i_j + \frac{1}{3} a^i_{jk} \underset{1}{\omega}{}^j \wedge (\underset{2}{\omega}{}^k - \underset{3}{\omega}{}^k), \\ d\underset{2}{\omega}{}^i &= \underset{2}{\omega}{}^j \wedge \overset{*}{\omega}{}^i_j + \frac{1}{3} a^i_{jk} \underset{2}{\omega}{}^j \wedge (\underset{3}{\omega}{}^k - \underset{1}{\omega}{}^k), \\ d\underset{3}{\omega}{}^i &= \underset{3}{\omega}{}^j \wedge \overset{*}{\omega}{}^i_j + \frac{1}{3} a^i_{jk} \underset{3}{\omega}{}^j \wedge (\underset{1}{\omega}{}^k - \underset{2}{\omega}{}^k), \end{aligned} \tag{4.64}$$

$$d\overset{*}{\omega}{}^i_j - \overset{*}{\omega}{}^k_j \wedge \overset{*}{\omega}{}^i_k = -\frac{1}{3} A^i_{jkl} (\underset{1}{\omega}{}^k \wedge \underset{2}{\omega}{}^l + \underset{2}{\omega}{}^k \wedge \underset{3}{\omega}{}^l + \underset{3}{\omega}{}^k \wedge \underset{1}{\omega}{}^l) \tag{4.65}$$

where

$$A^i_{jkl} = -A^i_{jlk} = \frac{2}{3} (a^m_{jk} a^i_{ml} + a^m_{lj} a^i_{mk} - a^m_{kl} a^i_{mj}) \tag{4.66}$$

is the curvature tensor of the connection $\overset{*}{\Gamma}$. We add equation (4.62) to these structure equations. If in (4.62) we substitute $\overset{*}{\omega}{}^i_j$, defined by (4.63), for ω^i_j, then equations (4.62) become

$$\overset{*}{\nabla} a^i_{jk} = 0. \tag{4.67}$$

From this and equations (4.61) and (4.66) we derive the equations $\overset{*}{\nabla} b^i_{jkl} = 0$ and

$$\overset{*}{\nabla} A^i_{jkl} = 0. \tag{4.68}$$

We recall that a manifold X with an affine connection Γ is called a *reductive space* if the torsion and curvature tensors of the connection Γ are covariantly constant. Thus, we have shown that *a manifold X carrying a Moufang three-web is a reductive space, and the middle connection $\overset{*}{\Gamma}$ is its canonical connection.*

Exterior differentiation of equations (4.67) and (4.68) leads to the relations:

$$a^i_{pk}A^p_{jlm} + a^i_{jp}A^p_{klm} - a^p_{jk}A^i_{plm} = 0 \tag{4.69}$$

and

$$A^i_{mjk}A^m_{lpq} + A^i_{lmk}A^m_{jpq} + A^i_{ljm}A^m_{kpq} - A^m_{ljk}A^i_{mpq} = 0, \tag{4.70}$$

whose meaning is clarified by the following theorem:

Theorem 4.19 *Local W-algebras, defined on a manifold X by a Moufang three-web $W = (X, \lambda_\alpha)$, are Mal'tsev algebras.*

Proof. According to the definition, given in Section 2.5, a local W-algebra of a three-web W has two operations, binary and ternary: the latter is defined by the torsion tensor and the former is defined by the curvature tensor. Since the torsion tensor of a Moufang web can be expressed in terms of its torsion tensor, then in a W-algebra of a Moufang web, there is only one independent operation:

$$[\xi, \eta]^i = -2a^i_{jk}\xi^j\eta^k. \tag{4.71}$$

We recall that a *Mal'tsev algebra* is an anti-commutative algebra in which the identity

$$[[\xi, \eta], [\xi, \zeta]] = [[[\xi, \eta], \zeta], \xi] + [[[\eta, \zeta], \xi], \xi] + [[[\zeta, \xi], \xi], \eta]. \tag{4.72}$$

holds.

Consider relations (4.69). If we substitute expressions (4.66) into them, we arrive at the relations:

$$\begin{aligned} &a^i_{jp}a^p_{mq}a^q_{lk} + a^i_{jp}a^p_{lq}a^q_{km} + a^i_{jp}a^p_{kq}a^q_{lm} + a^i_{kp}a^p_{mq}a^q_{jl} + a^i_{kp}a^p_{lq}a^q_{mj} \\ &+a^i_{kp}a^p_{jq}a^q_{ml} + a^i_{lp}a^p_{mq}a^q_{jk} + a^i_{mp}a^p_{lq}a^q_{kj} = a^i_{pq}a^p_{jk}a^q_{lm}. \end{aligned} \tag{4.73}$$

Define:

$$\begin{aligned} a^i_{jklm} &= a^i_{pq}a^p_{jk}a^q_{lm}, \\ A^i_{jklm} &= a^i_{jp}a^p_{kq}a^q_{lm} + a^i_{kp}a^p_{lq}a^q_{mj} + a^i_{lp}a^p_{mq}a^q_{jk} + a^i_{mp}a^p_{jq}a^q_{kl}. \end{aligned} \tag{4.74}$$

Then equations (4.73) can be written in the form:

$$A^i_{jmlk} + A^i_{jlkm} + A^i_{jklm} = a^i_{jklm}. \tag{4.75}$$

Moreover, the tensors A^i_{jklm} and a^i_{jklm} satisfy the following evident relations:

$$A^i_{j(klm)} = 0, \quad a^i_{jklm} = -a^i_{kjlm} = -a^i_{jkml} = -a^i_{lmjk}. \tag{4.76}$$

Cycling equations (4.75) in the indices i, k, l, and m adding the four series of relations so obtained and applying relations (4.76), we find:

$$A^i_{jklm} + A^i_{jmlk} = 0. \tag{4.77}$$

Subtracting equations (4.77) from equations (4.75), we get:

$$A^i_{jlkm} = a^i_{jklm},$$

or, in the notation of (4.74),

$$a^i_{jp}a^p_{lq}a^q_{km} + a^i_{kp}a^p_{mq}a^q_{jl} + a^i_{lp}a^p_{kq}a^q_{mj} + a^i_{mp}a^p_{jq}a^q_{lk} = a^i_{pq}a^p_{jk}a^q_{lm}. \tag{4.78}$$

These relations become more readable if we write them in vector form. Contracting relations (4.78) with the cordinates ξ^i, η^i, ζ^i and θ^i of vectors ξ, η, ζ and θ, we arrive at the identity:

$$[\xi,[\zeta,[\eta,\theta]]] + [\eta,[\theta,[\xi,\zeta]]] + [\zeta,[\eta,\theta,\xi]]] + [\theta,[\xi,[\zeta,\eta]]] = [[\xi,\eta],[\zeta,\theta]]. \tag{4.79}$$

This identity is called the *Sagle identity.* If we set $\theta = \xi$ in this identity, we arrive at the identity (4.72) characterizing Mal'tsev algebras. ■

2. Consider the system of equations (4.64), (4.65) and (4.67), defining a Moufang three-web. The tensor a^i_{jk} in these equations satisfies relations (4.69), (4.70), (4.72) and the Sagle identity (4.79). However, equations (4.70) follow from equations (4.69) since equations (4.68) are differential consequences of equations (4.67). Let us show that the systems (4.69) and (4.72) are equivalent. In one direction ((4.69) ⇒ (4.72)) this was proved in Theorem 4.18. Let us prove the converse. Relation (4.79) can be derived from relation (4.72) by linearization of the latter (see [Sag 62]). Next, if we symmetrize relations (4.78) (they are equivalent to the identity (4.79)) in the indices l and m, we obtain equations (4.77), which together with (4.78) give relations (4.75). But the latter differ from (4.73) and (4.69) only by the form in which they are written.

Thus, the structure equations of a Moufang web consist of equations (4.64), (4.65), (4.67) and the identity (4.72). This system is closed under the operation of exterior differentiation. As a result, we arrive at the following theorem:

Theorem 4.20 *A Moufang three-web is completely defined by its torsion tensor* a^i_{jk} *satisfying the identity* (4.72). ■

Using the fact that the torsion tensor is covariantly constant in the connection $\overset{*}{\Gamma}$, we prove the stronger statement: a Moufang web (as it was for a group web) is completely determined by a set of structure constants. Let $\mathcal{R}(W)$ be the fibration of adapted frames of the G_W-structure, defined by a Moufang three-web $W = (X, \lambda_\alpha)$ in the manifold X. We recall that the structure group of the G_W-structure is the group $\mathbf{GL}(r)$. The basis forms on $\mathcal{R}(W)$ are the forms $\underset{1}{\omega}^i, \underset{2}{\omega}^i$ and $\overset{*}{\omega}{}^i_j$ satisfying the structure equations (4.64) and (4.65). Consider in $\mathcal{R}(W)$ a distribution $\overset{*}{S}$ given by the equations:

$$\overset{*}{\omega}{}^i_j = \frac{1}{3} A^i_{jkl} \theta^{kl} \tag{4.80}$$

where, as in Section 4.1, θ^{kl} are parameters. Applying the same method as that used in Section 4.1, one can prove that

a) *The components of the torsion tensor a^i_{jk} are constant on the distribution $\overset{*}{S}$;*

b) *The distribution $\overset{*}{S}$ is involutive in $\mathcal{R}(W)$, and its integral manifolds are the holonomy fibrations defined by the connection $\overset{*}{\Gamma}$;*

c) *An arbitrary maximal integral manifold $\mathcal{R}'(W)$ of the distribution $\overset{*}{S}$ covers the manifold X, and as a result, a Moufang three-web, defined on X, is completely determined by a set of constants – values of the tensor a^i_{jk} in $\mathcal{R}'(W)$, i.e. by the structure tensor of some Mal'tsev algebra A;* and

d) *The holonomy algebra $\overset{*}{h}$ of the connection $\overset{*}{\Gamma}$ is generated by the linear transformations*

$$\overset{*}{d}{}^i_j = A^i_{jkl} \xi^k \eta^l,$$

which are differentiations of the Mal'tsev algebra A.

Consider the integral manifold $\mathcal{R}'(W)$ of the distribution $\overset{*}{S}$. The basis forms on this manifold are the forms $\underset{1}{\omega}{}^i, \underset{2}{\omega}{}^i$ and $\overset{*}{\theta}{}^i_j = \overset{*}{\omega}{}^i_j \big|_{\overset{*}{S}}$. They satisfy the structure equations (4.64)–(4.65), and the quantities a^i_{jk} in these equations are constant and form the structure tensor of the Mal'tsev algebra mentioned above. The system of structure equations (4.64)–(4.65) (we denote it also by $\overset{*}{S}$) is closed under the operation of exterior differentiation. In fact, exterior differentiation of equations (4.64) and (4.65) gives only relations (4.80) since equations (4.67) in $\mathcal{R}'(W)$ become identities. By (4.67), exterior differentiation of equations (4.80) also leads to identities. Thus, the system $\overset{*}{S}$ represents the structure equations (the Maurer–Cartan equations) of some Lie group G, and the Sagle identities (4.78) are equivalent to the Jacobi identities in the corresponding Lie algebra.

Equations (4.64) show that the subsystem $\underset{1}{\omega}{}^i = \underset{2}{\omega}{}^i = 0$ is completely integrable in $\overset{*}{S}$. Hence, it distinguishes in G a subgroup $\overset{*}{H}$ whose structure equations can be obtained from (4.65) and have the form:

$$d\pi^i_j = \pi^k_j \wedge \pi^i_k,$$

where $\pi^i_j = \overset{*}{\omega}{}^i_j \big|_{\underset{1}{\omega}{}^i = \underset{2}{\omega}{}^i = 0}$. The forms $\overset{*}{\pi}{}^i_j$ are expressed by the same equation (4.80) as the forms $\overset{*}{\omega}{}^i_j$. Since these very forms generate the holonomy algebra $\overset{*}{h}$ discussed above, we find that the group $\overset{*}{H}$ coincides with the holonomy group of the connection $\overset{*}{\Gamma}$.

On the other hand, the system $\underset{1}{\omega}^i = \underset{2}{\omega}^i = 0$ fixes a point in the manifold X carrying a Moufang web M. Thus, the manifold X is a homogeneous space $G/\overset{*}{H}$.

3. Let us show how the group G and the three-web M can be constructed provided that a Mal'tsev algebra is given. For this, we transfer from the connection $\overset{*}{\Gamma}$ to the connection $\widetilde{\Gamma}$ considered in Section 4.1. The relation between the corresponding connection forms $\overset{*}{\omega}{}^i_j$ and $\widetilde{\omega}^i_j$ can be found from equations (4.2) and (4.63):

$$\overset{*}{\omega}{}^i_j = \widetilde{\omega}^i_j - \frac{1}{3} a^i_{jk}(\underset{1}{\omega}^k - \underset{2}{\omega}^k).$$

Note that since the quantities a^i_{jk} are constant in G, the latter equations give a transformation of the basis of invariant forms of the group G. Comparing these equations with equations (4.80), we find that, in terms of the connection $\overset{*}{\Gamma}$, the distribution $\overset{*}{S}$ can be given by the equations:

$$\widetilde{\omega}^i_j = \frac{1}{3} A^i_{jkl}\theta^{kl} + \frac{1}{3} a^i_{jk}(\underset{1}{\omega}^k - \underset{2}{\omega}^k). \tag{4.81}$$

Since a Moufang web is a Bol web, its structure equations have the form (4.3) and (4.5). The structure equations of the group G can be obtained by restriction of these equations to the distribution (4.63) and can be written as follows:

$$\begin{aligned} d\underset{1}{\omega}^i &= \underset{1}{\omega}^j \wedge \widetilde{\theta}^i_j + a^i_{jk}\underset{1}{\omega}^j \wedge \underset{2}{\omega}^k, \\ d\underset{2}{\omega}^i &= \underset{2}{\omega}^j \wedge \widetilde{\theta}^i_j - a^i_{jk}\underset{1}{\omega}^j \wedge \underset{2}{\omega}^k, \\ d\widetilde{\theta}^i_j &= \widetilde{\theta}^k_j \wedge \widetilde{\theta}^i_k + R^i_{jkl}(\underset{1}{\omega}^k + \underset{2}{\omega}^k) \wedge (\underset{1}{\omega}^l + \underset{2}{\omega}^l) \end{aligned} \tag{4.82}$$

Here $\widetilde{\theta}^i_j$ is the notation for the restriction of the forms $\widetilde{\omega}^i_j$ to G.

The torsion tensor R^i_{jkl} of the connection $\widetilde{\Gamma}$ for the Moufang web can be expressed in terms of the tensor a^i_{jk}. From (4.6) and (4.61) we have:

$$R^i_{jkl} = \frac{1}{6}(a^m_{jk}a^i_{ml} + a^m_{lj}a^i_{mk} - 2a^i_{mj}a^m_{kl}). \tag{4.83}$$

As a result, the ternary operation (4.22)–(4.23), defining the triple Lie system, can be expressed in terms of the binary operation (4.71) of the Mal'tsev algebra as follows[1]:

$$d_{\eta,\zeta} =< \xi,\eta,\zeta >= R(\xi,\eta,\zeta) = \frac{1}{24}(2[\xi,[\eta,\zeta]] - [\eta,[\zeta,\xi]] - [\zeta,[\xi,\eta]]). \tag{4.84}$$

In this case, the triple Lie system is denoted by T_A.

[1]In the book [Lo 69] by O. Loos the coefficient $\frac{1}{24}$ is missing.

Consider the space

$$g = A_1 \dot{+} A_2 \dot{+} h,$$

where A_1 and A_2 are algebras isomorphic to the given Mal'tsev algebra A and h is the algebra of interior differentiations of the triple system A (see Section 4.1, p. 144). Let an isomorphism $\varphi : A_1 \to A_2$ be established in such a way that the corresponding vectors have equal coordinates, and let $\xi_1, \eta_1 \in A_1$, $\xi_2, \eta_2 \in A_2$ and $x, y \in h$. Define a commutator $(\; , \;)$ in g as follows:

$$\begin{array}{l} (\xi_\nu, \eta_\nu) = -2d_{\xi_\nu, \eta_\nu} \in h, \quad \nu = 1, 2, \\ (x, \xi_\nu) = -(\xi_\nu, x) = x(\xi_\nu) \in A_\nu, \\ (x, y) = x \circ y - y \circ x \in h, \\ (\xi_1, \eta_2) = -(\eta_2, \xi_1) = \frac{1}{2}[\xi_1, \varphi^{-1}(\eta_2)] - \frac{1}{2}[\varphi(\xi_1), \eta_2] - 2d_{\xi_1, \eta_2}. \end{array} \tag{4.85}$$

Lemma 4.21 *The space g with the operation $(\; , \;)$ forms a Lie algebra tangent to the group G.*

Proof. Let the structure equations of a Lie group be written in the form:

$$d\omega^u = -\frac{1}{2} c^u_{vw} \omega^v \wedge \omega^w,$$

Then, for any pair of invariant vector fields X and Y in this group, we have the relations:

$$d\omega^u(X, Y) = -c^u_{vw} X^v Y^w = -(X, Y)^u, \tag{4.86}$$

where (X, Y) is the commutator in the Lie algebra of this group. Calculate the commutator of the group G using its structure equations (4.82). Let $X = \xi_1 + \xi_2 + x$ and $Y = \eta_1 + \eta_2 + y$ be the invariant vector fields in G dual to the co-basis of invariant forms $\underset{1}{\omega}^i, \underset{2}{\omega}^i$ and $\widetilde{\theta}^i_j$. Then, by means of (4.71) and (4.84), we find from (4.82):

$$\begin{array}{rcl} d\underset{1}{\omega}^i(X, Y) & = & \underset{1}{\omega}^j \wedge \widetilde{\theta}^i_j + a^i_{jk} \underset{1}{\omega}^j \wedge \underset{2}{\omega}^k (\xi_1 + \xi_2 + x, \eta_1 + \eta_2 + y) \\ & = & \xi^j_1 y^i_j - \eta^j_1 x^i_j + a^i_{jk}(\xi^j_1 \eta^k_2 - \eta^j_1 \xi^k_2) \\ & = & \left(y(\xi_1) - x(\eta_1) - \frac{1}{2}[\xi_1, \varphi^{-1}(\eta_2)] + \frac{1}{2}[\eta_1, \varphi^{-1}(\xi_2)] \right)^i; \\ d\underset{2}{\omega}^i(X, Y) & = & \xi^j_2 y^i_j - \eta^j_2 x^i_j - (a^i_{jk} \xi^j_1 \eta^k_2 - \eta^j_1 \xi^k_2) \\ & = & \left(y(\zeta_2) - x(\eta_2) + \frac{1}{2}[\varphi(\xi_1), \eta_2] - \frac{1}{2}[\varphi(\eta_1), \xi_2] \right)^i; \\ d\widetilde{\theta}^i_j(X, Y) & = & x^k_j y^i_k - y^k_j x^i_k + R^i_{jkl}(\xi^k_1 \eta^l_1 - \eta^k_1 \xi^l_1) + R^i_{jkl}(\xi^k_1 \eta^l_2 - \eta^k_1 \xi^l_2) \\ & & + R^i_{jkl}(\xi^k_2 \eta^l_1 - \eta^k_2 \xi^l_1) + R^i_{jkl}(\xi^k_2 \eta^l_2 - \eta^k_2 \xi^l_2) \\ & = & (y \circ x)^i_j - (x \circ y)^i_j + 2(d_{\xi_1, \eta_1})^i_j + 2(d_{\xi_1, \eta_2})^i_j + 2(d_{\xi_2, \eta_1})^i_j + 2(d_{\xi_2, \eta_2})^i_j \\ & = & (y \circ x - x \circ y + 2d_{\xi_1, \eta_1} + 2d_{\xi_1, \eta_2} + 2d_{\xi_2, \eta_1} + 2d_{\xi_2, \eta_2})^i_j. \end{array}$$

Next, we calculate the commutator (X, Y) using definition (4.84):

$$
\begin{aligned}
(X,Y) = \ & (\xi_1+\xi_2+x, \eta_1+\eta_2+y) \\
= \ & -2d_{\xi_1,\eta_1} - 2d_{\xi_2,\eta_2} + x\circ y - y\circ x \\
& +\tfrac{1}{2}[\xi_1, \varphi^{-1}(\eta_2)] - \tfrac{1}{2}[\varphi(\xi_1), \eta_2] - 2d_{\xi_1,\eta_2} \\
& -\Big(\tfrac{1}{2}[\eta_1, \varphi^{-1}(\xi_2)] - \tfrac{1}{2}[\varphi(\eta_1), \xi_2] - 2d_{\eta_1,\xi_2}\Big) \\
& +x(\eta_1) + x(\eta_2) - y(\xi_1) - y(\xi_2) \\
= \ & -\Big(y(\xi_1) - x(\eta_1) - \tfrac{1}{2}[\xi_1, \varphi^{-1}(\eta_2)] + \tfrac{1}{2}[\eta_1, \varphi^{-1}(\xi_2)]\Big) \\
& -\Big(y(\xi_2) - x(\eta_2) + \tfrac{1}{2}[\varphi(\xi_1), \eta_2)] - \tfrac{1}{2}[\varphi(\eta_1), \xi_2]\Big) \\
& -(2d_{\xi_1,\eta_1} + 2d_{\xi_2,\eta_2} + 2d_{\xi_1,\eta_2} + 2d_{\xi_2,\eta_1} + y\circ x - x\circ y).
\end{aligned}
$$

Comparing these equations with previous series of equations, we see that the commutators, calculated by means of formulas (4.82) and (4.85), satisfy relation (4.85), i.e. they define the same Lie algebra. ■

	A_1	A_2	h
A_1	h		A_1
A_2		h	A_2
h	A_1	A_2	h

Figure 4.7

The multiplication table in the algebra g is given in Figure 4.7. One can see from this table that there are two subalgebras $A_1 \dot{+} h$ and $A_2 \dot{+} h$ in g. The corresponding subgroups G_1 and G_2 are defined in the group G by the equations: $\underset{2}{\omega}^i = 0$ and $\underset{1}{\omega}^i = 0$. In addition, the system $\underset{1}{\omega}^i + \underset{2}{\omega}^i = 0$ is also completely integrable on the three-web M. This system defines leaves of the third foliation. Since the forms $\underset{1}{\omega}^i$ and $\underset{2}{\omega}^i$ are invariant forms of the group G, the system $\underset{1}{\omega}^i + \underset{2}{\omega}^i = 0$ defines a subgroup G_3 in G. The tangent algebra to this subgroup is determined by the equation $\xi_1 + \xi_2 = 0$.

The intersection of the subgroups G_α, $\alpha = 1, 2, 3$ is the subgroup $\overset{*}{H}$ (we recall that the latter satisfies the equations $\underset{1}{\omega}^i = \underset{2}{\omega}^i = 0$), and since the subgroups G_α are defined in the group G by the same equations as leaves of the Moufang web in the manifold G/H, we arrive at the following theorem:

Theorem 4.22 *Let A be a real r-dimensional Mal'tsev algebra and h be the algebra of interior differentiations of the triple Lie system associated with the Mal'tsev algebra. Denote by A_1 and A_2 two isomorphic samples of the algebra A and by φ the isomorphism from A_1 to A_2. Suppose that the Lie algebra structure is defined in the space $g = A_1 \dot{+} A_2 \dot{+} h$ by formulas* (4.85). *Let G be the Lie group corresponding to this algebra,* $\dim\, G = 2r + \rho, \dim\, H = \rho$. *Then, on the homogeneous space $X = G/H$, where H is a subgroup with the tangent algebra h, a Moufang three-web* *of dimension $2r$ arises. Its leaves are the factor manifolds gG_α/gH, where the G_α are $(r+\rho)$-dimensional subgroups of the group G, corresponding to the subalgebras $g_1 = A_1 \dot{+} h$, $g_2 = A_2 \dot{+} h$ and the diagonal subalgebra g_3 defined in g by the equation $\xi_1 + \xi_2 = 0$.* ■

4. The canonical connection $\tilde{\gamma}$ arising in the base X_3 of the third foliation of a Bol three-web (see Section 4.1), in the case of a Moufang web, acquires additional properties.

Theorem 4.23 *Let a Moufang three-web M be given as in Theorem 4.20 by three $(r+\rho)$-dimensional subgroups G_α in the group G of dimension $2r+\rho$. Then, G/G_3 is a symmetric space.*

Proof. Consider in the algebra g a linear transformation σ_3 such that

$$\sigma_3(\xi_1 + \xi_2 + x) = -\varphi^{-1}(\xi_2) - \varphi(\xi_1) + x.$$

Applying formulas (4.85), we check that σ_3 is an automorphism in g (Problem **10**). Since $(\sigma_3)^2 = \mathrm{id}$, σ_3 is an involutive automorphism in g. Further, since the subalgebra g_3 is defined by the equation $\xi_1 + \xi_2 = 0$, this subalgebra, and only this subalgebra, is invariant relative to σ_3: $\sigma_3(g_3) = g_3$. Therefore, (g, g_3) is a symmetric pair. By well-known properties of symmetric spaces, it follows from this that G/G_3 is a symmetric space. Since the subgroup G_3 is determined by the equations $\underset{1}{\omega}^i + \underset{2}{\omega}^i = 0$, the points of this symmetric space are the leaves of the third foliation of the web M. Thus, this space is the space with the canonical connection $\tilde{\gamma}$ which was considered in Section 4.1. ■

The involutive automorphism σ_3 of the algebra g generates an automorphism τ_3 in the corresponding group G such that $\tau_3(G_3) = G_3$. Moreover, it follows from the definition of an automorphism that $\sigma_3(g_2) = g_1$ and $\sigma_3(g_1) = g_2$. In turn, this implies that $\tau_3(G_2) = G_1$ and $\tau_3(G_1) = G_2$. Next, since the foliations λ_α generating a Moufang three-web M, are of the same structure, we obtain that in the group G generating this web, there are three involutive automorphisms τ_α, $\alpha = 1,2,3$, such that $\tau_\alpha(G_\alpha) = G_\alpha$, $\tau_\alpha(G_\beta) = G_\gamma, \beta \neq \alpha \neq \gamma$. The automorphisms τ_α generate the symmetric group S_3 of (external) automorphisms of the group G.

The properties of the Moufang webs, listed above, also indicate a way to classify them.

4.5 The Moufang Three-Web of Minimal Dimension

1. A Moufang three-web of minimal dimension, which is not reduced to a group web, is eight-dimensional since the minimal dimension of a non-Lie Mal'tsev algebra (see

[Ku 70]) is four. Denote this non-Lie Mal'tsev algebra by A_4. In an appropriate basis, the structure tensor of the algebra A_4 has the following nonzero components:

$$a^1_{14} = a^2_{24} = -a^3_{34} = 1, \quad a^3_{12} = 1. \tag{4.87}$$

Substituting these components into formulas (4.61), we are able to calculate the nonzero components of the curvature tensor of the eight-dimensional Moufang three-web M:

$$b^3_{124} = b^3_{241} = b^3_{412} = -2, \quad b^3_{214} = b^3_{142} = b^3_{421} = 2. \tag{4.88}$$

From (4.61) and (4.66) we derive the formula:

$$A^i_{jkl} = b^i_{jkl} - \frac{4}{3} a^m_{kl} a^i_{mj}.$$

Substituting the values of the tensors a and b into this formula, we find the nonzero components of the tensor A^i_{jkl}:

$$\begin{array}{ll} A^1_{414} = A^2_{424} = A^3_{434} = -\frac{4}{3}, & A^1_{441} = A^2_{442} = A^3_{443} = \frac{4}{3}, \\ A^3_{124} = A^3_{241} = A^3_{412} = -\frac{2}{3}, & A^3_{142} = A^3_{214} = A^3_{421} = \frac{2}{3}, \end{array} \tag{4.89}$$

From relations (4.61), (4.66) and (4.83) we obtain the following expression for the tensor R^i_{jkl}:

$$R^i_{jkl} = \frac{3}{8} A^i_{jkl} - \frac{1}{8} b^i_{jkl}. \tag{4.90}$$

For the web under consideration, we have:

$$R^1_{414} = R^2_{424} = R^3_{434} = -\frac{1}{2}, R^1_{441} = R^2_{442} = R^3_{443} = \frac{1}{2}.$$

The other components of the tensor R^i_{jkl} are zero.

Equations (4.81), defining the group G in the bundle frame, in this case have the form:

$$\begin{array}{l} \overset{*}{\omega}{}^1_4 = -\frac{8}{3}\theta^{14}, \quad \overset{*}{\omega}{}^2_4 = -\frac{8}{3}\theta^{24}, \quad \overset{*}{\omega}{}^3_4 = -\frac{4}{3}\theta^{12} - \frac{8}{3}\theta^{34}, \\ \overset{*}{\omega}{}^3_1 = -\frac{4}{3}\theta^{24}, \quad \overset{*}{\omega}{}^3_2 = \frac{4}{3}\theta^{14}, \end{array}$$

and the other forms $\overset{*}{\omega}{}^i_j$ are zero. Eliminating the parameters θ^{kl}, we obtain 13 equations in 16 forms $\overset{*}{\omega}{}^i_j$:

$$\begin{array}{lll} \overset{*}{\omega}{}^1_4 + 2\overset{*}{\omega}{}^3_2 = 0, & \overset{*}{\omega}{}^2_4 - 2\overset{*}{\omega}{}^3_1 = 0, & \overset{*}{\omega}{}^4_i = 0, \\ \overset{*}{\omega}{}^1_1 = \overset{*}{\omega}{}^1_2 = \overset{*}{\omega}{}^1_3 = 0, & \overset{*}{\omega}{}^2_1 = \overset{*}{\omega}{}^2_2 = \overset{*}{\omega}{}^2_3 = 0, & \overset{*}{\omega}{}^3_3 = 0. \end{array} \tag{4.91}$$

(In this section all Latin indices take on the values 1, 2, 3 and 4.) Therefore, the group G, generating a Moufang web (see Section 4.4), is of dimension 11, and the basis forms in it are the forms $\underset{1}{\omega}{}^i, \underset{2}{\omega}{}^i, \overset{*}{\omega}{}^1_4, \overset{*}{\omega}{}^2_4$ and $\overset{*}{\omega}{}^3_4$. The stationary subgroup H, which is defined by the equations $\underset{1}{\omega}{}^i = \underset{2}{\omega}{}^i = 0$, is of dimension three, and its basis forms are the forms $\overset{*}{\omega}{}^1_4, \overset{*}{\omega}{}^2_4$ and $\overset{*}{\omega}{}^3_4$.

To get closed form equations of the Moufang web under consideration, we must write the system (4.64)–(4.65), taking into account relations (4.87) and (4.91), and integrate this system. However, before doing this, we make an important remark on a generalization of the results of Section 4.4, which allows us to construct an algorithm for solving problems similar to the problem formulated above.

As follows from Section 4.4, the study of Moufang webs is significantly simplified by the fact that a manifold X, carrying the three-web, is the homogeneous space G/H. Moreover, the groups G and H were intrinsically defined in Section 4.4, i.e. they were defined in terms of the corresponding Mal'tsev algebra. However, it is not at all necessary to choose them so precisely like this. In general, any two groups G and H such that $X = G/H$ can be selected. The group G must satisfy only the following condition: the components a^i_{jk} must be constant in the subbundle of frames defined by this group. The equations, defining the group G of maximal dimension with this property, can be obtained by substituting the given values of constants into equations (4.67):

$$a^m_{jk}\overset{*}{\omega}{}^i_m = a^i_{mk}\overset{*}{\omega}{}^m_j + a^i_{jm}\overset{*}{\omega}{}^m_k. \tag{4.92}$$

It follows from equations (4.92) that the forms $\overset{*}{\omega}{}^i_j$ are differentiations (in the algebraic sense as in Sections 4.1 and 4.4) of the algebra A. Therefore, these forms are invariant forms of the representation of the group Aut M of all automorphisms of the Moufang loop M for which the algebra A is the tangent algebra. We can define the structure of a Lie algebra in the direct sum $g = A_1 \dot{+} A_2 \dot{+}$Der A, where Der A is the algebra of all differentiations in A, proceeding in the same way as previously in Section 4.4, where we used the subalgebra of interior differentiations rather than the entire algebra Der A. Elements of the form $(0, 0, d)$, $d \in$ Der A from g constitute a subalgebra in g. The Moufang three-web is defined in the homogeneous space $X = G/H$ where G and H are the Lie groups corresponding to the algebras g and Der A.

In the four-dimensional case under consideration, equations (4.92) take the form:

$$\begin{gathered}\overset{*}{\omega}{}^4_i = 0, \quad \overset{*}{\omega}{}^1_3 = \overset{*}{\omega}{}^2_3 = 0, \quad -\overset{*}{\omega}{}^1_1 - \overset{*}{\omega}{}^2_2 + \overset{*}{\omega}{}^3_3 = 0, \\ 2\overset{*}{\omega}{}^3_1 - \overset{*}{\omega}{}^2_4 = 0, \quad 2\overset{*}{\omega}{}^3_2 + \overset{*}{\omega}{}^1_4 = 0.\end{gathered} \tag{4.93}$$

System (4.92) contains nine equations in 16 forms $\overset{*}{\omega}{}^i_j$. Thus, the group H is seven-dimensional. Denote this group by H_7.

As was indicated above, instead of the group H_7 we can take any of its subgroups. For example, we can take the three-dimensional subgroup defined by equations (4.91).

The optimal variant is in taking a subgroup of minimal dimension. We will indicate how this can be done for the three-web under consideration.

To simplify our considerations, we transfer from the connection $\overset{*}{\Gamma}$ to the Chern connection Γ, since the curvature tensor b^i_{jkl} of the latter has fewer non-zero components than the tensor A^i_{jkl}. Applying (4.87) and formula (4.63) connecting the forms $\overset{*}{\omega}{}^i_j$ and ω^i_j, we can write system (4.93), defining the subgroup H_7, in the form:

$$\begin{aligned} &\omega_2^4 = 0,\ \ \omega_3^1 = \omega_3^2 = 0, \qquad -\omega_1^1 - \omega_2^2 + \omega_3^3 = 2(\underset{1}{\omega}{}^4 - \underset{2}{\omega}{}^4),\\ &2\omega_1^3 - \omega_4^2 = -2(\underset{1}{\omega}{}^2 - \underset{2}{\omega}{}^2),\ \ 2\omega_2^3 + \omega_4^1 = 2(\underset{1}{\omega}{}^1 - \underset{2}{\omega}{}^1). \end{aligned} \tag{4.94}$$

The forms ω^i_j satisfy the structure equations (1.32) from which, by (4.88), in particular, we get:

$$\begin{aligned} d\omega_1^1 &= \omega_1^2 \wedge \omega_2^1, & d\omega_1^2 &= (\omega_1^1 - \omega_2^2) \wedge \omega_1^2,\\ d\omega_2^1 &= \omega_2^1 \wedge (\omega_1^1 - \omega_2^2), & d\omega_2^2 &= \omega_2^1 \wedge \omega_1^2,\\ d\omega_4^1 &= \omega_4^1 \wedge \omega_1^1 + \omega_4^2 \wedge \omega_2^1, & d\omega_4^2 &= \omega_4^1 \wedge \omega_1^2 + \omega_4^2 \wedge \omega_2^2. \end{aligned}$$

These equations show that the Pfaffian system

$$\omega_1^1 = \omega_1^2 = \omega_2^1 = \omega_2^2 = \omega_4^1 = \omega_4^2 = 0 \tag{4.95}$$

is completely integrable. Thus, this system defines a one-dimensional subgroup H_1 in H_7, and the manifold X of the web M is the homogeneous space G_9/H_1, $\dim\, G_9 = 9$. The invariant form in H_1 is the form $\pi_4^3 = \omega_4^3\big|_{\underset{1}{\omega}{}^i = \underset{2}{\omega}{}^i = 0}$. The tangent algebra g' of the group G_9 has the form $g' = A_4 \dotplus A_4 \dotplus h'$, $\dim\, h' = 1$.

Let us clarify how the algebra h' and the original algebra A_4 are related. The *J-kernel* of a Mal'tsev algebra A is said to be a set of such $\xi \in A$ for which the Jacobian

$$J(\xi, \eta, \zeta) = [[\xi, \eta], \zeta] + [[\eta, \zeta], \xi] + [[\zeta, \xi], \eta]$$

is zero for any $\xi, \eta \in A$. By relations (4.61), we obtain the following equations for finding coordinates of the vector ζ:

$$b^i_{jkl}\zeta^l = 0. \tag{4.96}$$

Let us find the J-kernel of the algebra A_4. For the values of the tensor b^i_{jkl} from (4.88) and (4.96), we get $\zeta^1 = \zeta^2 = \zeta^4 = 0$. Thus, the J-kernel (denote it $\mathcal{N}$) is one-dimensional.

Now let η be an arbitrary vector from A_4. To the pair η and ζ where $\zeta \in \mathcal{N}$, there corresponds the following one-parameter family of internal automorphisms of the triple system T_A:

$$(R^i_{jkl}\eta^k\zeta^l) = \begin{pmatrix} 0 & 0 & 0 & 0 \\ 0 & 0 & 0 & 0 \\ 0 & 0 & 0 & \frac{1}{2}\eta^4 \\ 0 & 0 & 0 & 0 \end{pmatrix}$$

Since the only nonzero element of this matrix is R^3_4, this gives the desired one-dimensional subalgebra h'.

Let us write the Maurer–Cartan equations of the group G_9. We apply relations (4.94) and (4.95) to the structure equations (1.26) and (1.32). As a result, we get:

$$\begin{array}{ll} d\underset{1}{\omega}^4 = 0, & d\underset{2}{\omega}^4 = 0, \\ d\underset{1}{\omega}^1 = 2\underset{1}{\omega}^1 \wedge \underset{1}{\omega}^4, & d\underset{2}{\omega}^1 = -2\underset{2}{\omega}^1 \wedge \underset{2}{\omega}^4, \\ d\underset{1}{\omega}^2 = 2\underset{2}{\omega}^2 \wedge \underset{2}{\omega}^4, & d\underset{2}{\omega}^2 = -2\underset{2}{\omega}^2 \wedge \underset{2}{\omega}^4, \\ d\underset{1}{\omega}^3 = \underset{1}{\omega}^4 \wedge \omega^3_4 - 2\underset{1}{\omega}^3 \wedge \underset{2}{\omega}^4 + \Omega, & d\underset{2}{\omega}^3 = \underset{2}{\omega}^4 \wedge \omega^3_4 + 2\underset{2}{\omega}^3 \wedge \underset{2}{\omega}^4 - \Omega, \\ d\omega^3_4 = 2\omega^3_4 \wedge (\underset{1}{\omega}^4 - \underset{2}{\omega}^4) - \Omega, & \end{array} \tag{4.97}$$

where $\Omega = \underset{1}{\omega}^1 \wedge \underset{2}{\omega}^2 - \underset{1}{\omega}^2 \wedge \underset{2}{\omega}^1$.

System (4.97) is of triangular form. This fact is connected with the solvability of the algebra A_4 (see Problem **13**). Successively integrating system (4.97), we find the invariant forms $\underset{1}{\omega}^i, \underset{2}{\omega}^i$ and ω^3_4 of the group G_9:

$$\begin{array}{ll} \underset{1}{\omega}^4 = dp^4, & \underset{2}{\omega}^4 = dq^4, \\ \underset{1}{\omega}^1 = e^{-2p^4} dp^1, & \underset{2}{\omega}^1 = e^{2q^4} dq^1, \\ \underset{1}{\omega}^2 = e^{-2p^4} dp^2, & \underset{2}{\omega}^2 = e^{2q^4} dq^2, \\ \underset{1}{\omega}^3 = -e^{2q^4}(e^{-2p^4}\omega + dp^3), & \underset{2}{\omega}^3 = e^{-2p^4}(e^{2q^4}\omega + dq^3), \\ \omega^3_4 = 2e^{2(q^4 - p^4)}\omega, & \end{array} \tag{4.98}$$

where

$$\omega = -\frac{1}{2}(q^2 dp^1 - p^1 dq^2 + p^2 dq^1 - q^1 dp^2) + d\tau,$$

and p^i, q^i and τ are parameters.

According to Theorem 4.21, the leaves of the web M are the factor manifolds gG_α/gH_1. The equations of the subgroups H_1 and G_α can be obtained by integration of the Pfaffian systems $\underset{1}{\omega}^i = \underset{2}{\omega}^i = 0$ and $\underset{1}{\omega}^i = \omega^3_4 = 0$, $\underset{2}{\omega}^i = \omega^3_4 = 0$, $\underset{1}{\omega}^i + \underset{2}{\omega}^i = \omega^3_4 = 0$. However, we will not look for these subgroups. We will proceed differently: we will indicate an eight-dimensional submanifold X' in the group G_9 that is transversal to the co-sets gH_1. This submanifold will be diffeomorphic to the factor space $X = G_9/H_1$, and the leaves of the web will be defined in X' by the systems $\underset{1}{\omega}^i\big|_{X'} = 0$, $\underset{2}{\omega}^i\big|_{X'} = 0$ and $(\underset{1}{\omega}^i + \underset{2}{\omega}^i)\big|_{X'} = 0$.

Define the manifold X' by the equation $d\tau = 0$. We find the first foliation of the web by integrating the system $\underset{1}{\omega}^i = 0$ subject to the condition $d\tau = 0$. Let x^i be the first integrals of the system. The integration leads to:

$$p^4 = x^4, \ p^1 = x^1, \ p^2 = x^2, \ \frac{1}{2}e^{2x^4}(x^2q^1 - x^1q^2) + p^3 = x^3. \tag{4.99}$$

Integration of the system $\underset{2}{\omega}^i\big|_{d\tau=0} = 0$, defining the second foliation, gives:

$$q^4 = y^4, \ q^1 = y^1, \ q^2 = y^2, \ \frac{1}{2}e^{2y^4}(y^2p^1 - y^1p^2) + q^3 = y^3, \tag{4.100}$$

and integration of the system $(\underset{1}{\omega}^i + \underset{2}{\omega}^i)\big|_{d\tau=0} = 0$, leads to the equations:

$$\begin{aligned} &e^{-z^4}p^1 + e^{z^4}q^1 = z^1, \\ &e^{-z^4}p^2 + e^{z^4}q^2 = z^2, \\ &-e^{z^4}p^3 + e^{-z^4}q^3 = z^3, \\ &p^4 + q^4 = z^4. \end{aligned} \tag{4.101}$$

Let us recall that the equations of a three-web or, equivalently, of its coordinate quasigroup, relate parameters of the leaves passing through the same point. Thus, the equations of a web can be obtained by eliminating the local coordinates p^i and q^i of the manifold X' from equations (4.99)–(4.101). As a result of this elimination, we arrive at the equations:

$$\begin{aligned} &z^1 = x^1e^{-z^4} + y^1e^{z^4}, \\ &z^2 = x^2e^{-z^4} + y^2e^{z^4}, \\ &z^3 = -x^3e^{z^4} + y^3e^{-z^4} - e^{y^4 - x^4}(x^1y^2 - x^2y^1), \\ &z^4 = x^4 + y^4. \end{aligned} \tag{4.102}$$

Let us find the coordinate loop of the web under study. Suppose that the identity element has zero coordinates in this loop. According to the general theory (see Section 4.2), we transfer to an LP-isotope of the coordinate quasigroup by means of translations, which, as follows from relations (4.102), have the form:

$$\begin{aligned} u^1 &= x^1e^{-x^4}, & v^1 &= y^1e^{y^4}, \\ u^2 &= x^2e^{-x^4}, & v^2 &= y^2e^{y^4}, \\ u^3 &= -x^4e^{x^4}, & v^3 &= y^3e^{-y^4}, \\ u^4 &= x^4, & v^4 &= y^4, \end{aligned}$$

As a result of these transformations, equations (4.102) take the form:

$$\begin{aligned} &z^1 = u^1e^{-v^4} + v^1e^{u^4}, \\ &z^2 = u^2e^{-v^4} + v^2e^{u^4}, \\ &z^3 = u^3e^{v^4} + v^3e^{-u^4} - (u^1v^2 - u^2v^1), \\ &z^4 = u^4 + v^4. \end{aligned} \tag{4.103}$$

We have obtained the closed form equations of the four-dimensional Moufang loop of minimal dimension. The very same equations define the eight-dimensional Moufang three-web in the standard parametrization.

PROBLEMS

1. Prove that if the torsion tensor of the Bol web B_m is zero, then the web is a web T. (*Hint*: apply equations (4.14).)

2. Prove that algebraically, relations (4.17) signify that the algebra, generated by the differentiations $d_{\xi,\eta}$, is the linear span of these differentiations.

3. Find closed form equations of a four-dimensional Bol three-web if a quadric and a plane generating the web are located as indicated below:

i) The quadric Q is oval and does not intersect the plane π;

ii) The quadric Q is oval and is tangent to the plane π; and

iii) The quadric Q is a ruled surface and is tangent to the plane π.

Solution. Consider, for example, the case i). By the definition of a Grassmann three-web, its equations relate coordinates of three points, in which a moving straight line l intersects surfaces generating this web. Choose an affine coordinate system in such a way that the plane π has the equation $x^3 = 0$, and the quadric Q has the equation:

$$x^3 = (x^1)^2 + (x^2)^2 + 1.$$

Let the straight line l intersect the quadric Q and the plane π at the points $x(x^i), y(y^i)$ and $z(z^1, z^2, 0)$, respectively. Then, the coordinates z^i can be expressed in terms of x^i and y^i as follows:

$$z^i = \frac{x^i y^3 - x^3 y^i}{y^3 - x^3}, \quad i = 1,2, \tag{4.104}$$

and in addition, the coordinates of the points x and y are connected by the equation of the quadric Q:

$$x^3 = (x^1)^2 + (x^2)^2 + 1, \quad y^3 = (y^1)^2 + (y^2)^2 + 1.$$

Substituting x^3 and y^3 into (4.104), we obtain

$$z^i = \frac{x^i\Big((y^1)^2 + (y^2)^2 + 1\Big) - y^i\Big((x^1)^2 + (x^2)^2 + 1\Big)}{\Big((y^1)^2 + (y^2)^2 - (x^1)^2 - (x^2)^2\Big)}.$$

Similar proofs can be given for the cases i) and ii).

4. Prove that in the moving frame associated with a Grassmann three-web as in Section 3.3, the equations of the hyperplane X_3 and the hyperquadric Q generating this web, can be written as follows:

$$a_i x^i - x^0 + x^{r+1} = 0,$$
$$b_{ij} x^i x^j - 2x^0 x^{r+1} = 0.$$

(*Hints*: a) The hyperplane X_3 can be defined by the point $A_0 + A_{r+1}$ and the points lying in the intersection of the tangent hyperplanes to the hyperquadric at the points A_0 and A_{r+1}; and b) the equation of the hyperquadric Q can be found from the condition for this hyperquadric to be stationary, in the same way as this was done in Section 4.2.)

5. Prove that the quadric $\widetilde{Q} = Q \bigcap X_3$, which is the absolute of the hyperplane X_3 (see Section 2.2), is non-degenerate if and only if the tensor $g_{ij} = b_{ij} + \frac{1}{2} a_i a_j$ is non-singular.

6. Let q be a second degree curve in a projective plane, AB be its chord, S be the pole of the straight line AB with respect to the curve q and l be an arbitrary straight line in the plane. Denote by x_1 and x_2 the intersection points of l and the curve q, by x_3 the intersection points of l and the line AB, and by y_1 and y_2 the projections of the points x_1 and x_2 respectively from the pole S onto the line AB (see Figure 4.8).

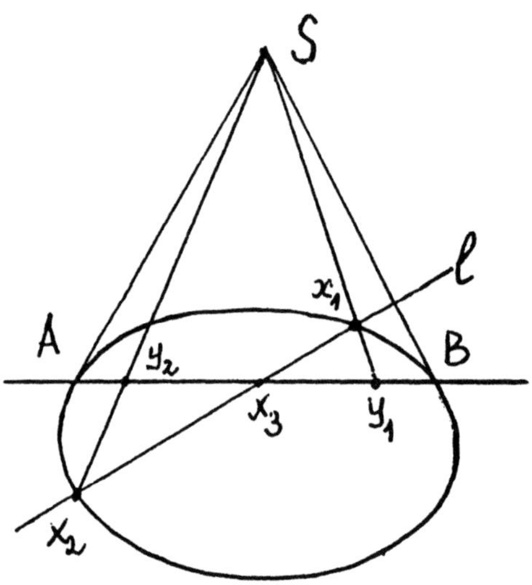

Figure 4.8

Prove that the cross-ratios $(A, B; y_2, x_3)$ and $(A, B; x_3, y_1)$ are equal.

(*Hint*: Apply a projective transformation of the plane which transforms the curve q into a circle, the straight line AB into its diameter and the point x_3 into its center.)

7. Prove that the group G, defined by structure equations (4.52), is the direct product of two simple three-dimensional groups each of which is isomorphic to the group of projective transformations on a straight line.

8. Give a geometric proof of the fact that the curve z_3, represented in Figure 4.6, does not depend on the choice of a triple of points A, S, T on the curve z_1.

9. Derive identity (4.79) from identity (4.72).

10. Prove that the linear transformation σ_3, introduced in the proof of Theorem 4.22, is an isomorphism of the algebra g.

11. In the algebra g considered in Problem **10**, find an invariant subspace of the transformation σ_3, corresponding to an eigenvalue of -1.

12. Establish a correspondence between the connections of a pencil $\gamma(W)$ of a Moufang three-web and the invariant bases in the group G, generating this web (see Sections 1.8 and 4.4).

13. Prove that the Mal'tsev algebra A_4, given by the structure tensor (4.87), is solvable.

14. Prove that the J-kernel of the Mal'tsev algebra A is an ideal in A.

Proof. Relation (4.72), characterizing the Mal'tsev algebras, can be written in the form:

$$J(x,y,xz) = J(x,y,z)x \quad \text{or} \quad J(x,xy,z) = J(x,y,z)x, \tag{4.105}$$

where $J(x,y,z)$ is the Jacobian of the algebra A (p. 79). By (4.105), the linearization $x \to x + w$ of the first of this equations leads to the identity:

$$J(x,y,wz) + J(w,y,xz) = J(x,y,z)w + J(w,y,z)x, \tag{4.106}$$

which is also satisfied in the algebra A. Now let z and w belong to the J-kernel, i.e. $J(x,y,z) = 0$ and $J(x,y,w) = 0$. Then, it follows from (4.106) that $J(x,y,wz) = 0$. ■

15. Prove that if a vector z belongs to the J-kernel of a Mal'tsev algebra A, then the interior differentiation $A^i_{jkl}\xi^k z^l$ of the algebra A is an interior differentiation of the Lie triple system associated with A. (*Hint*: apply relation (4.90).)

16. Find the general form of the matrices $A^i_{jkl}\xi^k z^l$ and $R^i_{jkl}\xi^k z^l$ of the four-dimensional Mal'tsev algebra A_4.

17. Find an admissible coordinate transformation which sends equations (4.103) into the equations:

$$\begin{aligned} z^1 &= x^1 + y^1 + y^1x^4, \\ z^2 &= x^2 + y^2 + y^2x^4, \\ z^3 &= x^3 + y^3 + x^3y^4 - x^1y^2 + x^2y^1, \\ z^4 &= x^4 + y^4 + x^4y^4. \end{aligned}$$

18. The *kernel* of a Moufang loop M is a set of elements z from M for which the identity of associativity $(xy)z = x(yz)$ holds for any x and y from M. Prove that:

i) The kernel is an associative subloop (i.e. a group) in M; and

ii) The tangent Lie algebra to the kernel of a local analytic Moufang loop M coincides with the J-kernel of the Mal'tsev algebra tangent to this loop.

19. A regular representation P of the Mal'tsev algebra is defined by the formula $P(x) = [x,a]$, where the commutator is defined by formula (4.71). Prove that the image of the J-kernel of the algebra A_4 under the regular representation is contained in the algebra of interior differentiations of the algebra A_4 itself as well as of its triple Lie system.

NOTES

4.1. Necessity of realization of relations (4.1) for the webs B_m and similar equations for the webs B_l and B_m (see Theorem 4.5) has been proved in the joint paper [AS 71a] by the authors of this book. Sufficiency of these relations (Theorem 4.4) has been proved in [F 78] by Fedorova who systematically investigated the multidimensional Bol three-webs. In particular, she proved Theorems 4.1–4.3 and 4.5. However, the Bianci identities (4.17) went unnoticed by her. Later, P. O. Mikheev and L. V. Sabinin found them during the study of analytic Bol loops [SM 82b].

The relationship between the locally symmetric spaces and the triple Lie systems is well-known (see, for example, [Lo 69], [Tr 89] and [MiS 88]). We study properties of the symmetric space associated with a web B_m following the scheme suggested in [S 89a].

The Bol algebra was introduced in [SM 82b].

4.2. The isoclinic Bol three-webs were first studied in [A 73a]. A detailed classification of four-dimensional Bol webs has been given by Ivanov (see [I 73], [I 74], [I 75] and [I 76]).

4.3. The results of Section 4.3 are due to Fedorova.

4.4. Relations (4.61), which the curvature tensor of a Moufang three-web satisfies, were obtained in [AS 71a]. Sufficiency of these relations and Theorems 4.18 and 4.19 was proved in [AS 71b]. Other properties of Moufang three-webs, indicated in Section 4.4, were obtained in [S 89a].

The relationship between the Mal'tsev algebras and the triple Lie algebras was investigated by many authors (see the survey [MiS 88]). The structure of the reductive space arising in a Moufang three-web was first described in [S 89a].

4.5. The closed form equations of the eight-dimensional Moufang three-web were found in Section 4.5 by the same procedure as in the paper [A 77a]. The interpretation of these results in terms of the corresponding Mal'tsev algebra was presented in [S 89a].

Chapter 5

Closed G-Structures Associated with Three-Webs

5.1 Closed G-Structures on a Smooth Manifold

1. In this chapter we will need some additional facts from the general theory of G-structures. Some basic information on G-structures has been given in Section 1.1. We noted there that a G-structure X_G in a smooth manifold X of dimension n is a subbundle of the frame bundle $\mathcal{R}(X)$, with the same base X as $\mathcal{R}(X)$, which is defined by a Lie group G – a subgroup of the general linear group $\mathbf{GL}(n)$. This subbundle can be given by differential equations (1.8). If a point $p \in X$ is fixed, the forms σ^α, $\alpha = 1, \ldots, \rho$ in these equations are invariant forms of the group G. The structure equations of the G-structure X_G can be written in the form:

$$d\omega^u = c^u_{v\alpha}\omega^v \wedge \sigma^\alpha + a^u_{vw}\omega^v \wedge \omega^w, \tag{5.1}$$

where ω^u, $u, v, w = 1, \ldots, n$, are basis forms of the manifold X, the coefficients $c^u_{v\alpha}$ define the imbedding of the group G into the group $\mathbf{GL}(n)$, and the quantities a^u_{vw} form the *first structure object* of the G-structure X_G. The latter is connected with the differential neighborhood of second order of the point p. The forms σ^α satisfy the equations:

$$d\sigma^\alpha = c^\alpha_{\beta\gamma}\sigma^\beta \wedge \sigma^\gamma + \omega^u \wedge \sigma^\alpha_u, \tag{5.2}$$

which, for $\omega^u = 0$, become the structure equations (1.6) of the Lie group G.

Exterior differentiation of equation (1.5) gives:

$$(\nabla a^u_{vw} - c^u_{v\alpha}\sigma^\alpha_w) \wedge \omega^v \wedge \omega^w + 2a^\kappa_{vw}a^u_{\kappa z}\omega^v \wedge \omega^w \wedge \omega^z = 0, \tag{5.3}$$

where

$$\nabla a^u_{vw} = da^u_{vw} - a^u_{zw}c^z_{v\alpha}\sigma^\alpha - a^u_{vz}c^z_{w\alpha}\sigma^\alpha + a^z_{vw}c^u_{z\alpha}\sigma^\alpha, \tag{5.4}$$

and $u, v, w, x, y, z = 1, \ldots, n$. It follows from equation (5.3) that the forms $\nabla a^u_{vw} - c^u_{v\alpha}\sigma^\alpha_w$ are principal forms, i.e. they can be expressed in terms of the basis form ω^u alone:

$$\nabla a^u_{vw} - c^u_{[v|\alpha|}\sigma^\alpha_{w]} = a^u_{vwz}\omega^z. \tag{5.5}$$

Substituting this expression into equation (5.3), we obtain the relations:

$$a^u_{[vwz]} + 2a^\kappa_{[vw}a^u_{|\kappa|z]} = 0, \tag{5.6}$$

which are called the *Bianci–Cartan identities* of the G-structure under consideration. The quantities a^u_{vwz} in equations (5.5) are related to the differential neighborhood of third order of a point of the manifold X. Together with the quantities a^u_{vw}, they form the *second structure object* of the G-structure X_G.

2. Consider relations (5.5) in more detail. If a point $p \in X$ is fixed, i.e. if $\omega^u = 0$, they have the form:

$$\nabla_\delta a^u_{vw} = c^u_{[v|\alpha|}\theta^\alpha_{w]},$$

where

$$\nabla_\delta a^u_{vw} = \nabla a^u_{vw}(\delta), \quad \theta^\alpha_w = \sigma^\alpha_w(\delta),$$

and the symbol δ, as in Section 1.1, denotes differentiation with respect to secondary parameters.

Set

$$\theta^u_{vw} = c^u_{[v|\alpha|}\theta^\alpha_{w]}$$

and denote by L the linear span of the forms θ^α_w and by L' its subspace spanned by the forms θ^α_{vw}. Let $\nu = \dim\ L, \ \ \nu' = \dim\ L'$. Two cases are possible:

(a) $\nu' = \nu$, i.e. $L' = L$, and

(b) $\nu' < \nu$, i.e. $L' \subset L$.

The forms θ^α_{vw} allow us to make a specialization of the structure object a^u_{vw}. In this specialization we can reduce ν' components of this object to zero. The remaining nonzero components of the object a^u_{vw} constitute the first structure tensor of the G-structure X_G under consideration.

As equations (5.5) show, after this specialization, the forms θ^α_{vw} are reduced to zero, and the forms $\sigma^u_{vw} = c^u_{[v|\alpha|}\sigma^\alpha_{w]}$ become linear combinations of the basis forms ω^u. If $L' = L$, then the forms σ^α_w will also be linear combinations of the forms ω^u. Then, equations (5.2) take the form:

$$d\sigma^\alpha = c^\alpha_{\beta\gamma}\sigma^\beta \wedge \sigma^\gamma + b^\alpha_{uv}\omega^u \wedge \omega^v, \tag{5.7}$$

and a connection arises in the G-structure X_G. This connection is called the *G-connection.* The quantities b^α_{uv} form the curvature tensor of this connection. In this case, the G-structure X_G is called the *G-structure of the first type.*

If $L' \subset L$, i.e. $\nu' < \nu$, then a part of the forms σ^α_v remains linearly independent in X_G. In this case, the consecutive prolongations of equations (5.2) will introduce a series of new forms $\sigma^\alpha_{v_1 v_2}, \sigma^\alpha_{v_1 v_2, v_3}, \ldots$. If $\omega^u = 0$, these forms are symmetric in the lower indices and they, together with the forms θ^α and θ^α_v, are invariant forms of the differential prolongations $G', G'', \ldots$ of the structure group G of the G-structure X_G under discussion. The consecutive prolongations of equations (5.5) will introduce new quantities $a^u_{v_1 v_2 v_3 v_4}, a^u_{v_1 v_2 v_3 v_4, v_5}, \ldots$ connected with the differential neighborhoods of fourth, fifth and higher orders of the point p of the manifold X. These quantities, together with a^u_{vw} and a^u_{vwz}, form the third, fourth and higher order structure objects of the G-structure X_G.

Two situations are possible:

(b$_1$) After specialization of the structure object $a^u_{vw}, a^u_{v_1 v_2 v_3}, \ldots, a^u_{v_1 \ldots v_s}$, all forms $\sigma^\alpha_{v_1 \ldots v_s}$ can be expressed in terms of the basis forms ω^u. In this case. the structure X_G is called the *structure of finite type s*. As to the forms $\sigma^\alpha_{v_1 \ldots v_q}, q < s$, for the structure of finite type s, a part of these forms or all of them can turn out to be independent on the basis forms ω^u of the manifold X. Then, the forms $\sigma^\alpha_v, \sigma^\alpha_{v_1 \ldots v_q}, q < s$, determine the group $\hat{G}$, which is a prolongation of the group G of the structure X_G, and in the manifold X a $\hat{G}$-connection arises.

(b$_2$) On each step, after the specialization of the structure object, a part of the forms $\sigma^\alpha_v, \sigma^\alpha_{v_1 v_2}, \ldots, \sigma^\alpha_{v_1 \ldots v_s}, \ldots$ will remain independent on the basis forms ω^u of the manifold X. In this case, the structure X_G is called the *structure of infinite type.*

For example, a Riemannian structure, defined in the manifold X by a non-degenerate invariant quadratic form $g(\xi, \xi)$, $\xi \in T_p(X)$, is a G-structure of finite type one, since it defines an invariant Levi–Civita connection in the frame bundle over X. A conformal structure, defined in the manifold X by a non-degenerate relatively invariant quadratic form $g(\xi, \xi)$, is a G-structure of finite type two. Similarly a Grassmann structure is a G-structure of finite type two. An almost complex structure, defined in the manifold X of even dimension by the field of an affinor J, satisfying the condition $J^2 = -\mathrm{id}$, is a G-structure of infinite type, since it is impossible to construct an invariant connection on it by means of the specialization of the series of the fundamental objects.

The G-structure, defined in the manifold X of dimension $n = 2r$ by a three-web $W = (X, \lambda_\alpha)$, where λ_α, $\alpha = 1, 2, 3$, are foliations of dimension r, is a G-structure of finite type one, since invariant connections in it are defined in a subbundle of frames. In Chapter 1 we named this structure the G_W-structure.

3. Let X_G be a G-structure of finite type one. Its structure equations have the form (5.1) and (5.7), where a^u_{vw} is the tensor obtained by means of the specialization of the first structure object. The tensor a^u_{vw} is called the *torsion tensor* of the G-structure under consideration, and the tensor b^α_{uv} in equations (5.7) is called its *curvature tensor*. As a result of covariant differentiation of these two tensors in the connection constructed in the structure X_G, we obtain two series of tensors:

$$a^u_{v_1 \ldots v_s}, \quad b^u_{v_1 \ldots v_{s-1}}, \quad s \geq 3. \tag{5.8}$$

They are connected with the differential neighborhood of order s of the structure X_G and satisfy a series of closed form equations of the same type as (5.6).

Definition 5.1 A G-structure X_G of finite type one is said to be *closed of order k* if the tensors (5.8) of order $s = k+1$ are comitants of the tensors of orders $s \leq k$, i.e. the following relations, invariant relative to the group G, hold:

$$\begin{aligned} a^u_{v_1 \ldots v_k v_{k+1}} &= F^u_{v_1 \ldots v_{k+1}}(a^v_{v_1 v_2}, b^\alpha_{v_1 v_2}, a^v_{v_1 v_2 v_3}, \ldots, b^v_{v_1 \ldots v_{k-1}}, a^v_{v_1 \ldots v_k}), \\ b^\alpha_{v_1 \ldots v_k} &= F^\alpha_{v_1 \ldots v_k}(a^v_{v_1 v_2}, b^\alpha_{v_1 v_2}, a^v_{v_1 v_2 v_3}, \ldots, b^v_{v_1 \ldots v_{k-1}}, a^v_{v_1 \ldots v_k}), \end{aligned} \tag{5.9}$$

The first example of closed G-structures is *parallelizable* G-structures (also called *locally flat*), which are characterized by the conditions:

$$a^u_{vw} = 0, \quad b^\alpha_{vw} = 0.$$

These closed G-structures are of order one.

A G-structure, defined by an r-parameter Lie group, is a closed G-structure of order two, since for it, we have $\omega^u = 0$ and $b^\alpha_{uv} = 0$, and the structure is defined by the structure equations:

$$d\sigma^\alpha = c^\alpha_{\beta\gamma}\sigma^\beta \wedge \sigma^\gamma, \quad \alpha, \beta, \gamma = 1, \ldots, r.$$

An example of closed G-structure of order three is provided by locally symmetric Riemannian spaces or spaces with an affine connection which are defined by the structure equations:

$$d\omega^u = \omega^v \wedge \sigma^u_v, \quad d\sigma^u_v = \sigma^w_v \wedge \sigma^u_w + R^u_{vwz}\omega^w \wedge \omega^z, \quad \nabla R^u_{vwz} = 0. \tag{5.10}$$

When we will continue our study of multidimensional three-webs, we will be able to discover new classes of closed G-structures.

A closed G-structure of order k is defined by equations (5.1), (5.7) and the Pfaffian equations

$$\begin{aligned} \nabla a^u_{v_1 \ldots v_s} &= a^u_{v_1 \ldots v_s v_{s+1}}\omega^{v_{s+1}}, \quad s = 2, \ldots, k, \\ \nabla b^\alpha_{v_1 \ldots v_{s-1}} &= b^\alpha_{v_1 \ldots v_{s-1} v_s}\omega^{v_s}, \quad s = 3, \ldots, k, \end{aligned} \tag{5.11}$$

where the tensors $a^u_{v_1 \dots v_k v_{k+1}}$ and $b^\alpha_{v_1 \dots v_k}$ can be expressed by formulas (5.9). Moreover, the tensors $a^u_{v_1 v_2}, b^\alpha_{v_1 v_2}, a^u_{v_1 v_2 v_3}, \dots, b^\alpha_{v_1 \dots v_{k-1}}, a^u_{v_1 \dots v_k}$ satisfy a series of closed form equations:

$$\Phi_\omega(a^u_{v_1 v_2}, b^\alpha_{v_1 v_2}, a^u_{v_1 v_2 v_3}, \dots, b^\alpha_{v_1 \dots v_{k-1}}, a^u_{v_1 \dots v_k}) = 0, \tag{5.12}$$

where the index ω runs over a certain range of values. System (5.12) contains relations (5.6) and similar relations of orders $s \leq k$. Moreover, the system (5.11)–(5.12) must be closed under the operation of exterior differentiation, i.e. its exterior differentiation should not create new closed form equations which are different from (5.12).

For example, for a symmetric space with an affine connection, the system of equations (5.12) can be written in the form:

$$\begin{array}{l} R^u_{v(wz)} = 0, \quad R^u_{[vwz]} = 0, \\ -R^w_{v_1 v_2 v_3} R^u_{w v_4 v_5} + R^u_{w v_2 v_3} R^w_{v_1 v_4 v_5} + R^u_{v_1 w v_3} R^w_{v_2 v_4 v_5} + R^u_{v_1 v_2 w} R^w_{v_3 v_4 v_5} = 0. \end{array} \tag{5.13}$$

These equations can be obtained as a result of exterior differentiation of equations (5.10). Applying the last of equations (5.10), one can easily prove that exterior differentiation of equations (5.10) leads to identities.

In the general case, the system of equations (5.1), (5.7), (5.11) and (5.12) of an arbitrary closed G-structure X_G is closed under the operation of exterior differentiation and does not contain any exterior quadratic equations except the structure equations (5.1) and (5.7). Such systems are called *formally completely integrable.* Their solution exists and depends on N arbitrary constants, where N is the number of linearly independent Pfaffian equations in system (5.11). Thus, we have proved the following theorem:

Theorem 5.2 *A closed G-structure of order k is defined by a formally completely integrable system of differential equations* (5.1), (5.7), (5.11) *and* (5.12). *A solution of this system exists and depends on N arbitrary constants, where N is the number of linearly independent Pfaffian equations in system (5.11).*

5.2 Closed G-Structures Defined by Multidimensional Three-Webs

Closed G-structures defined by three-webs are of special interest, in particular, because they lead to the classes of smooth loops that are a broad generalization of Lie groups. The property that the G_W-structure is closed, is reflected in the structure of the coordinate loops of the web W in the following manner: the canonical expansion of these loops is completely determined by the jet of finite order, i.e. by the first k terms of the expansion. For Lie groups, we have $k = 2$, and all terms of its canonical expansion – the Campbell–Hausdorff series can be expressed in terms of the commutator. The canonical expansion for Moufang loops has a similar structure. In this chaper we consider three-webs, defining closed G-structures of order $k = 3, 4$, and also discuss more general results, which were recently obtained.

1. In this section we will use the vector or index-free form of writing the structure equations of a web $W = (X, \lambda_\alpha)$, which is less cumbersome. We write equations (1.26) and (1.32) in the form:

$$d\underset{1}{\omega} = -\omega \wedge \underset{1}{\omega} + a\underset{1}{\omega} \wedge \underset{2}{\omega}, \; d\underset{2}{\omega} = -\omega \wedge \underset{2}{\omega} - a\underset{2}{\omega} \wedge \underset{2}{\omega}, \tag{5.14}$$

$$d\omega = -\omega \wedge \omega + b\underset{1}{\omega} \wedge \underset{2}{\omega}, \tag{5.15}$$

where $\underset{\alpha}{\omega} = (\underset{\alpha}{\omega}^i)$ are vector-valued 1-forms with their values in an r-dimensional space, $\omega = (\omega_j^i)$ is a matrix 1-form, and the symbol $\wedge$ denotes the matrix exterior multiplication. As earlier, a denotes the torsion tensor of the web W, and b denotes its curvature tensor.

The tensors a and b define the vector-valued forms $a(\xi, \eta)$ and $b(\xi, \eta, \zeta)$, the first of which is skew-symmetric:

$$a(\xi, \eta) = -a(\eta, \xi) \tag{5.16}$$

These forms allow us to write equations (1.31), which these forms satisfy, in the form:

$$\operatorname{alt}(b(\xi, \eta, \zeta) - 2a(a(\xi, \eta), \zeta)) = 0. \tag{5.17}$$

Consider further the differential prolongations of equations (5.14) and (5.15), i.e. consider equations (1.33) and (1.38):

$$\nabla a = \underset{1}{b}\underset{1}{\omega} + \underset{2}{b}\underset{2}{\omega}, \quad \nabla b = \underset{1}{c}\underset{1}{\omega} + \underset{2}{c}\underset{2}{\omega}, \tag{5.18}$$

where ∇ is the operator of covariant differentiation in the canonical connection Γ and

$$\underset{1}{b} = (b^i_{[j|l|k]}), \quad \underset{2}{b} = (b^i_{[jk]l}), \quad \underset{1}{c} = (\underset{1}{c}^i_{jklm}), \quad \underset{2}{c} = (\underset{2}{c}^i_{jklm})$$

The tensors $\underset{1}{c}$ and $\underset{2}{c}$ are connected with the tensors a and b by relations (1.39) and (1.40).

We introduce new notations: $c_2 = a$, $c_3 = b$, $c_4 = (\underset{1}{c}, \underset{2}{c})$ and write equations (5.18) in the form:

$$\nabla c_2 = c_3'\theta, \quad \nabla c_3 = c_4'\theta, \tag{5.19}$$

where θ denotes the 1-form $(\underset{1}{\omega}, \underset{2}{\omega})$ with values in the $(2r)$-dimensional space $T_p(X)$. The differential prolongation of equations (5.19) can be written in a similar way:

$$\nabla c_s = c_{s+1}\theta. \tag{5.20}$$

When we follow this procedure, we obtain an infinite set of differential equations if the web W is analytic. The tensor c_s of type $\binom{1}{s}$ is connected with the differential neighborhood of order s of a point p of the manifold X. In Chapter 1 we called the tensors c_s the fundamental tensors of the web W.

The fundamental tensors are connected by a sequence of closed form equations of type (5.12). We write these relations as follows:

$$\begin{array}{l} \varphi_2(c_2) = 0, \\ \varphi_3(c_2, c_3) = 0, \\ \dots\dots\dots\dots\dots\dots \\ \varphi_s(c_2, c_3, \dots, c_s) = 0, \\ \dots\dots\dots\dots\dots\dots \end{array} \tag{5.21}$$

The first of relations (5.21) is equivalent to identity (5.16), the second one equivalent to identity (5.17), and others can be obtained if we differentiate equations (5.20) and previous closed form equations. Relations (5.21) are invariant with respect to the structure group $G = \mathbf{GL}(r)$ of the G_W-structure, defined by the web W in the manifold X.

2. Let a three-web W be parametrized in a neighborhood of a point p in a standard way (Section 2.5). Then we can assume that the values $c_s(p)$ of the fundamental tensors of the web W at the point p are given in the tangent space T_e to the coordinate loop l_p at the identity element e. Since these tensors are of type $\binom{1}{s}$, in T_e, they determine the corresponding s-ary operations. The tensors $c_2 = a$ and $c_3 = b$ determine the binary and ternary operations, respectively. These two operations together generate the W-algebra (Section 2.5), the tensor $c_4 = (\underset{1}{c}, \underset{2}{c})$ determines two quaternary operations, etc.

Definition 5.3 The tangent space T_e to the coordinate loop l_p of a three-web W at the identity element e together with the set of binary, ternary, $\dots$, k-ary operations, determined by the fundamental tensors of this web, are called the *tangent* W_k*-algebra* of the web W at the point p.

For $k = 3$, we have the W algebra introduced in Secition 2.5.

On the other hand, in the tangent space T_e to the loop l_p, there is defined an additional sequence of algebras, namely Λ_k-algebras of this loop. Let us recall (Section 2.6) that the operations in these algebras are defined by the coefficients of the canonical expansion of the loop l_p, and there is one binary operation, two ternary operations, etc, and $k-1$ operations of k-ary operations.

The binary operations in the W_2-algebra and Λ_2-algebra are determined by the tensor a and the form $\underset{2}{\Lambda}$, respectively, and they differ only by sign. The ternary operations in the W_3-algebra and Λ_3-algebra are determined by the tensor b and the forms $\underset{2,1}{\Lambda}, \underset{1,2}{\Lambda}$, respectively, and these operations can be also expressed in terms of one another (Problem **37** to Chapter 2). It turns out that a similar result holds for any k:

Theorem 5.4 ([S 87a]) *For any k, the tangent W_k-algebra of a three-web W at a point p and the Λ_k-algebra of its coordinate loop l_p are equivalent, i.e. the operations of one of these algebras can be expressed in terms of the operations of another one and vice versa.*

Proof. The proof is obvious in one direction. As follows from the results of Section 1.6, all fundamental tensors of a web can be expressed in terms of the partial derivatives of the function $z = f(x, y)$, defining this web in local coordinates. On the other hand, we can assume that the equation $z = f(x, y)$ defines the operations in the coordinate loop l_p of the web in a neighborhood of the point $p(0, \ldots, 0)$ and the coordinates x^i are canonical. Expanding the function $f(x, y)$ in the Taylor series, we find the canonical expansion of the loop l_p. Its terms are the forms $\underset{s}{\Lambda}$ whose coefficients are partial derivatives of the function f (at the point p). Since the fundamental tensors of the web W can be expressed in terms of these partial derivatives, the theorem is proved in one direction.

The proof of the converse is connected with complicated calculations. We will not present this proof here. ■

Theorem 5.4 extends the correspondence between three-webs and their coordinate loops to the infinitesimal structures – the tangent algebras associated with them. In particular, Theorem 5.1 allows us to consider as equivalent two previously proved facts: the fact that a transversal vector ξ is an eigenvector of the tensor b^i_{jkl} and its covariant derivatives (Section 1.9) and the fact that a tangent vector to a one-parameter subloop is an eigenvector for all tensors $\underset{k,l}{\Lambda}$ (Theorem 2.26).

3. Suppose now that the G_W-structure, defined by a three-web W in the manifold X, is a closed G-structure of order k. Then, according to the general definition, its structure tensor c_{k+1} of order $k + 1$ is a comitant of the tensors $c_2, \ldots, c_k$:

$$c_{k+1} = F(c_2, c_3, \ldots, c_k). \tag{5.22}$$

The web under consideration is determined by the system of exterior equations (5.14) and (5.15), the system of Pfaffian equations (5.20), where $s = 2, \ldots, k$, and the tensor c_{k+1} has the expression (5.22), and the system of closed form equations:

$$\Phi_\omega(c_2, c_3, \ldots, c_k) = 0, \tag{5.23}$$

which can be obtained by differentiation of equations (5.14), (5.15) and (5.20). In addition, we assume that system (5.23) contains all closed form equations, obtained as a result of covariant differentiation of the equations of this system with application of formulas (5.20). So, this system, and along with it the whole system (5.14), (5.15), (5.20) and (5.23), is closed with respect to exterior differentiation. Finally, suppose that system (5.23) is consistent.

We will now prove the following important theorem:

Theorem 5.5 *Let, at a point p_0 of a C^{k+1}-manifold X of dimension $2r$, the following objects be given:*

a) *Three r-dimensional subspaces T_α, $\alpha = 1, 2, 3$, of the tangent space $T_{p_0}(X)$, which are in general position;*

b) *Vectorial frames $\{\underset{\alpha}{e}_i\}$ in the subspaces T_α, connected by the condition $\underset{1}{e}_i+\underset{1}{e}_i+\underset{1}{e}_i = 0$ and admitting the concordant transformations which constitute the group $G = \mathbf{GL}(r)$; and*

c) *The values $c_2^0, \ldots, c_k^0$ of the tensors $c_2, \ldots, c_k$ in the frames indicated in* b), *that satisfy closed form equations* (5.23).

Then, in a neighborhood of the point p_0 of the manifold X, there exists a unique three-web W, possessing a closed G_W-structure of finite type k, for which:

i) *The leaves, passing through the point p_0, are tangent to the subspaces T_α; and*

ii) *At the point p_0, the structure tensors $c_2, \ldots, c_k$ have the given values $c_2^0, \ldots, c_k^0$, and satisfy relations* (5.23) *in the neighborhood of the point p_0 indicated above.*

Proof. In fact, consider the system of equations (5.14), (5.15), (5.20) and (5.23) and suppose that the forms $\underset{1}{\omega}, \underset{2}{\omega}, \omega$ and the tensors $c_2, \ldots, c_k$ are unknown in this system. Since the system is closed with respect to exterior differentiiation, it is completely integrable. To find its solution, one should know the values of the unknown forms and functions at the point p_0 for an initial family of frames indicated in the condition of the theorem. Set $\underset{\alpha}{\omega}(T_\alpha)\Big|_{p_0} = 0$, $\omega\Big|_{p_0} = \pi$, where $\pi = (\pi_j^i)$ are the structure forms of the group $\mathbf{GL}(r)$ of admissible transformations of these frames. Finally, the initial values $c_2^0, \ldots, c_k^0$ of the tensors we are looking for are given in condition c) of the theorem. These initial conditions allow us to find a unique solution of the system (5.14), (5.15), (5.20) and (5.23) in a neighborhood of the point p_0, i.e. they allow us to find, in this neighborhood, a three-web possessing the properties indicated in the theorem. ■

Note that the values $c_2^0, \ldots, c_k^0$ of the tensors $c_2, \ldots, c_k$, satisfying relations (5.23), define the tangent W_k-algebra of the three-web at the point p_0 (see subsection **2**). Thus, Theorem 5.4 can be also formulated as follows:

The tangent W_k-algebra, given at the point p_0 of the manifold X and satisfying the closure conditions (5.23), *in a neighborhood U of the point p_0, determines a three-web W with a closed G-structure of order k.*

Next, consider the canonical expansion of the local loop l_p of a three-web W with a closed G-structure of order k. Since the G-structure is closed, the tensor c_{k+1} can be expressed in terms of the tensors $c_2, \ldots, c_k$ by formula (5.12). Differentiating this formula and eliminating the quantity c_{k+1} by means of equation (5.12) itself, we find an expression for the tensor c_{k+2} also in terms of the tensors $c_2, \ldots, c_k$. Continuing this procedure, we will be able to express all tensors c_s, $s \geq k+1$, in terms of the tensors $c_2, \ldots, c_k$. By Theorem 5.3, all coefficients of the forms $\underset{2}{\Lambda}, \underset{3}{\Lambda}, \ldots, \underset{k}{\Lambda}$ of the canonical expansion of the local loop l_p can be expressed in terms of the tensors $c_2, \ldots, c_k$ given at the point p. By the same Theorem 5.3, the tensors $c_2(p), \ldots, c_k(p)$ can be expressed in terms of the forms $\underset{2}{\Lambda}, \underset{3}{\Lambda}, \ldots, \underset{k}{\Lambda}$. Thus, we obtain, that all forms $\underset{s}{\Lambda}$ of the canonical expansion can be expressed in terms of its first $k-1$ forms $\underset{2}{\Lambda}, \underset{3}{\Lambda}, \ldots, \underset{k}{\Lambda}$.

The power series obtained in this way converges in a neighborhood of the identity element of the loop l_p, since this loop is a local loop of a three-web whose existence has been proved in Theorem 5.4.

Since the forms $\underset{2}{\Lambda}, \underset{3}{\Lambda}, \ldots, \underset{k}{\Lambda}$, satisfying the equations which are equivalent to equations (5.23), define a local $\underset{k}{\Lambda}$-algebra at the point p, the following theorem holds:

Theorem 5.6 *A local $\underset{k}{\Lambda}$-algebra completely determines a local loop l_p of a three-web W with a closed G_W-structure.* ■

This theorem is a far reaching generalization of the converse part of Lie's Third Fundamental Theorem according to which a local Lie group is completely determined by its Lie algebra.

4. Let us show that a three-web, on which one of the classical closure conditions indicated in Chapter 2, holds, possesses a closed G_W-structure.

Theorem 5.7 *Three-webs, on which one of the closure conditions T, R, M, B_l, B_r or B_m holds, determine in the manifold X, a closed G_W-structure. Moreover, for the web T, this structure is of order one, for the webs R and M, this structure is of order two, and for the webs B_l, B_r and B_m, this structure is of order three. The canonical expansions for local loops of the three-web are trivial, for the webs R and M, they are completely determined by their torsion tensor, and for the webs B_l, B_r and B_m, they are completely determined by their torsion and curvature tensors, given at a point and satisfying closed form equations of type* (5.23).

Proof. In fact, realization of condition T is equivalent to the parallelizability of a three-web W or vanishing of its torsion and curvature tensors. In this case, system (5.14) and (5.15) takes the form:

$$d\underset{\alpha}{\omega} = -\omega \wedge \underset{\alpha}{\omega}, d\omega = -\omega \wedge \omega,$$

and, as it is easy to show, it is closed under the operation of exterior differentiation. Because of this, the G_W-structure, determined in the manifold X by the web under consideration, is a closed G-structure of order one. As shown earlier, the local loops of the web T are abelian groups, and their canonical expansion has the form:

$$u \cdot v = u + v.$$

A three-web, on which the closure condition R holds, is characterized by the vanishing of its curvature tensor, which implies that the system of its structure equations consists of equations (5.14) and also the equations:

$$d\omega = -\omega \wedge \omega, \tag{5.24}$$

which can be obtained from equations (5.15) by substituting $b = 0$. By (5.18), the Pfaffian equations, defining this web, can be written in the form:

$$\nabla a = 0. \tag{5.25}$$

System (5.14), (5.24) and (5.25) is closed with respect to the operation of exterior differentiation. This implies that the G_W-structure, determined by a group three-web on the manifold X, is closed. Since the tensor a is defined in the second differential neighborhood, the order of this G_W-structure is two. In this case, closed form equations (5.23) are reduced to the Jacobi identities, which the torsion tensor of a group web satisfies (Section 1.5). As was proved earlier, the local W-algebras are Lie algebras, the local loops are Lie groups, and their canonical expansions coincide with the classical Campbell–Hausdorff formula:

$$u \cdot v = u + v + \frac{1}{2}[u, v] + \frac{1}{12}[u, [u, v]] + \frac{1}{12}[[u, v], v] + \dots, \tag{5.26}$$

where $[u, v] = -2a(u, v)$ is the commutator in the Lie algebra.

For the web M, the curvature tensor can be expressed in terms of its torsion tensor by formula (4.21). Thus, the structure equations of this web have the form (5.14) and (5.15), where

$$b = (2a^m_{[jk} a^i_{|m|l]}) = \frac{2}{3} J(a). \tag{5.27}$$

Therefore, the covariant differential ∇a of the torsion tensor a in the connection Γ is nonzero. However, as was proved in Section 4.4, it is zero in the middle connection $\overset{*}{\Gamma}$:

$$\overset{*}{\nabla} a = 0. \tag{5.28}$$

This implies that the system of equations (5.14) and (5.15), where the tensor b is given by formula (5.27), is closed with respect to the operation of exterior differentiation. The G-structure, determined by this system, is again of order two. The tensor a

of the web M satisfies the Sagle identity (4.79), to which closed form equations are reduced in the case under consideration. The local W-algebras of a Moufang web are Mal'tsev algebras, and the local loops are Moufang loops.

The canonical expansion for Moufang loops has the same form (5.26) as for Lie groups, since a Mal'tsev algebra is a binary Lie algebra, i.e. any two vectors in this algebra determine a Lie subalgebra.

For the webs B_l, B_r and B_m, the tensor c_4, connected with the fourth order differential neighborhood, can be expressed in terms of the tensors a and b. For the web B_m, this expression has the form (4.10). Since the tensor b is connected with the third order differential neighborhood of the web under discussion, the G_W-structure, determined by a Bol web, is a closed G-structure of order three.

The tensors a and b of the Bol web B_m allow us to construct the tensor R (see formula (4.6)), which is covariantly constant in the connection $\widetilde{\Gamma}$: $\widetilde{\nabla} R = 0$. The local W-algebras of the Bol web B_m, determined by the tensors a and b, are equivalent to the Bol algebras, for which the operations are defined by the tensors a and R and satisfy relations (4.23) and (4.25).

As was shown in Section 2.2, the local loops of a Bol three-web are Bol loops. Their canonical expansions can be expressed in terms of the tensors a and b or the tensors a and b, both satisfying the closed form equations indicated above. However, these canonical expansions in their explicit form have not yet been obtained.

5.3 Four-Dimensional Hexagonal Three-Webs

1. In this section we will prove that the G_W-structure, determined by a four-dimensional hexagonal three-web, is closed, and a four-dimensional hexagonal three-web itself is algebraizable.

As was indicated in Chapter 3 (p. 112), the torsion tensor of a four-dimensional three-web always has the form:

$$a^i_{jk} = a_{[j}\delta^i_{k]},$$

and this implies that the structure equations (1.26) of this web can be written in the form:

$$\begin{aligned} d\underset{1}{\omega}^i &= \underset{1}{\omega}^j \wedge \omega^i_j + a_j \underset{1}{\omega}^j \wedge \underset{1}{\omega}^k, \\ d\underset{2}{\omega}^i &= \underset{2}{\omega}^j \wedge \omega^i_j - a_j \underset{2}{\omega}^j \wedge \underset{2}{\omega}^k. \end{aligned} \tag{5.29}$$

Here and throughout this section the Latin indices i, j, k, l take the values 1 and 2. Exterior differentiation of equation (5.29) leads to the following exterior quadratic equation:

$$\begin{aligned} \Omega^i_j \wedge \underset{1}{\omega}^j - \nabla a_j \wedge \underset{1}{\omega}^j \wedge \underset{1}{\omega}^i = 0, \\ \Omega^i_j \wedge \underset{2}{\omega}^j + \nabla a_j \wedge \underset{2}{\omega}^j \wedge \underset{2}{\omega}^i = 0, \end{aligned} \tag{5.30}$$

where as earlier ∇ is the operator of covariant differentiation in the canonical connection Γ, and $\Omega^i_j = d\omega^i_j - \omega^k_j \wedge \omega^i_k$ is the curvature form of this connection. It follows from equations (5.30) that the forms Ω^i_j and ∇a_j have the form:

$$\Omega^i_j = b^i_{jkl}\underset{1}{\omega}^k \wedge \underset{2}{\omega}^l, \tag{5.31}$$

$$\nabla a_j = p_{jk}\underset{1}{\omega}^k + q_{jk}\underset{2}{\omega}^k. \tag{5.32}$$

Here p_{jk} and q_{jk}, unlike similar quantities in Section 3.2, in general, are not symmetric in the indices j and k. Substituting expansions (5.31) and (5.32) into equations (5.30), we obtain

$$b^i_{[jk]l} = q_{[j|l|}\delta^i_{k]}, \quad b^i_{[j|l|k]} = p_{[j|l|}\delta^i_{k]}. \tag{5.33}$$

Suppose further that the web W under consideration is hexagonal. As was proved in Section 3.1, such webs are characterized by the condition:

$$b^i_{(jkl)} = 0. \tag{5.34}$$

By means of formulas (5.33) and (5.34), expression (5.32) for the torsion tensor can be reduced to the form:

$$b^i_{jkl} = 2b^1_{[j|k|}\delta^i_{l]} + 2b^2_{[j|l|}\delta^i_{k]}, \tag{5.35}$$

where

$$b^1_{jk} = \frac{2}{3}p_{jk} - \frac{1}{3}q_{jk}, \quad b^2_{jk} = -\frac{1}{3}p_{jk} + \frac{2}{3}q_{jk}. \tag{5.36}$$

We can find the quantities p_{jk} and q_{jk} from equations (5.36):

$$p_{jk} = 2b^1_{jk} + b^2_{jk}, \quad q_{jk} = b^1_{jk} + 2b^2_{jk},$$

Set also

$$b^3_{jk} = -b^1_{jk} - b^2_{jk}. \tag{5.37}$$

Then, the previous equations can be written in the form:

$$p_{jk} = b^1_{jk} - b^3_{jk}, \quad q_{jk} = b^2_{jk} - b^3_{jk},$$

and relations (5.31) and (5.32) can be written as:

$$d\omega^i_j - \omega^k_j \wedge \omega^i_k = 2(b^1_{[j|k|}\delta^i_{l]} + b^2_{[j|l|}\delta^i_{k]})\underset{1}{\omega}^k \wedge \underset{2}{\omega}^l, \tag{5.38}$$

$$\nabla a_j = (b^1_{jk} - b^3_{jk})\underset{1}{\omega}^k + (b^2_{jk} - b^3_{jk})\underset{2}{\omega}^k. \tag{5.39}$$

Equations (5.29), (5.38) and (5.39) are the structure equations of four-dimensional hexagonal three-webs. Note also that equations (1.31), connecting the torsion and curvature tensors of a three-web, are automatically satisfied for four-dimensional three-webs, since the indices j, k and l, in which the alternation is performed, accept only the values 1 and 2.

2. The following theorem describes a G_W-structure, associated with a four-dimensional hexagonal three-web.

Theorem 5.8 *A G_W-structure, determined on a four-dimensional manifold X by a hexagonal three-web, is a closed G-structure of order four.*

Proof. First, we find differential prolongations of equations (5.38) and (5.39). Exterior differentiation of these equations leads to the following exterior equations:

$$\overset{\circ}{\nabla}(b^1_{[j|k|}\delta^i_{l]} + b^2_{[j|l|}\delta^i_{k]}) \wedge \underset{1}{\omega}^k \wedge \underset{2}{\omega}^l = 0, \tag{5.40}$$

$$\overset{\circ}{\nabla}(b^1_{jk} - b^3_{jk}) \wedge \underset{1}{\omega}^k + \overset{\circ}{\nabla}(b^2_{jk} - b^3_{jk}) \wedge \underset{2}{\omega}^k + (b^3_{j(k}a_{l)} - a_j b^1_{lk} - a_j b^2_{kl})\underset{1}{\omega}^k \wedge \underset{2}{\omega}^l = 0, \tag{5.41}$$

where $\overset{\circ}{\nabla}$ denotes the operator of covariant differentiation in the connection Γ, defined by the connection forms:

$$\overset{\circ}{\omega}{}^i_j = \omega^i_j - \frac{1}{2}\delta^i_j a_k(\underset{1}{\omega}^k - \underset{2}{\omega}^k)$$

(see Section 3.2).

It follows from equations (5.40) and (5.41) that the forms $\overset{\circ}{\nabla} b^\alpha_{ij}$ are linear combinations of the basis forms $\underset{1}{\omega}^k$ and $\underset{2}{\omega}^k$. After rather long but uncomplicated calculations, we obtain the following expansions for them:

$$\begin{aligned} \overset{\circ}{\nabla} b^1_{jk} &= -(c_{jkl} - 2b^1_{j(k}a_{l)} - \tfrac{2}{3}a_j b^1_{(kl)} - \tfrac{1}{3}a_j b^1_{lk})\underset{1}{\omega}^l - 2(c_{j[kl]} + \frac{1}{3}a_j b^2_{[kl]})\underset{2}{\omega}^l, \\ \overset{\circ}{\nabla} b^2_{jk} &= -2(c_{j[lk]} + \tfrac{1}{3}a_j b^1_{[lk]})\underset{1}{\omega}^l - (c_{jlk} + 2b^2_{j(k}a_{l)} + \frac{2}{3}a_j b^2_{(kl)} + \frac{1}{3}a_j b^2_{kl})\underset{2}{\omega}^l, \end{aligned} \tag{5.42}$$

where c_{jkl} is a tensor connected with the fourth order differential neighborhood.

Differential prolongation of equations (5.42) leads to the equations:

$$\overset{\circ}{\nabla} c_{ijk} = \underset{1}{c}_{ijkl}\underset{1}{\omega}^l + \underset{2}{c}_{ijkl}\underset{2}{\omega}^l, \tag{5.43}$$

where

$$\begin{aligned} \underset{1}{c}_{ijkl} = {} & 2a_k c_{i[lj]} + \tfrac{2}{3}a_i(c_{[l|j|k]} + c_{k[lj]}) + \tfrac{1}{2}c_{ijk}a_l - 3b^2_{(ij}b^1_{k)l} \\ & + \tfrac{2}{3}b^3_{[jk]}(b^1_{il} - b^3_{il}) + \tfrac{2}{3}(b^1_{ik} - b^2_{ik})(b^1_{[jl]} - b^3_{[jl]}) \\ & - 2b^2_{[ij]}b^1_{kl} + 2b^2_{ij}b^1_{[kl]} - \tfrac{8}{9}a_i b^1_{[jk]}a_l, \end{aligned} \tag{5.44}$$

and

$$\begin{aligned} \underset{2}{c}_{ijkl} = \; & -2a_j c_{i[kl]} - \tfrac{2}{3}a_i(c_{j[kl]} + c_{[lj]k}) - \tfrac{1}{2}c_{ijk}a_l + 3b^1_{(ij}b^2_{k)l} \\ & + \tfrac{2}{3}b^3_{[jk]}(b^2_{il} - b^3_{il}) + \tfrac{2}{3}(b^1_{ij} - b^2_{ij})(b^1_{[kl]} - b^3_{[kl]}) \\ & + 2b^1_{[ik]}b^2_{jl} - 2b^1_{ik}b^2_{[jl]} + \tfrac{8}{9}a_i b^2_{[jk]}a_l. \end{aligned} \tag{5.45}$$

In addition, differential prolongation of equations (5.42) gives an additional closed form equation, relating the tensors a_i, b^α_{ij} and c_{ijk}:

$$a_i a_j b^3_{[kl]} + 2a_j c_{i[kl]} + a_i c_{j[kl]} + 2b^2_{ij}b^1_{[kl]} - 2b^1_{ij}b^2_{[kl]} = 0. \tag{5.46}$$

Equations (5.44) and (5.45) show that the tensors $\underset{1}{c}_{ijkl}$ and $\underset{2}{c}_{ijkl}$, connected with the fifth order differential neighborhood, can be expressed in terms of the tensors a_i, b^α_{ij} and c_{ijk} connected with the differential neighborhoods of order two, three and four. This means that a G_W-structure, determined by a four-dimensional hexagonal three-web in the manifold X, is a closed G-structure of order four. ■

A four-dimensional hexagonal three-web is determined by the closed system of differential equations, consisting of exterior quadratic equations (5.29) and (5.39), Pfaffian equations (5.38), (5.42) and (5.43), where the tensors $\underset{\alpha}{c}_{ijkl}$ are expressed by formulas (5.44) and (5.45), and closed form equations (5.46), which are obtained as a result of differentiation of equations (5.43) and (5.46).

3. The study of the closed form equations, defining a four-dimensional hexagonal three-web, allows us to prove the following theorem:

Theorem 5.9 *Any four-dimensional hexagonal three-web is isoclinic.*

Proof. The complete proof of this theorem is very complicated and remains beyond the scope of this book. We outline here only the main stages of this proof.

As proved in Theorem 3.9, the condition for a four-dimensional three-web to be isoclinic is the symmetry of covariant derivatives p_{jk} and q_{jk} of the covector a_j. But, by (5.36) and (5.37), this condition is equivalent to the symmetry of the tensors b^α_{jk}. Let us prove that the symmetry of one of these tensors implies the symmetry of two others. In fact, suppose, for example, that

$$b^1_{[jk]} = 0. \tag{5.47}$$

Then, it follows from (5.37) that

$$b^2_{[jk]} = -b^3_{[jk]}. \tag{5.48}$$

Differentiate relation (5.47) by means of the operator $\overset{\circ}{\nabla}$. As a result, if we apply equations (5.42), we get

$$c_{j[kl]} - \frac{1}{3}a_j b^3_{[kl]}.$$

Substituting these expressions into equations (5.46), we arrive at the relations:

$$b^1_{ij} b^2_{[kl]} = 0.$$

These imply that either $b^2_{[kl]} = 0$, which completes the proof, or $b^1_{ij} = 0$. In the latter case, equations (5.42) imply that

$$c_{jkl} = 0, \quad a_j b^2_{[kl]} = 0.$$

This gives again $b^2_{[kl]} = 0$, which completes the proof, or $a_j = 0$. However, in the latter case $b^\alpha_{ij} = 0$, and the web under consideration is parallelizable, since its torsion and curvature tensors are zero. We exclude this case from consideration and assume that the covector a_j is not zero.

Thus, the quantities $b^\alpha_{[jk]}$, $\alpha = 1, 2, 3$, are either simultaneously zero, and then the web under consideration is isoclinic, or no one of these quantities is zero. Consider the latter case. Since the indices j and k take only two values, each of the tensors $b^\alpha_{[jk]} = 0$ has only one essential component, and we can set:

$$b^1_{[jk]} = f b^3_{[jk]}, \quad b^2_{[jk]} = g b^3_{[jk]}. \tag{5.49}$$

By (5.37), we find:

$$f + g + 1 = 0.$$

Similarly, the skew-symmetric tensor $c_{j[kl]}$ in equations (5.46) can be represented in the form:

$$c_{j[kl]} = c_j b^3_{[kl]}.$$

Substituting these expressions and relations (5.49) into equation (5.46) and dividing by $b^3_{[jk]}$, we obtain the equation:

$$-a_i a_j + 2a_j c_i + a_i c_j + 2 + b^2_{ij} - 2g b^1_{ij} = 0. \tag{5.50}$$

Alternating equation (5.50) in the indices i and j and applying equations (5.49), we arrive at the condition:

$$a_{[i} c_{j]} = 0,$$

implying the equations:

$$c_j = c a_j, \quad c_{j[kl]} = c a_j b^3_{[kl]}. \tag{5.51}$$

By (5.51), relation (5.50) takes the form:

$$(3c - 1) a_i a_j + 2f b^2_{ij} - 2g b^1_{ij} = 0.$$

This equation and relation (5.37) give the equations:

$$b^1_{ij} = \frac{1}{2}(1-3c)a_i a_j + f b^3_{ij}, \quad b^2_{ij} = \frac{1}{2}(1-3c)a_i a_j + g b^3_{ij}. \tag{5.52}$$

Next, by differentiating the second of relations (5.51) by means of the operator $\overset{\circ}{\nabla}$ and applying equations (5.52), we arrive at a contradiction, which proves that there are no three-webs on which at least one of the tensors $b^\alpha_{[ij]}$ is nonzero. ■

Further, any hexagonal three-web is transversally geodesic and, as we just proved, a four-dimensional hexagonal three-web is also isoclinic. Then, by Theorem 3.17, this web is algebraizable, i.e. it is equivalent to a four-dimensional Grassmann three-web, defined by a cubic surface in a projective space P^3. This proves the following theorem:

Theorem 5.10 *Any four-dimensional hexagonal three-web is equivalent to a Grassmann three-web, generated by a cubic surface in a three-dimensional projective space.* ■

To conclude this section, we return to the definition of a closed G-structure, given in Section 5.1.

The proof of Theorem 5.8 allows us to extend and generalize this notion. In fact, the property that the G-structure determined by a four-dimensional hexagonal three-web is closed follows from the fact that all fundamental tensors of this web can be expressed in terms of the tensors a_i, b^α_{ij} and c_{ijk}, whose covariant derivatives can be also expressed in terms of the same tensors. Thus, a G-structure X_G can also be called closed in the case when the torsion and curvature tensors, determined by this structure, and their covariant derivatives of all orders can be expressed in terms of a finite number of tensors defined in X_G.

5.4 The Closure of the G-Structure Defined by a Multidimensional Hexagonal Three-Web

1. The proof of the fact that the G_W-structure, determined by a four-dimensional hexagonal three-web, is closed, given in previous section, can not be extended to hexagonal three-webs of any dimension, since the fact that the web is four-dimensional, was essential to the proof. On the other hand, one can see from this proof what kind of computational difficulties we should expect in higher dimensions. In this section we propose another proof of closure, and this proof can be applied to any dimension. The proof is less constructive and does not allow us to write directly, as in Section 5.3, all equations characterizing a closed structure. However, this method allows wide generalizations (see [S 87a]).

Consider the structure equations of an arbitrary three-web, which were found in Section 1.4:

$$\nabla b^i_{jkl} = \underset{1}{c}^i_{jklm} \underset{1}{\omega}^m + \underset{2}{c}^i_{jklm} \underset{2}{\omega}^m, \tag{5.53}$$

$$\underset{1}{c}^i_{j[k|l|m]} = b^i_{jpl} a^p_{km}, \quad \underset{2}{c}^i_{jk[lm]} = -b^i_{jkp} a^p_{lm}, \tag{5.54}$$

$$\underset{1}{c}^i_{[jk]ml} - \underset{2}{c}^i_{j[l|k]m} = B^i_{jklm}, \tag{5.55}$$

where

$$B^i_{jklm} = -a^i_{jp} b^p_{klm} - a^i_{pk} b^p_{jlm} + a^p_{jk} b^i_{plm}. \tag{5.56}$$

Set

$$\begin{aligned} \nabla \underset{1}{c}^i_{jklm} &= X^i_{jklmn} \underset{1}{\omega}^n + \underset{2}{Z}^i_{jklmn} \underset{2}{\omega}^n, \\ \nabla \underset{2}{c}^i_{jklm} &= \underset{1}{Z}^i_{jklmn} \underset{1}{\omega}^n + Y^i_{jklmn} \underset{2}{\omega}^n. \end{aligned} \tag{5.57}$$

Differentiating equations (5.53) and applying equations (5.57), we arrive at the relations:

$$X^i_{jkl[mn]} = \underset{1}{c}^i_{jklp} a^p_{mn}, \quad Y^i_{jkl[mn]} = -\underset{2}{c}^i_{jklp} a^p_{mn}, \tag{5.58}$$

$$-\underset{2}{Z}^i_{jklmn} + \underset{1}{Z}^i_{jklnm} = B^i_{jklmn}, \tag{5.59}$$

where

$$B^i_{jklmn} = b^p_{jkl} b^i_{pmn} - b^i_{pkl} b^p_{jmn} - b^i_{jpl} b^p_{kmn} - b^i_{jkp} b^p_{lmn}. \tag{5.60}$$

The tensors $X, Y, \underset{1}{Z}$ and $\underset{2}{Z}$ are connected by additional sequence of relations, that can be obtained as a result of the differentiation of equations (5.54) and (5.55) by means of the operator ∇. Differentiating relations (5.54), we arrive at the equations:

$$\begin{aligned} X^i_{j[k|l|m]n} &= \underset{1}{c}^i_{jpln} a^p_{km} + b^i_{jpl} b^p_{[k|n|m]}, \\ \underset{2}{Z}^i_{j[k|l|m]n} &= \underset{2}{c}^i_{jpln} a^p_{km} + b^i_{jpl} b^p_{[km]n}, \\ Y^i_{jk[lm]n} &= -\underset{2}{c}^i_{jkpn} a^p_{lm} - b^i_{jkp} b^p_{[lm]n}, \\ \underset{1}{Z}^i_{jk[lm]n} &= -\underset{1}{c}^i_{jkpn} a^p_{lm} - b^i_{jkp} b^p_{[l|n|m]}, \end{aligned} \tag{5.61}$$

and differentiation of relations (5.55) gives:

$$\begin{aligned} X^i_{[jk]mln} - \underset{1}{Z}^i_{[j|l|k]mn} &= \underset{1}{B}^i_{jklmn}, \\ \underset{2}{Z}^i_{[jk]mln} - Y^i_{[j|l|k]mn} &= \underset{2}{B}^i_{jklmn}, \end{aligned} \tag{5.62}$$

where $\underset{1}{B}$ and $\underset{2}{B}$ are the covariant derivatives of the tensor B^i_{jklm}:

$$\nabla B^i_{jklm} = \underset{1}{B}^i_{jklmn} \underset{1}{\omega}^n + \underset{2}{B}^i_{jklmn} \underset{2}{\omega}^n.$$

Theorem 5.11 *A G_W-structure, determined by a multidimensional hexagonal three-web ($r > 1$), is a closed G-structure of order four.*

Proof. According to Chapter 3, a hexagonal three-web is characterized by the relation $b^i_{(jkl)} = 0$ among the components of its curvature tensor. This relation and equations (1.31) give:

$$b^i_{jkl} + b^i_{klj} + b^i_{ljk} = 2(a^m_{jk}a^i_{ml} + a^m_{kl}a^i_{mj} + a^m_{lj}a^i_{mk}) \overset{\text{def}}{=} 2J^i_{jkl}. \tag{5.63}$$

Differentiating equations (5.63) by means of the operator ∇ twice and applying equations (5.53) and (5.57), we obtain the relations:

$$\begin{aligned} X^i_{jklmn} + X^i_{kljmn} + X^i_{ljkmn} &= 2\underset{1}{\nabla}_n\underset{1}{\nabla}_m J^i_{jkl}, \\ \underset{2}{Z}^i_{jklmn} + \underset{2}{Z}^i_{kljmn} + \underset{2}{Z}^i_{ljkmn} &= 2\underset{2}{\nabla}_n\underset{1}{\nabla}_m J^i_{jkl}, \\ Y^i_{jklmn} + Y^i_{kljmn} + Y^i_{ljkmn} &= 2\underset{2}{\nabla}_n\underset{2}{\nabla}_m J^i_{jkl}, \\ \underset{1}{Z}^i_{jklmn} + \underset{1}{Z}^i_{kljmn} + \underset{1}{Z}^i_{ljkmn} &= 2\underset{1}{\nabla}_n\underset{2}{\nabla}_m J^i_{jkl}. \end{aligned} \tag{5.64}$$

Thus, the tensors $X, Y, \underset{1}{Z}$ and $\underset{2}{Z}$ of a hexagonal three-web W satisfy equations (5.58), (5.59), (5.61), (5.62) and (5.64). Let us prove that these equations allow us to express the tensors, indicated above, in terms of the tensors $a, b, \underset{1}{c}$ and $\underset{2}{c}$, connected with the differential neighborhood of at most fourth order. We will not look for the formulas for the tensors $X, Y, \underset{1}{Z}$ and $\underset{2}{Z}$, since they are very complicated (each of them occupies a few pages). Because of this, in what follows, instead of the system of equations (5.58), (5.59), (5.61), (5.62) and (5.64), we will consider the homogeneous system associated with this system, i.e. we replace by zeros the right-hand sides of all these equations which contain the tensors $a, b, \underset{1}{c}$ and $\underset{2}{c}$. If we prove that the homogeneous system, obtained in this manner, has only the trivial solution, this will imply that the G_W-structure is closed.

We make an additional simplification. Instead of the tensors $X, Y, \underset{1}{Z}$ and $\underset{2}{Z}$, we introduce the multilinear forms corresponding to these tensors. For example,

$$X_{\alpha\beta\gamma\delta\epsilon} = X^i_{jklmn}\xi^j_\alpha\xi^k_\beta\xi^l_\gamma\xi^m_\delta\xi^n_\epsilon\eta_i.$$

First, we consider equation (5.59). The homogeneous equation $\underset{1}{Z}_{\alpha\beta\gamma\delta\epsilon} = \underset{2}{Z}_{\alpha\beta\gamma\delta\epsilon}$ corresponds to this equation. Next, set $\underset{2}{Z} = Z$. Then, the homogeneous equations, corresponding to equations (5.58,1), (5.61,1) and (5.64,1) can be reduced to the form:

$$X_{\alpha\beta\gamma\delta\epsilon} = X_{\alpha\delta\gamma\beta\epsilon} = X_{\alpha\beta\gamma\epsilon\delta}, \quad X_{\alpha\beta\gamma\delta\epsilon} + X_{\beta\gamma\alpha\delta\epsilon} + X_{\gamma\alpha\beta\delta\epsilon} = 0. \tag{5.65}$$

If we do the same for equations (5.61,2), (5.61,4) and (5.64,2), we have:

$$Z_{\alpha\beta\gamma\delta\epsilon} = Z_{\alpha\delta\gamma\beta\epsilon} = Z_{\alpha\beta\epsilon\delta\gamma}, \quad Z_{\alpha\beta\gamma\delta\epsilon} + Z_{\beta\gamma\alpha\delta\epsilon} + Z_{\gamma\alpha\beta\delta\epsilon} = 0; \tag{5.66}$$

for equations (5.58,2), (5.61,3) and (5.64,3) we have:

$$Y_{\alpha\beta\gamma\delta\epsilon} = Y_{\alpha\beta\delta\gamma\epsilon} = Y_{\alpha\beta\gamma\epsilon\delta}, \quad Y_{\alpha\beta\gamma\delta\epsilon} + Y_{\beta\gamma\alpha\delta\epsilon} + Y_{\gamma\alpha\beta\delta\epsilon} = 0; \tag{5.67}$$

and equations (5.62) yield the system:

$$X_{\alpha\beta\gamma\delta\epsilon} - X_{\beta\alpha\gamma\delta\epsilon} = Z_{\alpha\delta\beta\epsilon\gamma} - Z_{\beta\delta\alpha\epsilon\gamma}, \quad Y_{\alpha\beta\gamma\delta\epsilon} - Y_{\gamma\beta\alpha\delta\epsilon} = Z_{\alpha\gamma\delta\beta\epsilon} - Z_{\gamma\alpha\delta\beta\epsilon}, \tag{5.68}$$

First, consider system (5.65). Since, by (5.65,1), the tensor X is symmetric in the indices β, δ and ϵ, we set $X_{\alpha\beta\gamma\delta\epsilon} = X_{\alpha\gamma}$. Then, from relations (5.65,2) it follows that $X_{(\alpha\beta)} + X_{(\beta\gamma)} + X_{(\gamma\alpha)} = 0$. For different sets of indices, we have:

$$X_{(\alpha\beta)} + X_{(\beta\gamma)} + X_{(\gamma\alpha)} = 0, \quad X_{(\alpha\beta)} + X_{(\beta\delta)} + X_{(\delta\alpha)} = 0,$$
$$X_{(\beta\gamma)} + X_{(\gamma\delta)} + X_{(\delta\beta)} = 0, \quad X_{(\gamma\delta)} + X_{(\delta\alpha)} + X_{(\alpha\gamma)} = 0.$$

If we add the first three equations and subtract the fourth one, we obtain:

$$X_{(\alpha\beta)} + X_{(\gamma\beta)} + X_{(\delta\beta)} = 0. \tag{5.69}$$

For another set of indices, we have:

$$X_{(\alpha\beta)} + X_{(\gamma\beta)} + X_{(\epsilon\beta)} = 0.$$

From this equation and equation (5.69), we find $X_{(\delta\beta)} = X_{(\epsilon\beta)}$, where ϵ, δ and β are all distinct. Therefore, $X_{(\alpha\beta)} = X_{(gamma\beta)} = X_{(\delta\beta)}$, and from (5.69) we find that $X_{(\alpha\beta)} = 0$. Since $X_{\alpha\beta} = X_{\alpha\gamma\beta\delta\epsilon}$, we get

$$X_{\alpha\gamma\beta\delta\epsilon} = -X_{\beta\gamma\alpha\delta\epsilon}. \tag{5.70}$$

Applying similar considerations, we obtain from (5.67) that the tensor Y satisfies the condition:

$$Y_{\alpha\beta\gamma\delta\epsilon} = -Y_{\beta\alpha\gamma\delta\epsilon}. \tag{5.71}$$

Next, it follows from (5.65,2) and (5.70) that

$$X_{\alpha\gamma\beta\delta\epsilon} - X_{\gamma\alpha\beta\delta\epsilon} = X_{\alpha\gamma\beta\delta\epsilon} + X_{\beta\gamma\alpha\delta\epsilon} + X_{\alpha\beta\gamma\delta\epsilon} = X_{\alpha\beta\gamma\delta\epsilon}.$$

Thus, the first of equations (5.68) takes the form:

$$X_{\alpha\beta\gamma\delta\epsilon} = Z_{\alpha\delta\gamma\epsilon\beta} - Z_{\gamma\delta\alpha\epsilon\beta}. \tag{5.72}$$

From (5.67,2), (5.71) and (5.68,2) we obtain the similar equations:

$$Y_{\alpha\beta\gamma\delta\epsilon} = Z_{\alpha\beta\delta\gamma\epsilon} - Z_{\beta\alpha\delta\gamma\epsilon}. \tag{5.73}$$

Since equation (5.65,1) implies $X_{\alpha\gamma\beta\delta\epsilon} = X_{\alpha\epsilon\delta\beta\gamma}$, from (5.72) we obtain the following equations:

$$Z_{\alpha\delta\beta\epsilon\gamma} - Z_{\beta\delta\alpha\epsilon\gamma} = Z_{\alpha\delta\beta\gamma\epsilon} - Z_{\beta\delta\alpha\gamma\epsilon}.$$

We write them in the form:

$$Z_{\alpha\delta\beta[\gamma\epsilon]} = Z_{\beta\delta\alpha[\gamma\epsilon]}. \tag{5.74}$$

In a similar manner, we obtain from (5.65,2) and (5.72) the following equations:

$$Z_{\alpha\beta\delta[\gamma\epsilon]} = Z_{\beta\alpha\delta[\gamma\epsilon]}. \tag{5.75}$$

Equations (5.74) and (5.75) show that the quantities $Z_{\alpha\beta\gamma[\delta\epsilon]}$ are symmetric in the first three indices.

Consider now relations (5.65,2). From these we obtain the equations:

$$Z_{\alpha\beta\gamma[\delta\epsilon]} + Z_{\gamma\alpha\beta[\delta\epsilon]} + Z_{\beta\gamma\alpha[\delta\epsilon]} = 0.$$

Since the quantities $Z_{\alpha\beta\gamma[\delta\epsilon]}$ are symmetric in the first three indices, it follows from these equations that $Z_{\alpha\beta\gamma[\delta\epsilon]} = 0$, i.e. the quantities $Z_{\alpha\beta\gamma\delta\epsilon}$ are also symmetric in the last two indices.

From this and the first relations of (5.66) we find that that the quantities Z are symetric in all indices. Hence, the second relations from (5.66) imply that $Z_{\alpha\beta\gamma\delta\epsilon} = 0$, and finally relations (5.72) and (5.73) give $X_{\alpha\beta\gamma\delta\epsilon} = 0$ and $Y_{\alpha\beta\gamma\delta\epsilon} = 0$. ■

5.5 Three-Webs and Identities in Loops

Each of the classes of the webs, T, R, M, B and H, which are associated with a closed G_W-structure, is characterized by a certain identity held in the coordinate loops of these webs. A natural question arises: what other identities lead to the closure of a G_W-structure?

An arbitrary identity in the loop $Q(\cdot)$ can be written in the form:

$$S_1(u, v, \ldots, w) = S_2(u, v, \ldots, w). \tag{5.76}$$

Here S_1 and S_2 are words, i.e. products of letters $u, v, \ldots, w$, where the parentheses are located in a certain way. We assume that the words S_1 and S_2 do not contain the inverse operations.

In smooth loops only balanced identities make a sense. A *balanced identity* is an identity of the type (5.76) in which each of the variables enters in the words S_1 and S_2 with the same multiplicity. In fact, suppose that the identity element e of the loop Q has coordinates which are zero. Then (see Section 2.6), in a neighborhood of the point e, the product $u \cdot v$ in the loop Q can be expanded in the Taylor series as follows:

$$u \cdot v = u + v + \underset{2}{\Lambda}(u,v) + \frac{1}{2}\underset{2,1}{\Lambda}(u,u,v) + \frac{1}{2}\underset{1,2}{\Lambda}(u,v,v) + \ldots \tag{5.77}$$

We assume that the second order terms of expansion (5.77) are reduced to their canonical form, i.e. the first of conditions (2.66) holds. This condition is equivalent to the skew-symmetry of the form $\underset{2}{\Lambda}$:

$$\underset{2}{\Lambda}(u,v) = -\underset{2}{\Lambda}(v,u). \tag{5.78}$$

Since an arbitrary word $S(u, v, \ldots, w)$ is a composition of products $u \cdot v$, its expansion has the form:

$$S(u,v,\ldots,w) = \alpha u + \beta v + \ldots + \gamma w + \{2\},$$

where $\alpha, \beta, \ldots, \gamma$ are multiplicities, with which the variables $u, v, \ldots, w$ enter in S, and $\{2\}$ denotes the terms of order higher than one.

Suppose that identity (5.76) holds in the loop Q. Comparing the first order terms in the expansions for S_1 and S_2, we find that the corresponding multiplicities are equal. In addition, it follows from this that the words S_1 and S_2 have the same length: $|S_1| = |S_2|$.

If the order of variables in the words S_1 and S_2 is the same, we will say that identity (5.76) does not have inversions. A balanced identity without inversions is called *regular*.

All non-trivial types of webs, R, M, B and H, are characterized by regular identities. It is known that the simplest irregular identity, the commutativity identity, leads to the parallelizable webs, i.e. to the trivial type of webs.

As to the other identities with inversions, it is easy to show by a simple argument that, as a rule, they also lead to the parallelizability.

Consider, for example, the identity of the form:

$$S_1(\ldots u \ldots v \ldots) = S_2(\ldots v \ldots u \ldots)$$

with one inversion $u \leftrightarrow v$. Suppose that the variables u and v have the multiplicities one in this identity. Then, setting other variables equal to the identity element of the loop, we obtain the commutativity identity, and therefore, we arrive at the class T of webs. It is clear that we obtain the same result if we make stronger assumptions about the original identity. Thus, the most interesting webs are those defined by regular identities.

We exclude from consideration the identities that can be reduced to equivalent identities with a smaller length. These are either the reducible identities of the form $S_1 \cdot S = S_2 \cdot S$, or the identities, whose reduction can be reached by a change of variables. For example, the identity $(uv)(wt) = u(v \cdot wt)$ is equivalent to the identity $(uv)w = u(vw)$ but the identities $u^2u^2 = u(uu^2)$ and $u^2v = u(uv)$ are not equivalent.

Regular identities can be classified, first of all, according to the length of the words, composing the identity, and the number of variables (the rank).

From the words of length three the elasticity identity $(uv)u = u(vu)$ was least studied. The class of webs defined by this identity will be considered in Chapter 7.

Consider regular identities of length four. There are only five words of length four with four different variables:

$$S_1 = (uv \cdot w)t, \quad S_2 = u(v \cdot wt), \quad S_3 = (u \cdot vw)t, \quad S_4 = u(vw \cdot t), \quad S_5 = (uv)(wt). \tag{5.79}$$

We apply (5.77) to find the Taylor expansion of these words, including the terms up to third order. Define:

$$\begin{aligned}
\theta^j &= u^j + v^j + w^j + t^j, \\
\theta^{jk} &= u^j v^k + u^j w^k + u^j t^k + v^j w^k + v^j t^k + w^j t^k, \\
\theta^{jkl} &= u^j v^k w^l + u^j v^k t^l + u^j w^k t^l + v^j w^k t^l, \\
\theta_1^{jkl} &= u^j u^k v^l + u^j u^k w^l + u^j u^k t^l + v^j v^k w^l + v^j v^k t^l + w^j w^k t^l, \\
\theta_2^{jkl} &= u^j v^k v^l + u^j w^k w^l + u^j t^k t^l + v^j w^k w^l + v^j t^k t^l + w^j t^k t^l, \\
\Xi^i &= \theta^i + \underset{2}{\Lambda}{}^i_{jk}\theta^{jk} + \frac{1}{2}\underset{2,1}{\Lambda}{}^i_{jkl}\theta_1^{jkl} + \frac{1}{2}\underset{1,2}{\Lambda}{}^i_{jkl}\theta_2^{jkl}.
\end{aligned} \tag{5.80}$$

Direct calculation with application of (5.79)–(5.80), leads to the following result:

Lemma 5.12 *For the words $S_1 - S_5$, the following expansions hold:*

$$\begin{aligned}
S_1^i(u,v,w,t) = {} & \Xi^i + (\underset{2,1}{\Lambda}{}^i_{jkl} + \underset{2}{\Lambda}{}^i_{pl}\underset{2}{\Lambda}{}^p_{jk})\theta^{jkl} + \{4\}, \\
S_2^i(u,v,w,t) = {} & \Xi^i + (\underset{1,2}{\Lambda}{}^i_{jkl} + \underset{2}{\Lambda}{}^i_{jp}\underset{2}{\Lambda}{}^p_{kl})\theta^{jkl} + \{4\}, \\
S_3^i(u,v,w,t) = {} & \Xi^i + (\underset{2,1}{\Lambda}{}^i_{jkl} + \underset{2}{\Lambda}{}^i_{pl}\underset{2}{\Lambda}{}^p_{jk})\theta^{jkl} - \beta^i_{jkl}u^j v^k w^l + \{4\}, \\
S_4^i(u,v,w,t) = {} & \Xi^i + (\underset{1,2}{\Lambda}{}^i_{jkl} + \underset{2}{\Lambda}{}^i_{jp}\underset{2}{\Lambda}{}^p_{kl})\theta^{jkl} + \beta^i_{jkl}v^j w^k t^l + \{4\}, \\
S_5^i(u,v,w,t) = {} & \Xi^i + (\underset{2,1}{\Lambda}{}^i_{jkl} + \underset{2}{\Lambda}{}^i_{pl}\underset{2}{\Lambda}{}^p_{jk})(u^j v^k w^l + u^j v^k t^l) \\
& + (\underset{1,2}{\Lambda}{}^i_{jkl} + \underset{2}{\Lambda}{}^i_{jp}\underset{2}{\Lambda}{}^p_{kl})(u^j w^k t^l + v^j w^k t^l) + \{4\},
\end{aligned} \tag{5.81}$$

where β^i_{jkl} is given by formula (2.32). ■

Any pair of words from list (5.79) gives a regular identity of length four and rank four. There are five identities, which can not be reduced to identities of smaller length by a contraction or a change of variables:

$$\begin{aligned}
S_1 = S_2 &: \quad (uv \cdot w)t = u(v \cdot wt); \\
S_1 = S_4 &: \quad (uv \cdot w)t = u(vw \cdot t); \\
S_2 = S_3 &: \quad u(v \cdot wt) = (u \cdot vw)t; \\
S_3 = S_5 &: \quad (u \cdot vw)t = (uv)(wt); \\
S_4 = S_5 &: \quad u(vw \cdot t) = (uv)(wt).
\end{aligned} \tag{5.82}$$

Applying expansions (5.81) and formula (2.49), we obtain:

$$\begin{aligned}
S_1^i - S_2^i &= -b_{kjl}^i(u^j v^k w^l + u^j v^k t^l + u^j w^k t^l + v^j w^k t^l) + \{4\},\\
S_1^i - S_4^i &= -b_{kjl}^i(u^j v^k w^l + u^j v^k t^l + u^j w^k t^l) + \{4\},\\
S_2^i - S_3^i &= b_{kjl}^i(u^j v^k t^l + u^j w^k t^l + v^j w^k t^l) + \{4\},\\
S_3^i - S_5^i &= b_{kjl}^i(u^j v^k w^l - u^j w^k t^l - v^j w^k t^l) + \{4\},\\
S_4^i - S_5^i &= b_{kjl}^i(u^j v^k w^l + u^j v^k t^l - v^j w^k t^l) + \{4\},
\end{aligned} \tag{5.83}$$

By Theorem 1.11, it immediately follows from these equations that a three-web W, in whose coordinate loops one of identities (5.82) holds, is the web R.

Regular identities of length four and rank three are called the *identities of Bol type*. Let us show that *a three-web W, in whose coordinate loops an identity of Bol type holds, is the web of one of the following types: B_r, B_l, M or R.*

In fact, any identity of Bol type can be obtained from a certain identity of length four and rank four by identifying some two of its four independent variables. Let, for example, in the coordinate loops of a web W, the first of identities (5.82) holds, where $u = v$. Then, the first formula from (5.83) gives:

$$b_{kjl}^i(u^j u^k w^l + u^j u^k t^l + 2u^j w^k t^l) = 0.$$

Since the variables u, w and t are independent, it follows from this that $b_{jkl}^i = 0$, and we arrive at the webs R.

Setting, in the same identity (5.82, 1), $w = u$ or $v = t$, we obtain the well-known Moufang identities (see Problem **7** of Chapter 2). If $u = t$, identity (5.82, 1) is transformed into the so-called extra-identity [Fe 69]:

$$(uv \cdot w)u = u(v \cdot wu).$$

For this identity, from identity (5.83, 1) we obtain:

$$b_{kjl}^i(u^j v^k w^l + v^j w^k u^l) + b_{kjl}^i u^j v^k u^l + b_{kjl}^i u^j w^k u^l = 0.$$

This implies the relations: $b_{k(jl)}^i = 0, \quad b_{kjl}^i + b_{lkj}^i = 0$. Subtracting the latter from the former, we find that $b_{klj}^i = b_{lkj}^i$. Thus, the tensor b is symmetric in the first two indices and is skew-symmetric in the last two indices. Therefore, this tensor is zero, and the class of webs under consideration is the class of webs R.

If $y = z$, identity (5.82, 1) becomes:

$$(uv \cdot v)t = u(v \cdot vt).$$

For $t = e$, we obtain from this the identity of right alternativity, characterizing the right Bol webs.

If $w = t$, from identity (5.82, 1), we find the identity:

$$(uv \cdot w)w = u(v \cdot ww).$$

which, for $u = e$, also gives the identity of right alternativity. Analyzing relations (5.83, 1), we can easily prove that the last two classes coincide with the class of webs R.

In the same manner, we can list the identities arising from the rest of identities (5.82). In all cases we arrive at one of the classes: B_r, B_l, M or R. In particular, the identity B_r can be obtained from (5.82, 2) if we set $v = t$, the identity B_l from (5.82, 3) if we set $u = v$, and the Moufang identity (see Table 2.1. p. 50) from (5.82, 5) if we set $u = t$.

In addition to identities (5.82), there are other identities of length four and rank four which, by a change of variables can be reduced to the identity of length three. An example is the identity:

$$(uv \cdot w)t = (uv)(wt).$$

However, it is easy to check that this kind of identity also leads to the same class of webs B_r, B_l, M or R.

Regular identities of length four with two different variables can be obtained from the identities of length four and rank four if we identify two pairs of variables. If we apply the method described above to the classes of webs obtained from these regular identities, we can easily find the corresponding tensorial condition. The identities of length four and rank two are considered in Problems **2** and **3**.

Consider now a class of webs defined by an identity of length four and rank one (i.e. with one variable) and show that this class coincides with the class of hexagonal webs.

In fact, if we identify all variables in any of identities (5.82), we arrive at the relations $b^i_{(jkl)}u^j u^k u^l = 0$ implying $b^i_{(jkl)} = 0$. As shown in Chapter 3, this tensor equation characterizes the webs H.

Conversely, in the coordinate loops of the web H the identity of power-associativity: $u^m \cdot u^n = u^{m+n}$ holds for any $m, n \in \mathbf{R}$. In particular, in these loops the identities of length four with one variable, which we indicated above, hold.

All the webs we considered so far are hexagonal. To obtain wider classes of webs with a closed G_W-structure, which include the webs H, we turn to the identities of order k with one variable generalizing the monoassociativity identity.

As above, we let $Q(\cdot)$ be an analytic loop, and $S(u)$ be a word of length n and rank one in Q. Suppose that the product $u \cdot v$ is written in the form of series (5.77), where the terms of order s have form (2.54):

$$\underset{s}{\Lambda}(u, v) = \sum_{i=1}^{s-1} \frac{1}{(s-i)!i!} \underset{s-i,i}{\Lambda}(u, \dots, u, v, \dots, v)$$

Then, the Taylor series of the word $S(u)$ can be written as follows:

$$S(u) = nu + \sum_{s=3}^{\infty}\Big(\sum_{i=1}^{\infty} \frac{1}{(s-i)!i!} \underset{s-i,i}{A} \underset{s-i,i}{\Lambda}(u) + \underset{s}{R}(u)\Big), \tag{5.84}$$

where $\underset{s-i,i}{A}$ are certain integers depending on the parenthesis structure of the word $S(u)$, and $\underset{s}{R}(u)$ denotes the comitants of those coefficients of expansion (5.77) whose order is less than s.

Let $S_1(u)$ and $S_2(u)$ be two words of length n and rank one in the loop Q. We will say that they are *k-equivalent* $\big(S_1(u) \overset{k}{\sim} S_2(u)\big)$ if their Taylor expansions coincide up to the kth order terms inclusive.

An identity

$$S: \quad S_1(u) = S_2(u)$$

is called an identity *of order k* if $S_1(u) \overset{k}{\sim} S_2(u)$.

Suppose that the expansions of the words $S_1(u)$ and $S_2(u)$ composing the identity S are written in the form (5.84). Subtracting these two equations, we obtain:

$$S_1(u) - S_2(u) = \sum_{i=1}^{k} \Big(\frac{\nu_{k+1-i,i}}{(k+1-i)!i!} \underset{s+1-i,i}{\Lambda}(u) + \underset{k+1}{R}'(u) \Big) + \{k+2\}, \tag{5.85}$$

where $\nu_{i,j} = \underset{i,j}{A_1} - \underset{i,j}{A_2}$.

We will say that a set of m identities of order k:

$$S_1^p(u) = S_2^p(u), \quad p = 1, 2, \ldots, m,$$

has the rank ρ, if the corresponding matrix, composed of the numbers $\nu^p_{i,j} = \underset{i,j}{A^p_1} - \underset{i,j}{A^p_2}$, has the rank ρ.

We will formulate without proof the following theorem (its proof is beyond the scope of this book):

Theorem 5.13 ([S 87a]) *If $k-1$ independent identities of order k hold in the coordinate loops of an analytic web W, then the G_W-structure, defined by this web, is a closed G-structure of finite order not higher than $2k$.* ■

Consider some particular cases of identities with one variable. The only identity of length three and rank one is the monoassociativity identity. To this identity there corresponds the class of the webs H which, as noted above, possess a closed G_W-structure of order four. This result follows from Theorem 5.13 if we set $k = 2$.

Identities of order three will appear if $n \geq 5$. For example, the following identities of order three are independent:

$$u^2(u^2u) = u((u^2u)u), \quad (u^2u)u^2 = (u(u^2u))u.$$

According to Theorem 5.13, to these identities there corresponds a closed G_W-structure of order not higher than six.

Identities of order four will appear if $n \geq 10$. To discover this fact, we used a computer. For example, the following three identities (one of length 10 and two of length 11) have order four and are independent:

$$\begin{aligned} u(u^2(u^2(u(u(uu^2))))) &= u^2(u(u(u(u^2(u^2u))))), \\ u(u^2(u((uu^2)(u(uu^2))))) &= u^2(u(u(u^2((u(uu^2))u)))), \\ u(u((u^2(u(uu^2)))(uu^2))) &= u^2((u^2(u((u(uu^2))u)))u). \end{aligned} \tag{5.86}$$

To these identities there corresponds a closed G_W-structure of order not higher than eight.

In conclusion, we remind the reader that according to Theorem 5.3, the canonical expansion of the coordinate loops of webs satisfying the conditions of Theorem 5.13, is completely determined by a partial sum of order not higher than $2k$.

PROBLEMS

1. Prove the following statement:

Let $S(u_1, u_2, \ldots, u_n)$ be a word in the loop Q, whose Taylor series is written in the form (5.77). Then the Taylor series for S has the form:

$$S(u_1, u_2, \ldots, u_n) = u_1 + u_2 + \ldots + u_n + \sum_{\substack{ij=1 \\ (i<j)}}^{n-1} \underset{2}{\Lambda}(u_i, u_j) + \{3\}, \tag{5.87}$$

where $\{3\}$ denotes the terms of order higher than two.

Proof. The proof is conducted by induction with respect to n. If $n = 1$, we obtain formula (5.77). Next, any word S can be always represented as a product od words of lower lengths: $S = S_1 \cdot S_2$, for which, by the induction assumption, formula (5.87) holds. Thus, by (5.77), we have

$$\begin{aligned} S = \; & S_1 + S_2 + \underset{2}{\Lambda}(S_1, S_2) + \{3\} = (u_1 + \ldots + u_k) + (u_{k+1} + \ldots + u_n) \\ & + \sum_{\substack{ij=1 \\ (i<j)}}^{k} \underset{2}{\Lambda}(u_i, u_j) + \sum_{i,j=k+1}^{n-1} \underset{2}{\Lambda}(u_i, u_j) + \underset{2}{\Lambda}(u_1 + \ldots + u_k, u_{k+1} + \ldots + u_n) + \{3\}. \end{aligned}$$

Grouping the summands in an appropriate way, we arrive at (5.87). ■

As a corollary, we note that for a word $S(u)$ of length n with one variable u (using the skew-symmetry of the form $\underset{2}{\Lambda}(u, v)$), we obtain

$$S(u) = nu + \{3\}. \tag{5.88}$$

2. You can see below all (up to the monoassociativity) regular identities of length four and rank two. Using (5.83), prove that if in the coordinate loops of a web W one of these identities holds, then the curvature tensor of the web W satisfies the conditions indicated in the right column:

H_1	$u^3v = u(u \cdot uv)$	$b^i_{(jk)l} = 0$
H_2	$(uv \cdot u)u = u(v \cdot u^2)$	$2b^i_{k(jl)} + b^i_{(j\vert k\vert l)} = 0$
H_3	$(u^2v)u = u(u \cdot vu)$	$b^i_{(jk)l} + 2b^i_{l(jk)} = 0$
H_4	$(vu^2)u = vu^3$	$b^i_{(j\vert k\vert l)} = 0$
H_5	$u^3v = u(u^2 \cdot v)$	$b^i_{(jk)l} = 0$
H_6	$(uv \cdot u)u = u(vu \cdot u)$	$b^i_{j(kl)} = 0$
H_7	$(u^2v)u = u(uv \cdot u)$	$b^i_{(jk)l} + b^i_{l(jk)} = 0$
H_8	$(vu \cdot u)u = vu^3$	$b^i_{(j\vert k\vert l)} = 0$
H_9	$u(vu^2) = (u \cdot vu)u$	$b^i_{k(jl)} + b^i_{(j\vert k\vert l)} = 0$
H_{10}	$u(u \cdot vu) = (u \cdot uv)u$	$b^i_{j(kl)} = 0$
H_{11}	$u^3v = u^2 \cdot uv$	$b^i_{(jk)l} = 0$
H_{12}	$(u \cdot vu)u = uv \cdot u^2$	$b^i_{k(jl)} - b^i_{(j\vert k\vert l)} = 0$
H_{13}	$(u \cdot uv)u = u^2 \cdot vu$	$b^i_{(jk)l} - b^i_{l(jk)} = 0$
H_{14}	$(vu^2)u = vu \cdot u^2$	$b^i_{(jkl)} = 0$
H_{15}	$u(u^2v) = u^2 \cdot uv$	$b^i_{(jkl)} = 0$
H_{16}	$u(vu \cdot u) = uv \cdot u^2$	$2b^i_{k(jl)} - b^i_{(j\vert k\vert l)} = 0$
H_{17}	$u(uv \cdot u) = u^2 \cdot vu$	$b^i_{(jk)l} - b^i_{(j\vert l\vert k)} = 0$
H_{18}	$vu^3 = vu \cdot u^2$	$b^i_{(j\vert k\vert l)} = 0$
H_{19}	$(uv \cdot u)u = uv \cdot u^2$	$b^i_{(j\vert k\vert l)} = 0$
H_{20}	$(u^2v)u = u^2(vu)$	$b^i_{j(kl)} = 0$
H_{21}	$(vu \cdot u)u = vu \cdot u^2$	$b^i_{(j\vert k\vert l)} = 0$
H_{22}	$u(u \cdot uv) = u^2 \cdot uv$	$b^i_{(jk)l} = 0$
H_{23}	$u(vu^2) = uv \cdot u^2$	$b^i_{j(kl)} = 0$
H_{24}	$u(u \cdot vu) = u^2 \cdot vu$	$b^i_{(jk)l} = 0$
H_{25}	$(u \cdot vu)u = u(vu \cdot u)$	$b^i_{j(kl)} = 0$
H_{26}	$(u \cdot uv)u = u(uv \cdot u)$	$b^i_{(jk)l} + b^i_{(j\vert l\vert k)} = 0$
H_{27}	$uv \cdot uv = u(v \cdot uv)$	$b^i_{(j\vert k\vert l)} = 0, b^i_{j(kl)} = 0$
H_{28}	$uv \cdot uv = u(vu \cdot v)$	$b^i_{j(kl)} = 0, b^i_{(j\vert k\vert l)} = 0$
H_{29}	$uv \cdot uv = (uv \cdot u)v$	$b^i_{(jk)l} = 0, b^i_{(j\vert k\vert l)} = 0$
H_{30}	$uv \cdot uv = (u \cdot vu)v$	$b^i_{(k\vert j\vert l)} = 0, b^i_{j(kl)} = 0$
H_{31}	$u(v \cdot uv) = (uv \cdot u)v$	$b^i_{(jk)l} + b^i_{l(jk)} = 0,$
		$b^i_{(k\vert j\vert l)} + b^i_{j(kl)} = 0$
H_{32}	$u(v \cdot uv) = (u \cdot vu)v$	$b^i_{(jk)l} = 0$
H_{33}	$u(vu \cdot v) = (uv \cdot u)v$	$b^i_{(j\vert k\vert l)} = 0$
H_{34}	$u(vu \cdot v) = (u \cdot vu)v$	$b^i_{k(jl)} = 0, b^i_{(j\vert k\vert l)} = 0$
H_{35}	$uv \cdot vu = u(v^2u)$	$b^i_{j(kl)} = 0, b^i_{(j\vert k\vert l)} = 0$
H_{36}	$uv \cdot vu = u(v \cdot vu)$	$b^i_{j(kl)} = 0, b^i_{(j\vert k\vert l)} = 0$
H_{37}	$uv \cdot vu = (uv \cdot v)u$	$b^i_{(jk)l} = 0, b^i_{(j\vert k\vert l)} = 0$
H_{38}	$uv \cdot vu = (uv^2)u$	$b^i_{(jk)l} = 0, b^i_{(j\vert k\vert l)} - b^i_{(jl)k} = 0$
H_{39}	$u(v^2u) = (uv \cdot v)u$	$b^i_{j(kl)} = 0, b^i_{(j\vert k\vert l)} = 0$

H_{40}	$u(v^2u) = (u \cdot v^2)u$	$b^i_{(j\|k\|l)} = 0, b^i_{j(kl)} = 0$
H_{41}	$u(v \cdot vu) = (uv \cdot v)u$	$b^i_{j(kl)} = 0$
H_{42}	$u(v \cdot vu) = (uv^2)u$	$b^i_{j(kl)} = 0, b^i_{(jk)l} = 0$
H_{43}	$u^2v^2 = (u^2v)v$	$b^i_{(j\|k\|l)} = 0$
H_{44}	$u^2v^2 = (u \cdot uv)v$	$b^i_{j(kl)} = 0, b^i_{(j\|k\|l)} = 0$
H_{45}	$u^2v^2 = u(uv^2)$	$b^i_{(jk)l} = 0$
H_{46}	$u^2 \cdot v^2 = u(uv \cdot v)$	$b^i_{(jk)l} = 0, b^i_{(j\|k\|l)} = 0$
H_{47}	$(u^2v)v = u(uv^2)$	$b^i_{(jk)l} = 0, b^i_{(j\|k\|l)} = 0$
H_{48}	$(u^2v)v = u(uv \cdot v)$	$b^i_{(j\|k\|l)} = 0, b^i_{(jk)l} = 0$
H_{49}	$(u \cdot uv)v = u(uv^2)$	$b^i_{(jk)l} = 0, b^i_{(j\|k\|l)} = 0$
H_{50}	$(u \cdot uv) = u(uv \cdot v)$	$b^i_{(jk)l} = 0, b^i_{(j\|k\|l)} = 0.$

3. Prove that the classes H_i (see Problem **2**) are connected with well-known classes B, M and E as follows:

1) $H_1 = H_5 = H_9 = H_{11} = H_{22} = H_{24} = H_{32} = B_l$.

2) $H_4 = H_7 = H_8 = H_{18} = H_{19} = H_{21} = H_{33} = H_{43} = B_r$.

3) $H_{25} = H_{26} = E$.

4) $H_{14} = B_r$.

5) $H_{15} = B_l$.

6) If $i = 2, 3, 12, 13, 16$ and 17, then $H_i \bigcap E = M$.

7) If $i = 6, 10, 20$ and 23, then $E \subset H_i \subset B_m$.

8) If $i = 27 - 31, 34 - 40, 42, 44$ and $46 - 50$, then $H_i = M$.

9) $M \subset H_{41} \subset B_m$, $H_{41} \bigcap B_l = M$ and $H_{41} \bigcap B_r = M$.

Solution: Let us take from Table 2.1, p. 50, and Problem **7** of Chapter 2 the identities characterizing the webs B_l, B_r and M:

$$
\begin{aligned}
B_l &: \quad (u \cdot vu)w = u(v \cdot uw), \\
B_r &: \quad u(vw \cdot v) = (uv \cdot w)v, \\
M &: \quad u(vw \cdot u) = (uv)(wu), \\
M_1 &: \quad u(v \cdot uw) = (uv \cdot u)w, \\
M_2 &: \quad (uv \cdot w)v = u(v \cdot wv).
\end{aligned}
\tag{5.89}
$$

Cases 1) *and* 2). Consider, for example, the webs H_9, whose curvature tensor satisfies the relations $b^i_{k(jl)} + b^i_{(j|k|l)} = 0$. Since all webs H_i are hexagonal, for the tensor b the conditions $b^i_{(jkl)} = 0$ hold. Together with previous relations, they give $b^i_{(jk)l} = 0$, by which the identities of this case are identically satisfied. But the relation $b^i_{(jk)l} = 0$ characterizes the webs B_l (Chapter 4).

Conversely: in the coordinate loops of any web B_l the identity B_l holds (see (5.89)), and the identity H_9 can be obtained from the identity B_l by setting $w = u$. For other classes, listed in 1) and 2), the proof is similar.

Case 3). If in the identity H_{25} (or H_{26}) we set $vu = w$ (repectively $uv = w$), we arrive at the identity of elasticity.

Cases 4) *and* 5). Applying the right alternativity identity: $vu \cdot u = vu^2$, which holds in the coordinate loops of the web B_r, we find that the identity H_{14} is satisfied. This implies the inclusion $B_r \subset H_{14}$. Conversely, consider now a web defined by the identity H_{14}, and prove that this web is the web B_r. To the identity H_{14} there corresponds the configuration represented in Figure 5.1 by solid lines:

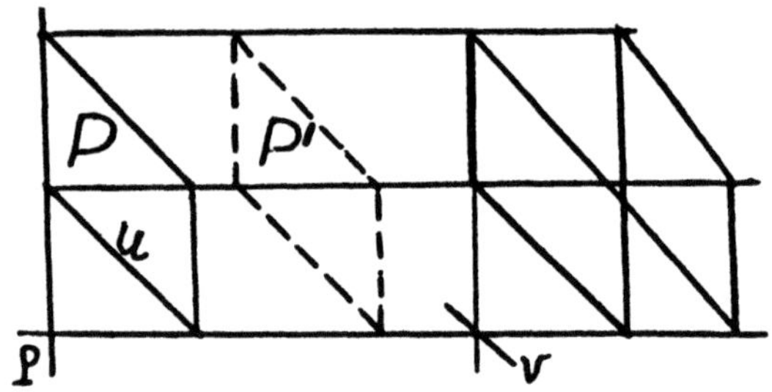

Figure 5.1

In this figure there are two parts which are independent from one another. If we vary v, the right part will be moving along the horizontal leaves. Since on the web under consideration all figures of the type represented in Figure 5.1 are closed, the figure, in which the parallelogram P is replaced by the parallelogram P' (represented by the dotted lines), is also closed. As a result, we obtain a closed right Bol figure PP'. It follows from this that all figures B_r are closed in the web H_{14}, i.e. this web is the web B_r. Thus, $H_{14} \subset B_r$. This and the inclusion $B_r \subset H_{14}$, obtained earlier, imply that $H_{14} = B_r$.

Case 6). The identity H_2, for example, can be obtained from the Moufang identity M_1 by setting $w = u$. On the other hand, the identity H_2, given the elasticity identity, can be transformed to the form $u(vu \cdot u) = u(vu^2)$, which implies $vu \cdot u = vu^2$. Thus, $H_2 \Rightarrow B_r$, i.e. in the coordinate loops of the webs H_2 the identity B_r holds. This and the elasticity identity imply the Moufang identity (see Problem 9 of Chapter 2).

Case 7). Consider, for example, the class H_6. The corresponding condition $b^i_{j(kl)} = 0$ for the curvature tensor characterizes the webs B_m. Hence, $H_6 \subset B_m$. On the other hand, by the elasticity identity, we have: $(uv \cdot u)u = (u \cdot vu)u = u(vu \cdot u)$, i.e. the identity H_6 holds. Therefore, $E \subset H_6$. For other classes, listed in the case 7), the proof is similar.

Case 8). Consider, for example, the class H_{27}. From the relations on the curvature tensor we obtain the relation $b = \text{alt } b$ characterizing the webs M (Section 4.4). Therefore, $H_{27} \subset M$. On the other hand, according to the Moufang theorem (see [Bel 66] or [Br 71], p. 117), any two elements in the loop M generate an associative subloop. In particular, the identity H_{27} holds in the loop M. As a result, we get $H_{27} = M$. In other cases the proof is similar.

Case 9). The condition $b^i_{j(kl)} = 0$ characterizes the webs B_m. On the other hand, the inclusion $M \subset H_{41}$ holds by the Moufang theorem.

4. Prove that the curvature tensor of the three-web, defined by the equations

$$\begin{aligned} z^1 = &\ x^1 + y^1 + 2x^2x^3(qy^3 - py^2) + 6y^2y^3(qx^3 - px^2) + 4q(x^3)^2y^2 - 4p(x^2)^2y^3, \\ z^2 = &\ x^2 + y^2, \\ z^3 = &\ x^3 + y^3, \end{aligned}$$

where p and q are constants, satisfies the conditions: $b^i_{(jkl)} = 0$, $b^i_{j[kl]} = 0$. This web is not isoclinic and can be smoothly deformed into a parallelizable three-web.

NOTES

5.1. The notion of a closed G-structure was introduced by Akivis in [A 75b], where, in particular, he stated that the symmetric spaces and the group, Moufang and Bol webs possess such a structure.

The notion of a formally completely integrable system introduced in [A 75b] is a particular case of pseudo-Kählerian systems of exterior differential equations, considered by L. N. Beskin in [Bes 58].

5.2. The problem of description of loops whose canonical expansion is completely determined by the first k terms, was posed in [HSa 80] but only for power associative loops.

Equations (4.10), giving the closure of the G-structure determined by a Bol web, were obtained by Fedorova [F 76].

5.3. Theorem 5.9 was proved by V. P. Botsu in [Bot 84]. A few other results on four-dimensional hexagonal Bol three-webs are also due to him (see [Bot 74] and [Bot 75]).

5.4. The results of this section and some other results on closed G_W-structures are due to Shelekhov (see [S 85b], [S 87a] and [S 89b]).

5.5. Many results of this section are due to Shelekhov (see [S 87a] and [S 89b]).

Classification of identities in abstract loops was performed by V.D. Belousov [Bel 66], F. Fenyves ([Fe 68] and [Fe 69]) and V. E. Vechtomov ([Ve 75] and [Ve 81]).

Classification of identities of order k with one variable in smooth loops is given in [BiS 87] and [BiS 90].

Six-dimensional hexagonal three-webs were studied by M. A. Shestakova ([Sh 88] and [Sh 90]). The results, formulated in Problem 4, are also due to her ([Sh 90]).

Chapter 6

Automorphisms of Three-Webs

6.1 The Autotopies of Quasigroups and Three-Webs

1. We remind the reader of the definition of the isotopy of quasigroups given in Section 2.1. Two 3-base quasigroups $q = (\cdot, X_\alpha)$ and $\widetilde{q} = (\circ, \widetilde{X}_\alpha)$, $\alpha = 1, 2, 3$, are called isotopic if there exists a triple $J = (J_1, J_2, J_3)$ of bijective mappings $J_\alpha : X_\alpha \to \widetilde{x}_\alpha$ such that, for any $x \in X_1$ and any $y \in X_2$, the relation

$$J_1(x) \circ J_2(y) = J_3(x \cdot y). \tag{6.1}$$

holds. An isotopic mapping of a quasigroup onto itself is called an autotopy.

Denote an autotopy $q \to q$ by $A = (A_\alpha)$, $\alpha = 1, 2, 3$. By the definition, we have:

$$A_1(x) \cdot A_2(y) = A_3(x \cdot y). \tag{6.2}$$

We will now consider some properties of an autotopy.

Proposition 6.1 *An autotopy is completely defined by preassigning two of its components.*

Proof. In fact, denote as usual by L_x and R_y a left translation and a right translation in $q = (\cdot, X_\alpha)$. Then equation (6.2) can be written as follows:

$$(L_{A_1(x)} A_2)(y) = (A_3 L_x)(y),$$

where LA_2 denotes the composition of the functions L and A_2. Since y is arbitrary, we obtain

$$L_{A_1} A_2 = A_3 L_x.$$

Similarly, we find that

$$R_{A_2(y)}A_1 = A_3R_y.$$

Using the unique solvability in the quasigroup q, we can express A_1 and A_2 in terms of two other components of the autotopy A:

$$A_1 = R^{-1}_{A_2(y)}A_3R_y, \quad A_2 = L^{-1}_{A_1(x)}A_3L_x. \blacksquare \tag{6.3}$$

Another proof of Proposition 6.1 will be proposed in Problem **2**.

The next two propositions are obvious.

Proposition 6.2 *Let $A = A_\alpha$ be an autotopy in the quasigroup $q = (\cdot, X_\alpha)$ and let the quasigroup $\widetilde{q} = (\circ, \widetilde{X}_\alpha)$ be isotopic to q, where the isotopy $q \to \widetilde{q}$ has the form $J = (J_\alpha)$. Then, in $\widetilde{q}$, the autotopy*

$$J^{-1}AJ = (J_1^{-1}A_1J_1,\ J_2^{-1}A_2J_2, J_3^{-1}A_3J_3)$$

arises. $\blacksquare$

Proposition 6.3 *All autotopies of the quasigroup q form a group.* $\blacksquare$

We will denote the group of autotopies of a quasigroup q by $\mathcal{A}_q$.

Let us recall one more definition from the theory of quasigroups. An autotopy $A = (A_\alpha)$ is called *regular* if one of the bijections A_β is the identity transformation. We will denote by $\mathcal{A}_q^\beta$ a set of regular autotopies of the form $A = (A_\alpha)$, where $A_\beta = \mathrm{id}$ for some fixed $\beta = 1, 2, 3$.

The next two propositions are also almost obvious.

Proposition 6.4 *Regular autotopies of the quasigroup q form normal subgroups in the group of all autotopies $\mathcal{A}_q$.* $\blacksquare$

Proposition 6.5 *The bijections A_α, forming a regular autotopy A of a 3-base quasigroup q, can be expressed in terms of translations of this quasigroup.*

Proof. In fact, if for example, $A_3 = \mathrm{id}$, then we find from (6.3) that

$$A_1 = R^{-1}_{A_2(y)}R_y, \quad A_2 = L^{-1}_{A_1(x)}L_x. \blacksquare \tag{6.4}$$

Suppose now that $Q(\cdot)$ is a one-base quasigroup with identity element, i.e. a loop. The right, left and middle kernels of the loop Q can be defined as follows:

$$\begin{aligned} N_r &= \{a \in Q | (ax)y = a(xy), x, y \in Q\}, \\ N_l &= \{b \in Q | (xy)b = x(yb), x, y \in Q\}, \\ N_m &= \{c \in Q | (xc)y = x(cy), x, y \in Q\}. \end{aligned}$$

It is easy to show that each of these kernels is closed under multiplication in Q. Therefore, each of these kernels is a subgroup in Q.

Proposition 6.6 *The groups of regular autotopies of a loop $Q(\cdot)$ are connected with its kernels as follows:*

$$\begin{aligned}\mathcal{A}_Q^{(1)} &= \{(id, R_a, R_a)\}, a \in N_r,\\ \mathcal{A}_Q^{(1)} &= \{(L_b, id, L_b)\}, b \in N_l,\\ \mathcal{A}_Q^{(3)} &= \{(R_c, L_c^{-1}, id)\}, c \in N_m.\end{aligned}$$

Proof. Consider, for example, the case when $A_3 = \mathrm{id}$. In equation (6.4) we set $x = e$ and take $y = c$ in such a way that $A_2(c) = e$. Then, we find from (6.4) that $A_1 = R_c$, $A_2 = L^{-1}_{A_1(e)}$. By definition (6.1), we have

$$A_1(e)A_2(c) = A_3(c) = c,$$

and since $A_2(c) = e$, we find from this that $A_1(e) = c$, $A_2 = L_c^{-1}$.

Let us now show that $c \in N_m$. By the definition of an autotopy, we have $R_c(x)L_c^{-1} = xy$. Setting $L_c^{-1} = y$, we write the last equation in the form $(xc)z = x(cz)$, which implies $c \in N_m$. For other groups $\mathcal{A}_Q^\alpha$, the proof is similar. ■

The kernels of loops are isomorphic to the corresponding groups $\mathcal{A}_Q^{(\alpha)}$.

Corollary 6.7 *A loop Q admits non-trivial autotopies if and only if it has the corresponsing non-trivial kernel.* ■

Taking into account Proposition 6.6, we will call the autotopies from $\mathcal{A}_Q^1$ and $\mathcal{A}_Q^2$ the *right* autotopies and the *left* autotopies, respectively.

We will call the autotopies from $\mathcal{A}_Q^3$ the *principal* isotopies – this corresponds to the term "the principal isotopy" (see Section 2.2).

2. As was indicated in Section 2.2, the notions of isotopy and autotopy of quasigroups, can be transferred to abstract three-webs. Consider three-webs $W = (X, \lambda_\alpha)$ and $\widetilde{W} = (\widetilde{X}, \widetilde{\lambda}_\alpha)$. Suppose that the coordinate quasigroups q and $\tilde{q}$ of these two quasigroups are isotopic, and the isotopy has the form $J = (J_\alpha)$. The bijection J_α is defined in the foliation λ_α. Thus, the triple of bijections gives a mapping of the web W onto the web $\widetilde{W}$. Condition (6.1) means that the incidence of leaves is preserved: three leaves of the web W passing through a point will be transferred into three leaves of the web $\widetilde{W}$ passing through a point. Such a transformation of webs is also called an isotopy. In particular, an autotopy of the coordinate quasigroup q generates an autotopy of the corresponding three-web W, i.e. a transformation of this web into itself. We will denote the group of autotopies of a web W by $\mathcal{A}_W$. In the case of geometric webs, $\mathcal{A}_W$ denotes a connected component of the identity element.

An autotopy $A = (A_\alpha)$ of a three-web W generates a bijective transformation φ of a set X, in which the web W is defined, in the following manner: $\varphi(x \cap y) = A_1(x) \cap A_2(y)$, $x \in \lambda_1$, $y \in \lambda_2$.

The transformation φ is called an automorphism of the three-web W. In particular, to a regular autotopy there corresponds a regular automorphism. The group of automorphisms of a web is isomorphic to the group of autotopies. Because of this, we will use the same notation $\mathcal{A}_W$ for both groups.

Suppose that an automorphism φ of a three-web W transfers a point p into a point p'. Since φ maps leaves of the web into leaves of the same foliation and preserves the incidence of leaves, the automorphism φ also maps the product of leaves, defined by multiplication in the loop l_p (see Figure 2.12), into the product (with respect to the operation in the loop $l_{p'}$) of the corresponding leaves. Thus, an automorphism of a three-web W induces an isomorphism of its coordinate loops in the corresponding points.

As a consequence, we find that all coordinate quasigroups of a three-web W are isomorphic to one another if this web admits a transitive group of automorphisms. It is easy to show that the converse also holds: if all coordinate loops of a three-web W are isomorphic, then the web admits a transitive group of automorphisms. Such webs are called *G-webs.* They will be considered in Section 6.4.

Another consequence is that if an automorphism φ of a three-web W keeps a point p fixed, then it induces an automorphism of the coordinate loop l_p.

Let us show that the converse holds: any automorphism of the coordinate loop l_p of a web W induces an autotopy of this web that keeps the point p fixed. In fact, suppose that A_3 is an isomorphism of the loop l_p. Since an isotopy of the coordinate loop l_p onto the coordinate quasigroup $q(\cdot)$ has the form $(R_b^{-1}, L_a^{-1}, \mathrm{id})$, where $p = (a, b)$ (see Section 2.3), according to Proposition 6.2, the isomorphism A_3, i.e. the autotopy (A_3, A_3, A_3), induces the autotopy (A_1, A_2, A_3) in $q(\cdot)$, where

$$A_1 = R_b^{-1} A_3 R_b, \quad A_2 = L_a^{-1} A_3 L_a. \tag{6.5}$$

By the definition of an autotopy, we have

$$A_1(a) A_2(y) = A_3(ay),$$

where y is an arbitrary element of the second foliation of the web W. Taking into account relations (6.5), we can write the last equation in the form:

$$A_1(a) L_a^{-1}(A_3(ay)) = A_3(ay).$$

Since y, and consequently ay and $A_3(ay)$, is an arbitrary element from λ_2, we find that $L_{A_1(a)} L_a^{-1} = \mathrm{id}$, i.e. $A_1(a) = A_1$.

Similarly, applying (6.5), we derive $A_2(b) = b$. Thus, the autotopy (A_1, A_2, A_3) keeps the leaves a and b, and subsequently the point p, fixed.

The following theorem combines the results obtained.

Theorem 6.8 *Any automorphism φ of a three-web $W = (X, \lambda_\alpha)$ induces an isomorphism of the coordinate loop l_p of this web onto the $\mathcal{A}_W(p)$ of automorphisms* *of the three-web W, keeping the point p fixed, is isomorphic to the group* Aut l_p *of automorphisms of the coordinate loop l_p of the web W.* ■

The group $\mathcal{A}_W(p)$ is called the *isotropy group* of the point p.

3. For a multidimensional three-web $W = (X, \lambda_\alpha)$, formed in a smooth manifold X by three foliations λ_α, the notion of an autotopy is replaced by the notion of a local autotopy, formed by three local diffeomorphisms $A_\alpha : \lambda_\alpha \rightarrow \widetilde{\lambda}_\alpha$, satisfying condition (6.2). Moreover, the mappings A_α are given in the bases X_α of the corresponding foliations λ_α.

The following natural question immediately arises: what is the relationship between the automorphisms of a three web W and the automorphisms of the Chern connection defined by the three-web? The following theorem answers this question.

Theorem 6.9 *Local automorphisms of a three-web $W = (X, \lambda_\alpha)$ are also automorphisms of the corresponding Chern connection. Conversely, let φ be automorphisms of the Chern connection of a three-web $W = (X, \lambda_\alpha)$ and let there exist a point p in X such that the linear transformation $d\varphi\big|_p$ maps the tangent planes to the leaves of the web W also to the tangent planes to the corresponding leaves. Then, φ is an automorphism of a three-web.*

Proof. Let us recall that an automorphism of an affine connection Γ, given in a manifold X, is a diffeomorphism of this manifold that preserves the law of parallel displacement, i.e. preserves the covariant differential relative to this connection.

The first part of Theorem 6.9 follows from the fact that an automorphism of a web preserves its foliations since these very foliations define the covariant differential in the Chern connection (see Section 1.7).

Conversely, suppose that an automorphism φ of the Chern connection of a three-web W possesses the property indicated in the condition of Theorem 6.9. Take a point p' in X, a smooth path from p to p', and let $s' = \varphi \cdot s$. Denote by τ and τ' the parallel displacements along s and s', respectively. Since φ is an automorphism of an affine connection, the linear transformation $d\varphi\big|_{p'} = \tau' \cdot d\varphi\big|_p \cdot \tau^{-1}$. By the condition of Theorem 6.9, the transformation φ transfers the tangent planes to the leaves of the web at the point p to the tangent planes to the corresponding leaves at the point $\varphi(p)$. By the properties of the Chern connection, the parallel displacements τ and τ' also transfer the tangent planes to the leaves of the web to the tangent planes to the leaves of the web. Thus, the the transformation $d\varphi\big|_p$ possesses this property. Therefore, the transformation $d\varphi$ preserves the distribution of the tangent planes to the leaves of the web. Consequently it preserves the leaves themselves. ■

It follows from Theorem 6.9 that all fundamental tensor fields $a^i_{jk}, b^i_{jkl}, \underset{\alpha}{c}{}^i_{jklm}, \ldots$ associated with a three-web W, are invariant with respect to an automorphism φ of this web, i.e.

$$d\varphi\big|_p(\mathcal{T}(p)) = \mathcal{T}(\varphi(p)), p \in X,$$

where $\mathcal{T}(p)$ denotes a value at the point p of some of the tensor fields listed above. In particular, if φ keeps the point p fixed, we obtain

$$d\varphi\big|_p(\mathcal{T}(p)) = \mathcal{T}(p). \tag{6.6}$$

We arrive at the following theorem:

Proposition 6.10 *If an automorphism φ of a three-web W keeps a point p fixed, then the fundamental tensor fields of this web, are invariant with respect to an automorphism φ.* ■

Suppose that in some local coordinates $d\varphi\big|_p = (\varphi^i_j)$. Then condition (6.6) becomes:

$$a^i_{jk}\varphi^j_{j'}\varphi^k_{k'} = a^{i'}_{j'k'}\varphi^i_{i'}, \quad b^i_{jkl}\varphi^j_{j'}\varphi^k_{k'}\varphi^l_{l'} = b^{i'}_{j'k'l'}\varphi^i_{i'}, \ldots \tag{6.7}$$

It follows from Proposition 6.10 that an arbitrary three-web does not admit non-trivial automorphisms since the quantities φ^i_j, from formulas (6.7), satisfy an infinite sequence of equations which, in general, are independent.

4. Consider autotopies of Grassmann three-webs. The latter can be given in the manifold $X = G(1, r+1)$ of straight lines of a projective space P^{r+1} by means of three smooth hypersurfaces X_α, $\alpha = 1, 2, 3$. The leaves of this web are the bundles of straight lines with vertices on X_α. Let $A = (A_\alpha)$ be an autotopy of a Grassmann three-web W and let φ be the corresponding automorphism of the manifold $X = G(1, r+1)$. Since φ is a local diffeomorphism $X \to X$, which transfers straight lines into straight lines, φ is a projective transformation of the space P^{r+1}. The transformations A_α, constituting the autotopy A, operate in the foliations λ_α, i.e. the bundles of straight lines with vertices on X_α are transferred also into the bundles of straight lines with vertices on X_α. Thus, the collineation $\varphi : P^{r+1} \to P^{r+1}$ maps the hypersurfaces X_α into themselves.

We proved the following proposition.

Proposition 6.11 *Let W be a Grassmann three-web, defined by hypersurfaces X_α of a projective space P^{r+1}, and let $\mathcal{A}_W$ be the group of autotopies of this web. Then:*

a) *The group $\mathcal{A}_W$ is a subgroup of the group $PGL(r+1)$ of collineations of the space P^{r+1}, and*

b) *The hypersurfaces X_α are invariant with respect to the group $\mathcal{A}_W$.*

■

Let, for example, X_1 and X_2 be domains in the same hyperquadric Q and X_3 be a hyperplane in P^{r+1}. Then the three-web defined by them is an isoclinic Bol web (Theorem 4.11, p. 146). This web admits the non-trivial group $\mathcal{A}_W$ of autotopies, which if we consider the hyperplane X_3 as the hyperplane at infinity, coincides with the group of transformations of an affine space A^{r+1} preserving the hyperquadric Q.

We indicate now the method of construction of Grassmann webs with a non-trivial local group of autotopies. Let G be a subgroup of dimension $\rho < r+1$ of the group $PGL(r+1)$. Fix three smooth surfaces $\widetilde{X}_\alpha$ of dimension $(r+1)-\rho$ in P^{r+1} such that their orbits $X_\alpha = G(\widetilde{X}_\alpha)$ are different and dim $X_\alpha = r$. Let there exist a straight line l intersecting the surfaces X_α in three different points A_α. Then, in a neighborhood of this straight line, a Grassmann web W is defined. By the construction, the group $\mathcal{A}_W$ of all autotopies of the web W contains the group G as a subgroup.

Consider regular autotopies of Grassmann three-webs. By the definition, a regular autotopy φ keeps one of the web foliations fixed. For a Grassmann web W, this means that each point of one of hypersurfaces, generating the web, remains fixed relative to the corresponding projective transformation φ in P^{r+1}. Thus, this hypersurface is a hyperplane.

A projective transformation with a pointwise fixed hyperplane Π is called a *homology*. The hyperplane Π is called the *douple hyperplane* of the homology. The homologies are characterized by the fact that the straight lines, joining the corresponding points, pass through a fixed point S, the *center* of the homology. The center S can lie in Π (the parabolic type) or does not belong to it (the hyperbolic type). Let $S \in \Pi$. If we consider the point S as "the point at infinity" and define the hyperplane Π by the equation $x^{r+1} = 0$, then the affine equivalent of the homology is the transformation $'x^{r+1} = kx^{r+1}$, $k \in \mathbf{R}\backslash 0$. In particular, if $k = -1$, we arrive at the *involutory homology*, or the *symmetry*.

Consider the homologies of parabolic type, when $S \in \Pi$. If the hyperplane Π is the hyperplane at infinity, then the corresponding affine homology is a parallel displacement in the direction of a straight line passing through the center of homology S.

Note an additional property of homologies. A hyperplane Π and a center S define not one homology but a one-parameter family (a pencil) of homologies. Assigning a pair of corresponding points, we distinguish a unique homology from this pencil.

Consider now a Grassmann web W admitting a regular automorphism φ. According to what was discussed above, φ is a homology. Moreover, one of hypersurfaces defining the three-web W, for example, the hypersurface X_3, is the double hyperplane of the homology φ, and two other hypersurfaces X_1 and X_2 are invariant relative to φ. If φ is a hyperbolic homology, then the affine analogies of the hypersurfaces X_1 and X_2 are cylinders whose generators are parallel to a certain axis. The same situation arises if a homology is parabolic. Thus, in both cases the hypersurfaces X_1 and X_2 are cones whose generators pass through the center S of homology.

Note that the symmetries relative to the hyperplane X_3 are not automorphisms of a locally defined Grassmann web, since they interchange two other hypersurfaces X_1 and X_2, i.e. the leaves of the first foliation are transferred into the leaves of the second one, and conversely.

We proved the following theorem.

Theorem 6.12 *Only non-involutory homologies can be regular automorphisms of a Grassmann three-web. A Grassmann three-web W admits a regular automorphism φ if and only if one of hypersurfaces defining the web is a double hyperplane of the homology φ and two other hypersurfaces are cones with generators passing through the center of this homology.* ■

6.2 Infinitesimal Automorphisms of Three-Webs

1. Consider a vector field ξ given in a smooth manifold X, $\dim X = n$. A curve $x(t)$ in X is called an *integral curve* of the field ξ if the tangent vector $\frac{dx}{dt}$ coincides with the vector $\xi(x(t))$ at each point of the curve $x(t)$. Integral curves are solutions of the following system of ordinary differential equations:

$$\frac{dx^u}{dt} = \xi^u, \quad u = 1, 2, \ldots, n.$$

A unique solution of this system passes through any point x_0 of the manifold X. This solution can be written in the form:

$$x(t) = \exp(t\xi)(x_0).$$

On the other hand, this solution determines a one-parameter group of local diffeomorphisms (translations) $x_t : X \to X$:

$$x_0 \to x_t(x_0) = \exp(t\xi)(x_0)$$

Because of this, the vector field ξ is called an *infinitesimal automorphism* of the manifold X. Note that a change of a parameter t on trajectories implies the multiplication of the vector ξ by a certain factor. Conversely, if we take a vector $\lambda\xi$ instead of the vector ξ, we obtain the same family of diffeomorphisms but with another parametrization.

If a manifold X carries a certain differential structure, then infinitesimal diffeomorphisms preserving this structure are called *infinitesimal automorphisms* of this structure.

Let X be a foliation with a base B and a projection π (see Section 1.1). Let us find conditions under which a vector field ξ of a manifold X is an infinitesimal diffeomorphism of the foliation (X, B, π), i.e. transfers a leaf of this foliation into a leaf of the same foliation.

As in Section 1.1, we take a co-frame $\{\omega^i, \omega^a\}$, $i, j = 1, 2, \ldots, m$; $a = m+1, \ldots, n$, in the manifold X, where ω^i are principal forms in the base B. We write the vector field ξ in the form:

$$\xi = \xi^i e_i + \xi^a e_a,$$

where the frame $\{e_i, e_a\}$ is dual to the co-frame $\{\omega^i, \omega^a\}$. The second component $\xi_2 = \xi^a e_a$ generates translations along a leaf of the foliation X, and the first component $\xi_1 = \xi^i e_i$ generates translations in the direction transversal to the leaf. Let a vector field ξ transfer a leaf into a leaf. Then this field operates in a set of leaves, i.e. in the base B. We will use the same notation ξ for the corresponding field in B. Since the basis forms ω^i of the base B satisfy the structure equations $d\omega^i = \omega^j \wedge \omega^i_j$ (see Section 1.1), the coordinates ξ^i of the vector ξ must satisfy the equations:

$$\nabla \xi^i \equiv d\xi^i + \xi^j \omega^i_j = \xi^i_j \omega^j. \tag{6.8}$$

Consider now a smooth three-web $W = (X, \lambda_\alpha)$, $\dim X = 2r$. The web W is formed in a manifold X by three foliations λ_α which are given by the forms $\underset{\alpha}{\omega}^i$. We will look for the differential equations of the vector fields defining infinitesimal automorphisms of this web. As in Section 1.3, we define a vector field ξ in X by the equations:

$$\xi = \underset{1}{\xi}^i \underset{2}{e}_i - \underset{2}{\xi}^i \underset{1}{e}_i = \underset{2}{\xi}^i \underset{3}{e}_i - \underset{3}{\xi}^i \underset{2}{e}_i = \underset{3}{\xi}^i \underset{1}{e}_i - \underset{1}{\xi}^i \underset{3}{e}_i, \tag{6.9}$$

where the vectors $\underset{\alpha}{e}_i$ are tangent to the leaf $\mathcal{F}_\alpha$ of the web W, $\mathcal{F}_\alpha \in \lambda_\alpha$. Since the forms $\underset{\alpha}{\omega}^i$ are connected by equations (1.16):

$$\underset{1}{\omega}^i + \underset{2}{\omega}^i + \underset{3}{\omega}^i = 0, \tag{6.10}$$

the coordinates of the vector ξ satisfy the equation:

$$\underset{1}{\xi}^i + \underset{2}{\xi}^i + \underset{3}{\xi}^i = 0. \tag{6.11}$$

A field ξ is an infinitesimal automorphism of a three-web W if and only if it preserves the foliations λ_α of this web, i.e. if each leaf of the web W is transferred into a leaf of the same foliation. According to formula (6.8), the components $\underset{\alpha}{\xi}^i$ must satisfy the equations:

$$d\underset{\alpha}{\xi}^i + \underset{\alpha}{\xi}^j \underset{\alpha}{\omega}^i_j = \underset{\alpha}{\xi}^i_j \underset{\alpha}{\omega}^j, \tag{6.12}$$

provided that the forms $\underset{\alpha}{\omega}^i$, defining the foliation λ_α, satisfy the equations:

$$d\underset{\alpha}{\omega}^i = \underset{\alpha}{\omega}^j \wedge \underset{\alpha}{\omega}^i_j, \quad \alpha = 1, 2, 3. \tag{6.13}$$

It follows from the structure equations (1.26) of the web that

$$\underset{1}{\omega}{}^i_j = \omega^i_j + a^i_{jk}\underset{1}{\omega}{}^k, \quad \underset{2}{\omega}{}^i_j = \omega^i_j - a^i_{jk}\underset{2}{\omega}{}^k, \quad \underset{3}{\omega}{}^i_j = \omega^i_j + a^i_{jk}(\underset{1}{\omega}{}^k - \underset{2}{\omega}{}^k).$$

Hence, equations (6.12) become

$$\begin{aligned} d\underset{1}{\xi}{}^i + \underset{1}{\xi}{}^j\omega^i_j &= \underset{1}{\xi}{}^i_j\underset{1}{\omega}{}^j - a^i_{jk}\underset{1}{\xi}{}^j\underset{1}{\omega}{}^k, \\ d\underset{2}{\xi}{}^i + \underset{2}{\xi}{}^j\omega^i_j &= \underset{2}{\xi}{}^i_j\underset{2}{\omega}{}^j + a^i_{jk}\underset{2}{\xi}{}^j\underset{2}{\omega}{}^k, \\ d\underset{3}{\xi}{}^i + \underset{3}{\xi}{}^j\omega^i_j &= \underset{3}{\xi}{}^i_j\underset{3}{\omega}{}^j - a^i_{jk}\underset{3}{\xi}{}^j(\underset{1}{\omega}{}^k - \underset{2}{\omega}{}^k). \end{aligned} \tag{6.14}$$

Adding equations (6.14) and applying equations (6.11), we obtain the relations:

$$\underset{1}{\xi}{}^i_j\underset{1}{\omega}{}^j + \underset{2}{\xi}{}^i_j\underset{2}{\omega}{}^j + \underset{3}{\xi}{}^i_j\underset{3}{\omega}{}^j = -a^i_{jk}(\underset{2}{\xi}{}^j\underset{1}{\omega}{}^k - \underset{1}{\xi}{}^j\underset{2}{\omega}{}^k). \tag{6.15}$$

Substitute in (6.15) for $\underset{3}{\omega}{}^i$ by means of (6.10) and equate to zero the coefficients in the independent forms $\underset{1}{\omega}{}^i$ and $\underset{2}{\omega}{}^i$. As a result, we find $\underset{1}{\xi}{}^i_j$ and $\underset{2}{\xi}{}^i_j$:

$$\underset{1}{\xi}{}^i_j = \underset{3}{\xi}{}^i_j + a^i_{jk}\underset{2}{\xi}{}^k, \quad \underset{2}{\xi}{}^i_j = \underset{3}{\xi}{}^i_j - a^i_{jk}\underset{1}{\xi}{}^k,$$

Define the quantities ξ^i_j as follows:

$$\underset{3}{\xi}{}^i_j = \xi^i_j - a^i_{jk}\underset{2}{\xi}{}^k + a^i_{jk}\underset{1}{\xi}{}^k. \tag{6.16}$$

Then, by (6.11), we obtain:

$$\underset{1}{\xi}{}^i_j = \xi^i_j + a^i_{jk}\underset{1}{\xi}{}^k, \quad \underset{2}{\xi}{}^i_j = \xi^i_j - a^i_{jk}\underset{2}{\xi}{}^k. \tag{6.17}$$

Thus, the first two equations of (6.14) take the form:

$$d\underset{1}{\xi}{}^i + \underset{1}{\xi}{}^j\omega^i_j = (\xi^i_j + 2a^i_{jk}\underset{1}{\xi}{}^k)\underset{1}{\omega}{}^j, \quad d\underset{2}{\xi}{}^i + \underset{2}{\xi}{}^j\omega^i_j = (\xi^i_j - 2a^i_{jk}\underset{2}{\xi}{}^k)\underset{2}{\omega}{}^j, \tag{6.18}$$

and the third equation of (6.14) follows from equations (6.18).

We have proven the following theorem.

Theorem 6.13 *A vector field $\xi = \underset{1}{\xi}{}^i\underset{2}{e}_i - \underset{2}{\xi}{}^i\underset{1}{e}_i$ defines an infinitesimal automorphism of a three-web W if and only if the coordinates of this field satisfy differential equations* (6.18). ■

2. Besides the quantities $\underset{1}{\xi}{}^i$ and $\underset{2}{\xi}{}^i$, equations (6.18) contain the function ξ^i_j which also should be found. The first of equations (6.17) shows that these quantities can be expressed in terms of the vector $\underset{1}{\xi}{}^i$, its covariant derivative $\underset{1}{\xi}{}^i_j$ and the tensor a^i_{jk}.

Therefore, the quantities ξ^i_j form a tensor, and subsequently its covariant differential is a linear combination of the principal forms $\underset{1}{\omega}^i$ and $\underset{2}{\omega}^i$:

$$\nabla\xi^i_j \equiv d\xi^i_j + \xi^m_j\omega^i_m - \xi^i_m\omega^m_j = \underset{1}{\xi}^i_{jk}\underset{1}{\omega}^k + \underset{2}{\xi}^i_{jk}\underset{2}{\omega}^k.$$

Now we apply exterior differentiation to equations (6.18) and substitute in the quadratic equations obtained, for the quantities $d\xi^i, d\underset{\alpha}{\omega}^i, d\omega^i_j$ and da^i_{jk} by means of equations (6.18), (1.26), (1.32) and (1.33). As a result, we obtain the equations containing the principal forms only. Equating to zero the coefficients in the independent products $\underset{1}{\omega}^k \wedge \underset{1}{\omega}^l, \underset{2}{\omega}^k \wedge \underset{2}{\omega}^l$ and $\underset{1}{\omega}^k \wedge \underset{2}{\omega}^l$, we obtain:

$$\underset{1}{\xi}^i_{jk} = b^i_{jkl}\underset{2}{\xi}^l, \quad \underset{2}{\xi}^i_{jk} = -b^i_{jlk}\underset{1}{\xi}^l; \tag{6.19}$$

$$\begin{aligned}
&\underset{1}{\xi}^i_{[jk]} + b^i_{[jk]l}\underset{1}{\xi}^l - b^i_{l[jk]}\underset{1}{\xi}^l + 6a^m_{[jl}a^i_{|m|k]}\underset{1}{\xi}^l - a^i_{lj}\xi^l_k - a^i_{kl}\xi^l_j + a^l_{kj}\xi^i_l = 0,\\
&\underset{2}{\xi}^i_{[jk]} - b^i_{[j|l|k]}\underset{2}{\xi}^l + b^i_{l[jk]}\underset{2}{\xi}^l + 6a^m_{[jl}a^i_{|m|k]}\underset{2}{\xi}^l + a^i_{lj}\xi^l_k + a^i_{kl}\xi^l_j - a^l_{kj}\xi^i_l = 0.
\end{aligned} \tag{6.20}$$

By means of (6.19), the expression for $\nabla\xi^i_j$ takes the form:

$$\nabla\xi^i_j = b^i_{jkl}(\underset{2}{\xi}^l\underset{1}{\omega}^k - \underset{1}{\xi}^k\underset{2}{\omega}^l), \tag{6.21}$$

and by (6.19) and (1.31), both equations (6.20) can be transformed to the same form:

$$\Phi^i_{jk}(\xi) \equiv b^i_{[j|m|k]}\underset{1}{\xi}^m + b^i_{[jk]m}\underset{2}{\xi}^m + a^i_{jm}\xi^m_k + a^i_{mk}\xi^m_j - a^m_{jk}\xi^i_m = 0. \tag{6.22}$$

Along with equations (6.21) and (6.22), their differential consequences must be satisfied on the web W. Exterior differentiation of equations (6.21) leads to the relations:

$$\Phi^i_{jkl}(\xi) \equiv \underset{1}{c}^i_{jklm}\underset{1}{\xi}^m + \underset{2}{c}^i_{jklm}\underset{2}{\xi}^m + b^i_{jkm}\xi^m_l + b^i_{jml}\xi^m_k + b^i_{mkl}\xi^m_j - b^m_{jkl}\xi^i_m = 0. \tag{6.23}$$

It is possible to show that differentiation of relations (6.22) leads to the equations $\Phi^i_{[jk]l} = 0$ and $\Phi^i_{[j|l|k]} = 0$ which are consequences of relations (6.23). Thus, the r^2+2r quantities $\underset{1}{\xi}^i, \underset{2}{\xi}^i$ and ξ^i_j satisfy the system of Pfaffian equations (6.18) and (6.21). In addition, they are connected by relations (6.22) and (6.23) which are the conditions of complete integrability of the system of equations (6.18) and (6.21). In general, the number of these relations far exceeds the number of the variables $\underset{1}{\xi}^i, \underset{2}{\xi}^i$ and ξ^i_j. This implies that for an arbitrary web, the system of equations (6.18) and (6.21) is inconsistent, i.e. an arbitrary three-web does not have non-trivial infinitesimal automorphisms. A similar result was obtained at the end of the subsection **1** of Section 6.1.

Consider the webs for which the rank of the system of equations (6.18) and (6.21) is equal to s, where $s < r^2 + 2r$. In this case the system of differential equations (6.18) and (6.21) has non-trivial solutions, and the set of its solutions depends on $N = r^2 + 2r - s$ parameters. To each solution $\underset{1}{\xi}{}^i, \underset{2}{\xi}{}^i$ (taken up to a factor) there corresponds a one-parameter family of infinitesimal automorphisms of the three-web. Thus, the set of all infinitesimal automorphisms, defined by the system of equations (6.18) and (6.21), depends on N parameters. Direct calculation shows that the set of solutions of system (6.18) is closed under the operation of commutation (Problem 7). This proves the following theorem.

Theorem 6.14 *If for a three-web $W = (X, \lambda_\alpha), \dim X = 2r$, the rank of the system of equations (6.22) and (6.23) is $s < r^2 + 2r$, then the system of equations (6.18) and (6.21) is completely integrable, and its solutions determine the group of automorphisms $\mathcal{A}_W$ of this web which depends on $N = r^2 + 2r - s$ parameters.* ■

3. Let us find a geometric meaning of the quantities ξ^i_j. Suppose that under an infinitesimal automorphism ξ, a point p remains fixed. Then $\xi(p) = 0$, or

$$\underset{1}{\xi}{}^i(p) = 0, \quad \underset{2}{\xi}{}^i(p) = 0, \tag{6.24}$$

and equations (6.18) take the form

$$d\underset{1}{\xi}{}^i\Big|_p = \xi^i_j(p)\underset{1}{\omega}{}^j\Big|_p, \quad d\underset{2}{\xi}{}^i\Big|_p = \xi^i_j(p)\underset{2}{\omega}{}^j\Big|_p, \tag{6.25}$$

Consider a point $p+dp$ close to the point p, where $dp = \underset{1}{\omega}{}^i\underset{2}{e}_i - \underset{2}{\omega}{}^i\underset{1}{e}_i$. By conditions (6.24), the value of the vector field ξ at the point $p + dp$ is $\xi(p + dp) = d\xi$. On the other hand, by relations (6.25), we obtain:

$$d\xi\Big|_p = d(\underset{1}{\xi}{}^i\underset{2}{e}_i - \underset{2}{\xi}{}^i\underset{1}{e}_i) = d\underset{1}{\xi}{}^i\Big|_p\underset{2}{e}_i - d\underset{2}{\xi}{}^i\Big|_p\underset{1}{e}_i = (\xi^i_j\underset{1}{\omega}{}^j)\underset{2}{e}_i - (\xi^i_j\underset{2}{\omega}{}^j)\underset{1}{e}_i.$$

The latter equations show that the vector $d\xi\Big|_p$ can be obtained from the vector dp by applying the operator:

$$\begin{pmatrix} \xi^i_j(p) & 0 \\ 0 & \xi^i_j(p) \end{pmatrix}.$$

Thus, the geometric meaning of the operator $\xi^i_j(p)$ is that the infinitesimal linear transformation, defined by this operator, transfers the vector dp into the vector $\xi(p+dp)$.

By relations (6.24), equations (6.22) and (6.23) imply the equations:

$$\xi^i_m a^m_{jk} = a^i_{mk}\xi^m_j + a^i_{jm}\xi^m_k, \quad \xi^i_m b^m_{jkl} = b^i_{mkl}\xi^m_j + b^i_{jml}\xi^m_k + b^i_{jkm}\xi^m_l,$$

which means that the linear transformation ξ^i_j is the differentiation of the tensors a^i_{jk} and b^i_{jki} in the algebraic sense (see Section 4.1).

Differentiating equations (6.22) and (6.23) and applying (6.25), we find that the operator $\xi^i_j(p)$ is also differentiation (in the algebraic sense) of the covariant derivatives of the curvature tensor. Using the terminology introduced in Section 5.2, we can say that the operator $\xi^i_j(p)$ is differentiation in any of W_k-algebras of the three-web. This is a generalization of the following well-known fact from Lie group theory: if ξ is an infinitesimal automorphism of a Lie group, then the derivative of ξ is differentiation in the tangent Lie algebra of this group.

4. Subsequently differentiating equations (6.23) by means of the operator ∇, we obtain a sequence of relations similar to relations (6.23). These relations contain the *kth* order covariant derivatives of the curvature tensor, $k = 2, 3, \ldots$ The next theorem clarifies the geometric meaning of all these equations.

Theorem 6.15 *Let $W = (X, \lambda_\alpha)$ be a multidimensional three-web. A vector field $\xi(\underset{1}{\xi^i}, \underset{2}{\xi^i})$ in X is an infinitesimal automorphism of the web W if and only if, along every integral curve of this field, there exists such a field of frames, in which the coordinates of the vector ξ, their covariant derivatives ξ^i_j and also the components of all fundamental tensors of the web W are constant.*

Proof. Let ξ be an infinitesimal automorphism of the web W. As earlier, we denote by $\exp(t\xi)$ a one-parameter group of transformations generated by the field ξ. Let us fix an adapted frame $\mathcal{R}_0$ at some point $x_0 \in X$. Then, at any point of an integral curve $x(t)$ of the field ξ passing through x_0, the frame $d\exp(t\xi)\mathcal{R}_0$ arises. Each of these frames is also an adapted frame of the web W since $\xi(t)$ is an automorphism of this web. In the field of frames constructed along $x(t)$, the coordinates of the vector ξ are constant along the integral curve. Substituting $\underset{1}{\xi^i} = \text{const}$, $\underset{2}{\xi^i} = \text{const}$ and the equations

$$\underset{1}{\omega^i} = \underset{1}{\xi^i}dt, \quad \underset{2}{\omega^i} = \underset{2}{\xi^i}dt \tag{6.26}$$

into equations (6.18), we obtain:

$$\omega^i_j = \xi^i_j dt. \tag{6.27}$$

Restricting equations (6.18) to the line $(x(t)$, we obtain $\xi^i_j = \text{const}$.

Consider now equations (1.33) which the curvature tensor of the web W satisfies:

$$da^i_{jk} + a^m_{jk}\omega^i_m - a^i_{mk}\omega^m_j - a^i_{jm}\omega^m_k = b^i_{[j|l|k]}\underset{1}{\omega^l} + b^i_{[jk]l}\underset{2}{\omega^l}.$$

Substituting the values of the forms from (6.26) and (6.27) into the last equations and applying equations (6.22), we find that $da^i_{jk} = 0$. Similarly, from equations (6.23), we obtain $db^i_{jkl} = 0$.

Conversely, let the tensors a^i_{jk} and b^i_{jkl} be constant along an integral curve $x(t)$ of a vector field $\xi(\underset{1}{\xi}^i, \underset{2}{\xi}^i)$ defined by equations (6.26) in a fixed family of frames $\mathcal{R}(t)$ given by equations (6.27). Substituting (6.26) and (6.27) into equations (1.33) and (1.38), provided that a^i_{jk} and b^i_{jkl} are constants, we obtain equations (6.22) and (6.23).

Consider equations (6.21). Conditions (6.23) of their complete integrability are satisfied. Thus, the quantities ξ^i_j must satisfy equations (6.21). But along the curve $x(t)$, these equations take the form $d\xi^i_j = 0$, which implies $\xi^i_j = \text{const}$.

Next, consider equations (6.18). Conditions (6.21) and (6.22) of their complete integrability are satisfied. Thus, the quantities $\underset{1}{\xi}^i$ and $\underset{2}{\xi}^i$ satisfy equations (6.18). But along the curve $x(t)$, these equations take the form $d\underset{1}{\xi}^i = 0$ and $d\underset{2}{\xi}^i = 0$. These imply $\underset{1}{\xi}^i = \text{const}$ and $\underset{2}{\xi}^i = \text{const}$. ■

Thus, equations (6.22), (6.23) and subsequent equations express the fact that the tensor fields $a^i_{jk}, b^i_{jkl}, \ldots$ defined in a three-web W are invariant along the trajectories of a vector field ξ with respect the group of automorphisms generated by this field.

Theorem 6.15 implies the following obvious corollary.

Corollary 6.16 *A three-web $W = (X, \lambda_\alpha)$ admits an N-parameter group of automorphisms if and only if there exists a fibration of the manifold X, in N-dimensional fibers of which, in a certain family of frames, the components of all fundamental tensors of the web W are constant.* ■

5. Let us find automorphisms of parallelizable and group three-webs. Suppose that a parallel three-web is given in an affine space A^{2r} of dimension $2r$ by the equations $x^i = \text{const}$, $y^i = \text{const}$ and $x^i + y^i = \text{const}$. Automorphisms of this web are linear transformations of the form:

$$'x^i = A^i_j x^j + B^i, \quad 'y^i = A^i_j y^j + C^i.$$

and only such transformations. In this case the group of automorphisms depends on $r^2 + 2r$ parameters, i.e. on the maximal number of parameters. It is not so difficult to prove the converse: if the group of automorphisms of some three-web W depends on the maximal number of parameters, then this web is parallelizable (see Problem 4).

Next, suppose that a three-web W is a group three-web, i.e. it is defined, as it was in Section 1.5, in the direct product $X = G \times G^{-1}$, where G is an r-dimensional Lie group. Then, according to Section 1.5, the curvature tensor b^i_{jkl} vanishes and the forms ω^i_j can be reduced to zero. As a result, the torsion tensor a^i_{jk} becomes constant and coincides with the structure tensor of the group G. In this case, we find from (6.21) that $\xi^i_j = c^i_j = \text{const}$, and equations (6.18) take the form:

$$d\underset{1}{\xi}^i = (c^i_j + 2a^i_{jk}\underset{1}{\xi}^k)\underset{1}{\omega}^j, \quad d\underset{2}{\xi}^i = (c^i_j - 2a^i_{jk}\underset{2}{\xi}^k)\underset{2}{\omega}^j, \qquad (6.28)$$

where $\underset{1}{\omega}^i$ and $\underset{2}{\omega}^i$ are invariant forms of the group X. Moreover, it is easy to show that equations (6.23) are identically satisfied, and relations (6.22) can be reduced to the following equations:

$$a^m_{jk}c^i_m = a^i_{mk}c^m_j + a^i_{jm}c^m_k. \tag{6.29}$$

Relations (6.29) mean that the linear operator c^i_j is a differentiation in the algebra g of the Lie group G. To each solution c^i_j of system (6.29) there corresponds a solution of the completely integrable system (6.28). In particular, if $c^i_j = 0$, system (6.28) determines a $(2r)$-parameter family of translations in the direct product $X = G \times G^{-1}$, preserving the foliations λ_α. Therefore, the group of all automorphisms of a group web is transitive and depends at least on $2r$ parameters.

We indicate another important solution of system (6.28). It is known that a Lie algebra possesses internal differentiations of the form $a^i_{jk}\xi^k$ where a^i_{jk} is the structure tensor of the algebra and ξ^k are constants. To these differentiations, there correspond internal automorphisms of the form $\Phi_a = a^{-1}xa$ of the corresponding Lie group (see, for example, [Po 82]). In this case, equations (6.82) take the form:

$$d\underset{1}{\xi}^i = a^i_{jk}(\xi^k + 2\underset{1}{\xi}^k)\underset{1}{\omega}^j, \quad d\underset{2}{\xi}^i = a^i_{jk}(\xi^k - 2\underset{2}{\xi}^k)\underset{2}{\omega}^j.$$

In particular, the first subsystem of this system admits a solution $\underset{1}{\xi}^i = -\frac{1}{2}\xi^i$. To this solution, there corresponds a vector field ξ in X whose horizontal component $\xi_2 = \underset{1}{\xi}^i \underset{2}{e}_i$ has constant coordinates, i.e. forms an invariant vector field in the subgroup G^{-1} of the direct product $X = G \times G^{-1}$. Similarly, to the solution $\underset{2}{\xi}^i = \frac{1}{2}\xi^i$ of the second subsystem there corresponds a vector field ξ in X whose vertical component $\xi_1 = \underset{2}{\xi}^i \underset{1}{e}_i$ is an invariant vector field in the subgroup G of X. The solution $\underset{1}{\xi}^i = -\frac{1}{2}\xi^i$, $\underset{2}{\xi}^i = \frac{1}{2}\xi^i$ determines a vector field ξ which is invariant in the whole group $X = G \times G^{-1}$. In this case, the coordinates of the field ξ are connected by the condition: $\underset{1}{\xi}^i + \underset{2}{\xi}^i = -\underset{3}{\xi}^i = 0$. Thus, the vector ξ is tangent to the leaf of the third foliation of the web W. We will consider this type of infinitesimal automorphism in the next section.

6.3 Regular Infinitesimal Automorphisms of Three-Webs

1. In Section 6.1 we called an autotopy $A = (A_\alpha)$ of a three-web $W = (X, \lambda_\alpha)$ regular if one of its mappings A_α is the identity transformation, i.e. if one of the foliations λ_α remains fixed under this isotopy. According to this definition, an infinitesimal automorphism of the web W is called *right, left or principal* if it transfers the leaves of the first, second or third foliation respectively into itself.

Consider, for example, a principal infinitesimal automorphism ξ. The vector ξ is tangent to a leaf of the third foliation . Thus, it follows from (6.9) that $\underset{3}{\xi^i} = 0$. Relations (6.11) imply that $\underset{1}{\xi^i} + \underset{2}{\xi^i} = 0$. By virtue of the last equations, we find from (6.18) that

$$(\xi^i_j - 2a^i_{jk}\xi^k)(\underset{1}{\omega^k} + \underset{2}{\omega^k}) = 0,$$

where we used the notation:

$$\xi^i \overset{\text{def}}{=} -\underset{1}{\xi^i} = \underset{2}{\xi^i}.$$

Since the forms $\underset{1}{\omega^k} + \underset{2}{\omega^k}$ are independent in the manifold X, we obtain:

$$\xi^i_j = 2a^i_{jk}\xi^k.$$

As a result, equations (6.18) are reduced to one equation:

$$\nabla\xi^i = 0. \tag{6.30}$$

Exterior differentiation of equation (6.30) leads to the relations:

$$b^i_{jkl}\xi^j = 0. \tag{6.31}$$

Performing similar calculations for the left and right infinitesimal automorphisms, we obtain the following proposition.

Proposition 6.17 *A vector field ξ generates the principal, right or left automorphism if it satisfies the equations:*

$$\xi = \xi^i \underset{3}{e_i}, \quad \underset{12}{\nabla}\xi^i = 0, \quad b^i_{jkl}\xi^j = 0; \tag{6.32}$$

$$\xi = \xi^i \underset{1}{e_i}, \quad \underset{23}{\nabla}\xi^i = 0, \quad b^i_{jkl}\xi^l = 0; \tag{6.33}$$

or

$$\xi = \xi^i \underset{2}{e_i}, \quad \underset{31}{\nabla}\xi^i = 0, \quad b^i_{jkl}\xi^k = 0, \tag{6.34}$$

respectively. ■

The third relations in (6.32), (6.33) and (6.34) can be interpreted in terms of the tangent W-algebras. As we did in Chapter 2, we define the operations of the tangent W-algebra in the point p of the web W in the tangent space T_e to the identity element of the coordinate loop l_p. We call the spaces, defined by the equations:

$$b^i_{jkl}(p)\xi^j = 0, \quad b^i_{jkl}(p)\xi^l = 0, \quad b^i_{jkl}(p)\xi^h = 0,$$

the *middle, right and left ternary kernel* of the tangent W-algebra of the three-web W at the point p, respectively. Denote these kernels by J_m, J_r and J_l, respectively. Note that if the web W is a Moufang web, then, by virtue of the skew-symmetry of the curvature tensor, all three kernels coincide with the J-kernel of the corresponding Mal'tsev algebra (Section 4.5).

Each of the kernels J_m, J_r and J_l forms a Lie algebra with respect to the binary operation [,] defined by the torsion tensor a^i_{jk}.

In fact, let vectors ξ, η and ζ from the space T_e satisfy, for example, relations (6.31). Contracting the coordinates of these vectors with equations (1.31), we obtain the Jacobi identity:

$$[\xi[\eta\zeta]] + [\eta[\zeta\xi]] + [\zeta[\xi\eta]] = 0.$$

The fact that the kernel is closed under the operation of commutation follows from the formula given in Problem **21** of Chapter 1. According to this formula, the commutator of two covariantly constant vector fields ξ and η can be calculated as follows:

$$[\xi, \eta]^i = -2a^i_{jk}\xi^j\eta^k.$$

Differentiating these equations by means of the operator ∇ and applying equation (1.33), we obtain:

$$\nabla[\xi, \eta]^i = -(b^i_{[j|l|k]}\underset{1}{\omega}^l + b^i_{[jk]l}\underset{2}{\omega}^l)\xi^j\eta^k.$$

By (6.31), the right-hand side of this equation vanishes. Thus, the commutator of two vector fields ξ and η satisfies equation (6.30). This completes the proof for the kernel J_m. For the kernels J_r and J_l, the proof is similar.

It follows from the above, that if ξ is a regular infinitesimal automorphism of a three-web W, then the vector $\xi(p), p \in X$, belongs to the corresponding ternary kernel of the tangent W-algebra of this web.

Suppose that at a point p, the rank of system (6.31) is equal to $r - s$, $s > 0$, i.e. the kernel J_m at this point is non-trivial. Then, in some neighborhood of this point, system (6.30) has s independent solutions ξ^i_a, $a = 1, \ldots, s$, which determine the s-dimensional group $\mathcal{A}^{(3)}_W$ of local regular automorphisms of the three-web W. It follows from Propsition 6.6 that the group $\mathcal{A}^{(3)}_W$ is locally isomorphic to the connected component of the identity element of the kernel N_m of the coordinate loop l_p of the web W. The tangent Lie algebra to the group $\mathcal{A}^{(3)}_W$ (or to the kernel N_m) coincides with the Lie algebra J_m.

Note that the choice of point p is not essential since the kernels, taken at different points, are isomorphic (see Problem **3**). Similar conclusions can be also made in two other cases.

2. We now consider in more detail the webs admitting a non-trivial group $\mathcal{A}^{(3)}_W$ of principal regular automorphisms. Independent solutions ξ^i_a, $a = 1, \ldots, s$ of system

(6.30) determine a distribution of $(2s)$-dimensional transversal subspaces. This distribution is involutive since equations (6.30) are completely integrable. Let V^{2s} be an integral transversally geodesic surface of the distribution indicated above. Denote by $\widetilde{W}$ the subweb which is cut on V^{2s} by leaves of the web (Section 1.9). Since the vectors ξ_a are parallel in the Chern connection, the web W is a group web (see Problem **15** of Chapter 1), and its coordinate loops are (at least locally) isomorphic to the kernel J_m of the coordinate loop l_p of the web W.

On the three-web W, to one-parameter subgroups there correspond transversally geodesic surfaces V^2 which cut the leaves of this web along geodesic lines (Section 1.9). Let l be one such geodesic line lying in a leaf $\mathcal{F}_3$ of the third foliation. Since the tangent bivector to the surface V^2 has the form $\xi^i \underset{1}{e_i} \wedge \xi^i \underset{2}{e_i}$ and the tangent r-plane to $\mathcal{F}_3$ can be defined by the vectors $\underset{3}{e_i}$, the tangent vector to their intersection line is $-\xi^i(\underset{1}{e_i} + \underset{2}{e_i}) = \xi^i \underset{3}{e_i}$. Comparing this with (6.32), we see that the line l is an integral line of the vector field ξ.

Varying the leaf $\mathcal{F}_3$, we find that an integral curve of the vector field ξ passes through each point of the surface V^2. Thus, the surface V^2 is invariant under the infinitesimal automorphism ξ. The same result holds for the transversally geodesic surface V^{2s} if system (6.30) has s independent solutions. Hence, we have proven the following theorem.

Theorem 6.18 *Let ξ be a principal infinitesimal automorphism of a three-web $W = (X, \lambda_\alpha)$. Then:*

i) *Integral lines of the vector field ξ are geodesic lines in the Chern connection.*

ii) *The manifold S is foliated into transversally geodesic submanifolds V^{2s} of dimension $2s$, where s is the dimension of the middle ternary kernel J_m of the W-algebra of the web W.*

iii) *Each transversally geodesic surface V^{2s} of the web contains integral lines of the field ξ, and subsequently, it is invariant relative to ξ.*

iv) *On transversally geodesic surfaces V^{2s}, the leaves of the web W cut group three-webs whose coordinate loops are isomorphic to the middle kernel N_m of the coordinate loop of the web W.*

■

3. Let us describe Grassmann three-web W admitting an s-parameter group of regular automorphisms ($s \geq 1$). By Theorem 6.12, such webs can be generated in a projective space P^{r+1} by a hyperplane X_3 and two hypercones X_1 and X_2 whose vertices have at least one common point S. Then, a one-parameter family of homologies with center S and the double hyperplabe X_3 is a desired regular infinitesimal automorphism of this web.

Suppose that the vertices of the hypercones X_1 and X_2 have an s-dimensional intersection Δ. Since each point of the s-plane Δ is the center of a one-parameter family of homologies with the double hyperplane X_3, the group of automorphisms $\mathcal{A}_W^{(3)}$ depends on $s+1$ parameters in this case.

Define: $\Delta' = \Delta \bigcap X_3$, $\dim \Delta' = s-1$. To the points of Δ' there correspond one-parameter homologies that form an s-dimensional subgroup $\tilde{\mathcal{A}}_W^{(3)}$ in $\mathcal{A}_W^{(3)}$. The group $\tilde{\mathcal{A}}_W^{(3)}$ is commutative. In fact, by Proposition 6.6, the coordinate loops of the three-web under consideration have non-trivial kernels N_m which are isomorphic to the group $\mathcal{A}_W^{(3)}$. Let us indicate subwebs corresponding to these kernels. Let p be a straight line in P^{r+1}. Any d-plane passing through p cuts in a Grassmann web W a subweb to ehich there corresponds a subloop of the coordinate loop l_p. Since the kernels are groups, subwebs formed by three hyperplanes correspond to them (Theorem 4.12). But such subwebs are cut by planes containing the plane Δ. Thus, to the kernels of the coordinate loops there correspond subwebs which are cut by $(s+2)$-dimensional planes containing the s-dimensional plane Δ. On the other hand, any $s+1$-dimensional plane Π' passing through the $s-1$-dimensional plane Δ' intersects the hypercones X_1 and X_2 and the hyperplane X_3 at s-dimensional planes containing Δ', i.e. belonging to a pencil. By Theorem 4.12, the corresponding three-web is parallelizable, and its coordinate quasigroup is a commutative group. We arrive at the following proposition.

Proposition 6.19 *A Grassmann three-web W admits an $(s+1)$-parameter group $\mathcal{A}_W^{(3)}$ of regular automorphisms if and only if one of the hypersurfaces X_α generating this web is a hyperplane (X_3) and two others (X_1 and X_2) are hypercones whose vertices have an intersection Δ of dimension s. The group $\mathcal{A}_W^{(3)}$ of regular automorphisms of such a web consists of homologies with centers in Δ and the double plane X_3. If $s \geq 1$, then there exists an $(s-1)$-dimensional intersection $\Delta' = \Delta \bigcap X_3$ to the points of which there correspond parabolic homologies. They form an s-dimensional commutative subgroup of the group $\mathcal{A}_W^{(3)}$.* ■

6.4 *G*-Webs

1. A three-web is said to be a *G-web* if it admits a transitive group of automorphisms $\mathcal{A}_W$. It follows from this definition that a manifold X of dimension $2r$, where the group of automorphisms of a G-web W operates, is a homogeneous space G/H where $G \equiv \mathcal{A}_W$ is the Lie group of dimension $2r+\rho$, and H is its subgroup of dimension ρ. As noticed in Section 6.1, the coordinate loops of a G-web are mutually isomorphic. But loops, isomorphic to all its isotopes, are called G-loops. Thus, the the coordinate loops of a G-web are G-loops.

Under an action of the group G, the leaves of a G-web W must be transferred into the leaves of the same web. Hence, an arbitrary G-web can be constructed by means of subgroups of the group G in the following way. Suppose that the group G has three subgroups G_α, $\alpha = 1, 2, 3$, each of dimension $r + \rho$, whose intersection is a subgroup H of dimension ρ. Leaves of the web W in the homogeneous space $X = G/H$ are r-dimensional factor manifolds gG_α/gH, where $g \in G$ and, as usual, gG_α denotes a co-set.

The group webs and the Moufang webs, which we considered earlier, are G-webs. In fact, a manifold X carrying a group web R is a direct product $G_1 \times G_1^{-1}$, where G_1 is a Lie group of dimension r (Section 1.5). The leaves of a group web are the co-sets of the group G relative to its normal subgroups G_1, G_1^{-1} and the subgroup generated by elements of the form (x, x^{-1}) (see p. 19). The fact that the Moufang webs are G-webs follows from Theorem 4.22 according to which a Moufang web can be realized in the group G precisely in the way we described above.

The Bol webs considered in Chapter 4 are not entirely contained in the class of G-webs. For example, the Grassmann Bol webs, generated in a projective space P^{r+1} by a hyperplane π and a hyperquadric Q, do not admit a transitive group of automorphisms. In fact, if we assume that the hyperplane π is the hyperplane at infinity and the hyperquadric Q is a hypersphere in some metric, then affine transformations, preserving the hypersphere Q, are rotations about the center of Q. However, the group of rotations operates intransitively in the set of straight lines.

Let us find the structure equations of a G-web. We take a co-basis of invariant forms $\underset{1}{\omega}^i, \underset{2}{\omega}^i, \sigma^\kappa$, $\kappa, \mu, \nu = 1, \ldots, \rho$, $\rho \geq 0$, in the initial group G in such a way that the subgroups G_1 and $G_2 = G_1^{-1}$ are defined by the equations $\underset{1}{\omega}^i = 0$ and $\underset{2}{\omega}^i = 0$, respectively. Then, the equations of the subgroup H are $\underset{1}{\omega}^i = \underset{2}{\omega}^i = 0$, and the third $(r + \rho)$-dimensional subgroup G_3 is defined by the system of the form:

$$\lambda^i_j \underset{1}{\omega}^j + \mu^i_j \underset{2}{\omega}^j = 0$$

with the constant coefficients λ^i_j and μ^i_j satisfying the conditions:

$$\det(\lambda^i_j) \neq 0, \quad \det(\mu^i_j) \neq 0.$$

We introduce a new co-basis of invariant forms by applying the substitutions $\lambda^i_j \underset{1}{\omega}^j \rightarrow \underset{1}{\omega}^i$ and $\mu^i_j \underset{2}{\omega}^j \rightarrow \underset{2}{\omega}^i$. As a result, the equations of the subgroup G_3 are reduced to the form: $\underset{1}{\omega}^i + \underset{2}{\omega}^i = 0$.

Consider the Maurer–Cartan equations but first only for the forms $\underset{1}{\omega}^i$ and $\underset{2}{\omega}^i$. By virtue of complete integrability of the systems $\underset{1}{\omega}^i = 0$ and $\underset{2}{\omega}^i = 0$, these equations can be written as follows:

$$\begin{aligned} d\underset{1}{\omega}^i &= \frac{1}{2} \underset{1}{c}{}^i_{jk} \underset{1}{\omega}^j \wedge \underset{1}{\omega}^k + \underset{1}{e}{}^i_{jk} \underset{1}{\omega}^j \wedge \underset{2}{\omega}^k + \underset{1}{c}{}^i_{j\kappa} \underset{1}{\omega}^j \wedge \sigma^\kappa, \\ d\underset{2}{\omega}^i &= \underset{2}{e}{}^i_{jk} \underset{2}{\omega}^j \wedge \underset{1}{\omega}^k + \frac{1}{2} \underset{2}{c}{}^i_{jk} \underset{2}{\omega}^j \wedge \underset{2}{\omega}^k + + \underset{2}{c}{}^i_{j\kappa} \underset{2}{\omega}^j \wedge \sigma^\kappa. \end{aligned} \tag{6.35}$$

All coefficients in these equations are constants satisfying the conditions: $\underset{1}{c}^i_{(jk)} = 0$ and $\underset{1}{c}^i_{(jk)} = 0$. The conditions of complete integrability of the system $\underset{1}{\omega}^i + \underset{2}{\omega}^i = 0$, defining the subgroup G_3, lead to the relations:

$$\underset{1}{e}^i_{[jk]} + \underset{2}{e}^i_{[jk]} = \frac{1}{2}(\underset{1}{c}^i_{jk} + \underset{2}{c}^i_{jk}), \qquad \underset{1}{c}^i_{j\kappa} = \underset{2}{c}^i_{j\kappa} \overset{\text{def}}{=} c^i_{j\kappa}. \tag{6.36}$$

To apply equations (6.36) to equations (6.35), we introduce the notation:

$$\frac{1}{2}\underset{1}{c}^i_{jk} - \underset{2}{e}^i_{[jk]} = -\frac{1}{2}\underset{2}{c}^i_{jk} + \underset{1}{e}^i_{[jk]} \overset{\text{def}}{=} a^i_{jk}, \tag{6.37}$$

and set:

$$\omega^i_j = \underset{2}{e}^i_{jk}\underset{1}{\omega}^k + \underset{1}{e}^i_{jk}\underset{2}{\omega}^k + c^i_{j\kappa}\sigma^\kappa. \tag{6.38}$$

As a result, equations (6.35) take the form:

$$d\underset{1}{\omega}^i = \underset{1}{\omega}^j \wedge \omega^i_j + a^i_{jk}\underset{1}{\omega}^j \wedge \underset{1}{\omega}^k, \quad d\underset{2}{\omega}^i = \underset{2}{\omega}^j \wedge \omega^i_j - a^i_{jk}\underset{2}{\omega}^j \wedge \underset{2}{\omega}^k,$$

i.e. their form coincides with that of structure equations (1.26) of a three-web. Hence, they define the G-web under consideration in the homogeneous space $X = G/H$. The quantities a^i_{jk} form the torsion tensor of this web.

Let us give a geometric meaning of the forms ω^i_j defined by equations (6.38). We fix a moving point p of the homogeneous space $X = G/H$ carrying the G-web under consideration, i.e. we set $\underset{1}{\omega}^i = \underset{2}{\omega}^i = 0$. Then equations (6.38) become

$$\pi^i_j = c^i_{j\kappa}\theta^\kappa,$$

where π^i_j and θ^κ denote the forms ω^i_j and σ^κ taken when the principal parameters are fixed. We recall that the forms π^i_j define concordant transformations of frames in the tangent spaces T_α to the leaves of the web passing though the point p, and these forms are invariant forms of the group $\mathbf{GL}(r)$ – the structure group of the G_W-structure defined by this web. On the other hand, the forms $\theta^\kappa = \sigma^\kappa(\delta)$ are invariant forms of the group H – the stationary subgroup of the point p of the homogeneous space X. Thus, the geometric meaning of formulas (6.38) is that the structure group of the G_W-structure defined by a G-web is reduced to the stationary group H of the homogeneous space G/H carrying a G-web. The adapted frames of this G_W-structure are connected by transformations with constant coefficients since they define a transformation from one invariant basis of the group G to another.

Next, we will find the curvature tensor of a G-web. To do this, we write the whole set of the structure equations of the group G. They consist of equations (6.35) and the equations which the forms θ^κ satisfy. By (6.36), we have:

$$
\begin{aligned}
d\underset{1}{\omega}{}^{i} &= \tfrac{1}{2}\underset{1}{c}{}^{i}_{jk}\underset{1}{\omega}{}^{j} \wedge \underset{1}{\omega}{}^{k} + \underset{1}{e}{}^{i}_{jk}\underset{1}{\omega}{}^{j} \wedge \underset{2}{\omega}{}^{k} + c^{i}_{j\kappa}\underset{1}{\omega}{}^{j} \wedge \sigma^{\kappa}, \\
d\underset{2}{\omega}{}^{i} &= \underset{2}{e}{}^{i}_{jk}\underset{2}{\omega}{}^{j} \wedge \underset{1}{\omega}{}^{k} + \frac{1}{2}\underset{2}{c}{}^{i}_{jk}\underset{2}{\omega}{}^{j} \wedge \underset{2}{\omega}{}^{k} + +c^{i}_{j\kappa}\underset{2}{\omega}{}^{j} \wedge \sigma^{\kappa}, \\
d\sigma^{\kappa} &= \tfrac{1}{2}\underset{3}{c}{}^{\kappa}_{\mu\nu}\sigma^{\mu} \wedge \sigma^{\nu} + \frac{1}{2}\underset{1}{c}{}^{\kappa}_{jk}\underset{1}{\omega}{}^{j} \wedge \underset{1}{\omega}{}^{k} + \bar{c}^{\kappa}_{jk}\underset{1}{\omega}{}^{j} \wedge \underset{2}{\omega}{}^{k} \\
&\quad + \tfrac{1}{2}\underset{2}{c}{}^{\kappa}_{jk}\underset{2}{\omega}{}^{j} \wedge \underset{2}{\omega}{}^{k} + \underset{1}{e}{}^{\kappa}_{j\nu}\underset{1}{\omega}{}^{j} \wedge \sigma^{\nu} + \underset{2}{e}{}^{\kappa}_{j\nu}\underset{2}{\omega}{}^{j} \wedge \sigma^{\nu}.
\end{aligned} \tag{6.39}
$$

The components of the structure tensor of the group G on the right-hand sides of equations (6.39) satisfy the Jacobi identity. The latter, because of the complicated structure of system (6.39), is broken up into a few sequences of relations. We will not write these relations in this book (see Problem **10**). However, the reader can find them by applying exterior differentiation to system (6.39).

To find the curvature tensor of a G-web, according to formula (1.32), we must calculate the difference $d\omega^{i}_{j} - \omega^{k}_{j} \wedge \omega^{i}_{k}$. Substitute for the forms ω^{i}_{j} in this difference their values from (6.38) and transform the expression obtained by means of the structure equations (6.39). As a result, we obtain an exterior quadratic form containing all kinds of the products of the basis forms $\underset{1}{\omega}{}^{i}, \underset{2}{\omega}{}^{i}$ and σ^{κ}. However, by virtue of the Jacobi identities, which the coefficients of equations (6.39) satisfy, all terms in this quadratic form vanish except the term $b^{i}_{jkl}\underset{1}{\omega}{}^{k} \wedge \underset{2}{\omega}{}^{l}$, and the curvature tensor b^{i}_{jkl} is found in the form:

$$
b^{i}_{jkl} = \underset{1}{e}{}^{m}_{kl}\underset{2}{e}{}^{i}_{jm} - \underset{2}{e}{}^{m}_{lk}\underset{1}{e}{}^{i}_{jm} - \underset{2}{e}{}^{m}_{jk}\underset{1}{e}{}^{i}_{ml} + \underset{1}{e}{}^{m}_{jl}\underset{2}{e}{}^{i}_{mk} + c^{i}_{j\kappa}\bar{c}^{\kappa}_{kl}. \tag{6.40}
$$

By means of formula (6.40), we get one unexpected result about group webs. As was noted in the beginning of this section, a group web is defined in the direct product of groups, i.e. it is a G-web in which two out of three subgroups G_{α}, generating the web, are normal subgroups of G. Formula (6.40) allows us to strengthen this result.

Theorem 6.20 *If at least one of subgroups G_{α} generating a G-web in the group G, is a normal subgroup of G, then this G-web is a group three-web.*

Proof. Since all subgroups G_{α} are of the same structure, it is sufficient to prove the theorem for one value of $\alpha = 1, 2, 3$. Suppose, for example, that the subgroup G_1 defined by the equations $\underset{1}{\omega}{}^{i} = 0$ is normal. Then, according to the well-known fact from Lie group theory, the quadratic forms $d\underset{1}{\omega}{}^{i}$ can be expressed in terms of the forms $\underset{1}{\omega}{}^{i}$ alone. This and equations (6.39) imply that

$$
\underset{1}{e}{}^{i}_{jk} = 0, \quad c^{i}_{j\kappa} = 0. \tag{6.41}
$$

Substituting (6.41) into (6.40), we find that $b^{i}_{jkl} = 0$. As shown in Section 1.5, the latter condition characterizes group three-webs. ■

The analytic proof we have performed is rather formal. We supplement it by geometric considerations clarifying the structure of the group G-web considered in

Theorem 6.20. First of all, it follows from relations (6.41) that the subgroup H is normal in G. Thus, the homogeneous space G/H is a group. We will use the same notation G for this group. The G-web under considearation is formed by the co-sets gG_α on G. Taking into account (6,38) and (6.41), we can write the corresponding structure equations in the form:

$$\begin{aligned} d\underset{1}{\omega}^i &= -\frac{1}{2}\underset{2}{c}^i_{jk}\underset{1}{\omega}^j \wedge \underset{1}{\omega}^k + \underset{2}{e}^i_{jk}\underset{1}{\omega}^j \wedge \underset{2}{\omega}^k, \\ d\underset{2}{\omega}^i &= \underset{2}{e}^i_{jk}\underset{2}{\omega}^j \wedge \underset{1}{\omega}^k + \frac{1}{2}\underset{2}{c}^i_{jk}\underset{2}{\omega}^j \wedge \underset{2}{\omega}^k. \end{aligned} \tag{6.42}$$

Since the subgroup G_1 is normal, multiplication can be defined in the set of co-sets gG_1, i.e. in the set of leaves of the third foliation of the web W. It follows immediately from this that the mapping $\varphi_{2,3} : G_2 \to G_3$ (see Section 1.3), established on the G-web under consideration by leaves of the third foliation, is an isomorphism. However, equations (6.42) show that the subgroups G_2 and G_3 are not normal in G and not isomorphic to the group G_1.

On the other hand, the equations of a group web can be written in the following form (see Section 1.5):

$$d\underset{1}{\bar{\omega}}^i = c^i_{jk}\underset{1}{\bar{\omega}}^j \wedge \underset{1}{\bar{\omega}}^k, \quad d\underset{2}{\bar{\omega}}^i = -c^i_{jk}\underset{2}{\bar{\omega}}^j \wedge \underset{2}{\bar{\omega}}^k. \tag{6.43}$$

As we have already noted many times, in this case, a manifold X in which a group three-web is defined, is a direct product of groups: $\bar{G}_1 \times \bar{G}_1$, $\dim \bar{G}_1 = r$. Denote this group by $\bar{G}$. Comparing systems (6.42) and (6.43), we see that, in general, the groups G and $\bar{G}$ are not isomorphic. Thus, a group three-web can be realized as a G-web on non-isomorphic groups.

However, since both systems (6.42) and (6.43) define the same three-web W, there exists an admissible transformation:

$$\underset{1}{\bar{\omega}}^i = A^i_j\underset{1}{\omega}^j, \quad \underset{2}{\bar{\omega}}^i = A^i_j\underset{2}{\omega}^j, \tag{6.44}$$

connecting the adapted co-frames $\mathcal{R} = \{\underset{1}{\omega}^i, \underset{2}{\omega}^i\}$ and $\bar{\mathcal{R}} = \{\underset{1}{\bar{\omega}}^i, \underset{2}{\bar{\omega}}^i\}$. Moreover, the functions A^i_j in equations (6.44) are variable since the groups G and $\bar{G}$ are not isomorphic to one another.

The functions A^i_j satisfy the equations:

$$dA^i_j - A^i_k\omega^k_j + A^k_j\bar{\omega}^i_k = 0$$

(cf. p. 21), where ω^i_j and $\bar{\omega}^i_j$ are the connection forms of the Chern connection with respect to the co-frames $\mathcal{R}$ and $\bar{\mathcal{R}}$. The forms $\bar{\omega}^i_j$ are equal to zero, and the forms ω^i_j can be found from (6.38) by applying (6.41):

$$\omega^i_j = \underset{2}{e}^i_{jk}\omega^k.$$

Substituting this into the previous equation, we obtain:

$$dA^i_j - A^i_k \underset{2}{e}^k_{jl} \underset{1}{\omega}^l = 0. \tag{6.45}$$

It is possible to check, using the Jacobi identities (see Problem **10**), that equations (6.45) are completely integrable. Thus, if a group web is given by equations (6.42), then from (6.45) we can find the quantities A^i_j and using them we transform equations (6.45) into "canonical" equations (6.43). The quantities A^i_j also connect the components of the torsion tensor given in different bases. As follows from equations (6.37) and (6.41), in the co-frame $\mathcal{R}$ the torsion tensor is: $a^i_{jk} = -\underset{2}{c}^i_{jk}$, and from equations (6.43) we find $\bar{a}^i_{jk} = c^i_{jk}$. According to the tensor law of transformation, we have:

$$c^i_{jk} = -A^i_l \widetilde{A}^p_j \widetilde{A}^q_k \underset{2}{c}^l_{pq} \tag{6.46}$$

where the matrix $\widetilde{A}^i_j$ is the inverse matrix of the matrix A^i_j. Note that although the quantities A^i_j are not constant, nevertheless the components c^i_{jk} defined by formulas (6.46) are constants (Problem **11**).

Problem **12** shows the existence of non-isomorphic groups G and $\bar{G}$ connected with a group three-web.

We now return to the general case and state two basic problems arising in the theory of G-webs:

a) Find a subclass of G-webs in a given class of three-webs, and

b) Find out whether a given three-web W is a G-web.

The way to solve the first problem is suggested by the fact that on a G-web, the components of all its fundamental tensors are constant since they can be expressed in terms of the structure tensor of the initial group G. Therefore, in order to find a subclass of G-webs in a given class $\mathcal{K}$ of three-webs, we must consider as constants the quantities $a^i_{jk}, b^i_{jkl}, \ldots$ in equations (1.26), (1.32) etc. Suppose that the system of equations obtained afterwards (denote it by Σ) does not imply new conditions for the fundamental tensors of the web and the forms $\underset{1}{\omega}^i$ and $\underset{2}{\omega}^i$. Then, from the system Σ we can express a part of the forms ω^i_j in terms of the principal forms $\underset{1}{\omega}^i$ and $\underset{2}{\omega}^i$, i.e. we reduce the family of adapted frames of the web. In this case, the structure equations form a system with constant coefficients closed with respect to the operation of exterior differentiation. This system defines a Lie group G on which the class $\mathcal{K}$ can be realized. Therefore, the whole class $\mathcal{K}$ is a part of the class of G-webs.

If the system Σ leads to new relations for the fundamental tensors but, in spite of these new relations, the forms $\underset{1}{\omega}^i$ and $\underset{2}{\omega}^i$ still remain independent, then these new relations define a subclass of G-webs in the class $\mathcal{K}$.

In the case when the system Σ implies relations among the forms $\underset{1}{\omega}^i$ and $\underset{2}{\omega}^i$, whether there are any relations for the fundamental tensors or not, there are no G-webs in the class $\mathcal{K}$.

We had similar considerations in Section 4.5 where we found the Moufang three-web of minimal dimension, and in Section 4.3 while discovering equations (6.38) of a six-dimensional Bol web with a symmetric matrix (a^{ij}) of rank one. Therefore, the latter web is also a G-web. Moreover, a six-dimensional Bol web with a non-symmetric matrix (a^{ij}) of rank three (see subsection **4** of Section 4.3) is also a G-web. The latter web is interesting due to the fact that the subgroup H is trivial for it: $H \equiv e$.

We now outline one of the methods to solve the problem b). Let a three-web W be given by closed form equations (1.3). Applying formulas of Section 1.5, we can find the structure equations of the web. Moreover, since a system of local coordinates is fixed, the forms ω_j^i can be expressed in terms of the local coordinates (the principal parameters) alone. In terms of the theory of G-structures, this means that the G_W-structure defined by the web W is the e-structure.

If the web W under consideration is a G-web, then the Lie group G on which this web can be realized, can be obtained by introducing new parameters. From this we obtain the following criterion: a three-web W under consideration is a G-web if by defining new basis forms $A_j^i \underset{1}{\omega}^i$ and $A_j^i \underset{2}{\omega}^i$, where the A_j^i are functions of the new and old parameters, we are able to write the structure equations of the web W in the form in which all functions are constants. The difficulty in using this approach is that there is no general argument allowing us to find the quantities A_J^i.

PROBLEMS

1. Prove that the groups of autotopies of isotopic quasigroups are isomorphic (see [Bel 67], p. 26).

2. Prove that an autotopy $A = (A_\alpha)$ is completely defined if two of its components are given.

Proof ([Bel 67], p. 26). In fact, if $A = (A_1, A_2, A_3)$ and $A = (A'_1, A'_2, A'_3)$ are autotopies, then $A'A^{-1} = (\mathrm{id}, A'_2 A_2^{-1}, \mathrm{id})$ is also an autotopy. Since two components of the latter autotopy are the identical transformations, the same is true for the third component, i.e. $A'_2 A_2^{-1} = \mathrm{id}, A'_2 = A_2^{-1}$. ■

3. Prove that the corresponding kernels of isotopic loops are isomorphic.

4. Prove that if the solution space of the system of equations (6.18) and (6.21) has maximal dimension, then the three-web is parallelizable.

5. Find under what conditions the basis vector $\underset{\alpha}{e}_i$ generates an infinitesimal automorphism of a three-web.

6. Write the structure equations of a three-web W on which all basis vectors $\underset{1}{e}_i$ generate infinitesimal automorphisms.

7. Prove that the set of vector fields ξ – solutions of system (6.18) – is closed with respect to the operation of commutation.

Proof. Let $\xi = (\underset{1}{\xi}{}^i, \underset{2}{\xi}{}^i)$ and $\eta = (\underset{1}{\eta}{}^i, \underset{2}{\eta}{}^i)$ be two solutions of system (6.18). According to Problem **21** of Chapter 1, the coordinates of their commutator $\zeta = [\xi, \eta]$ can be calculated by the following formulas:

$$\begin{aligned} \underset{1}{\zeta}{}^i &= \underset{1}{\xi}{}^j \eta^i_j - \underset{1}{\eta}{}^j \xi^i_j + 2a^i_{jk} \underset{1}{\xi}{}^j \underset{1}{\eta}{}^k, \\ \underset{2}{\zeta}{}^i &= \underset{2}{\xi}{}^j \eta^i_j - \underset{2}{\eta}{}^j \xi^i_j - 2a^i_{jk} \underset{2}{\xi}{}^j \underset{2}{\eta}{}^k. \end{aligned}$$

If we differentiate these two equations and transfor the result obtained by applying the same equations, relations (6.23) and (1.31), we arrive at the equations:

$$\nabla \underset{1}{\zeta}{}^i = (\zeta^i_j + 2a^i_{jk} \underset{1}{\zeta}{}^k) \underset{1}{\omega}{}^j, \nabla \underset{2}{\zeta}{}^i = (\zeta^i_j - 2a^i_{jk} \underset{2}{\zeta}{}^k) \underset{2}{\omega}{}^j,$$

which are also of form (6.18). In the last two equations we used the notation:

$$\zeta^i_j = \xi^k_j \eta^i_k - \eta^k_j \xi^i_k + b^i_{jkl} (\underset{1}{\xi}{}^k \underset{2}{\eta}{}^l - \underset{1}{\eta}{}^k \underset{2}{\xi}{}^l) ■$$

8. Prove that equations (6.28) define a linear representation of the group G which is homogeneous if $c^i_j = 0$.

9. Suppose that an infinitesimal automorphism ξ satisfies the condition $\xi(p) = 0$. Prove that the linear operator ξ^i_j is a differentiation for any of the tensors $\underset{1}{c}$ and $\underset{2}{c}$ – the covariant derivatives of the curvature tensor b^i_{jkl}.

10. Show that the Jacobi identities, which the coefficients of equations (6.39) satisfy provided that $\sigma^\kappa = 0$, have the following form:

$$\begin{aligned} & \underset{1}{c}{}^i_{m[j} \underset{1}{c}{}^m_{kl]} = 0, \quad \underset{2}{c}{}^i_{m[j} \underset{2}{c}{}^m_{kl]} = 0, \\ & \underset{1}{c}{}^i_{mk} \underset{1}{e}{}^m_{jl} - \underset{1}{c}{}^i_{mj} \underset{1}{e}{}^m_{kl} - \underset{1}{e}{}^i_{ml} \underset{1}{c}{}^m_{jk} + \underset{1}{e}{}^i_{km} \underset{2}{e}{}^m_{lj} - \underset{1}{e}{}^i_{jm} \underset{2}{e}{}^m_{lk} = 0, \\ & \underset{1}{e}{}^i_{mk} \underset{1}{e}{}^m_{jl} - \underset{1}{e}{}^i_{ml} \underset{1}{e}{}^m_{jk} - \underset{1}{e}{}^i_{jm} \underset{2}{c}{}^m_{lk} = 0, \\ & \underset{2}{c}{}^i_{mk} \underset{2}{c}{}^m_{jl} + \underset{2}{e}{}^i_{jm} \underset{1}{e}{}^m_{kl} - \underset{2}{e}{}^i_{lm} \underset{1}{e}{}^m_{kj} + \underset{2}{c}{}^i_{lm} \underset{2}{e}{}^m_{jk} - \underset{1}{c}{}^i_{jm} \underset{2}{e}{}^m_{lk} = 0, \\ & \underset{2}{e}{}^i_{mk} \underset{2}{e}{}^m_{lj} - \underset{2}{e}{}^i_{mj} \underset{2}{e}{}^m_{lk} + \underset{2}{e}{}^i_{lm} \underset{1}{c}{}^m_{kj} = 0. \end{aligned}$$

11. Prove that the quantities c^i_{jk}, defined by formulas (6.46), are constants.

Hint: First, show that the quantities $\widetilde{A}^i_j$ in formulas (6.46) satisfy equations $d\widetilde{A}^i_j + \widetilde{A}^i_k \underset{2}{e}{}^i_{jl} \underset{1}{\omega}{}^l = 0$; next, differentiate equations (6.46) and apply relations (6.45) and the Jacobi identities found in Problem **10**.

12. Prove that the systems $d\underset{1}{\omega} = 0$, $d\underset{2}{\omega} = 0$ and $d\underset{1}{\bar{\omega}} = 0$, $d\underset{2}{\bar{\omega}} = \underset{1}{\bar{\omega}} \wedge \underset{2}{\bar{\omega}}$ determine the same two-dimensional three-web. Find a transformation transferring the system of forms $\{\underset{1}{\omega}, \underset{2}{\omega}\}$ into the system of forms $\{\underset{1}{\bar{\omega}}, \underset{2}{\bar{\omega}}\}$. (*Hint*: Apply equations (6.45).)

NOTES

6.1. On autotopies of quasigroups and loops see, for example, [Bel 67] and [Bel 88]; on autotopies of abstract three-webs see [BS 83] and [S 90a].

Theorem 6.9 is due to P. Nagy [N 89].

6.2 – 6.3. The infinitesimal automorphisms of three-webs were studied by Gvozdovich in [Gv 81a], [Gv 81b], [Gv 82] and [Gv 85]. He found the basic equations (6.18), (6.21), (6.22) and (6.23). One of the important results which was obtained in [Gv 81b] but was not reflected in Chapter 6, is that analytic Bol three-webs admit an r-parameter group of autotopies.

6.4. G-loops were defined by R. H. Bruck in [Br 51]; see also Chapter X in the monograph [Bel 67]. Smooth G-loops have hardly been studied. We are aware only of one paper on analytic Bol G-loops [Mi 86].

G-webs were considered by A. Barlotti and K. Strambach [BS 83]. These webs were studied in more detail in [S 89a], [S 90a], [LS 89] and [S 91b].

Problems. Problem **7** is due to Gvozdovich [Gv 88].

Chapter 7

Geometry of the Fourth Order Differential Neighborhood of a Multidimensional Three-Web

The success in the study of web geometry on the level of the third order differential neighborhood has been achieved due to the two circumstances. First of all, each of the classes of webs listed in Table 2.2 (p. 83) can be characterized by a special structure of the torsion and curvature tensors a and b. Secondly, these tensors have a clear geometric meaning: they determine the principal part of the "deviation" from commutativity and associativity in the coordinate loop l_p of a web W, respectively (see Section 2.5). As a result, each of these classes has three characterizations: the closure condition, the algebraic identity and the tensor equation. For example, for the group three-webs these three characterizations are: a) the figures R are closed on them; b) their coordinate loops are associative; and c) the curvature tensor is zero. The same kind of characterizations can be given for other classes of webs.

In order to develop a similar theory in the differential neighborhood of order s, $s \geq 3$, we need to connect the fundamental tensors of type $\binom{1}{s}$ (the covariant derivatives of order $s-3$ of the curvature tensor b), arising in this neighborhood, with algebraic identities of s variables in the coordinate loops l_p. In other words, we must find pairs of words $S_1(u_1, u_2, \ldots, u_s)$ and $S_1(u_1, u_2, \ldots, u_s)$, $u_1, u_2, \ldots, u_s \in l_p$, such that the principal part of the expression $S_1 \cdot S_2^{-1}$ is defined by a tensor from the differential neighborhood of order s of a web W.

On the whole, this problem is very complicated, and its complete solution is known only for $s \leq 4$. In the present chapter we introduce the tensors L, R and M. Similar to the tensors a and b, each of the tensors $\mathcal{L}, \mathcal{R}$ and $\mathcal{M}$ determines the principal part of a certain polynomial composed of four elements of the loop l_p. Equating these polynomials to one, we distinguish special classes of webs in whose coordinate loops the corresponding identity with four variables holds. If we identify some of the variables, we obtain wider classes of webs. Some of these classes coincide with the classes of webs which were studied earlier. Analyzing the closure conditions, corresponding to the new classes of webs, we find solutions of some well-known problems of the theory of loops that are connected with the isotopic invariance of identities in loops.

7.1 Computation of Covariant Derivatives of the Curvature Tensor of a Three-Web

In this section we obtain an explicit expression for the tensors $\underset{1}{c} = (\underset{1}{c}{}^i_{jklm})$ and $\underset{2}{c} = (\underset{2}{c}{}^i_{jklm})$ in terms of the partial derivatives of the functions $z^i = f^i(x^j, y^k)$ defining a web, and we also calculate the values of these tensors at an arbitrary point p in terms of the coefficients of the Taylor expansions of the coordinate loop l_p.

1. The basis forms $\underset{1}{\omega}{}^i$ and $\underset{2}{\omega}{}^i$ of a manifold X carrying a three-web W can be expressed in terms of the partial derivatives $\bar{f}^i_j$ and $\tilde{f}^i_j$ of the function f as follows (see (1.51)):

$$\underset{1}{\omega}{}^i = \bar{f}^i_j dx^j, \quad \underset{2}{\omega}{}^i = \tilde{f}^i_j dy^j. \tag{7.1}$$

Substituting these values into equations (1.38) for the tensor ∇b^i_{jkl}, we obtain:

$$\nabla b^i_{jkl} = \underset{1}{c}{}^i_{jklm} \bar{f}^m_p dx^p + \underset{2}{c}{}^i_{jklm} \tilde{f}^m_p dy^p. \tag{7.2}$$

Applying the same relations (7.1), we can write the expression for the forms ω^i_j in the following way:

$$\omega^i_j = \Gamma^i_{nj} \bar{f}^n_p dx^p + \Gamma^i_{jn} \tilde{f}^n_p dy^p. \tag{7.3}$$

Applying (7.3), we can write in more detail the expression for the covariant differential ∇b^i_{jkl}:

$$\begin{aligned}\nabla b^i_{jkl} = db^i_{jkl} \quad & + b^m_{jkl}(\Gamma^i_{nm} \bar{f}^n_p dx^p + \Gamma^i_{mn} \tilde{f}^n_p dy^p) - b^i_{mkl}(\Gamma^m_{nj} \bar{f}^n_p dx^p + \Gamma^m_{jn} \tilde{f}^n_p dy^p) \\ & - b^i_{jml}(\Gamma^m_{nk} \bar{f}^n_p dx^p + \Gamma^m_{kn} \tilde{f}^n_p dy^p) - b^i_{jkm}(\Gamma^m_{nl} \bar{f}^n_p dx^p + \Gamma^m_{ln} \tilde{f}^n_p dy^p).\end{aligned}$$

Substituting this expression into (7.2) and equating the coefficients in dx^p and dy^p, we obtain:

$$\begin{aligned}\underset{1}{c}{}^i_{jklm} &= \bar{g}^p_m \frac{\partial}{\partial x^p}(b^i_{jkl}) + (-b^p_{jkl}\Gamma^i_{mp} + b^i_{pkl}\Gamma^p_{mj} + b^i_{jpl}\Gamma^p_{mk} + b^i_{jkp}\Gamma^p_{ml}), \\ \underset{2}{c}{}^i_{jklm} &= \tilde{g}^p_m \frac{\partial}{\partial y^p}(b^i_{jkl}) + (-b^p_{jkl}\Gamma^i_{pm} + b^i_{pkl}\Gamma^p_{jm} + b^i_{jpl}\Gamma^p_{km} + b^i_{jkp}\Gamma^p_{lm}).\end{aligned} \tag{7.4}$$

Here, as in Chapter 1, $\bar{g}^i_j$ and $\tilde{g}^i_j$ denote the inverse matrices of the matrices $\bar{f}^i_j$ and $\tilde{f}^i_j$, respectively.

Next, substitute in (7.4) for the quantities b^i_{jkl} their values taken from (1.58) and calculate the partial derivatives with respect to x^p and y^p in (7.4). After a long but uncomplicated computation, we arrive at the following expressions for the tensors $\underset{1}{c}$ and $\underset{2}{c}$:

$$
\begin{aligned}
\underset{1}{c}{}^i_{jklm} = \; & \left(\frac{\partial^4 f^i}{\partial x^s \partial x^q \partial y^n \partial y^p}\tilde{g}^n_j - \frac{\partial^4 f^i}{\partial x^s \partial x^q \partial x^n \partial y^p}\bar{g}^n_j\right)\bar{g}^s_k \bar{g}^q_m \tilde{g}^p_l \\
& - \frac{\partial^3 f^i}{\partial x^s \partial y^n \partial y^p}\left(\frac{\partial^2 f^{l_1}}{\partial y^{p_1} \partial x^{m_1}}\bar{g}^s_k \tilde{g}^n_{l_1} \tilde{g}^p_l \tilde{g}^{p_1}_j \bar{g}^{m_1}_m + \frac{\partial^2 f^{l_1}}{\partial x^{p_1} \partial x^{m_1}}\bar{g}^s_{l_1} \tilde{g}^n_j \tilde{g}^p_l \bar{g}^{p_1}_k \bar{g}^{m_1}_m \right.\\
& \left. + \frac{\partial^2 f^{l_1}}{\partial y^{p_1} \partial x^{m_1}}\bar{g}^s_k \tilde{g}^n_j \tilde{g}^p_{l_1} \tilde{g}^{p_1}_l \bar{g}^{m_1}_m\right) + \frac{\partial^3 f^i}{\partial x^s \partial x^n \partial y^p}\left(\frac{\partial^2 f^{l_1}}{\partial x^{p_1} \partial x^{m_1}}\bar{g}^s_k \bar{g}^n_{l_1} \tilde{g}^p_l \bar{g}^{p_1}_j \bar{g}^{m_1}_m \right.\\
& \left. + \frac{\partial^2 f^{l_1}}{\partial x^{p_1} \partial x^{m_1}}\bar{g}^s_{l_1} \bar{g}^n_j \tilde{g}^p_l \bar{g}^{p_1}_k \bar{g}^{m_1}_m + \frac{\partial^2 f^{l_1}}{\partial y^{p_1} \partial x^{m_1}}\bar{g}^s_k \bar{g}^n_j \tilde{g}^p_{l_1} \tilde{g}^{p_1}_l \bar{g}^{m_1}_m\right) \\
& - \frac{\partial^3 f^i}{\partial x^{m_1} \partial y^{n_1} \partial x^{p_1}}\left(\frac{\partial^2 f^p}{\partial y^s \partial y^n}\bar{g}^{m_1}_k \tilde{g}^{n_1}_p \tilde{g}^s_j \tilde{g}^n_l - \frac{\partial^2 f^p}{\partial x^s \partial x^n}\tilde{g}^{n_1}_l \bar{g}^{m_1}_p \bar{g}^s_j \bar{g}^n_k\right)\bar{g}^{p_1}_m \\
& - \frac{\partial^2 f^p}{\partial y^s \partial y^n}\bar{g}^{p_1}_m \tilde{g}^s_j \tilde{g}^n_l \left(\Gamma^i_{q_1 p}\frac{\partial^2 f^{q_1}}{\partial x^{m_1} \partial x^{p_1}}\bar{g}^{m_1}_k + \Gamma^i_{kl_1}\frac{\partial^2 f^{l_1}}{\partial y^{n_1} \partial x^{p_1}}\tilde{g}^{n_1}_p\right) \\
& - \Gamma^i_{kp}\frac{\partial^2 f^p}{\partial y^s \partial y^n}\frac{\partial^2 f^{l_1}}{\partial y^{p_1} \partial x^{m_1}}(\tilde{g}^n_l \tilde{g}^{p_1}_j \tilde{g}^s_{l_1} + \tilde{g}^s_j \tilde{g}^n_{l_1} \tilde{g}^{p_1}_l)\bar{g}^{m_1}_m \\
& + \frac{\partial^2 f^p}{\partial x^s \partial x^n}\bar{g}^{p_1}_m \bar{g}^s_j \bar{g}^n_k \left(\Gamma^i_{l_1 l}\frac{\partial^2 f^{l_1}}{\partial x^{m_1} \partial x^{p_1}}\bar{g}^{m_1}_p + \Gamma^i_{pl_1}\frac{\partial^2 f^{l_1}}{\partial y^{n_1} \partial x^{p_1}}\tilde{g}^{n_1}_l\right) \\
& + \Gamma^i_{pl}\frac{\partial^2 f^p}{\partial x^s \partial x^n}\frac{\partial^2 f^{l_1}}{\partial x^{p_1} \partial x^{m_1}}(\bar{g}^s_{l_1} \bar{g}^n_k \bar{g}^{p_1}_j + \bar{g}^s_j \bar{g}^n_{l_1} \bar{g}^{p_1}_k)\bar{g}^{m_1}_m \\
& + \left(\Gamma^i_{kp}\frac{\partial^3 f^p}{\partial x^q \partial y^s \partial y^n}\tilde{g}^s_j \tilde{g}^n_l - \Gamma^i_{pl}\frac{\partial^3 f^p}{\partial x^s \partial x^n \partial x^q}\bar{g}^s_j \bar{g}^n_k\right)\bar{g}^q_m \\
& - \frac{\partial^3 f^p}{\partial x^{m_1} \partial y^{n_1} \partial x^{p_1}}\bar{g}^{p_1}_m (\Gamma^i_{kp}\tilde{g}^{n_1}_l \bar{g}^{m_1}_j - \Gamma^i_{pl}\tilde{g}^{n_1}_j \bar{g}^{m_1}_k) \\
& - \frac{\partial^3 f^i}{\partial x^{m_1} \partial y^{n_1} \partial x^{p_1}}\bar{g}^{p_1}_m (\Gamma^p_{jl}\bar{g}^{m_1}_k \tilde{g}^{n_1}_p - \Gamma^p_{kj}\bar{g}^{m_1}_p \tilde{g}^{n_1}_l) \\
& - \Gamma^i_{ks}\left(\Gamma^s_{l_1 l}\frac{\partial^2 f^{l_1}}{\partial x^{m_1} \partial x^{p_1}}\bar{g}^{m_1}_j + \Gamma^s_{jl_1}\frac{\partial^2 f^{l_1}}{\partial y^{n_1} \partial x^{p_1}}\tilde{g}^{n_1}_l\right)\bar{g}^{p_1}_m \\
& - \Gamma^s_{jl}\left(\Gamma^i_{l_1 s}\frac{\partial^2 f^{l_1}}{\partial x^{m_1} \partial x^{p_1}}\bar{g}^{m_1}_k \bar{g}^{p_1}_m + \Gamma^i_{kl_1}\frac{\partial^2 f^{l_1}}{\partial y^{n_1} \partial x^{p_1}}\bar{g}^{p_1}_m \tilde{g}^{n_1}_s\right) \\
& + \Gamma^i_{sl}\left(\Gamma^s_{l_1 j}\frac{\partial^2 f^{l_1}}{\partial x^{m_1} \partial x^{p_1}}\bar{g}^{m_1}_k + \Gamma^s_{kl_1}\frac{\partial^2 f^{l_1}}{\partial y^{n_1} \partial x^{p_1}}\tilde{g}^{n_1}_j\right)\bar{g}^{p_1}_m \\
& + \Gamma^s_{kj}\left(\Gamma^i_{l_1 l}\frac{\partial^2 f^{l_1}}{\partial x^{m_1} \partial x^{p_1}}\bar{g}^{m_1}_s + \Gamma^i_{sl_1}\frac{\partial^2 f^{l_1}}{\partial y^{n_1} \partial x^{p_1}}\tilde{g}^{n_1}_l\right)\bar{g}^{p_1}_m \\
& + b^p_{jkl}\Gamma^i_{mp} - b^i_{pkl}\Gamma^p_{mj} - b^i_{jpl}\Gamma^p_{mk} - b^i_{jkp}\Gamma^p_{ml},
\end{aligned}
\tag{7.5}
$$

$$
\begin{aligned}
\underset{2}{c}{}^i_{jklm} = \; & \left(\frac{\partial^4 f^i}{\partial x^s \partial y^n \partial y^p \partial y^q}\tilde{g}^n_j - \frac{\partial^4 f^i}{\partial x^s \partial x^n \partial y^p \partial y^q}\bar{g}^n_j\right)\bar{g}^s_k \tilde{g}^p_l \tilde{g}^q_m \\
& - \frac{\partial^3 f^i}{\partial x^s \partial y^n \partial y^p}\left(\frac{\partial^2 f^{l_1}}{\partial y^{p_1} \partial y^{m_1}}\bar{g}^s_k \tilde{g}^n_{l_1} \tilde{g}^p_l \tilde{g}^{p_1}_j \tilde{g}^{m_1}_m + \frac{\partial^2 f^{l}}{\partial x^{p_1} \partial y^{m_1}}\bar{g}^s_{l_1} \tilde{g}^n_j \tilde{g}^p_l \bar{g}^{p_1}_k \tilde{g}^{m_1}_m \right.\\
& \left. + \frac{\partial^2 f^{l_1}}{\partial y^{p_1} \partial y^{m_1}}\bar{g}^s_k \tilde{g}^n_j \tilde{g}^p_{l_1} \tilde{g}^{p_1}_l \tilde{g}^{m_1}_m\right) + \frac{\partial^3 f^i}{\partial x^s \partial x^n \partial y^p}\left(\frac{\partial^2 f^{l_1}}{\partial x^{p_1} \partial y^{m_1}}\bar{g}^s_k \bar{g}^n_{l_1} \tilde{g}^p_l \bar{g}^{p_1}_j \tilde{g}^{m_1}_m\right.
\end{aligned}
$$

$$
\begin{aligned}
&+\frac{\partial^2 f^{l_1}}{\partial x^{p_1}\partial y^{m_1}}\bar{g}^s_{l_1}\bar{g}^n_j\tilde{g}^p_l\bar{g}^{p_1}_k\tilde{g}^{m_1}_m+\frac{\partial^2 f^{l_1}}{\partial y^{p_1}\partial y^{m_1}}\bar{g}^s_k\bar{g}^n_j\tilde{g}^p_{l_1}\tilde{g}^{p_1}_l\tilde{g}^{m_1}_m\Big)\\
&-\frac{\partial^3 f^i}{\partial x^{m_1}\partial y^{n_1}\partial y^{p_1}}\left(\frac{\partial^2 f^p}{\partial y^s\partial y^n}\bar{g}^{m_1}_k\tilde{g}^{n_1}_p\tilde{g}^s_j\tilde{g}^n_l-\frac{\partial^2 f^p}{\partial x^s\partial x^n}\bar{g}^{m_1}_p\tilde{g}^{n_1}_l\bar{g}^s_j\bar{g}^n_k\right)\tilde{g}^{p_1}_m\\
&-\frac{\partial^2 f^p}{\partial y^s\partial y^n}\tilde{g}^{p_1}_m\tilde{g}^s_j\tilde{g}^n_l\left(\Gamma^i_{l_1p}\frac{\partial^2 f^{l_1}}{\partial x^{m_1}\partial y^{p_1}}\bar{g}^{m_1}_k+\Gamma^i_{kl_1}\frac{\partial^2 f^{l_1}}{\partial y^{n_1}\partial y^{p_1}}\tilde{g}^{n_1}_p\right)\\
&-\Gamma^i_{kp}\frac{\partial^2 f^p}{\partial y^s\partial y^n}\frac{\partial^2 f^{l_1}}{\partial y^{p_1}\partial y^{m_1}}(\tilde{g}^s_{l_1}\tilde{g}^n_l\tilde{g}^{p_1}_j+\tilde{g}^s_j\tilde{g}^n_{l_1}\tilde{g}^{p_1}_l)\tilde{g}^{m_1}_m\\
&+\frac{\partial^2 f^p}{\partial x^s\partial x^n}\tilde{g}^{p_1}_m\bar{g}^s_j\bar{g}^n_k\left(\Gamma^i_{l_1l}\frac{\partial^2 f^{l_1}}{\partial x^{m_1}\partial y^{p_1}}\bar{g}^{m_1}_p+\Gamma^i_{pl_1}\frac{\partial^2 f^{l_1}}{\partial y^{n_1}\partial y^{p_1}}\tilde{g}^{n_1}_l\right)\\
&+\Gamma^i_{pl}\frac{\partial^2 f^p}{\partial x^s\partial x^n}\frac{\partial^2 f^{l_1}}{\partial x^{p_1}\partial y^{m_1}}(\bar{g}^s_{l_1}\bar{g}^n_k\bar{g}^{p_1}_j+\bar{g}^s_j\bar{g}^n_{l_1}\bar{g}^{p_1}_k)\tilde{g}^{m_1}_m\\
&+\left(\Gamma^i_{kp}\frac{\partial^3 f^p}{\partial y^s\partial y^n\partial y^q}\tilde{g}^s_j\tilde{g}^n_l-\Gamma^i_{pl}\frac{\partial^3 f^p}{\partial x^s\partial x^n\partial y^q}\bar{g}^s_j\bar{g}^n_k\right)\tilde{g}^q_m \qquad (7.6)\\
&-\frac{\partial^3 f^p}{\partial x^{m_1}\partial y^{n_1}\partial y^{p_1}}\tilde{g}^{p_1}_m(\Gamma^i_{kp}\bar{g}^{m_1}_j\tilde{g}^{n_1}_l-\Gamma^i_{pl}\bar{g}^{m_1}_k\tilde{g}^{n_1}_j)\\
&-\frac{\partial^3 f^i}{\partial x^{m_1}\partial y^{n_1}\partial y^{p_1}}\tilde{g}^{p_1}_m(\Gamma^p_{jl}\bar{g}^{m_1}_k\tilde{g}^{n_1}_p-\Gamma^p_{kj}\bar{g}^{m_1}_p\tilde{g}^{n_1}_l)\\
&-\Gamma^i_{kp}\left(\Gamma^p_{l_1l}\frac{\partial^2 f^{l_1}}{\partial x^{m_1}\partial y^{p_1}}\bar{g}^{m_1}_j+\Gamma^p_{jl_1}\frac{\partial^2 f^{l_1}}{\partial y^{n_1}\partial y^{p_1}}\tilde{g}^{n_1}_l\right)\tilde{g}^{p_1}_m\\
&-\Gamma^p_{jl}\left(\Gamma^i_{l_1p}\frac{\partial^2 f^{l_1}}{\partial x^{m_1}\partial y^{p_1}}\bar{g}^{m_1}_k+\Gamma^i_{kl_1}\frac{\partial^2 f^{l_1}}{\partial y^{n_1}\partial y^{p_1}}\tilde{g}^{n_1}_s\tilde{g}^{p_1}_m\right)\tilde{g}^{p_1}_m\\
&+\Gamma^i_{pl}\left(\Gamma^p_{l_1j}\frac{\partial^2 f^{l_1}}{\partial x^{m_1}\partial y^{p_1}}\bar{g}^{m_1}_k+\Gamma^p_{kl_1}\frac{\partial^2 f^{l_1}}{\partial y^{n_1}\partial y^{p_1}}\tilde{g}^{n_1}_j\right)\bar{g}^{p_1}_m\\
&+\Gamma^p_{kj}\left(\Gamma^i_{l_1l}\frac{\partial^2 f^{l_1}}{\partial x^{m_1}\partial y^{p_1}}\bar{g}^{m_1}_p+\Gamma^i_{pl_1}\frac{\partial^2 f^{l_1}}{\partial y^{n_1}\partial y^{p_1}}\tilde{g}^{n_1}_l\right)\tilde{g}^{p_1}_m\\
&+b^p_{jkl}\Gamma^i_{pm}-b^i_{pkl}\Gamma^p_{jm}-b^i_{jpl}\Gamma^p_{km}-b^i_{jkp}\Gamma^p_{lm}.
\end{aligned}
$$

2. Let us now find the values of the tensors $\underset{1}{c}=(\underset{1}{c}{}^i_{jklm})$ and $\underset{2}{c}=(\underset{2}{c}{}^i_{jklm})$ at an arbitrary point p in terms of the coefficients of the Taylor expansions of the functions f^i. Suppose that a three-web W is parametrized in a standard way in a neighborhood of the point p with coordinates equal to zero. Then (see Section 2.5) the equations $z^i=f^i(x^j,y^k)$ of the web W or the coordinate quasigroup q also define multiplication in the coordinate loop l_p of this web (the identity element e of this loop has zero coordinates). Thus, the Taylor expansion of the function $f(x,y)=x\cdot y$ can be written in form (2.52):

$$
\begin{aligned}
x\cdot y = {} & x+y+\underset{2}{\Lambda}(x,y)+\frac{1}{2}\underset{2,1}{\Lambda}(x,x,y)+\frac{1}{2}\underset{1,2}{\Lambda}(x,y,y)\\
&+\tfrac{1}{6}\underset{3,1}{\Lambda}(x,x,x,y)+\frac{1}{4}\underset{2,2}{\Lambda}(x,x,y,y)+\frac{1}{6}\underset{1,3}{\Lambda}(x,y,y,y)+\{5\}.
\end{aligned}
\qquad (7.7)
$$

In addition, we assume that expansion (7.7) is partially specialized by the condition:

$$
\underset{2}{\Lambda}(x,y)=-\underset{2}{\Lambda}(y,x) \qquad (7.8)
$$

which can be obtained from the general conditions of specialization (2.65) by setting $s=2$.

Applying expansion (7.7), we find the values of the partial derivatives, up to the fourth order inclusive, of the function f at the point $p(0,0,\ldots,0)$ (cf. (2.47)):

$$\frac{\partial f^i}{\partial x^j} = \frac{\partial f^i}{\partial y^j} = \delta^i_j, \quad \frac{\partial^2 f^i}{\partial x^j \partial y^k} = \underset{2}{\Lambda}{}^i_{jk}, \quad \frac{\partial^3 f^i}{\partial x^j \partial x^k \partial y^l} = \underset{2,1}{\Lambda}{}^i_{jkl}, \quad \frac{\partial^3 f^i}{\partial x^j \partial y^k \partial y^l} = \underset{1,2}{\Lambda}{}^i_{jkl},$$
$$\frac{\partial^4 f^i}{\partial x^j \partial x^k \partial x^l \partial y^m} = \underset{3,1}{\Lambda}{}^i_{jklm}, \quad \frac{\partial^4 f^i}{\partial x^j \partial x^k \partial y^l \partial y^m} = \underset{2,2}{\Lambda}{}^i_{jklm}, \quad \frac{\partial^4 f^i}{\partial x^j \partial y^k \partial y^l \partial y^m} = \underset{1,3}{\Lambda}{}^i_{jklm}.$$

(the rest of the partial derivatives are zero). Substituting the values of the partial derivatives from these equations into (7.5) and (7.6), we obtain:

$$\begin{aligned}
\underset{1}{c}{}^i_{jklm} = {} & \underset{2,2}{\Lambda}{}^i_{kmjl} - \underset{3,1}{\Lambda}{}^i_{kmjl} - \underset{2}{\Lambda}{}^i_{kp}\underset{1,2}{\Lambda}{}^p_{mjl} - \underset{2}{\Lambda}{}^i_{mp}\underset{1,2}{\Lambda}{}^p_{kjl} + \underset{2}{\Lambda}{}^p_{mk}\underset{1,2}{\Lambda}{}^i_{pjl} \\
& + \underset{2}{\Lambda}{}^p_{jl}\underset{2,1}{\Lambda}{}^i_{kmp} - \underset{2}{\Lambda}{}^p_{kj}\underset{2,1}{\Lambda}{}^i_{pml} - \underset{2}{\Lambda}{}^p_{mj}\underset{2,1}{\Lambda}{}^i_{pkl} - \underset{2}{\Lambda}{}^p_{mk}\underset{2,1}{\Lambda}{}^i_{pjl} + \underset{2}{\Lambda}{}^i_{mp}\underset{2,1}{\Lambda}{}^p_{jkl} \\
& + \underset{2}{\Lambda}{}^i_{kp}\underset{2,1}{\Lambda}{}^p_{jml} + \underset{2}{\Lambda}{}^i_{lp}\underset{2,1}{\Lambda}{}^p_{kmj} - \underset{2}{\Lambda}{}^i_{kp}\underset{2}{\Lambda}{}^p_{mq}\underset{2}{\Lambda}{}^q_{jl} - \underset{2}{\Lambda}{}^i_{mp}\underset{2}{\Lambda}{}^p_{kq}\underset{2}{\Lambda}{}^q_{jl} \\
& - \underset{2}{\Lambda}{}^i_{mp}\underset{2}{\Lambda}{}^p_{lq}\underset{2}{\Lambda}{}^q_{kj} - \underset{2}{\Lambda}{}^i_{kp}\underset{2}{\Lambda}{}^p_{lq}\underset{2}{\Lambda}{}^q_{mj} - \underset{2}{\Lambda}{}^i_{lp}\underset{2}{\Lambda}{}^p_{jq}\underset{2}{\Lambda}{}^q_{mk} + \underset{2}{\Lambda}{}^i_{pq}\underset{2}{\Lambda}{}^p_{mk}\underset{2}{\Lambda}{}^q_{jl};
\end{aligned} \tag{7.9}$$

$$\begin{aligned}
\underset{2}{c}{}^i_{jklm} = {} & \underset{1,3}{\Lambda}{}^i_{kjlm} - \underset{2,2}{\Lambda}{}^i_{kjlm} + \underset{2}{\Lambda}{}^i_{pm}\underset{2,1}{\Lambda}{}^p_{jkl} + \underset{2}{\Lambda}{}^i_{pl}\underset{2,1}{\Lambda}{}^p_{jkm} - \underset{2}{\Lambda}{}^p_{ml}\underset{2,1}{\Lambda}{}^i_{jkp} \\
& - \underset{2}{\Lambda}{}^p_{kj}\underset{1,2}{\Lambda}{}^i_{plm} + \underset{2}{\Lambda}{}^p_{jl}\underset{1,2}{\Lambda}{}^i_{kmp} + \underset{2}{\Lambda}{}^p_{jm}\underset{1,2}{\Lambda}{}^i_{klp} + \underset{2}{\Lambda}{}^p_{lm}\underset{1,3}{\Lambda}{}^i_{kjp} + \underset{2}{\Lambda}{}^i_{kp}\underset{1,2}{\Lambda}{}^p_{jlm} \\
& + \underset{2}{\Lambda}{}^i_{lp}\underset{1,2}{\Lambda}{}^p_{kjm} + \underset{2}{\Lambda}{}^i_{mp}\underset{1,2}{\Lambda}{}^p_{kjl} + \underset{2}{\Lambda}{}^i_{lp}\underset{2}{\Lambda}{}^p_{mq}\underset{2}{\Lambda}{}^q_{kj} + \underset{2}{\Lambda}{}^i_{mp}\underset{2}{\Lambda}{}^p_{kq}\underset{2}{\Lambda}{}^q_{jl} \\
& + \underset{2}{\Lambda}{}^i_{mp}\underset{2}{\Lambda}{}^p_{lq}\underset{2}{\Lambda}{}^q_{kj} + \underset{2}{\Lambda}{}^i_{lp}\underset{2}{\Lambda}{}^p_{kq}\underset{2}{\Lambda}{}^q_{jm} + \underset{2}{\Lambda}{}^i_{kp}\underset{2}{\Lambda}{}^p_{jq}\underset{2}{\Lambda}{}^q_{lm} - \underset{2}{\Lambda}{}^i_{pq}\underset{2}{\Lambda}{}^p_{kj}\underset{2}{\Lambda}{}^q_{lm}.
\end{aligned} \tag{7.10}$$

We simplify these expressions by introducing the tensors a and b of the web. It follows from (2.29) and (2.51) that $\underset{2}{\Lambda} = -a$. Moreover, applying formulas (2.30) where $N = \underset{1,2}{\Lambda}$, $M = \underset{2,1}{\Lambda}$, and relation (2.51), we eliminate the quantities $\underset{1,2}{\Lambda}$ from (7.9) and the quantities $\underset{1,1}{\Lambda}$ from (7.10). After uncomplicated calculations, we obtain:

$$\begin{aligned}
\underset{1}{c}{}^i_{jklm} = {} & \underset{2,2}{\Lambda}{}^i_{kmjl} - \underset{3,1}{\Lambda}{}^i_{kmjl} + a^i_{mp}b^p_{jkl} + a^i_{kp}b^p_{jml} - a^p_{mk}b^i_{jpl} \\
& + a^i_{pl}\underset{2,1}{\Lambda}{}^p_{kmj} - a^p_{jk}\underset{2,1}{\Lambda}{}^i_{pml} - a^p_{jm}\underset{2,1}{\Lambda}{}^i_{kpl} - a^p_{jl}\underset{2,1}{\Lambda}{}^i_{kmp};
\end{aligned} \tag{7.11}$$

$$\begin{aligned}
\underset{2}{c}{}^i_{jklm} = {} & \underset{1,3}{\Lambda}{}^i_{kjlm} - \underset{2,2}{\Lambda}{}^i_{kjlm} + a^i_{pm}b^p_{jkl} + a^i_{pl}b^p_{jkm} - a^p_{lm}b^i_{jkp} \\
& - a^i_{kp}\underset{1,2}{\Lambda}{}^p_{jlm} + a^p_{kj}\underset{1,2}{\Lambda}{}^i_{plm} + a^p_{lj}\underset{1,2}{\Lambda}{}^i_{kpm} + a^p_{mj}\underset{1,2}{\Lambda}{}^i_{klp}.
\end{aligned} \tag{7.12}$$

Thus, we have proved the following proposition.

Proposition 7.1 *Formulas* (7.5) *and* (7.6) *express the fundamental tensors* $\underset{1}{c}$ *and* $\underset{2}{c}$ *of order four of a web* W *in terms of the partial derivatives of order not higher than four of the functions* f^i *defining this web. Formulas* (7.9) *and* (7.10) *(or* (7.11) *and* (7.12)*) give the values of the tensors* $\underset{1}{c}$ *and* $\underset{2}{c}$ *at a point* p *of a manifold* X *carrying the web* W*, in terms of the coeficients of the Taylor series of the coordinate loop* l_p*.*

■

7.2 Internal Mappings in Coordinate Loops of a Three-Web

1. A sequence of groups is invariantly associated with an arbitrary loop Q. First of all, we note the stationary group J of the identity element. Next, although in general the right translations in Q do not form a group, they generate the group which is denoted by R_Q. The left translations also generate the group denoted by L_Q. The right and left translations together generate the group G_Q.

Consider, for example, the group L_Q. If Q is a group, the mapping $L : Q \to L_Q$ such that $L(x) = L_x$, is an isomorphism since, by associativity, $L_{xy} = L_x L_y$. Since, in general, there is no associativity in an arbitrary loop, the operator L is not an isomorphism, i.e. $L_{xy} \neq L_x L_y$. The latter condition can be written in the form: $l_{x,y} \neq \mathrm{id}$ where $l_{x,y} = L_{xy}^{-1} L_x L_y$.

The operators $l_{x,y}$ and $r_{x,y} = R_{xy}^{-1} R_x R_y$ generate the groups $as_l(Q)$ and $as_r(Q)$ which are called the *associants* of the loop Q. Moreover, $as_l(Q)$ is a subgroup in L_Q and $as_r(Q)$ is a subgroup in R_Q. The deviation from commutativity in Q is characterized by the operator $T_x = L_x^{-1} R_x$ which is called a *proper internal mapping.* The operators $l_{x,y}, r_{x,y}$ and T_x keep the identity element fixed, i.e. they belong to the group J. By the Albert–Bruck theorem, these operators generate the subgroup J^* of internal permutations of the loop Q which is a part of the group J lying in G_Q.

If Q is the coordinate loop l_p of a three-web $W = (X, \lambda_\alpha)$, the operators $l_{x,y}$ and $r_{x,y}$ are functions defined in the manifold X. In this section we calculate their Taylor expansions up to the terms of fourth order inclusive assuming that multiplication in the loop l_p is written in form (7.7)–(7.8). For convenience, following (2.24), we introduce the notation: $\underset{2}{\Lambda} = \Lambda$, $\underset{2,1}{\Lambda} = M$, $\underset{1,2}{\Lambda} = N$, $\underset{3,1}{\Lambda} = P$, $\underset{2,2}{\Lambda} = Q$ and $\underset{1,3}{\Lambda} = R$.

Proposition 7.2 *In the loop l_p, the following formulas hold:*

$$
\begin{aligned}
xy \cdot z = \ & K(x,y,z) + M(x,y,z) + \Lambda(\Lambda(x,y),z) \\
& + P_1(x,x,y,z) + P_2(x,y,y,z) + Q_2(x,y,z,z) + \{5\},
\end{aligned}
\tag{7.13}
$$

$$
\begin{aligned}
x \cdot yz = \ & K(x,y,z) + N(x,y,z) + \Lambda(x,\Lambda(y,z)) \\
& + Q_1(x,x,y,z) + R_1(x,y,y,z) + R_2(x,y,z,z) + \{5\},
\end{aligned}
\tag{7.14}
$$

where

$$
\begin{aligned}
K(x,y,z) = \ & x + y + z + \Lambda(x,y) + \Lambda(x,z) + \Lambda(y,z) \\
& + \tfrac{1}{2}(M(x,x,y) + M(x,x,z) + M(y,y,z)) \\
& + \tfrac{1}{2}(N(x,y,y) + N(x,z,z) + M(y,z,z)) \\
& + \tfrac{1}{6}(P(x,x,x,y) + P(x,x,x,z) + P(y,y,y,z)) \\
& + \tfrac{1}{4}(Q(x,x,y,y) + Q(x,x,z,z) + Q(y,y,z,z) \\
& + \tfrac{1}{6}(R(x,y,y,y) + R(x,z,z,z) + R(y,z,z,z));
\end{aligned}
\tag{7.15}
$$

$$\begin{aligned} P_1(x,x,y,z) = {} & \tfrac{1}{2}P(x,x,y,z) + M(x,\Lambda(x,y),z) + \tfrac{1}{2}\Lambda(M(x,x,y),z), \\ P_2(x,y,y,z) = {} & \tfrac{1}{2}P(x,y,y,z) + M(\Lambda(x,y),y,z) + \tfrac{1}{2}\Lambda(N((x,y,y),z), \\ Q_1(x,x,y,z) = {} & \tfrac{1}{2}Q(x,x,y,z) + \tfrac{1}{2}M(x,x,\Lambda(y,z)), \\ Q_2(x,y,z,z) = {} & \tfrac{1}{2}Q(x,y,z,z) + \tfrac{1}{2}N(\Lambda(x,y),z,z), \\ R_1(x,y,y,z) = {} & \tfrac{1}{2}R(x,y,y,z) + N(x,y,\Lambda(y,z)) + \tfrac{1}{2}\Lambda(x,M(y,y),z)), \\ R_2(x,y,z,z) = {} & \tfrac{1}{2}R(x,y,z,z) + N(x,\Lambda(y,z),z) + \tfrac{1}{2}\Lambda(x,M(y,z,z)). \end{aligned} \tag{7.16}$$

Proof. According to (7.7), we have:

$$\begin{aligned} xy\cdot z = {} & xy + z + \Lambda(xy,z) + \tfrac{1}{2}M(xy,xy,z) + \tfrac{1}{2}N(xy,z,z) \\ & + \tfrac{1}{6}P(xy,xy,xy,z) + \tfrac{1}{4}Q(xy,xy,z,z) + \tfrac{1}{6}R(xy,z,z,z) + \{5\}. \end{aligned}$$

Let $\underset{k}{L}(x,y)$ denote the terms of order k of the expansion for xy. As to the terms up to the third order inclusive, they were already found in Section 2.4:

$$\begin{aligned} \underset{3}{L}(xy\cdot z) = {} & \tfrac{1}{2}M(x,x,y) + \tfrac{1}{2}M(x,y,y) + \Lambda(\Lambda(x,y),z) + \tfrac{1}{2}M(x,x,z) + \tfrac{1}{2}M(y,z,z) \\ & + M(x,y,z) + \tfrac{1}{2}N(x,z,z) + \tfrac{1}{2}N(y,z,z). \end{aligned}$$

For the terms of fourth order, we have:

$$\begin{aligned} \underset{4}{L}(xyz) = {} & \underset{4}{L}(xy) + \Lambda(\underset{3}{L}(xy),z) + \frac{1}{2}N(\underset{1}{L}(xy),\underset{2}{L}(xy),z) \\ & + \tfrac{1}{2}N(\Lambda(x,y),z,z) + \tfrac{1}{6}P(\underset{1}{L}(xy),\underset{1}{L}(xy),\underset{1}{L}(xy),z) \\ & + \tfrac{1}{4}Q(\underset{1}{L}(xy),\underset{1}{L}(xy),z,z) + \frac{1}{6}R(\underset{1}{L}(xy),z,z,z). \end{aligned}$$

If we substitute

$$\underset{1}{L}(xy) = x + y,\ \underset{2}{L}(xy) = \Lambda(x,y),\ \underset{3}{L}(xy) = \frac{1}{2}M(x,x,y) + \frac{1}{2}N(x,y,y),$$

then after uncomplicated transformations, we arrive at formulas (2.13). The calculation of the product $x\cdot yz$ is similar. ■

2. The operators $l_{x,y}$ and $r_{x,y}$ can be defined by the equations:

$$(xy)l_{x,y}(u) = x(yu), \tag{7.17}$$

and

$$r_{x,y}(u)(xy) = (ux)y. \tag{7.18}$$

We also define the operator $m_{x,y}$ by the equation:

$$x(m_{x,y}(u)y) = (xu)y. \tag{7.19}$$

Next, we calculate the operator $l_{x,y}$. Let us denote $v = l_{x,y}$. According to (7.17), we have:

$$(xy)v = x(yu). \tag{7.20}$$

From formulas (7.13) and (7.14) we find that:

$$\begin{aligned}(xy)v = \; & K(x,y,v) + M(x,y,v) + \Lambda(\Lambda(x,y),v) \\ & + P_1(x,x,y,v) + P_2(x,y,y,v) + Q_2(x,x,v,v) + \{5\},\end{aligned} \tag{7.21}$$

and

$$\begin{aligned}x(yu) = \; & K(x,y,u) + N(x,y,u) + \Lambda(x,\Lambda(y,u)) \\ & + Q_1(x,x,y,u) + R_1(x,y,y,u) + R_2(x,y,u,u) + \{5\},\end{aligned} \tag{7.22}$$

where the polynomials $K, P_1, Q_1, R_1, P_2, Q_2$ and R_2 can be found from formulas (7.15) and (7.16). Comparing in (7.21) and (7.22) the terms up to second order inclusive, we obtain:

$$v = u + \{2\}, \quad v = u + \Lambda(x, u - v) + \Lambda(y, u - v) + \{3\};$$

and since the two middle terms in the last equation can be cancelled, we have:

$$v = u + \{3\}. \tag{7.23}$$

Comparison of the third order terms in (7.21) and (7.22) gives:

$$\begin{aligned}v = \; & u + \tfrac{1}{2}(M(x,x,u) + M(y,y,u))) + \tfrac{1}{2}(N(x,u,u) + M(y,u,u))) \\ & -\tfrac{1}{2}(M(x,x,v) + M(y,y,v))) - \tfrac{1}{2}(N(x,v,v) + M(y,v,v))) \\ & +N(x,y,u) + \Lambda(x,\Lambda(y,u)) - M(x,y,v) - \Lambda(\Lambda(x,y),v) + \{4\}.\end{aligned}$$

Applying (7.23), we obtain from this:

$$v = u + N(x,y,u) - M(x,y,u) + \Lambda(x,\Lambda(y,u)) - \Lambda(\Lambda(x,y),u) + \{4\},$$

or by formulas (2.30) and (2.51), we have:

$$v = u + b(y,x,u) + \{4\}. \tag{7.24}$$

From (7.24) and (7.15) we find that

$$\underset{3}{L}(K(x,y,u)) = \underset{3}{L}(K(x,y,v)), \quad \underset{4}{L}(K(x,y,u)) = \underset{4}{L}(K(x,y,v)),$$

where $\underset{s}{L}(K)$ denotes the terms of order s in the polynomial K. Thus, comparing the terms up to order four inclusive in (7.21) and (7.22) and taking into account equations (7.23) and (7.24), we find that

$$\begin{aligned} v = \ & u + \Lambda(x, u-v) + \Lambda(y, u-v) + b(y,x,u) + Q_1(x,x,y,u) + R_1(x,y,y,u) \\ & + R_2(x,y,u,u) - P_1(x,x,u,v) - P_2(x,y,y,v) - Q_2(x,y,u,v) + \{5\}. \end{aligned}$$

Let us transform the second and the fourth order terms in this expression by using new notation and applying (7.24) and (7.16), respectively. As a result, we obtain:

$$\begin{aligned} v = \ & l_{x,y}(u) = u + b(y,x,u) - \Lambda(x, b(y,x,u)) - \Lambda(y, b(y,x,u)) \\ & + \tfrac{1}{2}(Q(x,x,y,u) - P(x,x,y,u)) - \tfrac{1}{2}(Q(x,y,u,u) - P(x,y,u,u)) \\ & + \tfrac{1}{2}R(x,y,y,u) - \tfrac{1}{2}P(x,y,y,u) + \tfrac{1}{2}M(x,x,\Lambda(y,u)) \\ & - M(x,\Lambda(x,y),u) - M(\Lambda(x,y),y,u) + N(x,y,\Lambda(y,u)) \\ & + N(x,\Lambda(y,u),u) - \tfrac{1}{2}N(\Lambda(x,y),u,u) + \tfrac{1}{2}\Lambda(x, M(y,y,u)) \\ & + \tfrac{1}{2}\Lambda(x, N(y,u,u)) - \tfrac{1}{2}\Lambda(M(x,x,y),u)) - \tfrac{1}{2}\Lambda(N(x,y,y),u)) + \{5\}. \end{aligned} \tag{7.25}$$

For the operators $r_{x,y}$ and $m_{x,y}$, we obtain the following expressions:

$$\begin{aligned} r_{x,y}(u) = \ & u - b(x,u,y) - \Lambda(x, b(x,u,y)) - \Lambda(y, b(x,u,y)) \\ & + \tfrac{1}{2}(P(u,u,x,y) - Q(u,u,x,y)) + \tfrac{1}{2}(P(u,x,x,y) - R(u,x,x,y)) \\ & + \tfrac{1}{2}Q(u,x,y,y) - \tfrac{1}{2}R(u,x,y,y) + \tfrac{1}{2}M(u,\Lambda(u,x),y)) \\ & + M(\Lambda(u,x),x,y) - \tfrac{1}{2}M(u,u,\Lambda(x,y)) + \tfrac{1}{2}N(\Lambda(u,x),y,y)) \\ & - N(u,x,\Lambda(x,y)) - N(u,\Lambda(x,y),y) + \tfrac{1}{2}\Lambda(M(u,u,x),y) \\ & + \tfrac{1}{2}\Lambda(N(u,x,x),u) - \tfrac{1}{2}\Lambda(u, M(x,x,y)) - \tfrac{1}{2}\Lambda(u, N(x,y,y),u)) + \{5\}; \end{aligned} \tag{7.26}$$

and

$$\begin{aligned} m_{x,y}(u) = \ & u - b(u,x,y) + \Lambda(x, b(u,x,y)) + \Lambda(b(u,x,y),y) \\ & + \tfrac{1}{2}(P(x,u,u,y) - R(x,u,u,y)) + \tfrac{1}{2}(Q(x,u,y,y) - R(x,u,y,y)) \\ & + \tfrac{1}{2}P(x,x,u,y) - \tfrac{1}{2}Q(x,x,u,y) + M(x,\Lambda(x,u),y)) \\ & + M(\Lambda(x,u),u,y) - \tfrac{1}{2}M(x,x,\Lambda(u,y)) + \tfrac{1}{2}N(\Lambda(x,u),y,y)) \\ & - N(x,u,\Lambda(u,y)) - N(x,\Lambda(u,y),y) + \tfrac{1}{2}\Lambda(M(x,x,u),y) \\ & + \tfrac{1}{2}\Lambda(N(x,u,u),y) - \tfrac{1}{2}\Lambda(x, M(u,u,y)) - \tfrac{1}{2}\Lambda(x, N(u,y,y))) + \{5\}. \end{aligned} \tag{7.27}$$

We now express the operators $l_{x,y}, r_{x,y}$ and $m_{x,y}$ in terms of the fundamental tensors of the three-web W. We take formulas (7.11) and (7.12) and write them in new notations:

$$\begin{aligned}
-\underset{2}{c}(x,y,z,t) = {} & (Q-R)(y,x,z,t) - a(b(x,y,z),t) - a(b(x,y,t),z) \\
& + b(x,y,a(z,t)) + a(y,N(x,z,t)) - N(a(y,x),z,t) \\
& - N(y,a(z,x),t) - N(y,z,a(t,x)); \\
\underset{1}{c}(x,y,z,t) = {} & (Q-P)(y,t,x,z) + a(t,b(x,y,z)) + a(y,b(x,t,z)) \\
& - b(x,a(t,y),z) + a(M(y,t,x),z) - M(a(x,y),t,z) \\
& - M(y,a(x,t),z) - M(y,t,a(x,z)).
\end{aligned} \tag{7.28}$$

This implies:

$$\begin{aligned}
-\underset{2}{c}(y,x,u,u) = {} & (Q-R)(x,y,u,u) - 2a(b(y,x,u),u) \\
& + a(x,N(y,u,u)) - N(a(x,y),u,u) - 2N(x,u,a(u,y)); \\
-\underset{2}{c}(y,x,y,u) = {} & (Q-R)(x,y,y,u) - a(b(y,x,y),u) - a(b(y,x,u),y) \\
& + b(y,x,a(y,u)) + a(x,N(y,y,u)) - N(x,y,a(u,y)) \\
& - N(a(x,y),y,u) \\
\underset{1}{c}(y,x,u,x) = {} & (Q-P)(x,x,y,u) + 2a(x,b(y,x,u)) + a(M(x,x,y),u) \\
& - 2M(a(y,x),x,u)) - M(x,x,a(y,u)); \\
\underset{1}{c}(y,x,u,y) = {} & (Q-P)(x,y,y,u) + a(y,b(y,x,u)) + a(x,b(y,y,u)) \\
& - b(y,a(y,x),u) + a(M(x,y,y),u) \\
& - M(a(y,x),y,u)) - M(x,y,a(y,u)).
\end{aligned} \tag{7.29}$$

From (7.29) we find the differences $Q-R$ and $Q-P$ and substitute them into equations (7.25)–(7.27). Taking into account the relation $\Delta = -a$, after uncomplicated transformations, we obtain the following expression for $l_{x,y}$:

$$\begin{aligned}
l_{x,y}(u) = {} & u + b(y,x,u) + a(x,b(y,x,u)) + a(y,b(y,x,u)) \\
& + \tfrac{1}{2}(\underset{1}{c}(y,x,u,x) + \underset{1}{c}(y,x,u,y) + \underset{2}{c}(y,x,u,u) + \underset{2}{c}(y,x,y,u)) \\
& - a(b(y,x,u),u) - \tfrac{1}{2}(a(b(y,x,y),u) - \tfrac{1}{2}a(b(y,x,u),y) \\
& + \tfrac{1}{2}b(y,x,a(y,u)) - a(x,b(y,x,u)) - \tfrac{1}{2}a(y,b(y,x,u)) \\
& - \tfrac{1}{2}a(x,b(y,y,u)) + \tfrac{1}{2}b(y,a(y,x),u) - \tfrac{1}{2}a(x,M(y,y,u) \\
& - N(y,y,u)) + \tfrac{1}{2}M(a(x,y),y,u) - \tfrac{1}{2}N(a(x,y),y,u) \\
& + \tfrac{1}{2}M(x,y,a(y,u)) - \tfrac{1}{2}N(x,y,a(y,u)) + \tfrac{1}{2}a(u,M(x,y,y) \\
& - N(x,y,y)) + \{5\}.
\end{aligned} \tag{7.30}$$

By virtue of equations (2.30) and (2.51), we have:

$$\begin{aligned}
& -\tfrac{1}{2}a(x,M(y,y,u) - N(y,y,u)) = \tfrac{1}{2}a(x,b(y,y,u)) - \tfrac{1}{2}a(x,a(y,a(y,u))); \\
& \tfrac{1}{2}M(a(x,y),y,u) - \tfrac{1}{2}N(a(x,y),y,u) \\
& = -\tfrac{1}{2}b(y,a(x,y),u) + \tfrac{1}{2}a(a(x,y),a(y,u)) - \tfrac{1}{2}a(a(a(x,y),y),u); \\
& \tfrac{1}{2}M(x,y,a(y,u)) - \tfrac{1}{2}N(x,y,a(y,u)) \\
& = -\tfrac{1}{2}b(y,x,a(y,u)) + \tfrac{1}{2}a(x,a(y,a(y,u))) - \tfrac{1}{2}a(a(x,y),a(y,u)); \\
& \tfrac{1}{2}a(u,M(x,y,y) - N(x,y,y)) = -\tfrac{1}{2}a(u,b(y,x,y)) - \tfrac{1}{2}a(u,a(a(x,y),y).)
\end{aligned}$$

Substituting these equations into equations (7.30), after simple calculations we get:

$$\begin{aligned} l_{x,y}(u) = \;& u + b(y,x,u) + +\tfrac{1}{2}(\underset{1}{c}(y,x,u,x) + \underset{1}{c}(y,x,u,y) + \underset{2}{c}(y,x,u,u) \\ & + \underset{2}{c}(y,x,y,u)) + a(y,b(y,x,u)) + a(u,b(y,x,u)) - b(y,a(x,y),u) + \{5\}. \end{aligned} \tag{7.31}$$

We can make one more simplification. According to Section 1.3, the tensors $a, b, \underset{1}{c}$ and $\underset{2}{c}$ are connected by the relations:

$$\begin{aligned} & -\underset{2}{c}(x,y,z,t) + \underset{2}{c}(x,y,t,z) = 2b(x,y,a(z,t)), \\ & \underset{1}{c}(x,y,z,t) - \underset{1}{c}(x,t,z,y) = 2b(x,a(y,t),z), \\ & \underset{1}{c}(x,y,t,z) - \underset{1}{c}(y,x,t,z) - \underset{2}{c}(x,z,y,t) + \underset{2}{c}(y,z,x,t) \\ & = 2(a(b(y,z,t),x) - a(b(x,z,t),y)) + b(a(x,y),z,t)). \end{aligned} \tag{7.32}$$

By means of the second of equations (7.32), we can write formula (7.31) in the form:

$$\begin{aligned} l_{x,y}(u) = \;& u + b(y,x,u) + a(y,b(y,x,u)) + a(u,b(y,x,u)) + \tfrac{1}{2}(\underset{1}{c}(y,x,u,x) \\ & + \underset{1}{c}(y,y,u,x) + \underset{2}{c}(y,x,u,u) + \underset{2}{c}(y,x,y,u)) + \{5\}. \end{aligned} \tag{7.33}$$

Similar calculations lead to the following form of the operators $r_{x,y}$ and $m_{x,y}$:

$$\begin{aligned} r_{x,y}(u) = \;& u - b(x,u,y) + a(x,b(x,u,y)) + a(u,b(x,u,y)) - \tfrac{1}{2}(\underset{1}{c}(x,x,y,u) \\ & + \underset{1}{c}(x,u,y,u) + \underset{2}{c}(x,u,y,y) + \underset{2}{c}(x,u,x,y)) + \{5\}. \end{aligned} \tag{7.34}$$

and

$$\begin{aligned} m_{x,y}(u) = \;& u - b(u,x,y) + b(u,a(x,u),y) - \tfrac{1}{2}(\underset{1}{c}(u,x,y,x) \\ & + \underset{1}{c}(u,x,y,u) + \underset{2}{c}(u,x,y,y) + \underset{2}{c}(u,x,u,y)) + \{5\}. \end{aligned} \tag{7.35}$$

Thus, we have proved the following proposition.

Proposition 7.3 *The operators $l_{x,y}, r_{x,y}$ and $m_{x,y}$ can be expressed in terms of the tensors $a, b, \underset{1}{c}$ and $\underset{2}{c}$ up to the fifth order terms by formulas* (7.33) – –(7.35). ■

7.3 An Algebraic Characterization of the Tangent W_4-Algebra of a Three-Web

In an arbitrary analytic loop $Q(\cdot)$ the operators $l_{x,y}, r_{x,y}$ and $m_{x,y}$ are not automorphisms. This means, for example, that

$$l_{x,y}(uv) \neq l_{x,y}(u)l_{x,y}(v).$$

Define the functions $\mathcal{L}_{x,y}(u,v), \mathcal{R}_{x,y}(u,v)$ and $\mathcal{M}_{x,y}(u,v)$ by the equations:

$$\begin{aligned} \mathcal{L}_{x,y}(u,v) &= {}^{-1}(l_{x,y}(uv))(l_{x,y}(u)l_{x,y}(v)), \\ \mathcal{R}_{x,y}(u,v) &= {}^{-1}(r_{x,y}(uv))(r_{x,y}(u)r_{x,y}(v)), \\ \mathcal{M}_{x,y}(u,v) &= {}^{-1}(m_{x,y}(uv))(m_{x,y}(u)m_{x,y}(v)). \end{aligned} \tag{7.36}$$

If, for example, the operator $l_{x,y}(uv)$ is an automorphism, then $\mathcal{L}_{x,y} = \mathrm{id}$.

Lemma 7.4 *For any x, y, u and z from Q, the equation*

$$\mathcal{L}_{x,y}(u,v) = l_{x,y}(u)l_{x,y}(v)) - l_{x,y}(uv) + \{5\} \tag{7.37}$$

holds. Similar equations hold for the functions $\mathcal{R}_{x,y}$ and $\mathcal{M}_{x,y}$.

Proof. Define $\bar{u} = l_{x,y}(uv)$ and $\bar{v} = l_{x,y}(u)l_{x,y}(v)$ and expand the functions $\bar{u}$ and $\bar{v}$ in their series by applying formulas (7.33) and (7.11). The calculations show that the functions $\bar{u}$ and $\bar{v}$ differ only in terms of the fifth order:

$$\bar{v} = \bar{u} + \{5\}. \tag{7.38}$$

Next, applying expansion (7.7), we obtain:

$$\begin{aligned} \mathcal{L}_{x,y}(u,v) = \; & {}^{-1}\bar{u} \cdot \bar{v} = -\bar{u} + \bar{v} + \Lambda({}^{-1}\bar{u}, \bar{v}) + \tfrac{1}{2}M({}^{-1}\bar{u}, {}^{-1}\bar{u}, \bar{v}) + \tfrac{1}{2}N({}^{-1}\bar{u}, \bar{v}, \bar{v}) \\ & + \tfrac{1}{6}P({}^{-1}\bar{u}, {}^{-1}\bar{u}, {}^{-1}\bar{u}, \bar{v}) + \tfrac{1}{4}Q({}^{-1}\bar{u}, {}^{-1}\bar{u}, \bar{v}, \bar{v}) + \tfrac{1}{6}R({}^{-1}\bar{u}, \bar{v}, \bar{v}, \bar{v}) + \{5\}. \end{aligned}$$

The expression for ${}^{-1}\bar{u}$ in terms of $\bar{u}$ can be found in Problem **1**. Substituting this expression in the previous formula and taking into account equation (7.38), after uncomplicated calculations we arrive at formula (7.37). ■

Theorem 7.5 *Let l_p be the coordinate loop of a three-web W. Then the following formulas hold:*

$$\begin{aligned} \mathcal{L}_{x,y}(u,v) &= \mathcal{L}(x,y,u,v) + \{5\}, \\ \mathcal{R}_{x,y}(u,v) &= \mathcal{R}(x,y,u,v) + \{5\}, \\ \mathcal{M}_{x,y}(u,v) &= \mathcal{M}(x,y,u,v) + \{5\}, \end{aligned} \tag{7.39}$$

where $\mathcal{L}, \mathcal{R}$ and $\mathcal{M}$ are multilinear forms which can be expressed in terms of the fundamental tensors of the web as follows:

$$\begin{aligned}\mathcal{L}(x,y,u,v) &= -\underset{2}{c}(y,x,u,v) - 2a(u,b(y,x,v)),\\ \mathcal{R}(x,y,u,v) &= -\underset{1}{c}(x,v,y,u) + 2a(v,b(x,u,y)),\\ \mathcal{M}(x,y,u,v) &= \underset{1}{c}(v,u,y,x) + \underset{2}{c}(u,x,v,y).\end{aligned} \tag{7.40}$$

Proof. We will prove this for the operator $\mathcal{L}_{x,y}(u,v)$. By (7.33), we have:

$$\begin{aligned} l_{x,y}(u)l_{x,y}(v) = \;& l_{x,y}(u) + l_{x,y}(v) + \Lambda(u + b(y,x,u), v + b(y,x,v)) \\ & + \tfrac{1}{2}M(u,u,v) + \tfrac{1}{2}N(u,v,v) + \tfrac{1}{6}P(u,u,u,v) \\ & + \tfrac{1}{4}Q(u,u,v,v) + \tfrac{1}{6}R(u,v,v,v) + \{5\} \\ = \;& u\cdot v + b(y,x,u) + \Lambda(u, b(y,x,v)) + \Lambda(b(y,x,u),v) + \\ & \underset{4}{L}(l_{x,y}(u)) + \underset{4}{L}(l_{x,y}(v)), \end{aligned}$$

where $\underset{4}{L}(\ldots)$ denotes the fourth order terms in the expansion of the function in parentheses. On the other hand, up to terms of fourth order inclusive, we have:

$$\begin{aligned} l_{x,y}(uv) &= uv + b(y,x,uv) + \underset{4}{L}(l_{x,y}(uv)) + \{5\} \\ &= uv + b(y,x,u+v+\Lambda(u,v)) + \underset{4}{L}(l_{x,y}(uv)) + \{5\}. \end{aligned}$$

Substituting this and previous equations into (7.37) and taking into account that $\Lambda = -a$, we obtain:

$$\begin{aligned} \mathcal{L}_{x,y}(u,v) = \;& \underset{4}{L}(l_{x,y}(u)) + \underset{4}{L}(l_{x,y}(v)) - \underset{4}{L}(l_{x,y}(u\cdot v)) \\ & + \Lambda(u, b(y,x,v)) + \Lambda(b(y,x,u),v) - b(y,x,\Lambda(u,v)) + \{5\} \\ = \;& a(u,b(y,x,u)) + \tfrac{1}{2}\underset{2}{c}(y,x,u,u) + a(v,b(y,x,v)) + \frac{1}{2}\underset{2}{c}(y,x,v,v) \\ & - a(u+v, b(y,x,u+v)) - \tfrac{1}{2}\underset{2}{c}(y,x,u+v,u+v) - a(u,b(y,x,v)) \\ & - a(b(y,x,u),v) - b(y,x,a(u,v)) + \{5\} \\ = \;& \tfrac{1}{2}(-\underset{2}{c}(y,x,u,v) - \underset{2}{c}(y,x,v,u) - 2a(u,b(y,x,v)) \\ & + b(y,x,a(u,v)) + \{5\}. \end{aligned}$$

Finally, substituting the values taken from (7.32) for the last terms in this equation, we arrive at (7.39). ∎

According to (7.39), the tensors $\mathcal{L}, \mathcal{R}$ and $\mathcal{M}$ have the following interpretation in terms of the coordinate loops of the web W: they characterize the principal part of the deviation of the operators $l_{x,y}, r_{x,y}$ and $m_{x,y}$ from an automorphism. Applying the same method as that used in Section 2.4 for the W-algebra, we can, by means of the tensors $\mathcal{L}, \mathcal{R}$ and $\mathcal{M}$, define 4-ary operations in the tangent space of the loop Q. Let $x(t), y(t), u(t)$ and $v(t)$ be smooth curves in Q passing through the identity element e, where the parameter t is chosen in such a way that $x(0) = y(0) = u(0) = v(0) = e$.

Denote the tangent vectors to these four lines at e by ξ, η, ζ and θ. Consider another three lines $\mathcal{L}(t), \mathcal{R}(t)$ and $\mathcal{M}(t)$ where for example,

$$\mathcal{L}(t) = {}^{-1}\Big(l_{x(t),y(t)}\big(u(t), v(t)\big)\Big)\Big(l_{x(t),y(t)}\big(u(t)\big)l_{x(t),y(t)}\big(v(t)\big)\Big).$$

Define a new parameter on these curves by setting $t = \sqrt[4]{s}$. Then, by the same method as that used in Chapter 3, we can prove that the tangent vectors λ, ρ and μ to the lines $\mathcal{L}(\sqrt[4]{s}), \mathcal{R}(\sqrt[4]{s})$ and $\mathcal{M}(\sqrt[4]{s})$ at the point e are determined by the formulas:

$$\lambda = \mathcal{L}(\xi, \eta, \zeta, \theta), \quad \rho = \mathcal{R}(\xi, \eta, \zeta, \theta), \quad \mu = \mathcal{M}(\xi, \eta, \zeta, \theta). \tag{7.41}$$

Formulas (7.41) define three 4-ary operations in the tangent space T_e of the loop Q. These operations are connected with the binary and ternary operations, defined by the tensors a and b in the W-algebra, by the relations which can be obtained by eliminating the tensors $\underset{1}{c}$ and $\underset{1}{c}$ from equations (7.32) and (7.40):

$$\begin{aligned}
&\mathcal{L}(x,u,v,y) + \mathcal{M}(x,y,u,v) + \mathcal{R}(v,y,x,u) = 2a(u,b(v,x,y)) - 2a(v,b(u,x,y))\\
&\tfrac{1}{2}\big(\mathcal{L}(x,y,u,v) - \mathcal{L}(x,y,v,u)\big) = b(y,x,a(u,v) - a(b(y,x,u),v) - a(u,b(y,x,v)),\\
&\tfrac{1}{2}\big(\mathcal{R}(x,y,u,v) - \mathcal{R}(x,y,v,u)\big) = b(x,a(u,v),y) - a(b(x,u,y),v) - a(u,b(x,v,y)),\\
&\tfrac{1}{2}\big(\mathcal{M}(x,y,u,v) - \mathcal{M}(x,y,v,u)\big) = b(a(v,u),x,y) - a(b(v,x,y),u) - a(v,b(u,x,y)).
\end{aligned} \tag{7.42}$$

We can now give another definition of the W_4-algebra, which is more symmetric than that given in Section 5.2.

Definition 7.6 Let T be a vector space in which the binary, ternary and three quaternary operations are defined by the forms $a, b, \mathcal{L}, \mathcal{R}$ and $\mathcal{M}$, respectively. The set $(T, a, b, \mathcal{L}, \mathcal{R}, \mathcal{M})$ is called the W_4-algebra if the ternary and binary operations are connected by the generalized Jacobi identity (2.31), and all five operations satisfy relations (7.42).

Thus, the local W_4-algebra of this type is associated with each point of C^4-manifold X carrying a three-web W. Note that this W_4-algebra and the W_4-algebra considered in Chapter 5 are equivalent, i.e. the operations on one of them can be expressed in terms of the operations of the other.

7.4 Classification of Three-Webs in the Fourth Order Differential Neighborhood

The results of Section 7.3 allow us to continue and deepen the well-known classification of three-webs based on the closure conditions or the identities in their coordinate loops.

The loops, in which the operators $l_{x,y}$ are automorphisms, are called the *left special loops* or *A_l-loops.* In the same way the *right special loops* or *A_r-loops* and the *middle special loops* or *A_m-loops* can be defined. Accordingly, the webs in whose coordinate loops the identities:

$$A_l: \quad l_{x,y}(uv) = l_{x,y}(u)l_{x,y}(v), \tag{7.43}$$

$$A_r: \quad r_{x,y}(uv) = r_{x,y}(u)r_{x,y}(v), \tag{7.44}$$

$$A_m: \quad m_{x,y}(uv) = l_{x,y}(u)m_{x,y}(v), \tag{7.45}$$

hold, are called the *webs* A_l, A_r and A_m, respectively. We will use the same notations for the corresponding figures.

Let us describe a construction of the configuration determined, for example, by equation (7.43). First, we construct the element $w = l_{x,y}(u)$ which is the solution of the equation $xy \cdot w = x \cdot yu$. Its construction is represented in Figure 7.1.

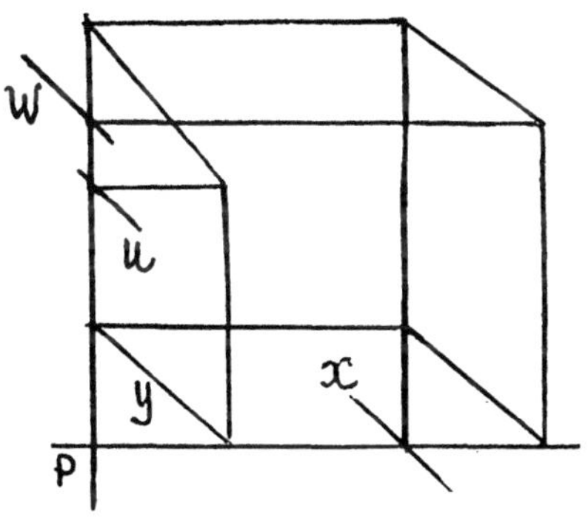

Figure 7.1

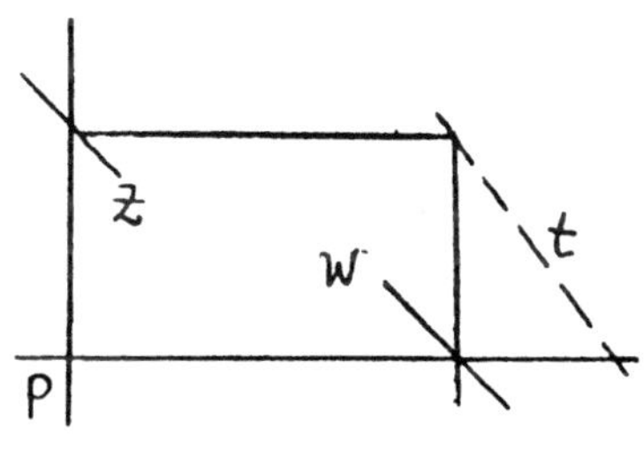

Figure 7.2

Suppose that the elements $l_{x,y}(u) = w$, $l_{x,y}(v) = z$ and $l_{x,y}(uv) = t$ have been already constructed. Then, equation (7.43) is equivalent to the fact that the leaves wz and t coincide (Figure 7.2). If we make all constructions in the same figure, we obtain the shape of the figure A_l. We do not reproduce this picture here since it is rather complicated and moreover, it will not be used in our further considerations. The figures A_r and A_m can be constructed by the same method. Figure 7.3 reproduces the construction of the element $w = r_{x,y}(u)$.

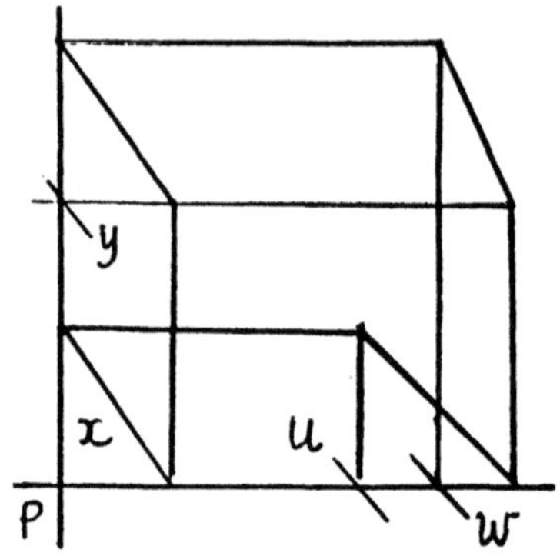

Figure 7.3

Identifying in equations (7.43)–(7.45) some or all of the variables, we obtain the new identities. We will call them and the corresponding webs the *identities* (*webs*) of type A_l, A_r and A_m. The corresponding configurations can be obtained from the figures A_l, A_r and A_m by identification of certain leaves (for example, the Bol figures or the figure H can be obtained from the figure R – see Figures 2.3, 2.5 – 2.8, pp. 53–55).

Consider some of the classes of the types A_l and A_r. We will see that some of them coincide with the known classes of webs B_l, B_r or R. We will consider only identity (7.43) since for identity (7.44) all considerations and constructions are similar.

Setting $u = y$, $v = \underbrace{y(y(\ldots(yy))}_{n-1}\ldots)$ in (7.43), we obtain the equation:

$$O_l(u): \quad l_{x,y}(\underbrace{y(y(\ldots(yy))}_{n}\ldots) = l_{x,y}(y)\cdot l_{x,y}((\underbrace{y(y(\ldots(yy))}_{n-1}\ldots). \tag{7.46}$$

Theorem 7.7 *On a three-web W, the closure conditions $O_l(n)$ and $O_l(n+1)$ simultaneously hold if and only if this webs is the right Bol web.*

Proof. Consider the case $n = 2$. Let, in the coordinate loops of a web W, the identities $O_l(2)$ and $O_l(3)$ hold:

$$O_l(2): \quad l_{x,y}(y^2) = (l_{x,y}(y))^2, \tag{7.47}$$

$$O_l(3): \quad l_{x,y}(y\cdot y^2) = l_{x,y}(y)\cdot l_{x,y}(y^2). \tag{7.48}$$

The figures $O_l(2)$ and $O_l(3)$ are closed on this web. Figures 7.4 and 7.5, where the notation $\tilde{y} = l_{x,y}(y)$ is used, reproduce these two figures.

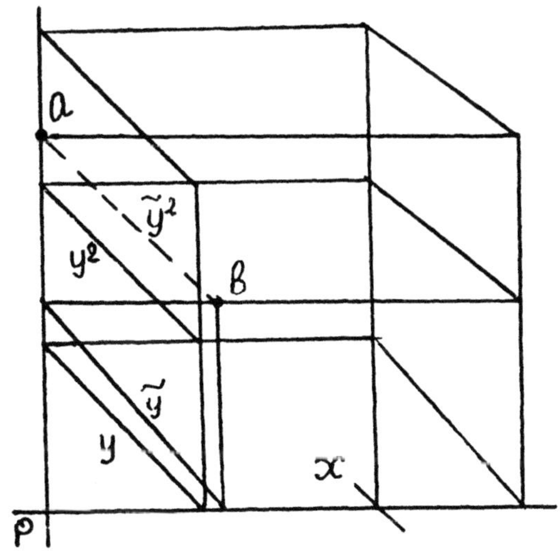

Figure 7.4

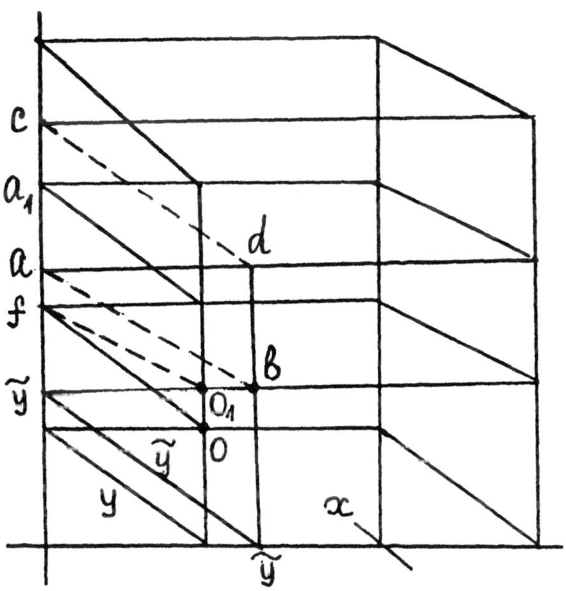

Figure 7.5

These figures can be constructed following the rule described above (see Figures 7.1 and 7.2). The closure of the figure $O_l(2)$ means that the points a and b belong to one leaf of the third foliation, i.e. we have $a3b$ (see the notation on p. 58). The closure of the figure $O_m(2)$ in Figure 7.6 means $c3d$.

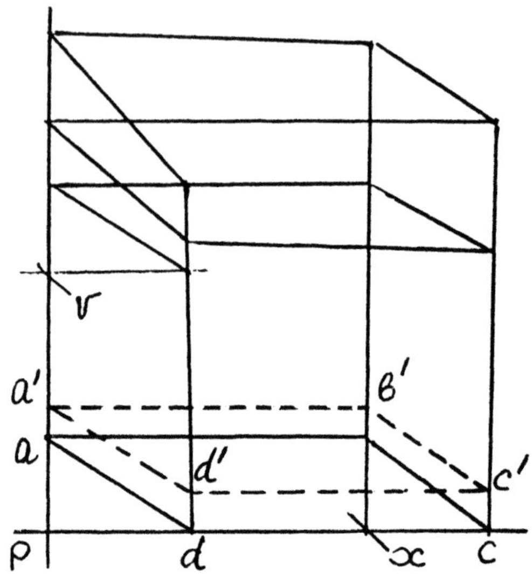

Figure 7.6

Comparing Figures 7.5 and 7.6, we discover the figure $O_l(2)$ in Figure 7.6 which is located below the horizontal leaf passing through the point a_1. According to the condition of the theorem, this figure must be closed. Thus, the points a and b belong to one leaf of the third foliation.

Consider now the part of the figure $O_l(3)$ (see Figure 7.5) which is located above the horizontal leaf passing through the point b. Comparing this part with Figure 7.4, we find that this part is the figure $O_l(2)$. Since this figure must be closed, we have $f3o_1$. However, according to the definition of a three-web, only one leaf of each foliation passes through a point. Thus, $o \equiv o_1$ which implies $\tilde{y} = y \Rightarrow l_{x,y}(y) = y \Rightarrow xy \cdot y = x \cdot y^2$. Since x and y were taken arbitrarily, we conclude that in the loop l_p the right alternativity identity holds. These considerations are valid for any coordinate loop of the web W. Therefore, the latter is the right Bol web. If $n > 2$, the proof is similar. The only difference is that the number of "levels" in Figures 7.6 and 7.5 will be higher.

The converse: $B_r \Rightarrow O_l(n)\&O_l(n+1)$ follows from the known fact that in a loop B_r the identity $(xy)^m \cdot y^n = x \cdot y^{m+n}$ holds. The latter is equivalent to the identity $l_{x,y}(y^n) = y^n$, which implies that equation (7.46) is identically satisfied. ■

Theorem 7.7 allows us to solve the following problem posed by L.V. Sabinin in [Sa 88]: *find the class of smooth power-associative loops where the so-called identity of right geometricity:*

$$l_{x,y}(y^t) = (l_{x,y}(y))^t, \quad t \in \mathbf{R} \tag{7.49}$$

is universal.

Theorem 7.8 *The identity of right geometricity is universal in the class of right Bol loops.*

Proof. Setting in (7.49) $t = 2$ and $t = 3$, we get identities (7.47) and (7.48). Suppose that identity (7.49) is universal in a certain class of isotopic loops. To this class, there corresponds a three-web in whose coordinate loops the closure conditions $O_l(2)$ and $O_l(3)$ hold. Then, by Theorem 7.7, we find that this web is the web B_r and subsequently its coordinate loops are right Bol loops. Thus, the class of loops under consideration is the class of right Bol loops. The converse can be proved as in Theorem 7.7. ■

Setting $u = y$ in identity (7.43), we arrive at the identity:

$$l_{x,y}(yv) = l_{x,y}(y)l_{x,y}(v). \tag{7.50}$$

Theorem 7.9 *Let identity* (7.50) *hold in each coordinate loop of a three-web* W*. Then this web is the web* R*.*

Proof. Setting first $v = y$ and next $v = y^2$ in (7.50), we get identities (7.47) and (7.48). By Theorem 7.7, we derive that the web under consideration is the web B_r. As noted earlier, for the latter web, $l_{x,y}(y) = y$. Thus, identity (7.50) takes the form:

$$l_{x,y}(y \cdot v) = y \cdot l_{x,y}(v). \tag{7.51}$$

If we disregard the dotted lines in Figure 7.6, Figure 7.6 reproduces the corresponding configuration. Denote it by F. The figure F consists of two parts, upper and lower. The former is defined by the choice of the element v alone, and the latter by the choice of the element y alone. Changing y, we get a new figure F' with the same upper part and the lower part $a'b'c'd'$. By the condition of the theorem, the figure F' must be closed. Thus, the figure $R = (abcda'b'c'd')$ is closed. Since, the figures F and F' can be attached to any figure R in the web W under consideration by the procedure outlined above, we conclude that all figures R are closed on the web W. ■

Since identity (7.50) is a particular case of identity (7.43) which characterizes the webs A_l, Theorem 7.9 implies the obvious corollary.

Corollary 7.10 *Any three-web* A_l *is the web* R*. The same is true for the webs* A_r*.* ■

In the paper [BP 56] R.H. Bruck and I.J. Paige posed the following problem: *describe the class of loops in which identities* (7.43) *and* (7.44) *are isotopically invariant.* Corollary 7.10 gives the solution of this problem: such loops are groups. Moreover, note that in our proof we did not use either the smoothness or even continuity of the loops under consideration. Thus, this proof is also valid for the abstract loops considered in [BP 56]. Note also that Theorem 7.9 shows that the much weaker assumption of the universality of one of identities (7.50) or (7.51) (instead of the universality of both identities, A_l and A_r) leads to the class of groups.

Consider the equation:

$$H_l(n): \quad l_{x,x}(\underbrace{x(x(\ldots(xx)}_{n}\ldots) = l_{x,x}(x)(l_{x,x}(\underbrace{x(x(\ldots(xx)}_{n-1}\ldots), \tag{7.52}$$

which can be obtained from (7.46) by setting $y = x$.

Theorem 7.11 *The closure conditions $H_l(n)$ and $H_l(n+1)$ hold on a web W if and only if the web is hexagonal.*

Proof. The proof precisely follows the proof of Theorem 7.7 with the only difference being that we should set $y = x$ everywhere. Then, we find that the figures B_r in which two vertical leaves x and y coincide, are closed. But these are the figures H. ■

Theorem 7.12 *Let the identity*

$$l_{x,x}(xv) = l_{x,x}(x)l_{x,x}(v) \tag{7.53}$$

hold in each coordinate loop of a web W. Then this web is the left Bol web.

Proof. Setting first $v = x$ and next $v = x^2$ in (7.53), we get identities $H_l(2)$ and $H_l(3)$. Thus, by Theorem 7.11, the web under consideration is the web H, and subsequently we have $l_{x,x}(x) = x$. As a result, identity (7.53) takes the form:

$$l_{x,x}(xv) = xl_{x,x}(v). \tag{7.54}$$

Since identity (7.54) can be obtained from identity (7.51) by setting $x = y$, to this identity there corresponds the figure F in which the vertical leaves x and y coincide. Applying the same considerations as those used in the proof of Theorem 7.9, we arrive at the fact that on the web under consideration, all figures R in which the vertical leaves x and y coincide, are closed. But these are the figures B_l. ■

Corollary 7.13 *If in the coordinate loops of a three-web W the mappings $l_{x,x}$ are automorphisms, then the web W is a left Bol web.*

Proof. In fact, let the identity

$$l_{x,x}(uv) = l_{x,x}(u)l_{x,x}(v) \tag{7.55}$$

hold in each coordinate loop of a web W. This identity implies identity (7.53). The latter means that the web under consideration is the web B_l. ■

7.5 Three-Webs with Elastic Coordinate Loops

1. We denoted by E, those three-webs in whose coordinate loops the elasticity identity

$$E: \quad x \cdot yx = xy \cdot x, \tag{7.56}$$

holds. Since relation (7.56) implies the monoassociativity identity $x^2 \cdot x = x \cdot x^2$, every web E is hexagonal. Moreover, the following theorem holds.

Theorem 7.14 *Any three-web E is the middle Bol web of special type: its torsion and curvature tensors a and b are connected by the relation:*

$$b(x, y, a(x, y)) = 0. \tag{7.57}$$

Proof. We write identity (7.56) in the equivalent form: $x = l_{x,y}(x)$. Setting $u = x$ in formula (7.33) and taking into account the last identity, we obtain three relations connecting the fundamental tensors of a web E:

$$\begin{aligned} &b(y, x, x) = 0, \quad \underset{1}{c}(y, x, x, x) + \underset{2}{c}(y, x, x, x) = 0, \\ &\underset{1}{c}(y, y, x, x) + \underset{2}{c}(y, x, y, x) = 0. \end{aligned} \tag{7.58}$$

The first equation in (7.58) means that the curvature tensor is skew-symmetric in the last two indices. As we already know (Theorem 4.4, p. 138), the latter condition characterizes the middle Bol webs B_m. This completes the proof of the first part of Theorem 7.14.

Since the three-web under consideration is the web B_m, the covariant derivatives of its curvature tensor can be expressed in terms of the torsion and curvature tensors by formulas (4.10). The latter formulas can be written in the index-free notation as follows:

$$\begin{aligned} &\underset{1}{c}(x, y, z, t) = -\underset{2}{c}(x, y, z, t) \\ &= -b(x, a(z, t), y) + b(x, a(y, z), t) + b(x, a(y, t), z). \end{aligned}$$

By virtue of these relations, the second of equations (7.58) is identically satisfied, and the third one takes the form:

$$3b(y, a(y, x), x) + b(y, a(x, y), x) = 0.$$

By the skew-symmetry of the tensors a and b, this implies (7.57). ■

The meaning of identity (7.56) is that the triples of elements of the form (x, y, x) in the coordinate loops l_p of an elastic web are associative. Let us show that some other elements from l_p possess the same property.

Theorem 7.15 *In the coordinate loops of a web E the identities of the form:*

$$(x^l y)x^n = x^l(yx^n), \quad l, n \in \mathbf{Z} \tag{7.59}$$

hold.

Proof. If l and n are natural numbers, we apply a generalized induction method. Denote by E_{ln} the figures in the web E corresponding to identity (7.59). To identity (7.56) there corresponds the figure $E \equiv E_{11}$ shown in Figure 7.7.

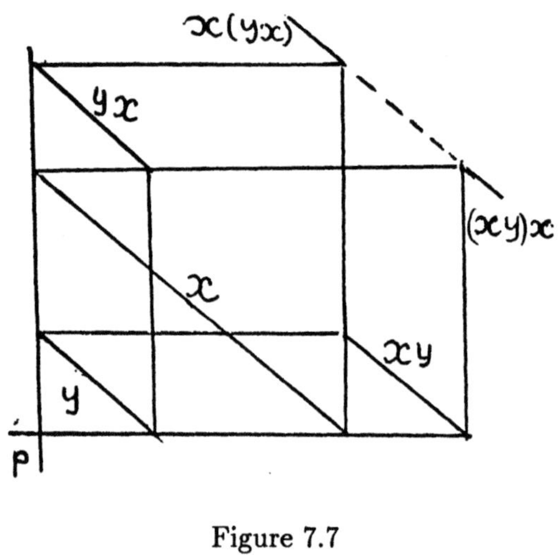

Figure 7.7

First, we prove that in the coordinate loops l_p the identity:

$$(x^l y)x = x^l(yx), \tag{7.60}$$

holds (this identity can be obtained from (7.59) if we take $n = 1$). If $l = 2$, identity (7.60) implies the identity $(x^2y)x = x^2(yx)$ to which there corresponds the figure E_{21}

shown in Figure 7.8.

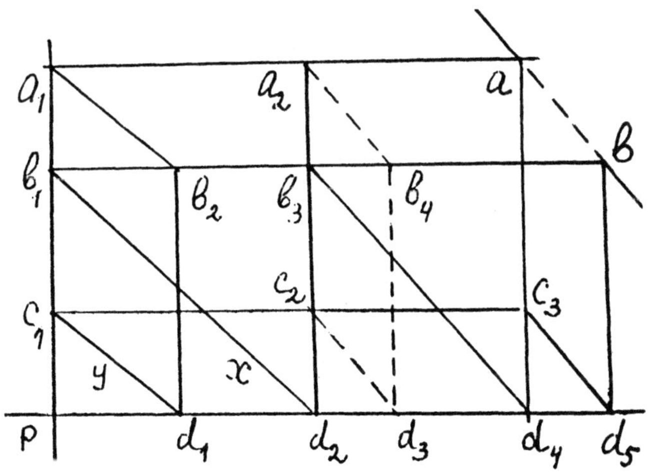

Figure 7.8

We prove that this figure is closed, i.e. the points a and b belong to one leaf of the third foliation.

We construct the points a_2, d_3 and b_4 as shown in Figure 7.8, and consider the figure $E^1 = (c_1d_1b_1d_2a_1b_2c_2d_3a_2b_4)$. Since this figure is closed, we have a_23b_4. Next, we consider the figure $E^2 = (c_2d_3b_3d_4a_2b_4c_3d_5ab)$. Since this figure is closed, we have $a3b$.

Suppose now that identity (7.60) holds for $l = k$. We will prove that it also holds for $l = k + 1$. In the loop l_p we consecutively construct the following leaves of the web W: $x^2, x^3, \ldots, x^k, x^{k+1}, x^ky, yx(x^ky)x = x^k(yx), x^{k+1}y, (x^{k+1}y)x, x^{k+1}(yx)$. We obtain the figure shown in Figure 7.9 by the solid lines.

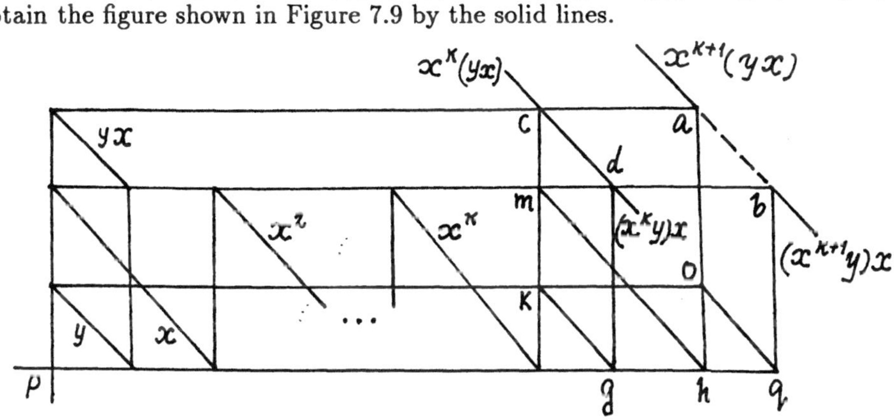

Figure 7.9

We need to prove that $(x^{k+1}y)x = x^{k+1}(yx)$, i.e. that we have $a3b$. By the induction assumption, the figures E_{k1} are closed, i.e. we have $c3d$. Next, consider the figure $E = (kgmhcdoqab)$. Since this figure is closed, we get $a3b$. Thus, identity (7.60) holds for all natural numbers l.

We now consider the general case. Suppose that identity (7.59) holds for any l and for $n = 1, 2, \ldots, m$, i.e. all figures E_{lm} are closed on the web under consideration. We prove that identity (7.59) also holds for $n = m + 1$. Consider an arbitrary figure E_{lm+1} shown in Figure 7.10 by the solid lines.

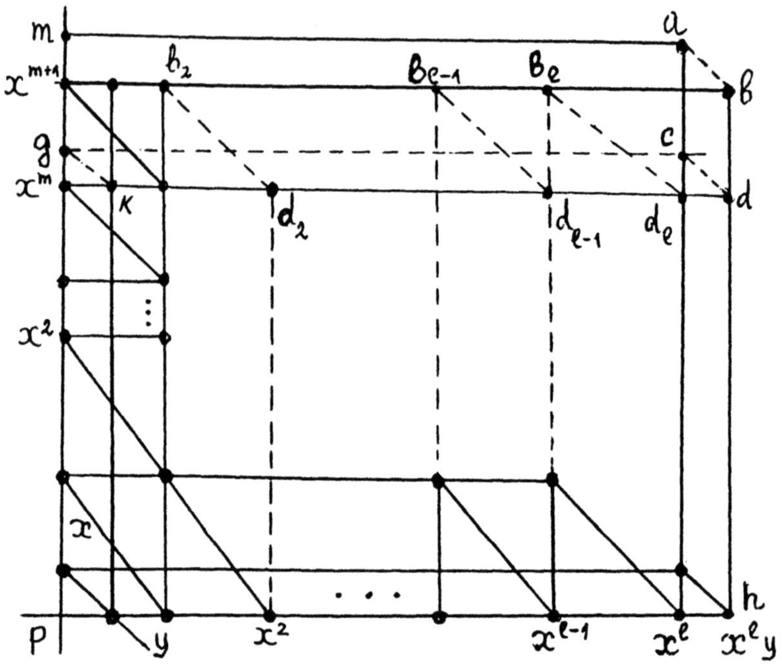

Figure 7.10

We will prove that this figure is closed, i.e. we have $a3b$. Construct the points d, g and c as shown in Figure 7.10. As a result, we obtain the figure $E_{lm} = (pgcdh)$. Since, by the induction assumption, this figure is closed, we have $c3d$. Note further that since a web E is the web H, the identity of power-associativity holds in its coordinate loops. We write this identity in the form: $x^i \cdot x^m = x^{i-1} \cdot x^{m+1}$. Considering the figures corresponding to this identity for $i = 2, 3, \ldots, l$, we arrive at the conclusion that the relation $b_i 3 d_i$ holds for any i.

The points a and b belong to the figure $E_{l1} = (dkgmab)$ which is similar to the figure shown in Figure 7.10. We have already proved that this figure is closed. Thus, we have $a3b$, i.e. the figure E_{lm+1} under consideration is closed, and identity (7.59) holds for any natural numbers l and n.

For the negative integers l and n, the proof can be reduced to the previous proof since, by virtue of the power-associativity mentioned above, we have the identities: $x^{-1} = {}^{-1}x$ and $(x^{-1})^n = (x^n)^{-1}$. ■

Note that Theorem 7.15 also holds for abstract three-webs E since we did not make any assumptions on smoothness. In the case when the web E is a geometric web, Theorem 7.15 can be strengthened.

Theorem 7.16 *In the coordinate loops of an analytic web E, triples of the form (x, y, z), where y is an arbitrary element of the loop and x and z belong to the same one-parametric subgroup of this loop, are associative.*

Proof. In fact since a web E is hexagonal, its coordinate loops possess the maximal number of one-parametric subgroups (Theorem 3.5, p. 109). The elements x^m and x^n in identity (7.59) belong to the one-parametric subgroup g of the loop l_p, and this subgroup is generated by the element x (sufficiently close to the identity element of the loop l_p). In the subgroup g, the equation $x^m = z$, $m \in \mathbf{N}$ can be (locally) uniquely solved. Hence, identity (7.59) implies the more general identity with a fractional exponent:

$$(x^{\frac{l}{m}} \cdot y) \cdot x^{\frac{n}{m}} = x^{\frac{l}{m}}(y \cdot x^{\frac{n}{m}}).$$

Since multiplication is continuous, it follows from this that identity (7.59) holds for any real l and n. But then x^l and x^n can be any two elements from the subgroup g. ■

2. Since the webs E are middle Bol webs, a classification of the former is based on a classification of the latter. The isoclinic and hexagonal webs B_m have been most studied. In the following considerations we will separate the class of subwebs E in each of these classes – the class of isoclinic webs B_m and the class of hexagonal webs B_m.

Theorem 7.17 *The class of isoclinic webs E coincides with the class of isoclinic webs R.*

Proof. Let W be an isoclinic web E. Then, by Theorem 7.15, this web is an isoclinic web B_m, and its structure tensors a and b have form (4.26) and (4.29):

$$a^i_{jk} = a_{[j}\delta^i_{k]}, \quad b^i_{jkl} = b_{jk}\delta^i_l - b_{jl}\delta^i_k, \quad b_{ij} = b_{ji}. \tag{7.61}$$

Consider relations (7.57) which we write in the coordinate form:

$$b^i_{jkm}a^m_{pq} + b^i_{pkm}a^m_{jq} + b^i_{jqm}a^m_{pk} + b^i_{pqm}a^m_{jk} = 0. \tag{7.62}$$

Substituting for a and b in (7.62) their values from (7.61), after some transformations we obtain:

$$-b_{jk}a_q\delta^i_p - b_{jq}a_k\delta^i_p + 2b_{jp}a_q\delta^i_k + 2b_{jp}a_k\delta^i_q - b_{pk}a_q\delta^i_j - b_{pq}a_k\delta^i_j = 0.$$

Contracting this equation in the indices i and p, we arrive at the equation: $b_{jk}a_q + b_{jq}a_k = 0$ which means that the tensor $t_{jkq} = b_{jk}a_q$ is skew-symmetric in the indices k and q. But it is also symmetric in the indices j and k. Therefore, the quantities t_{jkq} are equal to zero, i.e. $b_{jk}a_q = 0$. If at least one of the quantities a_q is not zero, we have $b_{jk} = 0$, and it follows from (7.61) that $b^i_{jkl} = 0$. This tensor equation characterizes the webs R.

Suppose now that all a_q are zero. As shown in Section 4.3, the tensor a_q of an isoclinic web B_m satisfies the differential equation:

$$\nabla a_q = b_{ql}(\underset{1}{\omega}^l - \underset{2}{\omega}^l)$$

Substituting $a_q = 0$ into this equation, by the independence of the forms $\underset{1}{\omega}^l - \underset{2}{\omega}^l$, we obtain that $b_{ql} = 0$. Thus, we again arrive at the webs R. ■

Note that if $r = 2$, the tensors a and b of a web B_m can be always written in the form (7.61) (see subsection **6** of Section 4.2). It is easy to derive from this that *there are no four-dimensional webs E except the group webs.*

3. However, six-dimensional non-trivial (i.e. non-Moufang) three-webs E do exist. One of such webs, namely, the web E_1, was considered in Theorem 4.17 and Problems **26** and **27** of Chapter 2. Before we indicate other examples, we recall that six-dimensional Bol webs B_m, which include the class of six-dimensional webs E, can be classified by means of the tensor a^{ij} and the latter can be expressed in terms of the torsion tensor by formulas (4.36). Moreover, according to Lemma 4.14, a field of frames on a six-dimensional web E can be chosen in such a way that the components of the tensor a^{ij} become constants. The tensor a^{ij} satisfies relations (4.31) and (4.32) by means of which the components of the curvature tensor can be calculated. The tensors of the web E also satisfy relations (7.57) which, by formulas (4.36), take the form:

$$\Big(b^{is}_j(\epsilon_{kps}\epsilon_{lmq} + \epsilon_{mps}\epsilon_{lkq}) + b^{is}_l(\epsilon_{kps}\epsilon_{jmq} + \epsilon_{mps}\epsilon_{jkq})\Big)a^{pq} = 0, \qquad (7.63)$$

where ϵ_{ijk} is the discriminant tensor (see Section 4.3). Thus, for finding six-dimensional webs E, in each of the classes of six-dimensional Bol webs B_m, we should separate the subclass of webs whose tensors satisfy relations (7.63). Since the number of relations (7.63) is very high, it is impossible to list all cases in this book. However, if we accomplished this, we would get the following result.

Theorem 7.18 *The are no six-dimensional webs E with a matrix (a^{ij}) of rank two or three.* ■

A web E with a symmetric matrix (a^{ij}) of rank one is the web E_1. Let us consider a web B_m with a non-symmetric matrix (a^{ij}) of rank one and find out whether this web is the web E.

A family of frames on a manifold of such a web can be chosen in such a way that, at each point, the matrix (a^{ij}) has the form:

$$(a^{ij}) = \begin{pmatrix} 0 & 0 & 0 \\ 0 & 0 & 1 \\ 0 & 0 & 0 \end{pmatrix}$$

i.e. $a^{23} = 1$ and other a^{ij} are zero. Substituting these values into equations (4.32) and (7.63), we find that the tensor b_k^{ij} has only three essential components: b_3^{32}, b_1^{22} and b_1^{32}.

Let us substitute the values of the tensors we found into equations (4.30) and (4.31). In particular, if $i = 3$ and $j = 2$, we obtain from equation (4.31) the equation:

$$b_1^{32}(\underset{1}{\omega}^1 - \underset{2}{\omega}^1) + b_3^{32}(\underset{1}{\omega}^3 - \underset{2}{\omega}^3) = 0.$$

By the independence of the basis forms, it follows that $b_1^{32} = b_3^{32} = 0$. As a result, the system of equations (4.30) and (4.31) takes the following form:

$$\begin{aligned} &\omega_1^1 = \omega_2^1 = \omega_3^1 = \omega_2^3 = 0, \quad \omega_3^2 = \tfrac{1}{2} b_{22}^1(\underset{1}{\omega}^1 - \underset{2}{\omega}^1), \\ &db_{22}^1 = b_{22}^1(\omega_3^3 - \omega_2^2). \end{aligned} \tag{7.64}$$

It follows from the last equation of (7.64) that the quantity b_1^{22} is a relative invariant. If $b_1^{22} \neq 0$, we reduce the family of adapted frames by setting $b_1^{22} = 2$. Then we find from (7.64) that

$$\omega_2^2 - \omega_3^3 = 0, \quad \omega_3^2 = \underset{1}{\omega}^1 - \underset{2}{\omega}^1. \tag{7.65}$$

Since now $d\omega_2^2 = d\omega_3^3 = 0$, we can one more time reduce the family of frames by setting $\omega_2^2 = \omega_3^3 = 0$.

By virtue of all equations obtained so far, the system of structure equations (4.37) and (4.38) of the Bol web under consideration takes the form:

$$\begin{aligned} &d\underset{1}{\omega}^1 = 0, \quad d\underset{1}{\omega}^3 = 0, \quad d\underset{1}{\omega}^2 = \underset{1}{\omega}^1 \wedge \omega_1^2 + 2\underset{1}{\omega}^1 \wedge \underset{1}{\omega}^2 - \underset{1}{\omega}^1 \wedge \underset{1}{\omega}^3 - \underset{1}{\omega}^3 \wedge \underset{2}{\omega}^1, \\ &d\underset{2}{\omega}^1 = 0, \quad d\underset{2}{\omega}^3 = 0, \quad d\omega_1^2 = 2\underset{1}{\omega}^3 \wedge \underset{2}{\omega}^1 - 2\underset{1}{\omega}^1 \wedge \underset{2}{\omega}^3, \\ &d\underset{2}{\omega}^2 = \underset{2}{\omega}^1 \wedge \omega_1^2 + \underset{2}{\omega}^1 \wedge \underset{2}{\omega}^3 + \underset{2}{\omega}^3 \wedge \underset{1}{\omega}^1 - 2\underset{2}{\omega}^1 \wedge \underset{2}{\omega}^3. \end{aligned} \tag{7.66}$$

In addition to the forms $\underset{1}{\omega}^i, \underset{2}{\omega}^i$ and ω_1^2, this system contains only constants the components of the torsion and curvature tensors. Moreover, this system is closed with respect to the operation of exterior differentiation. Therefore, system (7.66) determines a seven-dimensional Lie group G, and the corresponding Bol web is a G-web (see Section 6.4).

Consecutively integrating equations (7.66), we find the invariant forms of the group G:

$$
\begin{aligned}
&\underset{1}{\omega}^1 = du^1, \quad \underset{1}{\omega}^3 = du^3, \quad \underset{2}{\omega}^3 = dv^3, \\
&\omega_1^2 = u^3dv^1 - v^1du^3 - u^1dv^3 + v^3du^1 + d\tau, \\
&\underset{1}{\omega}^2 = e^{-2u^1}du^2 - u^3dv^1 + (u^3 + u^3v^1 + u^1v^3)du^1 - \tau du^1, \\
&\underset{2}{\omega}^2 = e^{-2v^1}dv^2 - u^1dv^3 + (-v^3 - v^3u^1 + v^1u^3)dv^1 - \tau dv^1.
\end{aligned}
\tag{7.67}
$$

The foliations of the web are determined by the equations:

$$
\underset{1}{\omega}^i\Big|_{\tau=0} = 0, \quad \underset{2}{\omega}^i\Big|_{\tau=0} = 0, \quad \underset{1}{\omega}^i\Big|_{\tau=0} + \underset{2}{\omega}^i\Big|_{\tau=0} = 0.
$$

Denoting the parameters of the leaves by x^i, y^i and z^i, as a result of integration of the systems indicated above, we find the equations of the foliations in the form:

$$
\begin{aligned}
&\lambda_1: \quad u^1 = x^1, \quad u^2e^{2x^1} - v^1x^3 = x^2, \quad u^3 = x^3; \\
&\lambda_2: \quad v^i = y^i; \\
&\lambda_3: \quad -u^1 + v^1 = z^1, \quad u^2 + v^2e^{-2z^1} - z^3u^1e^{-2u^1} + u^1u^3e^{-2u^1} = z^2, \quad u^3 + v^3 = z^3.
\end{aligned}
$$

If we eliminate the local coordinates u^i and v^i, we arrive at the closed form equations of the web (or its coordinate quasigroup):

$$
\begin{aligned}
&z^1 = x^1 + y^1, \quad z^3 = x^3 + y^3, \\
&z^2 = x^2e^{-2x^1} + y^2e^{-2z^1} - 2z^3e^{-2x^1} + (y^1x^3 - x^1y^3)e^{-2x^1}.
\end{aligned}
$$

Making the isotopic transformation:

$$
(x^2 - x^3)e^{-2x^1} \to x^2, \quad y^2e^{-2y^1} - y^3 \to y^2,
$$

we get the equations of the coordinate loop with the identity element $e(0,0,0)$:

$$
\begin{aligned}
&z^1 = x^1 + y^1, \\
&z^2 = x^2 + y^2e^{-x^1} + (y^1x^3 - x^1y^3)e^{-2x^1}, \\
&z^3 = x^3 + y^3.
\end{aligned}
\tag{7.68}
$$

It can be directly verified that the elasticity identity holds in this loop. Since the web we found is a G-web, all its coordinate loops are isomorphic (Section 6.4), and therefore, they are also elastic. Thus, the three-web determined by equation (7.68) is a web E. Denote it by E_2.

We have proved the following theorem.

Theorem 7.19 *There exists only two six-dimensional non-Moufang webs E. These are the Bol webs E_1 and E_2 whose tensor a^{ij} is of rank one. Each of these webs is a web G.* ■

PROBLEMS

1. Apply expansion (7.7) to find the first four terms of the Taylor series for the elements ${}^{-1}x$ and x^{-1}.

Solution. Since ${}^{-1}x \cdot x = e$, we find from (7.7) that

$$\begin{aligned} 0 = \; & {}^{-1}x + x + \Lambda({}^{-1}x, x) + \tfrac{1}{2}M({}^{-1}x, {}^{-1}x, x) + \tfrac{1}{2}N({}^{-1}x, x, x) \\ & + \tfrac{1}{6}P({}^{-1}x, {}^{-1}x, {}^{-1}x, x) + \tfrac{1}{4}({}^{-1}x, {}^{-1}x, x, x) + \tfrac{1}{6}R({}^{-1}x, x, x, x) + \{5\}. \end{aligned}$$

(Here we used the notations introduced on p. 248.) From (2.37) and (2.38) we find the expansion for ${}^{-1}x$ up to the fourth order terms:

$${}^{-1}x = -x + \frac{1}{2}b(x, x, x) + \{4\}.$$

Substituting this value of ${}^{-1}x$ into the terms of order higher than one in the previous equation, we find that

$$\begin{aligned} {}^{-1}x = \; & -x + \tfrac{1}{2}b(x, x, x) \\ & + \tfrac{1}{2}a(b(x, x, x), x) + \tfrac{1}{6}P(x, x, x, x) - \tfrac{1}{4}Q(x, x, x, x) + \tfrac{1}{6}R(x, x, x, x) + \{5\}. \end{aligned}$$

For x^{-1}, the proof is similar.

2. Prove that in the coordinate loops of a web E the triples (x, y, z) of elements x, y and z, where x is the solution of the equation $z \cdot yz = xy$, are associative.

3. Prove directly that the web defined by equations (7.68) is a web E. (*Hint*: See Problem **27** of Chapter 2.)

NOTES

7.1. The results of this section are due to Shelekhov [S 86b]. For the generalization of these results see [S 90b] and [S t.a.].

7.2. The results of this section are due to Shelekhov [S 88a].

Properties of the associants of a loop Q were considered by Sabinin in [Sa 72].

7.3. The results of this section are due to Shelekhov [S 88a].

7.4. The results of this section are due to Shelekhov [S 89b].

7.5. The existence of analytic non-Moufang webs E or, equivalently, of the class of isotopically invariant elastic loops different from the Moufang loops, was in doubt for a long time. Their existence (Theorem 7.19) has been proved not long ago in [S t.a.]. Theorem 7.14 has been proved in [S 87c] and Theorem 7.15 in [SS 84].

Note also that the webs E possess a closed G-structure of finite type not higher than three since they are the webs B_m (Theorem 7.14). However, the description of the complete system of relations connecting the fundamental tensors of the webs E is still unknown.

Chapter 8

d-Webs of Codimension r

Along with three-webs, other types of webs are also studied in differential geometry. As far back as the first papers of Blaschke and his school, webs formed by four or more families of curves in a plane as well as webs formed by families of surfaces in a three-dimensional space were investigated. These constructions were generalized to higher dimensions. In this way, the theory of webs formed by d foliations of codimension r in a smooth manifold X of dimension nr arose. We will denote such webs by $W(d,n,r)$. Of course, we assume that foliations λ_α, $\alpha = 1,\ldots,d$, forming a d-web, are in general position, i.e. n leaves of a web $W(d,n,r)$ belonging to different foliations have at most one common point.

In particular, we will now use the notation $W(3,2,r)$ for three-webs, which were studied in previous chapters and denoted there by $W = (X, \lambda_\alpha)$, $\alpha = 1,2,3$.

Two webs $W_1(d,n,r)$ and $W_2(d,n,r)$ defined in the manifolds X_1 and X_2, both of dimension nr, are said to be *equivalent* if there exists a local diffeomorphism $\varphi : X_1 \rightarrow X_2$ which transfers the foliations of the first web W_1 into the foliations of the second web W_2.

A web $W(d,n,r)$ is said to be *parallelizable* if it is equivalent to a web consisting of d families of parallel planes of codimension r. If $d \leq n$, a web $W(d,n,r)$ is always parallelizable. Because of this, we will assume that $d \geq n+1$.

In this chapter we will study the basic properties of webs $W(d,n,r)$, the geometric structures associated with these webs and special types of these webs defined by certain closure conditions. Special attention will be given to the Grassmannization problem, the linearization problem and the algebraization problem. The problem of web rank will be also discussed.

Many matters that will be discussed in this chapter generalize the contents of previous chapters. Because of this, our considerations in this chapter will be more concise. On some occasions, we will give only a review of the main results. The reader can find a more detailed exposition of the theory of d-webs in the book [G 88].

8.1 $(n+1)$-Webs on a Manifold of Dimension nr

1. Consider an $(n+1)$-web $W(n+1,n,r)$ defined on a manifold X of dimension nr. The foliations $\lambda_\alpha, \alpha = 1,\ldots,n+1$, forming this web, can be defined by the following completely integrable systems of Pfaffian equations:

$$\underset{\alpha}{\omega}^i = 0, \quad i = 1,\ldots,r; \quad \alpha = 1,\ldots,n+1. \tag{8.1}$$

Since the number of 1-forms on the left-hand sides of equations (8.1) is $(n+1)r$ and $\dim X = nr$, the forms $\underset{\alpha}{\omega}^i$ are connected by linear equations. Applying the same method as that used in Chapter 1, we can show that these equations can be reduced to the following form:

$$\underset{1}{\omega}^i + \underset{2}{\omega}^i + \ldots + \underset{n+1}{\omega}^i = 0. \tag{8.2}$$

Relations (8.2) remain invariant under the transformations:

$$'\underset{\alpha}{\omega}^i = A^i_j \underset{\alpha}{\omega}^j, \quad \det(A^i_j) \neq 0. \tag{8.3}$$

forming the group $G = \mathbf{GL}(r)$ – the structure group of a web $W(n+1,n,r)$. Thus, the structure group of a web $W(n+1,n,r)$ does not depend on the number n.

By virtue of condition (8.2), the structure equations of a web $W(n+1,n,r)$ can be reduced to the form:

$$\begin{aligned} d\underset{u}{\omega}^i &= \underset{u}{\omega}^j \wedge \omega^i_j + \sum_{v \neq u} \underset{uv}{a}{}^i_{jk} \underset{u}{\omega}^j \wedge \underset{v}{\omega}^k, \\ d\underset{n+1}{\omega}^i &= \underset{n+1}{\omega}^j \wedge \omega^i_j, \end{aligned} \tag{8.4}$$

where $u, v = 1,\ldots,n$ and $\underset{uv}{a}{}^i_{jk}$ are the *torsion tensors* of the web satisfying the conditions:

$$\underset{uv}{a}{}^i_{jk} = \underset{vu}{a}{}^i_{kj}, \quad \sum_{u,v} \underset{uv}{a}{}^i_{jk} = 0. \tag{8.5}$$

In addition, we assume that $\underset{uu}{a}{}^i_{jk} = 0$. The forms ω^i_j satisfy the structure equations:

$$d\omega^i_j = \omega^k_j \wedge \omega^i_k + \sum_{u,v} \underset{uv}{b}{}^i_{jkl} \underset{u}{\omega}^k \wedge \underset{v}{\omega}^l, \tag{8.6}$$

and define an affine connection $\tilde{\Gamma}$ on the manifold X which is similar to the connection $\tilde{\Gamma}$ for a three-web (see Section 1.8. p. 34). The tensors $\underset{uv}{a}{}^i_{jk}$ and $\underset{uv}{b}{}^i_{jkl}$ are the *torsion tensors* and the *curvature tensors* of this connection, respectively.

Equations (8.6) are differential prolongations of equations (8.4). In addition, as another result of exterior differentiation of equations (8.4), we obtain the Pfaffian equations:

$$\nabla \underset{uv}{a}{}^{i}_{jk} = \sum_{w}^{n} \left(\underset{uvw}{a}{}^{i}_{jkl} + \underset{uv}{a}{}^{i}_{mk} \underset{wu}{a}{}^{m}_{lj} + \underset{uv}{a}{}^{i}_{jm} \underset{vw}{a}{}^{m}_{kl} \right) \underset{w}{\omega}{}^{l}, \quad u \neq v, \tag{8.7}$$

and the closed form relations:

$$\underset{uv}{b}{}^{i}_{jkl} = \frac{1}{2} \left(\underset{wuv}{a}{}^{i}_{jkl} - \underset{vwu}{a}{}^{i}_{ljk} \right), \quad w \neq u, v, \tag{8.8}$$

$$\underset{uv}{b}{}^{i}_{[jkl]} = 0, \tag{8.9}$$

$$\underset{uu}{b}{}^{i}_{ljk} + 2 \underset{uv}{b}{}^{i}_{[jk]l} = 0. \tag{8.10}$$

Relation (8.8) shows that if $n > 2$, the curvature tensor of a web $W(n+1, n, r)$ is completely determined by the covariant derivatives of the torsion tensor of this web. This implies the following theorem.

Theorem 8.1 *A web $W(n+1, n, r)$ is parallelizable if and only if its torsion tensor vanishes, i.e.* $\underset{uv}{a}{}^{i}_{jk} = 0$.

Proof. In fact, by (8.7) and (8.8), the condition $\underset{uv}{a}{}^{i}_{jk} = 0$ implies $\underset{uv}{b}{}^{i}_{jkl} = 0$. As a result, the structure equations of the web under consideration take the form:

$$d\underset{\alpha}{\omega}{}^{i} = \underset{\alpha}{\omega}{}^{i} \wedge \omega^{i}_{j}, \quad d\omega^{i}_{j} = \omega^{k}_{j} \wedge \omega^{i}_{k}, \quad \alpha = 1, \ldots, n+1, \quad i, j, k = 1, \ldots, r.$$

But these equations determine an $(n+1)$-web in an affine space A of dimension nr formed by foliations of parallel planes of codimension r (cf. Section 1.5). The converse immediately follows from the previous equations. ■

Note that for a multidimensional three-web $W(3, 2, r)$, as follows from relations (1.31) and (1.33), the symmetric part of the curvature tensor can not be expressed in terms of the covariant derivatives of the torsion tensor. Thus, its parallelizability condition can be written in the form:

$$a^{i}_{jk} = 0, \quad b^{i}_{(jkl)} = 0.$$

2. On a manifold X carrying an $(n+1)$-web $W(n+1, n, r)$, we now consider a submanifold of dimension kr that is the intersection of the leaves $\mathcal{F}_{\alpha_{k+1}}, \ldots, \mathcal{F}_{\alpha_n}$ belonging to the foliations $\lambda_{\alpha_{k+1}}, \ldots, \lambda_{\alpha_n}$ of the web under consideration. This submanifold can be defined by the following system of Pfaffian equations:

$$\underset{\alpha_{k+1}}{\omega}{}^{i} = 0, \ldots, \underset{\alpha_n}{\omega}{}^{i} = 0.$$

The foliations $\lambda_{\alpha_{n+1}}, \lambda_{\alpha_1}, \ldots, \lambda_{\alpha_k}$ cut a $(k+1)$-web $W(k+1, k, r)$ on the manifold M. This web is called a *subweb* of the initial web $W(n+1, n, r)$ and is denoted by $[\alpha_{n+1}, \alpha_1, \ldots, \alpha_k]$. In particular, if $k = 2$, then $\dim M = 2r$ and the foliations $\lambda_{\alpha_{n+1}}, \lambda_{\alpha_1}, \lambda_{\alpha_2}$ cut a 3-web $[\alpha_{n+1}, \alpha_1, \alpha_2]$ on M.

Consider three-subwebs $[n+1, u, v]$ of a web $W(n+1, n, r)$. It follows from relations (8.3) – (8.9) that the torsion and curvature tensors of this subweb can be calculated by the following formulas:

$$\underset{uv}{\bar{a}}{}^{i}_{jk} = \underset{uv}{a}{}^{i}_{[jk]}, \tag{8.11}$$

$$\underset{uv}{\bar{b}}{}^{i}_{jkl} = 2\underset{uv}{b}{}^{i}_{jkl} - \underset{vuv}{a}{}^{i}_{jkl} + \underset{uvu}{a}{}^{i}_{jlk} - \underset{vu}{a}{}^{m}_{jk}\underset{uv}{a}{}^{i}_{ml} + \underset{uv}{a}{}^{m}_{jl}\underset{vu}{a}{}^{i}_{mk}. \tag{8.12}$$

In the same way we find that the torsion and curvature tensors of a subweb $[u, v, w]$ can be calculated by the formulas:

$$\underset{uvw}{\bar{a}}{}^{i}_{jk} = \underset{uv}{a}{}^{i}_{[jk]} + \underset{vw}{a}{}^{i}_{[jk]} + \underset{wu}{a}{}^{i}_{[jk]} \tag{8.13}$$

$$\begin{aligned} \underset{uvw}{\bar{b}}{}^{i}_{jkl} = {} & \underset{uv}{\bar{b}}{}^{i}_{jkl} + \underset{vw}{\bar{b}}{}^{i}_{ljk} + \underset{wu}{\bar{b}}{}^{i}_{klj} - 2\underset{wu}{b}{}^{i}_{jkl} - \underset{uvw}{a}{}^{i}_{jlk} + \underset{vuw}{a}{}^{i}_{jkl} \\ & + (\underset{vu}{a}{}^{m}_{jk} - \underset{vw}{a}{}^{m}_{jk})\underset{uw}{a}{}^{i}_{ml} + (\underset{vw}{a}{}^{m}_{jk} - \underset{uw}{a}{}^{m}_{jk})\underset{uv}{a}{}^{i}_{ml} + (\underset{vu}{a}{}^{m}_{lk} - \underset{vw}{a}{}^{m}_{lk})\underset{uw}{a}{}^{i}_{jm}. \end{aligned} \tag{8.14}$$

It follows from equation (8.14) that

$$\underset{uvw}{\bar{b}}{}^{i}_{(jkl)} = \underset{uv}{\bar{b}}{}^{i}_{(jkl)} + \underset{vw}{\bar{b}}{}^{i}_{(ljk)} + \underset{wu}{\bar{b}}{}^{i}_{(klj)}. \tag{8.15}$$

We will now give the following definition.

Definition 8.2 An $(n+1)$-web $W(n+1, n, r)$ is said to be *hexagonal* if all its three-subwebs are hexagonal.

Formula (8.15) implies the following theorem.

Theorem 8.3 *A web $W(n+1, n, r)$ is hexagonal if and only if all its three-subwebs $[\alpha, \beta, \gamma]$, α fixed, are hexagonal.*

Proof. First, we renumber the foliations in order to have $\alpha = n + 1$. As shown in Section 3.1, the hexagonality of three-subwebs $[n + 1, u, v]$ leads to the conditions:

$$\underset{uv}{\bar{b}}{}^{i}_{(jkl)} = 0. \tag{8.16}$$

But by virtue of (8.15), it follows from equation (8.16) that $\underset{uvw}{\bar{b}}{}^{i}_{(jkl)} = 0$ for any u, v, and w. ∎

Theorem 8.3 is a generalization of the classical Dubourdieu theorem ([Bl 55]) for a web $W(4, 3, 1)$ formed by two-dimensional foliations in a three-dimensional manifold.

Theorem 8.3 implies the following theorem.

Theorem 8.4 *Relations (8.16) are necessary and sufficient conditions of hexagonality of an $(n + 1)$-web $W(n + 1, n, r)$.* ∎

8.2 $(n+1)$-Webs on a Grassmann Manifold

1. An important example of a web $W(n+1, n, r)$ can be constructed in the Grassmannian $G(n-1, r+n-1)$ of $(n-1)$-planes of a projective space P^{n+r-1} of dimension $n+r-1$. First of all, note that $\dim G(n-1, r+n-1) = nr$. Next, consider a submanifold in $G(n-1, r+n-1)$ formed by $(n-1)$-planes passing through a fixed point $x \in P^{n+r-1}$. We will call this manifold a *bundle* of $(n-1)$-planes and denote by S_x. It is easy to see that a bundle S_x is isomorphic to the Grassmannian $G(n-2, r+n-2)$ and that $\dim S_x = (n-1)r$.

Using bundles of $(n-1)$-planes, we can construct foliations and webs of codimension r in the Grassmannian $G(n-1, r+n-1)$. In fact, in the space P^{n+r-1}, let us consider a smooth manifold X of dimension r and a set of the bundles S_x with vertices belonging to X: $x \in X$. If we exclude the $(n-1)$-planes tangent to the manifold X and the $(n-1)$-planes intersecting X more than at one point from each of the bundles S_x, the remaining parts $\widetilde{S}_x$ of S_x form a foliation in an open domain D of the Grassmannian $G(n-1, r+n-1)$.

In the space P^{n+r-1}, we further consider submanifolds X_α, $\alpha = 1, \ldots, n+1$, of dimension r in general position. Each generates a foliation in an open domain $D_\alpha \subset G(n-1, r+n-1)$ described above. All foliations constructed in this manner generate an $(n+1)$-web of codimension r in the domain $D = \bigcap_{\alpha=1}^{n+1} D_\alpha$. We call this web a *Grassmann* $(n+1)$-*web* and denote it by $GW(n+1, n, r)$.

Next, denote by L a moving $(n-1)$-plane of a Grassmann $(n+1)$-web and by A_α the points of intersection of L and the submanifolds X_α. Since the submanifolds X_α are in general position, n of those points, for example, $A_1, \ldots, A_n$, can be taken as the vertices of a projective frame of the $(n-1)$-plane L. We also take the vertex A_{n+1} as the unit point of this frame. Thus, $A_{n+1} = A_1 + A_2 + \ldots + A_n$. Let us take the points $A_i, i = n+2, \ldots, n+r+1$, that supplement the points $A_u, u = 1, \ldots, n$, to a complete frame of the space P^{n+r-1}. As usual, the equations of infinitesimal displacement of this moving frame can be written in the form:

$$dA_\sigma = \omega_\sigma^\rho A_\rho, \quad \sigma, \rho = 1, \ldots, n, n+2, \ldots, n+r+1, \tag{8.17}$$

and the structure equations, which the forms ω_σ^ρ satisfy, in the form:

$$d\omega_\sigma^\rho = \omega_\sigma^\tau \wedge \omega_\tau^\rho, \quad \sigma, \rho, \tau = 1, \ldots, n, n+2, \ldots, n+r+1. \tag{8.18}$$

Since the $(n-1)$-plane L is not tangent to any of the submanifolds X_u, the 1-forms ω_u^i can be taken as co-basis forms on these submanifolds. This implies that

$$\omega_u^v = \lambda_{ui}^v \omega_u^i, \quad v \neq u, \tag{8.19}$$

and

$$dA_u = \omega_u^u A_u + \omega_u^i \Big(A_i + \sum_{v \neq u} \lambda_{ui}^v A_v\Big). \tag{8.20}$$

Here and in what follows, the summation is carried over the indices i, j, k according to the usual rule while the summation is carried over the indices u, v, w only if there is the summation sign.

Let us locate the points A_i in the space $T_{A_{n+1}}$ tangent to the manifold X_{n+1} generated by the point A_{n+1}. Then, we have:

$$dA_{n+1} = \omega A_{n+1} + A_i \sum_u \omega_u^i, \tag{8.21}$$

where

$$\omega = \sum_v \omega_v^u. \tag{8.22}$$

Define:

$$\sum_u \omega_u^i = -\omega_{n+1}^i. \tag{8.23}$$

Since the point A_{n+1} generates an r-dimensional manifold X_{n+1}, the forms ω_{n+1}^i are linearly independent. In the frame which we have constructed, the equations:

$$\omega_1^i = 0, \ldots, \omega_n^i = 0, \qquad \sum_u \omega_u^i = 0$$

determine $n+1$ foliations in the Grassmannian $G(n-1, r+n-1)$, and these foliations form a Grassmann $(n+1)$-web $GW(n+1, n, r)$. The forms ω_u^i are the co-basis forms of this web, and equations (8.19) and (8.23) are its fundamental equations.

2. Let us find the torsion and curvature tensors of a Grassmann web $GW(n+1, n, r)$. For this, we first prolong equations (8.17), i.e. we take the exterior derivatives of these equations and apply the Cartan lemma to the exterior quadratic equations obtained as the result of exterior differentiation. As a result, we obtain:

$$\nabla\lambda_{ui}^v + \lambda_{ui}^v \lambda_{uj}^v \omega_{n+1}^j + \sum_{w \neq u,v} (\lambda_{ui}^v - \lambda_{ui}^w)(\lambda_{uj}^v - \lambda_{wj}^v)\omega_w^j + \omega_i^v = \lambda_{uij}^v \omega_u^j, \tag{8.24}$$

where u, v and w are distinct and $\lambda_{uij}^v = \lambda_{uji}^v$. In equations (8.24) we used the notations:

$$\begin{aligned} \nabla\lambda_{ui}^v &= d\lambda_{ui}^v - \lambda_{uj}^v \theta_i^j; \\ \theta_i^j &= \omega_j^j - \delta_i^j \omega, \end{aligned} \tag{8.25}$$

and the forms ω_{n+1}^j are determined by formula (8.23).

Exterior differentiation of relation (8.22) leads to the equation:

$$(\omega_i^v - \omega_i^u) \wedge \omega_{n+1}^i = 0.$$

The solution of the latter equation can be represented in the form:

$$\omega_i^u = \omega_i^{n+1} + p_{ij}^u \omega_{n+1}^j, \tag{8.26}$$

where

$$p^u_{ij} = p^u_{ji}, \qquad \sum_u p^u_{ij} = 0,$$

and

$$\omega^{n+1}_i = \frac{1}{n}\sum_u \omega^u_i \tag{8.27}$$

are secondary Pfaffian forms, i.e. they are not linear combinations of the co-basis forms ω^i_u.

Let us fix the $(n-1)$-plane L of the web under consideration, i.e. we set $\omega^i_u = 0$. Then equations (8.24) take the form:

$$\nabla_\delta \lambda^v_{ui} + \pi^v_i = 0, \tag{8.28}$$

where as usual δ denotes the differentiation symbol with respect to the secondary parameters and $\pi^v_i = \omega^v_i(\delta)$. Define the quantities:

$$\lambda_i = \frac{1}{n(n-1)}\sum_{u\neq v} \lambda^v_{ui}. \tag{8.29}$$

It follows from equations (8.28) that when the principal parameters are fixed, the quantities λ_i satisfy the equations:

$$\nabla_\delta \lambda_i + \pi^{n+1}_i = 0.$$

On the other hand, equations (8.17) and (8.27) imply that

$$\delta A_i = \pi^j_i A_j + \pi^{n+1}_i A_{n+1}.$$

The last two relations lead to the equations:

$$\delta(A_i + \lambda_i A_{n+1}) = \pi^j_i (A_j + \lambda_j A_{n+1}),$$

which show that the plane π, spanned by the points $A_i + \lambda_i A_{n+1}$, is invariant.

Let us locate the points A_i in π. Then the quantities λ_i vanish and (8.29) gives:

$$\sum_{u\neq v} \lambda^v_{ui} = 0. \tag{8.30}$$

The forms π^{n+1}_i become zero, and the forms ω^{n+1}_i become principal forms:

$$\omega^{n+1}_i = \sum_v q^v_{ij}\omega^j_v. \tag{8.31}$$

Equations (8.31), (8.26) and (8.23) imply that

$$\omega^u_i = \sum_v \left(q^v_{ij} - p^u_{ij}\right)\omega^j_v. \tag{8.32}$$

We can now find the torsion and curvature tensors of a web $GW(n+1,n,r)$. Applying exterior differentiation to its co-basis forms ω^i_u, we get:

$$d\omega^i_u = \omega^j_u \wedge \omega^i_j + \omega^v_u \wedge \omega^i_v.$$

Next, from equation (8.22) we obtain:

$$\omega^u_u = \omega - \sum_{v \neq u} \omega^u_v.$$

Substituting these expressions in the previous formulas and applying relations (8.18), we obtain:

$$d\omega^i_u = \omega^j_u \wedge \theta^i_j + \sum_{v \neq u} \Big(\delta^i_k \lambda^v_{uj} + \delta^i_j \lambda^u_{vk}\Big) \omega^j_u \wedge \omega^k_v, \tag{8.33}$$

where the 1-form θ^i_j is determined by formula (8.25). By (8.30), the expressions $\delta^i_k \lambda^v_{uj} + \delta^i_j \lambda^u_{vk}$ satisfy equations (8.5) of Section 8.1. Therefore, equations (8.33) are the structure equations of a Grassmann web $GW(n+1,n,r)$, and the torsion tensor of this web has the form:

$$\underset{uv}{a}{}^i_{jk} = \delta^i_k \lambda^v_{uj} + \delta^i_j \lambda^u_{vk}. \tag{8.34}$$

Since, if $n > 2$, the torsion tensor completely defines the geometry of a web $W(n+1,n,r)$ (see p. 273), we arrive at the following theorem.

Theorem 8.5 *A web $W(n+1,n,r), n > 2$, is Grassmannizable, i.e. equivalent to a Grassmann web $GW(n+1,n,r)$ if and only if its torsion tensor has the form* (8.34). ■

The forms θ^i_j determined by equations (8.25) define an affine connection $\tilde{\Gamma}$ on a web $GW(n+1,n,r)$. It follows from the previous equations that

$$d\omega^i_j - \omega^k_j \wedge \omega^i_k = \sum_u \omega^u_j \wedge \omega^i_u, \quad d\omega = -\omega^k_{n+1} \wedge \omega^{n+1}_k.$$

By virtue of formulas (8.25), (8.31) and (8.32), we find from this equation that

$$d\theta^i_j - \theta^k_j \wedge \theta^i_k = \sum_{u,v} (\delta^i_l q^u_{jk} + \delta^i_j q^u_{lk} + \delta^i_k p^u_{jl}) \omega^k_u \wedge \omega^l_v.$$

This gives the following expression for the curvature tensor of the Grassmann $(n+1)$-web under consideration:

$$\underset{uv}{b}{}^i_{jkl} = \frac{1}{2}(\delta^i_l q^u_{jk} + \delta^i_j q^u_{lk} + \delta^i_k p^u_{jl}) - \frac{1}{2}(\delta^i_k q^v_{jl} + \delta^i_j q^v_{kl} + \delta^i_l p^v_{jk}). \tag{8.35}$$

3. Consider now hexagonal Grassmann webs $GW(n+1,n,r)$. By Definition 8.2, all three-subwebs of such webs are hexagonal. These subwebs are cut on the $(2r)$-dimensional intersections of leaves $\mathcal{F}_{\alpha_1},\ldots,\mathcal{F}_{\alpha_{n-2}}$ of the foliations $\lambda_{\alpha_1},\ldots,\lambda_{\alpha_{n-2}}$ by the remaining three foliations $\lambda_{\alpha_{n-1}},\lambda_{\alpha_n}$ and $\lambda_{\alpha_{n+1}}$. For simplicity, we assume that $\alpha_\beta=\beta$.

For a Grassmann $(n+1)$-web generated by manifolds $X_1,\ldots,X_{n+1}$ of a projective space P^{n+r-1}, the leaves $\mathcal{F}_\kappa, \kappa=1,\ldots,n-2$, are the bundles S_κ of $(n-1)$-planes with vertices at points $x_\kappa \in X_\kappa$. The bundles S_κ have a common $(2r)$-bundle S of $(n-1)$-planes whose vertex is the $(n-3)$-plane $[x_1,\ldots,x_{n-2}]$. The bundle S projects the space P^{n+r-1} on the subspace P^{r+1} which is complimentary for $[x_1,\ldots,x_{n-2}]$. Under this projection, the $(n-1)$-plane L, belonging to the bundle S, is projected onto the straight line $l=\pi(L)$ in P^{r+1}, and the submanifolds $X_\alpha \subset P^{n+r-1}, \alpha=n-1,n,n+1$, are projected onto the r-dimensional manifolds $\widetilde{X}_\alpha=\pi(X_\alpha)$. Thus, the three-web, generated by the bundles S_κ, is isomorphic to a Grassmann three-web $GW(3,2,r)$ generated by three submanifolds $\widetilde{X}_\alpha$ in the space P^{r+1}.

If a web $GW(n+1,n,r)$ is hexagonal, then all its three-subwebs are also hexagonal. Therefore, all Grassmann three-webs $GW(3,2,r)$, which are obtained by projection from the subspace $[x_1,\ldots,x_{n-2}]$, are hexagonal as well. However, as was proved in Section 3.3, the submanifolds $\widetilde{X}_\alpha$ generating such three-webs, belong to one hypersurface of third order. Therefore, for any points $x_{\alpha_1},\ldots,x_{\alpha_{n-2}}$, belonging to the submanifolds $X_{\alpha_1},\ldots,X_{\alpha_{n-2}}$, the projections of the submanifolds $X_{\alpha_{n-1}},X_{\alpha_n}$ and $X_{\alpha_{n+1}}$ from the plane $[x_{\alpha_1},\ldots,x_{\alpha_{n-2}}]$ belong to one cubic hypersurface. This can be the case if and only if all submanifolds $X_1,\ldots,X_{n+1}$ of the space P^{n+r-1}, which generate a Grassmann web $GW(n+1,n,r)$, belong to one algebraic manifold of order $n+1$.

Since the converse is obvious, we have proved the theorem.

Theorem 8.6 *A Grassmann web $GW(n+1,n,r)$ is hexagonal if and only if all r-dimensional submanifolds $X_\alpha \subset P^{n+r-1}, \alpha=1,\ldots,n+1$, generating this web, belong to one r-dimensional algebraic manifold of order $n+1$.* ■

We write the analytic conditions of hexagonality of a Grassmann $(n+1)$-web by means of the curvature tensors $\underset{u,v}{\bar{b}}{}^i_{jkl}$ of its three-subwebs $[n+1,u,v]$, $u,v=1,\ldots,n$. Formula (8.12), giving the expressions for these tensors, contains the quantities $\underset{uvu}{a}{}^i_{jkl}$ which arise after differentiation of the torsion tensor. Differentiating (8.34), we obtain

$$\underset{uvu}{a}{}^i_{jkl} = \delta^i_k(\lambda^v_{ujl}+p^v_{jl}-q^u_{jl}) - \delta^i_j(q^u_{kl}-p^u_{kl}+\lambda^v_{ul}\lambda^u_{vk}) - \delta^i_l\lambda^u_{vk}\lambda^v_{uj}.$$

Substituting these values and expressions (8.34) and (8.35) for the torsion and curvature tensors of a web $GW(n+1,n,r)$ into (8.12), we get the expression for the curvature tensor of a Grassmann web:

$$\underset{uv}{\bar{b}}{}^i_{jkl} = \delta^i_j(p^u_{kl}-p^v_{kl}) + \delta^i_l\lambda^v_{ujk} - \delta^i_k\lambda^u_{vjl}. \tag{8.36}$$

As was proved in Theorem 8.4, the condition of hexagonality of a web $W(n+1,n,r)$ can be written in form (8.16). By (8.35), for a Grassmann web $GW(n+1,n,r)$, this condition takes the form:

$$\delta^i_{(j}(p^u_{kl)} - p^v_{kl)} + \lambda^v_{|u|kl)} - \lambda^u_{|v|kl)}) = 0.$$

This implies the equations:

$$p^u_{kl} - p^v_{kl} + \lambda^v_{ukl} - \lambda^u_{vkl} = 0. \tag{8.37}$$

Let us find a geometric meaning for relations (8.37). For this, we find the 2nd fundamental forms of submanifolds X_α at the points x_α of their intersection with the moving $(n-1)$-plane L. Using expressions (8.20) and (8.21), for the first differentials of the points A_u and A_{n+1}, we find that

$$\begin{aligned} d^2A_u &\equiv \textstyle\sum_{v\neq u} \lambda^v_{uij}\omega^i_u\omega^j_u A_v && \bmod T_{A_u}(X_u), \\ d^2A_{n+1} &\equiv \textstyle\sum_{v\neq u}(p^u_{ij} - p^v_{ij})\omega^i_{n+1}\omega^j_{n+1}A_v && \bmod T_{A_{n+1}}(X_{n+1}). \end{aligned}$$

These relations give the following structure of the 2nd fundamental forms of the submanifolds X_u and X_{n+1}:

$$\begin{aligned} h^v_u &= \lambda^v_{uij}\omega^i_u\omega^j_u \quad (v \neq u), \\ h^v_{n+1} &= (p^u_{ij} - p^v_{ij})\omega^i_{n+1}\omega^j_{n+1} \quad (v \neq u). \end{aligned}$$

Thus, the tensors in equations (8.37) are the 2nd fundamental tensors of the submanifolds X_u, X_v and X_{n+1} belonging to the system of submanifolds X_α generating a Grassmann web $GW(n+1,n,r)$. We have proved the following theorem.

Theorem 8.7 *A Grassmann web $GW(n+1,n,r)$ generated by r-dimensional submanifolds $X_\alpha \subset P^{n+r-1}, \alpha = 1,\ldots,n+1$, is hexagonal if and only if the 2nd fundamental tensors of the submanifolds X_α satisfy conditions* (8.37) *for any u and v.* ■

Note that under the condition of Theorem 8.6, the submanifolds X_α belong to a one r-dimensional algebraic manifold of order $n+1$. This implies that conditions (8.37) are the conditions of algebraizability of the submanifolds X_α as well as of the web $GW(n+1,n,r)$ generated by these submanifolds. This allows us to combine Theorems 7.6 and 8.7 in the following theorem.

Theorem 8.8 *The following three conditions are equivalent:*

1) *A Grassmann web $GW(n+1,n,r)$ is hexagonal.*
2 *A system of submanifolds $X_\alpha \subset P^{n+r-1}, \alpha = 1,\ldots,n+1$, generating this web is algebraizable.*
3) *The 2nd fundamental tensors of the submanifolds X_α satisfy conditions* (8.37). ■

Note also that if $n=2$ conditions (8.37) become the condition (3.52) of hexagonality and algebraizability of a Grassmann three-web $GW(3,2,r)$.

8.3 $(n+1)$-Webs and Almost Grassmann Structures

1. Consider the Plucker mapping of the Grassmannian $G(n-1, n+r-1)$ of $(n-1)$-planes of a projective space P^{n+r-1} onto an algebraic manifold $\Omega(n-1, n+r-1)$ of dimension nr of a projective space P^N where $N = \binom{n+r}{n} - 1$. This mapping can be constructed by means of the Grassmann coordinates of an $(n-1)$-plane L in P^{n+r-1} which are the determinants of order n of the matrix:

$$\begin{pmatrix} x_1^1 & \ldots & x_1^n & x_1^{n+1} & \ldots & x_1^{n+r} \\ \multicolumn{6}{c}{\dotfill} \\ x_n^1 & \ldots & x_n^n & x_n^{n+1} & \ldots & x_n^{n+r} \end{pmatrix}$$

composed of the coordinates of the basis points $x_1, \ldots, x_n$ of the $(n-1)$-plane L (cf. Section 3.3). The Grassmann coordinates are connected by a set of quadratic relations that define the manifold $\Omega(n-1, n+r-1)$ in the space P^N (see [HP 47]). We will say that this manifold carries the *Grassmann structure*. We will denote this manifold shortly by Ω. We will study in more detail the structure of the manifold Ω.

Let L_1 and L_2 be two $(n-1)$-planes in P^{n+r-1} meeting in the $(n-2)$-plane K. They generate a linear pencil S of $(n-1)$-planes $\lambda L_1 + \mu L_2$. A rectilinear generator of the manifold Ω corresponds to this pencil. All the $(n-1)$-planes of the pencil S belong to an n-plane M. This pencil, and consequently the corresponding straight line in Ω, is completely determined by a pair of planes K and M, $K \subset M$.

We consider a bundle of $(n-1)$-planes, i.e. a set of all $(n-1)$-planes passing through a fixed $(n-2)$-plane K. On the manifold Ω, to this bundle there corresponds an r-dimensional plane generator ξ^r. On the manifold Ω, to a family of $(n-1)$-planes belonging to a fixed n-plane M, there corresponds an n-dimensional plane generator η^n. Thus, the manifold Ω carries two families of plane generators of dimensions r and n, respectively.

If the planes K and M are incident, i.e. $K \subset M$, the plane generators ξ^r and η^n defined by these planes, meet along a straight line. If they are not incident, then the generators ξ^r and η^n have no common points.

Let us consider a fixed $(n-1)$-plane L in P^{r+n-1}. It contains an $(n-1)$-parameter family of $(n-1)$-planes K. Therefore, the $(n-1)$-parameter family of generators ξ^r passes through the point $p \in \Omega$ corresponding to L. On the other hand, an $(r-1)$-parameter family of n-planes M passes through the same plane M. Consequently, an $(r-1)$-parameter family of generators η^n passes through the point p. Furthermore, any two generators ξ^r and η^n passing through p meet along a straight line. It follows from this that all the plane generators ξ^r and η^n passing through the point p, form a cone whose projectivization is the Segre manifold $S(r-1, n-1)$ in the projective space P^{nr-1}. This Segre manifold carries two familes of plane generators of dimensions $r-1$ and $n-1$ and can be considered as the projective embedding of the Cartesian product of two projective spaces P^{r-1} and P^{n-1} into the space P^{nr-1}. The above described cone, whose projectivization is the Segre manifold $S(r-1, n-1)$, is called the *Segre cone* and is denoted by $C_p(r,n)$. This cone is the intersection of the manifold Ω and its tangent space $T_p(\Omega)$, whose dimension is the same as that of Ω, namely nr. In the space P^{r+n-1}, the set of all $(n-1)$-dimensional planes intersecting a fixed $(n-1)$-plane L in $(n-2)$-planes corresponds to the cone $C_p(r,n)$.

Therefore, with each point p of the algebraic manifold $\Omega \subset P^N$, there is connected the Segre cone $C_p(r,n)$ with vertex p located in the tangent space $T_p(\Omega)$, and the generators of this cone are generators of the manifold Ω. The Segre cone $C_p(r,n)$ in the space $T_p(\Omega)$ can be defined by the parametric equations:

$$z_u^i = \eta_u \xi^i, \quad u = 1, \ldots, n, \quad i = n+1, \ldots, n+r. \tag{8.38}$$

The same equations define the Segre manifold $S(r-1, n-1)$ in the space $P^{nr-1} = PT_p(\Omega)$.

2. Now we can define the notion of an almost Grassmann structure, a particular case of which is the Grassmann structure on the algebraic manifold Ω considered above.

Let X be a differentiable manifold of dimension nr, let p be an arbitrary point of X and let $T_p(X)$ be the tangent space to X at the point p. In each space $T_p(X)$, we consider the Segre cone $C_p(r,n)$ with vertex p. We will assume that the field of the Segre cones on X is differentiable.

The differential geometric structure on X^{nr} defined by the field of Segre cones is called an *almost Grassmann structure* and is denoted by $AG(n-1, n+r-1)$. Its structural group is a subgroup of the general linear group $\mathbf{GL}(nr)$ of transformations of the space $T_p(X)$, and the cone $C_p(r,n)$ is invariant under transformations of this group. We denote this group by $GL(r,n)$. One can see from equations (8.38) defining the Segre cone that this group is isomorphic to the product $\mathbf{GL}(r) \times \mathbf{SL}(n)$.

The Segre cone $C_p(r,n)$ attached to a point $p \in X$ determines in $T_p(X)$ an $(n-1)$-parameter family of r-planes ξ^r and an $(r-1)$-parameter family of n-planes η^n.

An almost Grassmann structure $AG(n-1, r+n-1)$ is called *r-semi-integrable* if on X there is an $(r+1)(n-1)$-parameter family of subvarieties V^r such that $T_p(V^r) = \xi_p^r$ for any point $p \in V^r$, and each r-plane ξ^r is tangent to one and only one subvariety V^r. Similarly, a structure $AG(n-1, r+n-1)$ is called *n-semi-integrable*, if on X there is an $(r-1)(n+1)$-parameter family of subvarieties V^n such that $T_p(V^n) = \eta_p^n$ for any point $p \in V^n$, and each n-plane η_p^n is tangent to one and only one subvariety V^n.

An almost Grassmann structure which is both r- and n-semi- integrable is called *integrable*. The following theorem holds: *an integrable almost Grassmann structure is locally Grassmann*, i.e., a neighborhood of every point p of the manifold X admits a differentiable mapping into the algebraic manifold $\Omega(n-1, r+n-1)$ of the projective space P^N, and the plane generators ξ^r and η^n of the manifold $\Omega(n-1, n+r-1)$ correspond to subvarieties V^r and V^n of the manifold X, respectively.

3. We now consider an $(n+1)$-web $W(n+1,n,r)$ on a smooth manifold X of dimension nr. Let $T_p(X)$ be the tangent space to X at the point p. The co-basis forms $\underset{u}{\omega}^i$, $u=1,\ldots,n$; $i=1,\ldots,r$, of the $(n+1)$-web introduced in Section 8.1 can be taken as coordinates in the space $T_p(X)$. Then, the equations of the subspaces T_α of this space which are tangent to the leaves of the web passing through the point p can be written in the form (8.1). By virtue of (8.2), we can see that the relations

$$\underset{n+1}{\omega}{}^i = -\sum_{u=1}^{n} \underset{u}{\omega}^i$$

hold. Equations (8.1) and (8.2) are invariant under transformations of the group $\mathbf{GL}(r)$ of the web $W(n+1,n,r)$.

Let $(v,w,u_1,\ldots,u_{n-1})$ be a permutation of the indices $(1,\ldots,n+1)$. In $T_p(X)$, we consider the intersection of the subspaces $T_{u_k}, k=1,\ldots,n-1$. Denote this intersection by T_{vw}. Its dimension is r, and it is defined by the equations $\underset{u_k}{\omega}^i=0$. The number of such subspaces is $\binom{n+1}{2}=\frac{n(n+1)}{2}$. If $n=2$, this number is equal to 3, and the subspaces T_{uv} coincide with the subspaces T_α tangent to the leaves of the web passing through the point p.

In the space $T_p(X)$, there exists a unique Segre cone $C_p(r,n)$ containing all subspaces T_{uv}. This cone can be defined by parametric equations (8.38) where $z_u^i=\underset{u}{\omega}^i$. By (8.2), it follows from these equations that

$$\underset{n+1}{\omega}{}^i = -\xi^i \eta_{n+1},$$

where $\eta_{n+1} = -\sum_{u=1}^{n} \eta_u$. The subspaces T_{uv} belonging to the Segre cone can be given on this cone by the equations $\eta_{u_k}=0$, where the indices u_k take the values indicated above. These subspaces belong to the family of the r-dimensional plane generators ξ^r of the Segre cone $C_p(r,n)$.

Since the family of Segre cones $C_p(r,n)$ given in the tangent spaces $T_p(X)$, defines an almost Grassmann structure $AG(n-1,r+n-1)$ in the manifold X, the following theorem holds.

Theorem 8.9 *An almost Grassmann structure $AG(n-1,r+n-1)$ is invariantly connected with an $(n+1)$-web $W(n+1,n,r)$ given on a smooth manifold X of dimension nr. The structural group of this web is a normal subgroup of the structural group of the almost Grassmann structure.* ■

Note that the last statement of Theorem 8.9 follows from the fact that the structural group of an $(n+1)$-web is the group $\mathbf{GL}(r)$ and the structural group of the almost Grassmann structure is the group $\mathbf{GL}(r)\times\mathbf{SL}(n)$.

The r-dimensional plane generators ξ^r of the Segre cones $C_p(r,n)$ associated with a web $W(n+1,n,r)$ are called its *isoclinic subspaces.* In addition to them, the Segre cones $C_p(r,n)$ carry the n-dimensional plane generators η^n. They are called the *transversal subspaces* of this web.

8.4 Transversally Geodesic and Isoclinic $(n+1)$-Webs

1. A web $W(n+1,n,r)$ is called *transversally geodesic* if the almost Grassmann structure $AG(n-1,r+n-1)$ associated with this web is n-semi-integrable. A web $W(n+1,n,r)$ is called *isoclinic* if the structure $AG(n-1,r+n-1)$ is r-semi-integrable.

We consider first transversally geodesic $(n+1)$-webs W and find analytic conditions characterizing them. The semi-integrability of the almost Grassmann structure associated with a web means the existence of a family of subvarieties V^n on the manifold X which are tangent to the n-planes η^n. The equations of these subvarieties have the form:

$$\underset{u}{\omega}^i = \xi^i\theta_u, \quad u=1,\ldots,n; \quad i=1,\ldots,r, \tag{8.39}$$

where θ_u are 1-forms independent on V^n and ξ^i are the coordinates of a vector determining the location of the transversal subspace of the web. By means of formulas (8.4), exterior differentiation of equations (8.39) leads to the following exterior quadratic equations:

$$(\nabla\xi^i + \sum_{v\neq u} a^i_{uv}\theta_v)\wedge\theta_u = -\xi^i d\theta_u, \tag{8.40}$$

where we used the notations

$$\nabla\xi^i = d\xi^i + \xi^j\omega^i_j, \quad \underset{uv}{a}^i = \underset{uv}{a}{}^i_{jk}\xi^j\xi^k, \tag{8.41}$$

and the quantities a^i_{uv} satisfy the equations:

$$\underset{uv}{a}^i = \underset{vu}{a}^i, \quad \sum_{u,v}\underset{uv}{a}^i = 0. \tag{8.42}$$

If we add up equations (8.40) written for all $u=1,\ldots,n$, and use conditions (8.42), we find that

$$\nabla\xi^i\wedge(\sum_u\theta_u) = -\xi^i d(\sum_u\theta_u). \tag{8.43}$$

Equations (8.43) show that

$$d(\sum_u\theta_u) = (\sum_u\theta_u)\wedge\theta,$$

where θ is a 1-form. Substituting the last expression into equations (8.43), we obtain the equation

$$(\nabla\xi^i - \xi^i\theta)\wedge(\sum_u\theta_u) = 0.$$

By Cartan's lemma, we find that

$$\nabla \xi^i = \xi^i \theta + \sigma^i \sum_u \theta_u. \tag{8.44}$$

On a submanifold V^n, the foliations of a web $W(n+1,n,r)$ cut out an $(n+1)$-web $W(n+1,n,1)$ of codimension 1. The forms θ_u are the basis forms of this web, and the leaves of its $(n-1)$-dimensional foliations are determined by the following systems of Pfaffian equations:

$$\theta_u = 0, \ \sum_u \theta_u = 0.$$

By (8.4), the structure equations of a web $W(n+1,n,1)$ have the form:

$$d\theta_u = \theta_u \wedge \omega + \sum_{v \neq u} \underset{uv}{a} \theta_u \wedge \theta_v, \tag{8.45}$$

where

$$\sum_{u,v} \underset{uv}{a} = 0.$$

Substituting expressions (8.44) and (8.45) into equations (8.40), we find that $\omega = \theta$ and

$$\sigma^i + \underset{uv}{a}{}^i = \xi^i \underset{uv}{a}.$$

Summing up all these equations in u and v, we obtain

$$\sigma^i = 0, \ \underset{uv}{a}{}^i = \xi^i \underset{uv}{a}. \tag{8.46}$$

By (8.46), equations (8.44) and (8.41) take the form:

$$\nabla \xi^i = \xi^i \theta, \tag{8.47}$$

and

$$\underset{uv}{a}{}^i_{jk} \xi^j \xi^k = \underset{uv}{a} \xi^i. \tag{8.48}$$

The following theorem gives a geometric meaning of relation (8.47).

Theorem 8.10 *The subvarieties V^n defined on X by equations* (8.39), *are totally geodesic in the connection $\widetilde{\Gamma}$ induced by a web $W(n+1,n,r)$.*

Proof. In fact, denote by $\{\underset{u}{e}_i\}$ the frame which is dual to the co-frame $\{\underset{u}{\omega}{}^i\}$ consisting of the co-basis forms of the web $W(n+1,n,r)$, and consider the vectors $\xi_u = \xi^i \underset{u}{e}_i$. By (8.39), these vectors are tangent to V^n. By (8.47), they as well as all vectors of the form $c^u \xi_u$ where c^u are constants, can be displaced in a parallel way along V^n. Therefore, V^n is a totally geodesic submanifold. ■

It follows from this that the submanifolds V^n cut out the leaves of the $(n+1)$-web (which are themselves are totally geodesic submanifolds on X) along geodesic lines. This is the reason that the submanifolds V^n are called *transversal* submanifolds of a web $W(n+1,n,r)$, and the web itself is called *transversally geodesic.*

Next, we will study equations (8.48). First of all, note that if $n=2$, the left-hand side of (8.48) is identically zero since the torsion tensor a^i_{jk} of a web $W(3,2,r)$ is skew-symmetric in the indices j and k. We can see from relations (8.41) and (8.46) that in this case $\underset{12}{a}{}^i = 0$ and equation (8.48) becomes an identity. Therefore, if $n=2$, the property of a three-web to be transversally geodesic can be expressed in terms of the curvature tensor (cf. Section 3.1).

Suppose further that $n>2$. Since equation (8.48) must be identically satisfied with respect to ξ^i, the expression $\underset{uv}{a}$ on its right-hand side is linear in ξ^i, i.e. the relations

$$\underset{uv}{a} = \underset{uv}{a}{}_k \xi^k. \tag{8.49}$$

hold. Substituting (8.49) into (8.47), we obtain the equation

$$(\underset{uv}{a}{}^i_{jk} - \delta^i_j \underset{uv}{a}{}_k)\xi^j\xi^k = 0,$$

from which it follows that

$$\underset{uv}{a}{}^i_{(jk)} = \delta^i_{(j}\underset{uv}{a}{}_{k)}, \tag{8.50}$$

Theorem 8.11 *For a web $W(n+1,n,r)$, $n>2$, to be transversally geodesic, it is necessary and sufficient that the symmetric part of its torsion tensor has the form* (8.50).

Proof. The necessity of the condition of Theorem 8.11 was proved above. Its sufficiency follows from the fact that by (8.50), the system of Pfaffian equations (8.39) and (8.47) defining the transversally geodesic submanifolds of a web $W(n+1,n,r)$ is completely integrable. ■

2. A web $W(n+1,n,r)$ is called *polyhedral* if it is transversally geodesic and all its subwebs $W(n+1,n,1)$ cut out on its transversally geodesic submanifolds, are parallelizable.

It follows from equation (8.45) that the condition for the transversal subwebs to be parallelizable can be written in the form: $\underset{uv}{a} = 0$. By (8.49) and (8.50), it follows that the tensor $\underset{uv}{a}{}^i_{(jk)}$ vanishes on a polyhedral web:

$$\underset{uv}{a}{}^i_{(jk)} = 0. \tag{8.51}$$

Conversely, if on a web $W(n+1,n,r)$ condition (8.51) holds, then by (8.50), this web is transversally geodesic. Then equations (8.49) imply that $\underset{uv}{a} = 0$. By virtue of this, its transversal subwebs are parallelizable. We arrive at the theorem.

Theorem 8.12 *For a web $W(n+1,n,r)$, $n>2, r>2$, to be polyhedral, it is necessary and sufficient that its torsion tensor satisfies condition* (8.51). ■

The name "polyhedral" for the $(n+1)$-web under consideration is connected with the fact that on this web, the figures of the form of a $(2n+2)$-edron, whose faces are web leaves, are closed. In the case $n=3$, these figures are octahedrons. If $n=3$ and $r=1$, then $\dim X=3$. Such an octahedron is presented in Figure 8.1.

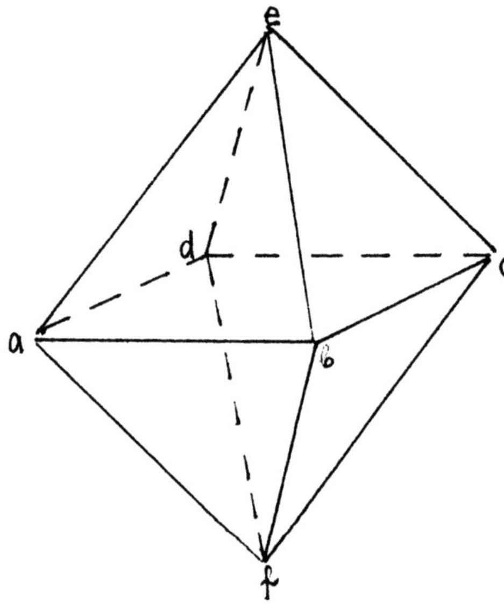

Figure 8.1

In this figure the faces *ade* and *bcf* belong to the first foliation of the web, the faces *dce* and *abf* belong to the second foliation, the faces *cbe* and *daf* belong to the third foliation, and the faces *bae* and *cdf* belong to the fourth foliation.

3. We now consider isoclinic webs $W(n+1,n,r)$. The almost Grassmann structure associated with this web must be r-semi-integrable. Therefore, on the manifold X, there exists a family of submanifolds V^r tangent to the isoclinic subspaces ξ^r. The equations of these submanifolds can be written in the form:

$$\underset{u}{\omega}^i = \eta_u \theta^i, \quad u=1,\ldots,n; \quad i=1,\ldots,r, \tag{8.52}$$

where the θ^i are 1-forms, which are linearly independent on V^r, and η_u are parameters determining the location of the isoclinic subspace of the web.

Theorem 8.13 *For a web $W(n+1,n,r)$, $n \geq 3, r > 2$, to be isoclinic, it is necessary and sufficient that the skew-symmetric part of its torsion tensor has the form:*

$$\underset{uv}{a}{}^i_{[jk]} = \underset{uv}{b}{}_{[j}\delta^i_{k]}. \tag{8.53}$$

■

The proof of this theorem is similar to that of Theorem 8.11, and we will not give it here.

The following important theorem follows from Theorems 8.11 and 8.13.

Theorem 8.14 *A web $W(n+1,n,r)$ is Grassmannizable if and only if it is both isoclinic and transversally geodesic.*

Proof. In fact, suppose that a web $W(n+1,n,r)$ is Grassmannizable. Then it is equivalent to a Grassmann web $G(n+1,n,r)$. As was proved in Section 8.2 (see formula (8.34)), the torsion tensor of a Grassmann web has the form:

$$\underset{uv}{a}{}^i_{jk} = \delta^i_k \lambda^v_{uj} + \delta^i_j \lambda^u_{vk},$$

This implies the relations:

$$\underset{uv}{a}{}^i_{[jk]} = (\lambda^v_{u[j} - \lambda^u_{v[j})\delta^i_{k]}, \quad \underset{uv}{a}{}^i_{(jk)} = (\lambda^v_{u(j} + \lambda^u_{v(j})\delta^i_{k)},$$

i.e. the torsion tensor of the web under consideration satisfies the conditions of Theorems 8.11 and 8.13. Thus, a Grassmannizable web is both isoclinic and transversally geodesic.

Conversely, suppose that a web $W(n+1,n,r)$ is both isoclinic and transversally geodesic. Then, relations (8.50) and (8.53) hold on this web. So we have

$$\underset{uv}{a}{}^i_{jk} = \underset{uv}{a}{}^i_{(jk)} + \underset{uv}{a}{}^i_{[jk]} = \underset{uv}{a}{}_{(j}\delta^i_{k)} + \underset{uv}{b}{}_{[j}\delta^i_{k]} = (\underset{uv}{a}{}_j + \underset{uv}{b}{}_j)\delta^i_k + (\underset{uv}{a}{}_k - \underset{uv}{b}{}_k)\delta^i_j.$$

Thus, the torsion tensor of this web has the form (8.34), and therefore the web is Grassmannizable. ■

Hence, relations (8.50) and (8.53) are analytic conditions of the Grassmannizability of an $(n+1)$-web $W(n+1,n,r)$.

8.5 d-Webs on a Manifold of Dimension nr

1. In the previous sections of this chapter, we studied the webs $W(n+1,n,r)$ formed by $n+1$ foliations of codimension r on a manifold X of dimension nr. In this section, we consider the webs $W(d,n,r)$, $d > n+1$ on a manifold X of the same dimension nr.

As in Section 8.1, we define the foliations λ_α forming the web $W(d,n,r)$ on the manifold X, by the following completely integrable systems of equations:

$$\underset{\alpha}{\omega}^i = 0, \quad i = 1, \ldots, r; \quad \alpha = 1, \ldots, d. \tag{8.54}$$

Since the foliations λ_α are in general position, each of subsystems $\underset{\alpha_1}{\omega}^i, \ldots, \underset{\alpha_n}{\omega}^i$ of system (8.54) is linearly independent. We take the forms $\underset{u}{\omega}^i$, $u = 1, \ldots, n$ as basis forms of the manifold X. Then other forms of system (8.54) are their linear combinations:

$$\underset{\sigma}{\omega}^i = \sum_u \underset{\sigma u}{\lambda}{}^i_j \underset{u}{\omega}^j, \quad \sigma = n+1, \ldots, d,$$

where all matrices $\underset{\sigma u}{\lambda}{}^i_j$ are non-singular. By the change of the co-bases in the foliations λ_α, we can reduce the latter equations to the form:

$$-\underset{n+1}{\omega}{}^i = \underset{1}{\omega}^i + \underset{2}{\omega}^i + \ldots + \underset{n}{\omega}^i, \tag{8.55}$$

$$-\underset{a}{\omega}^i = \underset{a1}{\lambda}{}^i_j \underset{1}{\omega}^j + \underset{a2}{\lambda}{}^i_j \underset{2}{\omega}^j + \ldots + \underset{an-1}{\lambda}{}^i_j \underset{n-1}{\omega}{}^j + \underset{n}{\omega}^i, \tag{8.56}$$

where $a = n+2, \ldots, d$. Now all the forms $\underset{\alpha}{\omega}^i$ admit only the concordant transformations of the form:

$$'\underset{\alpha}{\omega}^i = A^i_j \underset{\alpha}{\omega}^j.$$

These transformations form the structural group $\mathbf{GL}(r)$ of the web $W(d,n,r)$, and the matrices $(\underset{au}{\lambda}{}^i_j)$ become tensors with respect to these transformations. Denote these tensors by $\underset{au}{\Lambda}$ and note that $\underset{an}{\lambda}{}^i_j = \delta^i_j$.

In addition to be non-singular, the tensors $\underset{au}{\Lambda}$ satisfy additional conditions: their differences $\underset{au}{\Lambda} - \underset{au}{\Lambda}$ as well as some other their combinations must be non-singular. All these conditions follow from the fact that the foliations λ_α are in general position.

The structure equations of a web $W(d,n,r)$ consist of equations (8.4) (which are the conditions of integrability of the systems of equations $\underset{u}{\omega}^i = 0$, $u = 1, \ldots, n$ and $\underset{n+1}{\omega}{}^i = 0$) and the conditions of integrability of the systems of equations $\underset{a}{\omega}^i = 0$, $a = n+2, \ldots, d$, which contain the differentials of the tensors $\underset{au}{\Lambda}$. We will not write the general form of all these equations.

2. Consider a four-web $W(4,2,r)$ in more detail. For this web, the system of equations (8.53) and (8.54) takes the form:

$$-\underset{3}{\omega}^i = \underset{1}{\omega}^i + \underset{2}{\omega}^i, \quad -\underset{4}{\omega}^i = \lambda^i_j \underset{1}{\omega}^j + \underset{2}{\omega}^i,$$

where the tensors λ^i_j and $\delta^i_j - \lambda^i_j$ must be non-singular. The tensor λ^i_j is called the *basis affinor* of the web $W(4,2,r)$ and is denoted by Λ. Let us clarify its geometric meaning.

Let p be an arbitrary point of a manifold X of dimension $2r$ carrying a web $W(4,2,r)$, and let $T_p(X)$ be the tangent space to X at this point. Denote by T_α, $\alpha = 1,2,3,4$, the subspaces of $T_p(X)$ tangent to the leaves $\mathcal{F}_\alpha$ of the web W passing through the point p. Consider in the space $T_p(X)$ the frame $\{\underset{1}{e}_i, \underset{2}{e}_i\}$ which is dual to the co-frame $\{\underset{1}{\omega}^i, \underset{2}{\omega}^i\}$ and such that any vector $\xi \in T_p(X)$ can be written in the form:

$$\xi = \underset{2}{\omega}^i(\xi)\underset{1}{e}_i - \underset{1}{\omega}^i(\xi)\underset{2}{e}_i$$

(cf. Section 1.3). Then $\underset{1}{e}_i \in T_1$, $\underset{2}{e}_i \in T_2$, $\underset{1}{e}_i + \underset{2}{e}_i \in T_3$ and $\lambda^i_j \underset{1}{e}_j + \underset{2}{e}_i \in T_4$, and each of these systems of vectors form a basis in the corresponding subspace T_α.

The triples of subspaces $\{T_1, T_2, T_3\}$ and $\{T_1, T_2, T_4\}$ define in the space $T_p(X)$ two systems of transversal bivectors. Let $\xi_1 = \xi^i \underset{1}{e}_i$ be an arbitrary vector from T_1. The bivector $H = \xi_1 \wedge \xi_2$, where $\xi_2 = \xi^i \underset{2}{e}_i$, passes through ξ_1, and this bivector H is transversal to the first triple of subspaces. Similarly, the bivector $H' = \xi'_1 \wedge \xi_2$, where $\xi'_1 = \xi^i \lambda^j_i \underset{1}{e}_j$, passes through $\xi_2 = \xi^i \underset{2}{e}_i$, and this bivector H' is transversal to the second triple of subspaces. This defines the linear transformation $\Lambda : T_1 \to T_1$ which can be written in the form $\xi'_i = \lambda^i_j \xi^j$. The linear transformations defined by the affinor Λ in other spaces T_α have a similar meaning.

Consider the eigenvectors of the operator $\Lambda : T_1 \to T_1$. Since for them $\xi'_i = \lambda \xi_1$, the transversal bivectors H and H', defined by the eigenvector ξ_1, belong to the common transversal subspace of the quadruple of subspaces T_α. We arrive at the following theorem.

Theorem 8.15 *At each point $p \in X$, the basis affinor $\Lambda = (\lambda^i_j)$ of a four-web $W(4,2,r)$ defines the linear operators $\Lambda_\alpha : T_\alpha \to T_\alpha$ in the subspaces $T_\alpha \subset T_p(X)$ that are constructed by means of the transversal bivectors of these subspaces in the manner described above and can be naturally expressed in terms of the affinor Λ. To the eigenvectors of the operators Λ_α, there corresponds the common transversal subspace of the quadruple of subspaces T_α.* ■

3. We now return to the study of the general webs $W(d,n,r)$. Each subsystem of foliations $\lambda_{\alpha_1}, \ldots, \lambda_{\alpha_{n+1}}$ of this web form an $(n+1)$-subweb on the manifold X. We denote this subweb by $[\alpha_1, \ldots, \alpha_{n+1}]$. The total number of such subwebs is $\binom{d}{n+1} = \frac{d(d-1)\ldots(d-n)}{(n+1)!}$. In the tangent space $T_p(X)$, each of these subwebs determines (see Section 8.3) the Segre cone $C_p(r,n)$ and consequently, the almost Grassmann structure $AG(n-1, n+r-1)$ in the manifold X. Thus, a system of almost Grassmann structures arises in the manifold X. However, the most interesting case is indicated in the following definition.

Definition 8.16 A web $W(d,n,r)$ is said to be *almost Grassmannizable* if all almost Grassmann structures, defined by its $(n+1)$-subwebs, coincide.

Theorem 8.15 proved in subsection **2** of this section, implies that the web $W(4,2,r)$ is almost Grassmannizable if and only if its basis affinor is scalar: i.e. $\Lambda = \lambda E$. In fact, in this case, all transversal bivectors of the subweb $[1,2,3]$ are also transversal bivectors of the subweb $[1,2,4]$, and consequently, for all of its other three-subwebs, $[1,3,4]$ and $[2,3,4]$. But the transversal bivectors constitute one of the families of the plane generators of the Segre cones $C(2,r)$ associated with the web W. Therefore, if $\Lambda = \lambda E$, the Segre cones defined by different subwebs of the web $W(4,2,r)$, coincide. The converse is obvious.

In the general case we have the following theorem.

Theorem 8.17 *For a web $W(d,n,r)$ to be almost Grassmannizable, it is necessary and sufficient that all its basis affinors $\underset{au}{\Lambda}$ are scalar, i.e. proportional to the identity affinor $E = (\delta^i_j)$.*

Proof. In fact, the almost Grassmann structures, determined on the manifold X by two $(n+1)$-subwebs of a web $W(d,n,r)$, coincide if and only if, at each point $p \in X$, the Segre cones located in the tangent space $T_p(X)$ and determined by the tangent subspaces to the leaves of these subwebs, coincide. Consider the subwebs $[1,\ldots,n,n+1]$ and $[1,\ldots,n,a]$ on X. As shown in Section 8.3, the Segre cone, determined in $T_p(X)$ by the first subweb, can be given by the equations (8.38). In a similar way, we can show that the Segre cone, determined in $T_p(X)$ by the second subweb, can be given by the equations:

$$z^i_u = \eta_u(\underset{au}{\lambda}{}^i_j \xi^j). \tag{8.57}$$

If

$$\underset{au}{\lambda}{}^i_j = \underset{au}{\lambda}\delta^i_j, \tag{8.58}$$

then equations (8.57) take the form:

$$z^i_u = \underset{au}{\lambda}\eta_u \xi^i \tag{8.59}$$

and determine the same Segre cone as equations (8.38). Conversely, if equations (8.57) define the same Segre cone as equations (8.38), the tensors $\underset{au}{\lambda}{}^i_j$ have the form (8.58), i.e. proportional to the identity tensor. ■

It follows from Theorem 8.17 that for an almost Grassmannizable web $W(d,n,r)$, equations (8.56) take the form:

$$-\underset{a}{\omega}{}^i = \underset{a1}{\lambda}\underset{1}{\omega}{}^i + \ldots + \underset{an-1}{\lambda}\underset{n-1}{\omega}{}^i + \underset{n}{\omega}{}^i. \tag{8.60}$$

Since the foliations λ_α are in general position, in the matrix

$$\begin{pmatrix} 1 & \dots & 1 & 1 \\ \underset{n+1,1}{\lambda} & \dots & \underset{n+1,n-1}{\lambda} & 1 \\ \dots & \dots & \dots & \dots \\ \underset{d,1}{\lambda} & \dots & \underset{d,n-1}{\lambda} & 1 \end{pmatrix},$$

composed of the coefficients of equations (8.55) and (8.60), all the minors of any order are different from zero.

Denote an almost Grassmannizable web $W(d,n,r)$ by $AGW(d,n,r)$ and consider the almost Grassmann structure $AG(n-1,n+r-1)$ associated with this web. If this structure is r-semi-integrable, then as in Section 8.4, the web $AGW(d,n,r)$ is called *isoclinic*. If the almost Grassmann structure $AG(n-1,n+r-1)$ is n-semi-integrable, then the web $AGW(d,n,r)$ is called *transversally geodesic*.

A web $AGW(d,n,r)$ is called *Grassmannizable* if it is equivalent to a Grassmann web $GW(d,n,r)$ formed on the Grassmannian $G(n-1,n+r-1)$ by d foliations λ_α whose structure has been described in Section 8.2. It follows from Theorem 8.14 that a web $AGW(d,n,r)$ is Grassmannizable if and only if it is both isoclinic and transversally geodesic.

However, these conditions of Grassmannizability can be weakened since the following theorem holds.

Theorem 8.18 *If $d \geq n+2$ and $r \geq 3$, an almost Grassmannizable web $AGW(d,n,r)$ is isoclinic.*

Proof. We write the system of Pfaffian forms defining the foliation λ_{n+2} on the web $AGW(d,n,r)$ in the form:

$$-\underset{n+2}{\omega}{}^i = \underset{1}{\lambda}\underset{1}{\omega}{}^i + \dots + \underset{n}{\lambda}\underset{n}{\omega}{}^i. \tag{8.61}$$

In equation (8.61) we omitted the index $n+2$ in the coefficients $\underset{n+2,u}{\lambda}$ and assumed that $\underset{n}{\lambda}$ is not necessarily equal to one. By the Frobenius theorem, the condition of complete integrability of system (8.61) can be written in the form:

$$d\underset{n+2}{\omega}{}^i = \underset{n+2}{\omega}{}^j \wedge \theta^i_j. \tag{8.62}$$

By virtue of formulas (8.4), exterior differentiation of (8.61) leads to the quadratic equations:

$$-d\underset{n+2}{\omega}{}^i = -\underset{n+2}{\omega}{}^j \wedge \omega^i_j + \sum_u d\underset{u}{\lambda} \wedge \underset{u}{\omega}{}^i + \sum_{u,v} \underset{u}{\lambda}\underset{uv}{a}{}^i_{jk}\underset{u}{\omega}{}^j \wedge \underset{v}{\omega}{}^k. \tag{8.63}$$

In these equations, the coefficients $\underset{u}{\lambda}$ are relative invariants which implies

$$d\underset{u}{\lambda} = \underset{u}{\lambda}\theta + \sum_{v} \underset{uv}{\lambda}_j \underset{v}{\omega}^j.$$

Substituting these expansions into (8.63), we obtain

$$-d\underset{n+2}{\omega}^{i} = -\underset{n+2}{\omega}^{j} \wedge (\omega_j^i - \delta_j^i\theta) + \sum_{u,v}(\underset{uv}{\lambda}_j\delta_k^i + \underset{u}{\lambda}\underset{uv}{a}^i_{jk})\underset{u}{\omega}^j \wedge \underset{v}{\omega}^k.$$

From condition (8.62) it follows that the second term on the right-hand side of the latter equation must have the form $-\underset{n+2}{\omega}^{j} \wedge \sigma_j^i$ where $\sigma_j^i = \sum_u \underset{u}{\mu}^i_{jk}\underset{u}{\omega}^k$. Equating these two expressions and applying (8.62), we get

$$\sum_{u,v}(\underset{uv}{\lambda}_j\delta^i_k + \underset{u}{\lambda}\underset{uv}{a}^i_{jk})\underset{u}{\omega}^j \wedge \underset{v}{\omega}^k = \sum_{u,v}\underset{u}{\lambda}\underset{v}{\mu}^i_{jk}\underset{u}{\omega}^j \wedge \underset{v}{\omega}^k.$$

Comparing the alternated coefficients and applying relations (8.5), we arrive at the equations:

$$(\underset{u}{\lambda} - \underset{v}{\lambda})\underset{uv}{a}^i_{jk} = \underset{uv}{\lambda}_k\delta^i_j - \underset{vu}{\lambda}_j\delta^i_k + \underset{u}{\lambda}\underset{v}{\mu}^i_{jk} - \underset{v}{\lambda}\underset{u}{\mu}^i_{kj}. \tag{8.64}$$

Setting $u = v$ in (8.64), we find that

$$\underset{u}{\mu}^i_{[jk]} = \frac{1}{\underset{u}{\lambda}}\underset{uu}{\lambda}_{[j}\delta^i_{k]}.$$

By virtue of these equations, the alternation of relations (8.64) in the indices j and k gives

$$\underset{uv}{a}^i_{[jk]} = \underset{uv}{b}_{[j}\delta^i_{k]}, \tag{8.65}$$

where we used the notation

$$\underset{uv}{b}_k = \frac{1}{\underset{u}{\lambda} - \underset{v}{\lambda}}\Big(\frac{\underset{v}{\lambda}}{\underset{u}{\lambda}}\underset{uu}{\lambda}_k + \frac{\underset{u}{\lambda}}{\underset{v}{\lambda}}\underset{vv}{\lambda}_k - 2\underset{(uv)}{\lambda}_k\Big).$$

From relations (8.65) and Theorem 8.13 it follows that if $r \geq 3$, the $(n+1)$-subweb $[1, \ldots, n, n+1]$ of the web $AGW(d,n,r)$ is isoclinic. This immediately implies the isoclinicity of the web $AGW(d,n,r)$. ■

Theorems 8.14 and 8.18 give another result.

Theorem 8.19 *If $d \geq n+2$ and $r \geq 3$ and a web $W(d,n,r)$ is almost Grassmannizable and transversally geodesic, then it is Grassmannizable.* ■

4. We now consider a d-web of codimension one on a manifold X of dimension n, i.e. a web $W(d,n,1)$, provided that $d \geq n+1$. Through each point of this web, there pass d of its leaves belonging to the different foliations. Any $n+1$ out of d subspaces

T_α, $\alpha = 1, \ldots, d$, tangent to the leaves $\mathcal{F}_\alpha$ of the web W passing through the point $p \in X$, determine the Segre cone $C_p(1, n)$, which carries the $(n-1)$-parameter family of one-dimensional generators, and only one n-dimensional generator coinciding with the space $T_p(X)$. Thus, this cone is a bundle of straight lines with center p in $T_p(X)$. Because of this, all Segre cones determined by the $(n+1)$-subwebs of a web $W(d, n, 1)$ coincide, and this web is almost Grassmannizable.

Suppose further that a web $W(d, n, 1)$ is Grassmannizable. This means that this web admits a mapping into the Grassmannian $G(n-1, n)$. The latter is a set of hyperplanes of a projective space P^n, i.e. the conjugate projective space P^{n*}. In P^n to each leaf of the web $W(d, n, 1)$, there corresponds the bundle of hyperplanes S_x with center at a point $x \in P^n$, and to the foliation λ_α, $\alpha = 1, \ldots, d$, there corresponds one-parameter family of the bundles S_x whose centers form a curve X_α in P^n. Thus, in P^n to a Grassmannizable web $W(d, n, 1)$, there corresponds a web $GW(d, n, 1)$ generated by the system of curves X_α, $\alpha = 1, \ldots, d$. The elements of this web are hyperplanes, and its leaves are the bundles of hyperplanes with centers located on the curves X_α.

We consider a correlation $\kappa : P^n \rightarrow P^n$ which maps hyperplanes of the space P^n into points and bundles of hyperplanes into hyperplanes. The correlation κ maps a Grassmann web $GW(d, n, 1)$ into a web whose foliations λ_α are one-parameter families of hyperplanes in the space P^n. The latter web is called *hyperplanar* and is denoted by $LW(d, n, 1)$. A web $W(d, n, 1)$ is called *linearizable* if it is equivalent to a hyperplanar web $LW(d, n, 1)$. The following theorem holds.

Theorem 8.20 *A web $W(d, n, 1)$ is linearizable if and only if it is Grassmannizable.*

Proof. The sufficiency of the condition of Theorem 8.20 was proved in preceding considerations. Its necessity follows from the fact the correlation κ^{-1} maps a hyperplanar web $LW(d, n, 1)$ into a Grassmann web $GW(d, n, 1)$. ■

5. The following theorem is of interest since it is related to the linearizability problem.

Theorem 8.21 *Every Grassmannizable web $W(d, 2, r)$ is linearizable, i.e. admits a mapping onto a d-web $LW(d, 2, r)$ formed in a projective space P^{2r} by r-parametric families of r-planes.*

Proof. We perform three stages to construct a mapping $W(d, 2, r) \rightarrow LW(d, 2, r)$. First, since the web $W(d, 2, r)$ is Grassmannizable, it is equivalent to a Grassmann d-web $GW(d, 2, r)$ which is defined in the Grassmannian $G(1, r+1)$ of straight lines of a projective space P^{r+1} by the system of hypersurfaces X_α, $\alpha = 1, \ldots, d$.

Second, consider the Plucker mapping π of the Grassmannian $G(1, r+1)$ onto $(2r)$-dimensional point manifold $\Omega(1, r+1)$ in a projective space P^N where $N = \frac{1}{2}(r+1)(r+2) - 1$. Since the leaves of the Grassmann web $GW(d, 2, r)$ are bundles of straight lines, on the manifold $\Omega(1, r+1)$, r-dimensional plane generators of this manifold $\Omega(1, r+1)$ correspond to these leaves. Thus, on the manifold $\Omega(1, r+1)$ to the Grassmann web $GW(d, 2, r)$, there corresponds the web $\Omega W(d, 2, r)$ whose leaves are r-dimensional plane generators of the manifold $\Omega(1, r+1)$.

Finally, the third mapping can be constructed as follows. Let p be a point on the manifold which is the image of a straight line of the Grassmann web $GW(d, 2, r)$ under the Plucker mapping, and let $T_p(\Omega)$ be the tangent space to $\Omega(1, r+1)$ at the point p. Its dimension as well as the dimension of the manifold $\Omega(1, r+1)$ is equal to $2r$. In P^N, we consider a subspace Z which is complementary to $T_p(\Omega)$, and suppose that $\dim Z = \frac{1}{2}r(r-1) - 1$. We project $\Omega(1, r+1)$ from Z onto $T_p(\Omega)$. This projection maps a sufficiently small neighborhood $U_p \subset \Omega(1, r+1)$ of the point p onto its neighborhood in the space $T_p(\Omega)$. Moreover, r-dimensional plane generators of the manifold $\Omega(1, r+1)$ will be projected onto r-planes on $T_p(\Omega)$, and the web $\Omega W(d, 2, r)$ will be projected onto the web $LW(d, 2, r)$ formed by planar foliations of codimension r in the projective space $T_p(\Omega)$. Thus, the sequence of local diffeomorphisms

$$W(d, 2, r) \to GW(d, 2, r) \to \Omega W(d, 2, r) \to LW(d, 2, r)$$

maps a Grassmannizable web $W(d, 2, r)$ into a planar web $LW(d, 2, r)$. ■

Note that the converse does not hold, i.e. not every linearizable web $LW(d, 2, r)$ is Grassmannizable.

Theorem 8.21 does not hold for Grassmannizable webs $W(d, n, r), n > 2$, since to the leaves of such a web, there correspond submanifolds $G(n-2, n+r-2)$ on the Grassmannian $G(n-1, n+r-1)$, and in turn, to these submanifolds, there correspond submanifolds $\Omega(n-2, n+r-2)$ on the manifold $\Omega(n-1, n+r-1)$, and the latter submanifolds are not planar.

8.6 The Algebraization Problem for Multidimensional d-Webs

1. A web $W(d, n, r)$ is called *algebraizable* if it is Grassmannizable and r-dimensional submanifolds X_α of a projective space P^{n+r-1} generating this web, belong to an algebraic manifold V_d^r of dimension r and degree d. We will denote an algebraic d-web by the symbol $AW(d, n, r)$.

We first consider the case $n = 2$, i.e. $\dim X = 2r$. A Grassmann web $GW(d, 2, r)$ is generated by d hypersurfaces X_α, $\alpha = 1, \ldots, d$, in a projective space P^{r+1}. This web is algebraizable if all hypersurfaces X_α belong to an algebraic hypersurface V_d^r of degree d. We now find conditions of algebraizability of a system of hypersurfaces in a space P^{r+1}.

Denote by l_0 a straight line that intersects each of the hypersurfaces X_α at a single point, and consider a neighborhood D of the line l^0 on the Grassmannian $G(1, r+1)$. Let us fix affine coordinates $(y^1, \ldots, y^{r+1})$ in P^{r+1} by locating their origin at the point

$O \in l_0$ and taking the line l_0 as the axis y^{r+1}. Then any line l from the neighborhood D can be given by the equations:

$$y^i = m^i y^{r+1} + b^i. \tag{8.66}$$

The coefficients m^i and b^i are the coordinates of the line l on the $(2r)$-dimensional Grassmannian $G(1, r+1)$, and the line l^0 has line coordinates which are equal to zero: $m^i = 0, b^i = 0$ for all i. Define: $m = (m^1, \ldots, m^r)$, $b = (b^1, \ldots, b^r)$.

A line $l = l(m, b) \in D$ intersects the hypersurface X_α in a point $x_\alpha(m, b)$. We denote the $(r+1)$th coordinate of this point by $z_\alpha = z_\alpha(m, b)$.

Theorem 8.22 *A system of hypersurfaces X_α, $\alpha = 1, \ldots, d$, and along with it, the web $W(d, 2, r)$ generated by this system, are algebraizable if and only if the conditions*

$$\sum_{\alpha=1}^{d} \frac{\partial^2 z_\alpha}{\partial b^i \partial b^j} = 0 \tag{8.67}$$

hold for all $i, j = 1, \ldots, r$.

Proof. *Necessity.* Suppose that all hypersurfaces X_α belong to an algebraic hypersurface V_d^r of degree d defined by the equation:

$$P(y^1, \ldots, y^{r+1}) = 0, \tag{8.68}$$

whose left-hand side is the degree d polynomial. We find the intersection points of this hypersurface with the line l. Substituting the coordinates y^i from (8.66) into equation (8.68), we obtain the equation:

$$P(m^1 y^{r+1} + b^1, \ldots, m^r y^{r+1} + b^r, y^{r+1}) = 0,$$

which is of degree d and can be written in the form:

$$a_0 (y^{r+1})^d - a_1 (y^{r+1})^{d-1} + \ldots \pm a_d = 0.$$

The coordinates z_α of the points x_α are the roots of this equation, and consequently their sum can be expressed in terms of the coefficients of this equation by the formula:

$$\sum_{\alpha=1}^{d} z_\alpha = \frac{a_1}{a_0}.$$

However, the coefficient a_0 does not depend on the quantities b^i, and the coefficient a_1 is linear in the b's. Thus, the second b-partials of $\frac{a_1}{a_0}$ vanish, i.e. equation (8.67) holds.

Sufficiency. Suppose that the hypersurface X_α is defined by the equation:

$$\varphi_\alpha(y^1, \ldots, y^{r+1}) = 0.$$

Then the coordinate $y^{r+1} = z_\alpha = z_\alpha(m,b)$ of its intersection point with the line l satisfies the equation:

$$\varphi_\alpha(m^1 z_\alpha + b^1, \ldots, m^r z_\alpha + b^r, z_\alpha) = 0$$

for all (m,b). Differentiating this identity first with respect to m^j and next with respect to b^j, we get:

$$\sum_i \frac{\partial \varphi_\alpha}{\partial y^i}\left(m^i \frac{\partial z_\alpha}{\partial m^j} + \delta^i_j z_\alpha\right) + \frac{\partial \varphi_\alpha}{\partial y^{r+1}} \frac{\partial z_\alpha}{\partial m^j} = 0,$$
$$\sum_i \frac{\partial \varphi_\alpha}{\partial y^i}\left(m^i \frac{\partial z_\alpha}{\partial b^j} + \delta^i_j\right) + \frac{\partial \varphi_\alpha}{\partial y^{r+1}} \frac{\partial z_\alpha}{\partial b^j} = 0.$$

From these two equations we find that

$$\frac{\partial z_\alpha}{\partial m^j} = z_\alpha \frac{\partial z_\alpha}{\partial b^j}.$$

Relation (8.67) shows that $\sum_\alpha z_\alpha$ is linear in the b's, i.e. has the form:

$$\sum_\alpha z_\alpha = \sum_i \rho_i(m) b^i + \sigma(m),$$

where $\rho_i(m)$ and $\sigma(m)$ are functions of m alone.

Next, the expression

$$\frac{1}{2}\frac{\partial(\sum_\alpha z_\alpha^2)}{\partial b^j} = \sum_\alpha z_\alpha \frac{\partial z_\alpha}{\partial b^j} = \sum_\alpha \frac{\partial z_\alpha}{\partial m^j} = \sum_i \frac{\partial \rho_i}{\partial m^j} b^i + \frac{\partial \sigma}{\partial m^j}$$

is also linear in the b's. Thus, the expression $\sum_\alpha z_\alpha^2$ is of second degree in the b's. The proof that the expression $\sum_\alpha z_\alpha^q$ is of degree q in the b's is similar.

We next define the function $A(\zeta, m, b)$ by

$$A(\zeta, m, b) = \prod_\alpha (\zeta - z_\alpha(m,b)).$$

The function A is a polynomial of degree d in ζ, with coefficients which are functions of m and b. We write it as follows:

$$A(\zeta, m, b) = \zeta^d - A_1(m,b)\zeta^{d-1} + \ldots \pm A_d(m,b).$$

Since $\sum_\alpha z_\alpha^q$ is of degree q in the b's, by the Newton identities for elementary symmetric functions, the expressions $A_q(m,b)$ are also of degree q in the b's.

We introduce now new variables $\xi^i = m^i\zeta + b^i$ and let $\xi = (\xi^1, \ldots, \xi^r)$. Eliminating b from $A(\zeta, m, b)$, we define the new function

$$B(\xi, \zeta, m) = A(\zeta, m, \xi - m\zeta).$$

Since $A_q(m, b)$ is the polynomial of degree q in b, the function $B(\xi, \zeta, m)$ is a polynomial of degree d in ξ and ζ, with coefficients that are functions of m.

We will show that, if the point $(\xi, \zeta) \in X_\alpha$, then $B(\xi, \zeta, m) = 0$, for all m. We have

$$B(\xi, \zeta, m) = A(\zeta, m, \xi - m\zeta) = \prod_\alpha (\zeta - z_\alpha(m, \xi - m\zeta)).$$

But $z_\alpha(m, \xi - m\zeta)$ is the $(r+1)$th coordinate of the intersection point of the hypersurface X_α with the line defined by the equation: $y^i = m^i y^{r+1} + \xi^i - m^i\zeta$. It is obvious that this line passes through the point (ξ, ζ). Since the point (ξ, ζ) is the intersection of the hypersurface X_α and the line l, it follows that $z_\alpha(m, \xi - m\zeta) = \zeta$. The latter implies that $B(\xi, \zeta, m) = 0$.

Consider the hypersurface $X(m)$ defined by the equation $B(\xi, \zeta, m) = 0$. Since $B(\xi, \zeta, m)$ is the polynomial of degree d with coefficients depending on m, $X(m)$ is an algebraic hypersurface of degree d. But since for any m this hypersurface $X(m)$ contains each X_α and the latter does not depend on m, the coefficients of the equation $B(\xi, \zeta, m) = 0$ cannot actually depend on m. Thus, this equation defines the unique algebraic hypersurface V_d^r of degree d containing all hypersurfaces X_α. ■

We can write equation (8.67) in an invariant form. In fact, if $m^i = 0$, from (8.66) we obtain:

$$\frac{\partial^2 y_\alpha}{\partial b^i \partial b^j} = \frac{\partial^2 z_\alpha}{\partial y^i \partial y^j}.$$

In particular, the latter equation holds for the line l_0. But the partials $\frac{\partial^2 z_\alpha}{\partial y^i \partial y^j}\Big|_{y^i=0}$ coincide with the coefficients b^α_{ij} of the 2nd fundamental forms of the hypersurfaces X_α at the points of their intersection with the line l_0. Since the line l_0 is an arbitrary straight line of the domain D of regularity of a web $GW(d, 2, r)$, the equations

$$\frac{\partial^2 z_\alpha}{\partial b^i \partial b^j}\Big|_{\substack{m^i=0,\\ b^i=0}} = b^\alpha_{ij}.$$

hold. Thus, relations (8.67) can be written in the form:

$$\sum_{\alpha=1}^{d} b^\alpha_{ij} = 0, \tag{8.69}$$

which are invariant conditions of algebraizability of a web $GW(d, 2, r)$.

If $d = 3$, this condition is identical with the algebraizability condition for a Grassmann web $GW(3, 2, r)$ given in Section 3.3.

2. We further find the algebraizability condition for a web $W(d, n, r)$, $d \geq n + 1$. Since by the definition, this web must be Grassmannizable, it is generated by a system of d r-dimensional submanifolds X_α, $\alpha = 1, \ldots, d$, in a projective space P^{n+r-1}.

Let $L_0 \subset P^{r+n-1}$ be a subspace of dimension $n-1$ that intersects each of the manifolds X_α at the single point x^0_α, and let D be a neighborhood of the subspace L_0 in the Grassmannian $G(n-1, r+n-1)$. We take any $n-2$ of the points x^0_α. Suppose these are the points $x^0_{\alpha_1}, \ldots, x^0_{\alpha_{n-2}}$. Denote their linear span by $Z = [x^0_{\alpha_1}, \ldots, x^0_{\alpha_{n-2}}]$. We consider a subspace P^{r+1} complimentary to Z and the projection $\pi : P^{r+n-1} \backslash Z \rightarrow P^{r+1}$ from the center Z. Define: $l = \pi(L)$, $\widetilde{X}_{\alpha_m} = \pi(X_{\alpha_m})$, $m = n-1, \ldots, d$, where L is a subspace of dimension $n-1$ belonging to the neighborhood D, l is a straight line and $\widetilde{X}_{\alpha_m}$ are hypersurfaces in P^{r+1}. The number of these hypersurfaces is $d' = d-(n-2) \geq 3$. The algebraizability condition of hypersurfaces $\widetilde{X}_{\alpha_m}$ in P^{r+1} can be written in the form (8.69), where $\alpha = \alpha_m$ and $b^{\alpha_m}_{ij}$ are the coeeficients of the 2nd fundamental forms of the hypersurfaces $\widetilde{X}_{\alpha_m}$ at the points of their intersection with the straight line l. These coefficients are also the coefficients of the 2nd fundamental forms of the r-dimensional manifolds X_{α_m} at the points of their intersection with the subspace L, relative to the hyperplanes ξ_{α_m} that are the linear spans of the center Z of projection and the tangent subspaces $T(X_{\alpha_m})$ to the manifolds $X\alpha_m$ at the points $x_{\alpha_m} = X_{\alpha_m} \bigcap L$, $m = n-1, \ldots, d$. This gives the following theorem.

Theorem 8.23 *A system of submanifolds X_α, $\alpha = 1, \ldots, d$ of dimension r in a projective space P^{n+r-1} is algebraizable if and only if, for any subspace $L \subset D$ and any distinct values α_m, the condition*

$$\sum_{m=n-1}^{d} b^{\alpha_m}_{ij} = 0 \tag{8.70}$$

is satisfied where $b^{\alpha_m}_{ij}$ are the coefficients of the 2nd fundamental forms of the manifolds X_{α_m} defined above. ■

8.7 The Rank Problem for d-Webs

1. The problems of Grassmannizability, algebraizability and linearizability of webs $W(d, n, r)$ considered in two previous sections, are closely connected with the rank problem for a web. There are many papers on this subject. In this book we cannot give a complete exposition of this problem and give only a short review of it.

The definition of the rank of a web $W(d, n, r)$ given on a manifold X of dimension nr, can be given as follows. In the base X_α of the foliation λ_α belonging to the web W, we consider an r-form:

$$\Omega_\alpha = \underset{\alpha}{\omega^1} \wedge \underset{\alpha}{\omega^2} \wedge \ldots \wedge \underset{\alpha}{\omega^r}, \quad \alpha = 1, \ldots, d,$$

which is called a *normal* of the foliation λ_α. Next, on the web we construct the forms

$$\Omega^k = \sum_{\alpha=1}^{d} f^k_\alpha(p)\Omega_\alpha, \quad p \in X, \quad f^k_\alpha(p) \neq 0, \tag{8.71}$$

which are closed: $d\Omega^k = 0$. The equations $\Omega^k = 0$ are called the ***abelian equations*** of the web $W(d,n,r)$. The maximal number of linearly independent abelian equations which the web $W(d,n,r)$ admits, is called the *rank of the web*.

Any linear combination of equations (8.71) with constant coefficients is an equation of the same type. For this reason, the abelian equations of the web $W(d,n,r)$ form a linear space over the field of constants. The rank of the web is the dimension of this linear space.

The rank problem for a web $W(d,n,r)$ consists of finding the upper bound for its rank, the determination of the maximal rank webs, and the study of the properties of such webs.

2. For a web $W(d,n,1)$ of codimension one, we have $\Omega_\alpha = \underset{\alpha}{\omega}$, and its abelian equations take the form:

$$\Omega^k = \sum_\alpha f_\alpha^k(p)\underset{\alpha}{\omega}.$$

If $d = n+1$, the co-basis forms of the web are connected by condition (8.2) which for $r = 1$ has the form:

$$\underset{1}{\omega} + \underset{2}{\omega} + \ldots + \underset{n+1}{\omega} = 0.$$

Because of this, in the abelian equation (8.71) of the web $W(d,n,1)$, all the functions f_α are equal. If a web $W(d,n,1)$ is not parallelizable, then there exists no function $f(p)$ on X such that the form

$$\Omega = f(p)\sum_{\alpha=1}^{d}\underset{\alpha}{\omega}$$

is closed. Thus, in this case the rank of the web is zero. If a web $W(d,n,1)$ is parallelizable, then Theorem 8.1 implies the existence of a function $f(p)$ such that the form Ω is closed. Therefore, the rank of a parallelizable web $W(d,n,1)$ is equal to one. It is easy to see that the converse is also true: if the rank of a web $W(d,n,1)$ is equal to one, then this web is parallelizable.

In the paper [CG 78a] Chern and Griffiths established the relationship between the linearizability and the algebraizability of the web $W(d,n,1)$. It is proved there that *if a web $W(d,n,1)$ is linearizable and its rank $\rho \geq 1$, then this web is algebraizable, i.e. all its leaves belong to an algebraic curve of degree d in the dual projective space P^{n*}.* This theorem follows from the converse of Abel's theorem proved by S. Lie. The latter can be stated as follows. *Let $X_1, \ldots, X_d$ be analytic curves in P^n, let $\underset{\alpha}{\omega}$ be holomorphic forms in X_α, $\alpha = 1, \ldots, d$, and let there exist a hyperplane L_0 which intersects each curve at a single point. If in a neighborhood of the hyperplane L_0 the abelian equation*

$$\sum_\alpha \underset{\alpha}{\omega}(X_\alpha) = 0$$

holds, then all the curves X_α belong to one algebraic curve X of degree d and $\underset{\alpha}{\omega} = \omega\big|_{X_\alpha}$*, where ω is an 1-form in X.*

Note that Theorem 8.23 proved in Section 8.6 is a multidimensional analogue of this theorem.

As far back as 1936, Chern (see [C 36b]) found the least upper bound for the rank of a web $W(d,n,1)$. Namely, he proved that its rank satisfies the inequality

$$\rho \leq \pi(d,n),$$

where

$$\pi(d,n) = \frac{1}{2(n-1)}\{(d-1)(d-n) + t(n-t-1)\} \tag{8.72}$$

and

$$t \equiv -d+1 \bmod n-1, \ 0 \leq t \leq n-2.$$

It turns out that the integer $\pi(d,n)$ is connected with the theory of algebraic curves. It represents the *Castelnuovo bound* (see for example [H 77]) for the genre g of a non-degenerate algebraic curve of degree d in a projective space P^n. Algebraic curves whose genre is $\pi(d,n)$, are called *extremal.* A web $W(d,n,1)$ whose rank is equal to this number, is called a *web of maximum rank.*

We now state a sufficient condition of algebraizability of a web $W(d,n,1)$ proved in [CG 78a] [CG 81]. *If for $n \geq 3$ and $d \geq 2n$, the rank of a normal web $W(d,n,1)$ is maximal, i.e. $\rho = \pi(d,n)$, then the web is algebraizable, and all its foliations are generated by the same extremal algebraic curve of degree d in the dual projective space P^{n*}.* (For the definition of normality we refere to [CG 81].) For $n=3$, the proof of this theorem can be found in the book [BB 38]. The case $n=2$ was also considered in [BB 38].

To prove this theorem, Chern and Griffiths applied the Poincaré mapping approach. We will outline how this mapping can be constructed. We take the coefficients $f_\alpha^k(p)$, $k = 1,\ldots,\rho$, of abelian equations (8.71) as homogeneous coordinates of a point $z_\alpha(p)$ in an auxiliary projective space $P^{\rho-1}$ of dimension $\rho-1$. Thus, we have

$$z_\alpha(p) = \{f_\alpha^1(p),\ldots,f_\alpha^\rho(p)\}, \quad \alpha = 1,\ldots,d.$$

Since there are precisely n linearly independent forms among the forms $\underset{\alpha}{\omega}(X_\alpha)$ in equations (8.71), there will be only $d-n$ linearly independent points among d points $z_\alpha(p)$. Thus, the mapping $X \to G(d-n-1,\rho-1)$ arises. This mapping sends a point $p \in X$ into a subspace $P^{d-n-1}(p) = z_1(p) \wedge \ldots \wedge z_d(p)$ of the space $P^{\rho-1}$. This mapping is called the *Poincaré mapping.*

For $d = 2n$, the Poincaré mapping allows us to give a simple proof of the theorem stated above. In fact, in this case $\pi(2n,n) = n+1$ and $P^{\pi-1} = P^n$. Thus, the Poincaré mapping takes the form: $X \to G(n-1,n) = P^{n*}$. It sends a point $p \in X$

into a hyperplane in P^n and a web leaf into a bundle of hyperplanes. This shows that the linearizability of a web follows from the principle of duality in P^n.

In particular, note that the maximum rank of the web $W(4,2,1)$ formed by four families of curves in the plane is equal to three. Each web $W(4,2,1)$ of maximum rank is rectifiable and formed by the tangents to a curve of third class.

If $d > 2n$, the proof of the algebraization theorem for the web $W(d,n,1)$ is much more complicated. This proof is given in detail in [CG 78a], and more briefly in [CG 77] and [Be 80]. If $n+1 < d-2n, n \geq 3$, the maximal rank of a web $W(d,n,r)$ is equal $\pi(d,n) = d-n$, and there are examples of non-linearizable webs of maximum rank. For $n=2, d=5$, an example of such a web was constructed by Bol as far back as 1936 (see [B 36] and also the book [Bl 55], p. 124). The web of this Bol example consists of four pencils of straight lines whose vertices are in general position in the plane P^2, and a pencil of conics passing through these vertices (see Figure 8.2).

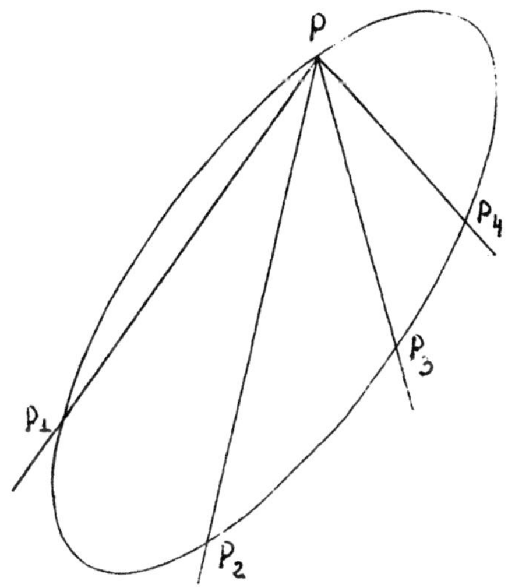

Figure 8.2

This Bol example was generalized by D. Damiano for higher dimensions in the papers [D 80] and [D 83]. Damiano studied a curvilinear web on an n-dimensional manifold. A curvilinear d-web in X^n is called *quadrilateral* if quadrilateral figures formed by any two families of curves composing the d-web, are closed (see Figure

8.3).

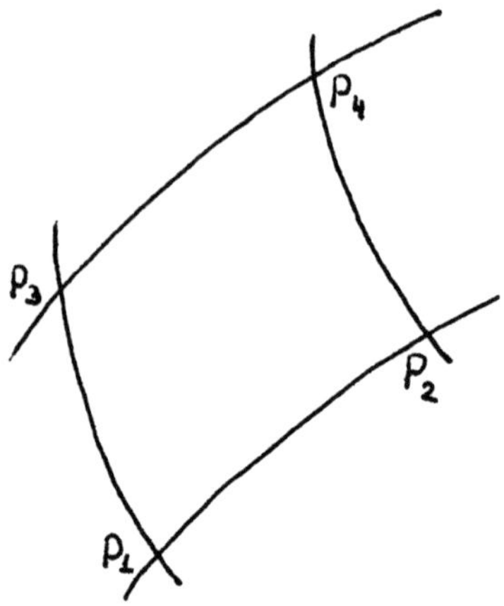

Figure 8.3

In [D 80] it was proved that if $d \geq n \geq 3$ and $d \neq n+3$, any quadrilateral curvilinear d-web on a manifold X^n is equivalent to a web formed by bundles of straight lines in a projective space P^n. In the exceptional case $d = n+3$, such web is equivalent to a web formed in P^n by $n+2$ bundles of straight lines whose vertices p_α, $\alpha = 1, \ldots, n+2$, are in general position, and an $(n-1)$-parametric family of rational normal curves passing through the points p_α. This web is called the *exceptional web* and is a web of maximum rank among all curvilinear $(n+3)$-webs. In the case $n = 2$, the exceptional web constructed by Damiano, is reduced to the Bol 5-web represented in Figure 8.2.

Note also that the webs $W(2n, n, 1)$ of maximum rank are closely connected with the theory of hypersurfaces of double translation in an affine space A^{n+1}. For $n = 2$ these surfaces were the subject of study of such famous geometers as S. Lie [Lie 37], H. Poincaré [P 01], N.G. Chebotarev [Ch 27] and Ya. P. Blank [Bla 52]. The main theorem here is that the tangent lines to two nets of translation on a hypersurface V^n in the space A^{n+1} intersect the hyperplane at infinity of the space A^{n+1} in the points belonging to one curve of degree $d = 2n$. As shown by Chern [C 82], this theorem can be easily derived from the algebraizability of the web $W(2n, n, 1)$.

3. In the paper [CG 78b] Chern and Griffiths gave an estimation for the rank of a web $W(d, n, r)$ of arbitrary codimension r on a manifold X of dimension nr provided that $r \leq n$ and $n \geq 2$. They showed that this rank possesses the least upper bound

$$\pi(d, n, r) = \sum_{\mu \geq 0} \max\{\binom{r+\mu-1}{\mu}\Big(d - (r+\mu)(n-1) - 1\Big),\ 0\}. \tag{8.73}$$

In [CG 78b] this formula was proved for the case $r = 2$. For $r > 2$, it was proved by Little in [Lit 89].

Goldberg in [G 90a] gave a few other expressions for the number $\pi(d,n,r)$ that are more convenient for computations. In particular, he represented this number in the form:

$$\pi(d,n,r) = \frac{1}{(r+1)!(n-1)!}(d-rt-1)\prod_{\mu=1}^{r}\Big(d+t-1-\mu(n-1)\Big), \qquad (8.74)$$

where, as in formula (8.72),

$$t \equiv -d+1 \bmod n-1,\ 0 \leq t \leq n-2.$$

This expression can be further simplified by setting

$$t+d-1 = \alpha(n-1). \qquad (8.75)$$

By (8.75), formula (8.74) takes the form:

$$\pi(d,n,r) = \frac{d-rt-1}{(r+1)!(n-1)!}(n-1)^r(\alpha-1)(\alpha-2)\dots(\alpha-r). \qquad (8.76)$$

We consider some particular cases of the previous formulas. Since the first term in the sum of formula (8.73) is equal to $d-r(n-1)-1$, we have $\pi(d,n,r)=0$ if and only if $d<r(n-1)$. Next, if $r=2$, formula (8.74) takes the form:

$$\pi(d,n,2) = \frac{1}{6(n-1)!}(d-2t-1)(d+t-n)(d+t-2n-1),$$

which was found by Chern and Griffiths in [CG 78b]. By substitution (8.75), the latter formula can be presented in the form:

$$\pi(d,n,2) = \frac{1}{6(n-1)!}(d-2t-1)(n-1)^2(\alpha-1)(\alpha-2).$$

In particular, if $n+1 \leq d \leq 2n$, we find from (8.75) that $\alpha=2$ and $\pi(d,n,2)=0$. If $r=n=2$, we find that $t=0$, $\alpha=d-1$ and

$$\pi(d,2,2) = \frac{1}{6}(d-1)(d-2)(d-3). \qquad (8.77)$$

This number coincides with the maximum geometric genus of a smooth algebraic surface of degree d in a projective space P^3.

In the paper [CG 78b] Chern and Griffiths proved the following geometric theorem: *If $d \geq 2n+1$ and $n \geq 3$, the projectivization of all normals $\Omega_\alpha(p)$ of a web $W(d,n,2)$ of maximum rank belongs to a one ruled Segre variety $S_p(1,n-1) \subset PT_p^*(X)$.* Since the tangent subspaces $T_p(F_\alpha)=T_\alpha$ to leaves of the web under consideration are dual to its normals $\Omega_\alpha(p)$, this theorem is equivalent to the fact that all subspaces T_α, $\alpha = 1,\dots,d$, determine a unique Segre cone $C_p(2,n) \subset T_p(X)$. Thus, *for $d \geq 2n+1$ and*

$n \geq 3$, a web $W(d,n,2)$ of maximum rank is almost Grassmannizable in the sense of Section 8.5 (see Definition 8.16, p. 290). On the manifold X, this web determines an almost Grassmann structure $AG(n-1, n+1)$. J. Little [Lit 89] extended this result to webs $W(d,n,r)$. He proved that *if $r \geq 2$ and $d > r(n-1)+2$, every web $W(d,n,r)$ of maximum rank is almost Grassmannizable.*

4. The rank problem for a web $W(d,2,r)$ on a manifold X of dimension $2r$ has been studied in a series of papers by Goldberg. First of all, it follows from formula (8.76) that

$$\pi(d,2,r) = \frac{1}{(r+1)!}(d-1)(d-2)\dots(d-r-1). \tag{8.78}$$

This shows that if $d \leq r+1$, we have $\pi(d,2,r) = 0$.

It follows from Little's result mentioned above that a web $W(d,2,r)$ of maximum rank is almost Grassmannizable if $r \geq 2$ and $d > r+2$. Since it follows from the last two inequalities that $d > 4$, the case $d = 4$ should be studied separately. This case was considered in the paper [G 83]. In this paper it is shown that if $r \geq 2$ and a web $W(4,2,r)$ admits at least one abelian equation, the web is almost Grassmannizable. It is also proved in this paper (see also [G t.a.] and [G 88]) that a web $W(d,2,2), d > 4$, is of maximum rank if and only if it is algebraizable.

Therefore, the webs $W(4,2,2)$ are exceptional among webs $W(d,2,2)$ of maximum rank since they are not necessarily algebraizable. It follows from formula (8.77) that the rank of such webs is equal to one. There are isoclinic and non-isoclinic 4-webs among these exceptional 4-webs of maximum rank. Examples of both kinds have been presented in [G 85], [G 86] and [G 87] (see Problems **5, 6** and **8** of this chapter).

The results of the papers [G 83], [G 85], [G 86], [G 87] and [G t.a.] are given in more detail in the book [G 88] by Goldberg.

PROBLEMS

1. Apply formulas of Sections 8.1 and 8.4 to prove the following theorems:

i) If a web $W(n+1,n,r)$ is transversally geodesic, then its curvature tensor $\underset{uv}{b}{}^i_{jkl}$ satisfies the condition:

$$\underset{uv}{b}{}^i_{(jkl)} = \delta^i_{(j}\,\underset{uv}{b}{}^{\,i}_{kl)},$$

and the tensors $\underset{uvw}{a}{}^i_{jkl}$ (covariant derivatives of the torsion tensor $\underset{uv}{a}{}^i_{jk}$) satisfy similar conditions.

ii) A web $W(n+1,n,r)$ is transversally geodesic if and only if all its three-subwebs $[u,v,n+1]$ are transversally geodesic.

iii) Any hexagonal web $W(n+1,n,r)$ is transversally geodesic.

2. Derive the structure equations of a web $W(n+1,n,1)$ by applying the structure equations of a web $W(n+1,n,r)$.

3. Prove that the Segre cone $C(r,n)$ is the intersection of the finite number of cones of second order and derive from this that the Segre cones defined by equations (8.38) and (8.59), coincide.

4. Prove that the 4-web $W(4,2,2)$ defined on a four-dimensional manifold X^4 with coordinates x^1, x^2, y^1, y^2 by the equations:

$$\begin{aligned}
\lambda_1 &: \; x^1 = \text{const.}, \quad x^2 = \text{const.};\\
\lambda_2 &: \; y^1 = \text{const.}, \quad y^2 = \text{const.};\\
\lambda_3 &: \; u_3^1 = x^1 + y^1 = \text{const.}, \quad u_3^2 = (x^2+y^2)(y^1-x^1) = \text{const.};\\
\lambda_4 &: \; u_4^1 = (x^1-y^1)^2(x^2+y^2)^2/[(x^2+C)(y^2-C)] = \text{const.},\\
&\quad u_4^2 = x^1 + y^1 + [(y^1-x^1)(x^2+y^2)/\sqrt{(x^2+C)(y^2-C)}]\cdot\\
&\quad \cdot \arctan\sqrt{(y^2-C)(x^2+C)} = \text{const},
\end{aligned}$$

is an isoclinic web of maximum rank one. Its only abelian equation has the form:

$$\left(\frac{1}{x^2}\right)dx^1\wedge dx^2 + \left(\frac{1}{y^2}\right)dy^1\wedge dy^2 - \left(\frac{1}{u_3^2}\right)du_3^1\wedge du_3^2 - \left(\frac{1}{2u_4^1}\right)du_4^1\wedge du_4^2 = 0.$$

5. Prove that the 4-web $W(4,2,2)$ defined on a four-dimensional manifold X^4 with coordinates x^1, x^2, y^1, y^2 by the equations:

$$\begin{aligned}
\lambda_1 &: \; x^1 = \text{const.}, \quad x^2 = \text{const.};\\
\lambda_2 &: \; y^1 = \text{const.}, \quad y^2 = \text{const.};\\
\lambda_3 &: \; u_3^1 = x^1 + y^1 = \text{const.}, \quad u_3^2 = -x^1y^2 + x^2y^1 = \text{const.};\\
\lambda_4 &: \; u_4^1 = (u_3^2 + C_1u_3^1)/(x^2+y^2+C_1u_3^2) = \text{const.},\\
&\quad u_4^2 = -u_4^1 \ln|(y^2+C_1y^1-C_2)/(x^2+C_1x^1+C_2)| - u_3^1 = \text{const.}
\end{aligned}$$

is an isoclinic web of maximum rank one. Its only abelian equation has the form:

$$\begin{aligned}
-\frac{1}{x^2+C_1x^1+C_2}dx^1\wedge dx^2 - \frac{1}{y^2+C_1y^1-C_2}dy^1\wedge dy^2\\
+\frac{1}{u_3^2+C_2u_3^1}du_3^1\wedge du_3^2 - \frac{1}{u_4^1}du_4^1\wedge du_4^2 = 0.
\end{aligned}$$

6. Prove that the 4-web $W(4,2,2)$ defined on a four-dimensional manifold X^4 with coordinates x^1, x^2, y^1, y^2 by the equations:

$$\begin{aligned}
\lambda_1 &: \; x^1 = \text{const.}, \quad x^2 = \text{const.};\\
\lambda_2 &: \; y^1 = \text{const.}, \quad y^2 = \text{const.};\\
\lambda_3 &: \; u_3^1 = x^1 + y^1 + (x^1)^2y^2/2 = \text{const.}, \quad u_3^2 = x^2 + y^2 - x^1(y^2)^2/2 = \text{const.}\\
\lambda_4 &: \; u_4^1 = -x^1 + y^1 + (x^1)^2y^2/2 = \text{const.}, \quad u_4^2 = x^2 - y^2 - x^1(y^2)^2/2 = \text{const},
\end{aligned}$$

is a non-isoclinic web of maximum rank one. Its only abelian equation has the form:

$$2dx^1 \wedge dx^2 + 2dy^1 \wedge dy^2 - du_3^1 \wedge du_3^2 + du_4^1 \wedge du_4^2 = 0.$$

The webs indicated in Problems **4**, **5** and **6** are not Grassmannizable. These webs are considered in more detail in the papers [G 85], [G 86] and [G 87] (see also Chapter 8 of the book [G 88]).

NOTES

8.1. The basic equations of the theory of $(n+1)$-webs $W(n+1, n, r)$ as well as the connection $\tilde{\Gamma}$ were obtained by Goldberg in the papers [G 73] and [G 74a] (see also his [G 88]). Theorems 8.1, 8.3 and 8.4 can be also found in these papers and the book. For the webs $W(3, 2, r)$ the connection $\tilde{\Gamma}$ was substantially used in [F 78].

8.2. Grassmann webs $GW(n+1, n, r)$ for $n = 2$ were considered by Akivis in [A 73a], for $n = 3$ in [AG 74], and for any n in [G 75c].

8.3. A geometric definition of an almost Grassmann structure on a differentiable manifold by means of the field Segre cones was given by Akivis in the papers [A 80] and [A 82b]. Theorem 8.9 was also proved in these papers. An analytic definition of an almost Grassmann structure can be found in the papers [Ha 60] and [M 78]. The torsion tensor of an almost Grassmann structure was found in [M 78] and [G 75d]. In the latter paper this tensor was calculated in the general frame.

8.4. Transversal geodesic webs $W(n+1, n, r)$ were introduced in [G 73] and [G 74a], and isoclinic webs $W(n+1, n, r)$ in [G 74b]. In connection with the theory of almost Grassmann structures these webs were considered in [G 75d], [A 80] and [A 82b]. Polyhedral webs $W(n+1, n, r)$ were investigated in [G 76] (in this paper they are called $(2n+2)$-hedral).

The problem of Grassmannizability for webs $W(3, 2, r)$ was solved in [A 74] and for webs $W(n+1, n, r)$ in the papers [A 80], [A 82b] and [G 82c] (see Theorem 8.14).

8.5. The theory of webs $W(4, 2, r)$ was constructed in [G 77] and [G 80]. The geometric meaning of the basis affinor λ^i_j (Theorem 8.15) was also established there.

A geometric definition of almost Grassmannizability for webs $W(d, n, r)$ in the case $d \geq n+1$ was introduced by Akivis in [A 83a]. But actually this kind of webs was considered in [A 81b] where the analytical characterization of these webs was given. Analytically, a definition of almost Grassmannizable webs was given in [G 84] (in this paper they are called scalar webs). Theorem 8.18 supplements Theorem 8.1.10 on almost Grassmann webs $AGW(d, 2, r)$ from the book [G 88].

The problem of linearizability for a web $W(3, 2, 1)$ is connected with the problem of anamorphosis in nomography (see notes **3.3** on p. 134). For webs $W(d, n, 1)$, the problem of linearizability was considered in [CG 77]. Theorem 8.21 was first proved in this book.

8.6. The problem of algebraizability of d-webs $W(d, n, r)$ was posed by Chern and Griffiths in [CG 78a]. For webs $W(3, 2, r)$ this problem was solved by Akivis as far back as 1974 (see [A 73a] and [A 74]). For webs $W(4, 2, r)$ this problem was solved by Goldberg in [G 82d] by the same method as that which was indicated in Section 3.3. Applying this method, J. Wood solved the algebraizability problem for a web $W(d, 2, r), d \geq 4$, in his Ph. D. thesis [W 82]. It turns out that for an arbitrary d, this proof involves complicated calculations. Later on, Wood in [W 84] gave a much simpler proof of Theorem 8.20 given in Section 8.6. The method applied by Wood in [W 84], was essentially used by Lie and

Scheffers for their proof of algebraizability of quadruples of curves in a plane, i.e. in the case $r = 1$ and $d = 4$ (see [Lie 37] and [Sc 04]).

8.7. The notion of rank was introduced in 1933 by Blaschke in the paper [Bl 33] for d-webs of codimension one on manifolds of dimensions two and three. In 1936 Chern in [C 36b] extended this notion to webs $W(d, n, 1)$. When at the end of the 1970's, Chern and Griffiths noticed that the least upper bound for the rank of a web $W(d, n, 1)$ coincides with the Castelnuovo boundary for the genus g of non-degenerate algebraic curve of degree d, Griffiths also became interested the rank problem for webs. As a result, Chern and Griffiths published a series of papers: [CG 77], [CG 78a], [CG 78b] and [CG 81]. Their results are presented in Chapter 8.

The rank problem for webs $W(d, 2, r)$ was studied by Goldberg in more detail (see [G 83], [G 84], [G 85], [G 86], [G 87], [G 88], [GR 88] and [G t.a.]). He found the upper least bound for these webs and investigated webs of maximum rank.

Appendix A

Web Geometry and Mathematical Physics

E.V. Ferapontov

In spite of numerous and deep connections of the theory of webs with algebraic geometry, differential geometry and the theory of non-associative structures, until recently, there were practically no applications of the web theory to physics. In this appendix we will present some such applications. One of them arises in the study of geometry of characteristics of solutions of systems of differential equations of hydrodynamic type (see [Kil 71], [Kil 75], [Fer 88], [Fer 89], [Fer 90] and [Fer 91]), and the second one appears when one constructs invariants of the wave systems (see [BF 92]).

A.1 Web Geometry and Systems of Hydrodynamic Type

A system of hydrodynamic type is a quasi-linear system of first order differential equations of the form:

$$u^i_t + v^i_j(u)u^j_x = 0, \quad u = (u^1, \ldots, u^n), \tag{A.1}$$

where t is the time, x is an one-dimensional coordinate, $u^i = u^i(x,t)$, $i = 1, \ldots, n$, are unknown functions, and the entries of the coefficient matrix (v^i_j) are functions of u alone, i.e. they do not depend on t and x. Usually systems of form (A.1) arise in problems of gas dynamics, chemical kinetics, in averaging method of G. B. Whitham etc. (see, for example, [Ts 90]). It is convenient to represent solutions $u^i(x,t)$ of system (A.1) as two-dimensional surfaces in an n-dimensional space $\mathbf{R}^n(u^1, \ldots, u^n)$ which are parameterized by the coordinates x and t. These integral manifolds can intersect one another along curves which are called *characteristics.*

Denote the eigenvalues of the matrix $v^i_j(u)$ by $\lambda^i(u)$ and suppose that all these eigenvalues are real and distinct, i.e. system (A.1) is strongly hyperbolic. Then, the characteristics can be defined by the equations: $dx - \lambda^i dt = 0$. On each solution of system (A.1), these curves form n one-parameter systems of curves, i.e. a curvilinear n-web. Since the eigenvalues λ^i explicitly depend on u, in general, the properties of this web depend on a solution which has been chosen. Thus, two n-webs of characteristics on two different solutions of system (A.1) can be not equivalent.

A natural problem arises: to separate those solutions in the complete set of solutions of system (A.1) on which the web of characteristics is sufficiently simple, for example, hexagonal. The first results in this direction are due to H. Kilp. In [Kil 71] and [Kil 75]) she studied the geometry of characteristics by the Cartan method of exterior forms. In [Kil 75] she found all those solutions of the equations of the one-dimensional non-isentropic flow of a polytropic gas:

$$\begin{aligned} \rho_t + u\rho_x + \rho u_x &= 0, \\ u_t + uu_x + p_x/\rho &= 0, \\ p_t + up_x + \gamma p u_x &= 0, \end{aligned}$$

(where ρ is the density, u is the velocity, p is the pressure and $\gamma = \text{const}$ is the polytropic constant) on which the characteristics form a hexagonal three-web. It turned out that for $\gamma = -1$ (in this case the equations of gas dynamics are known under the name "Chaplygin's gas"), the three-web of characteristics is hexagonal on any solution while for $\gamma \neq -1$, the solutions with a hexagonal three-web of characteristics form a finite-parameter family and are characterized by the fact that the velocity u is a linear function of the coordinate x: $u = a(t)x + b(t)$. In gas dynamics these solutions are known as solutions with "a linear profile of the velocity field" (see [Se 81], p. 318).

In what follows, we restrict ourselves by consideration of such systems (A.1) that can be reduced to the diagonal form:

$$R^i_t + \lambda^i(R) R^i_x = 0, \tag{A.2}$$

and for definiteness, we will assume that the number of equations in system (A.1) is equal to three: $i = 1, 2, 3$. Of course, reduction of system (A.1) to the diagonal form is possible only in the case when the net of the eigendirections of the matrix (v^i_j) is holonomic. The coordinates R^i along the eigendirections of (v^i_j) are called the *Riemann invariants* of system (A.1). In general, the existence of Riemann invariants is not essential for studying the geometry of characteristics. We made this assumption only for simplification of further calculations.

Since system (A.2) is equivalent to the exterior quadratic equations $dR^i \wedge (dx - \lambda^i dt) = 0$, the characteristics (their equations are $dx - \lambda^i dt = 0$) are the level lines of the Riemann invariants: $R^i = \text{const}$. While studying the geometry of characteristics, it is very natural to investigate the following problems:

1. Describe all systems (A.2) consisting of three equations, for which the characteristics form a hexagonal three-web on any solution.

2. For a system of type (A.2) which is not in the first class, find all solutions with a hexagonal three-web of characteristics (in the general case these solutions depend on a finite number of parameters).

Before we o discuss these problems, we give some necessary definitions.

Definition A.1 System (A.2) is called *weakly non-linear* if for any i the condition $\frac{\partial \lambda^i}{\partial R^i} = 0$ holds.

The weak non-linearity is connected with such well-known phenomena as overthrowning (the gradient catastrophe) which is typical for systems of hydrodynamic type. It turns out that the condition for system (A.2) to be weakly non-linear is an obstacle to the phenomena of shock formation, and the solutions $R^i(x,t)$ which are single-valued at the initial moment $t = 0$, will stay the same in the process of evolution [RS 67].

Definition A.2 System (A.2) is said to be *semi-Hamiltonian* if

$$\partial_k\left(\frac{\partial_j \lambda^i}{\lambda^j - \lambda^i}\right) = \partial_j\left(\frac{\partial_k \lambda^i}{\lambda^k - \lambda^i}\right), \quad \partial_i = \frac{\partial}{\partial R^i}$$

for any triple of indices $i \neq j \neq k \neq i$.

The condition for a system (A.2) to be semi-Hamiltonian is equivalent to the existence of infinitely many conservation laws [Ts 85].

Theorem A.3 ([Fer 88], [Fer 89]) *The three-web of characteristics is hexagonal on any solution of system (A.2) if and only if this system is both weakly non-linear and semi-Hamiltonian.*

Proof We will outline the main stages of the proof. Since the characteristics are defined by the equations $dx - \lambda^i dt = 0$, $i = 1,2,3$, it is convenient to introduce the 1-forms:

$$\omega^1 = (\lambda^3 - \lambda^2)(dx - \lambda^1 dt), \quad \omega^2 = (\lambda^1 - \lambda^3)(dx - \lambda^2 dt).$$

Then the equations of the characteristics of the first, second and the third families can be written in the form:

$$\omega^1 = 0, \quad \omega^2 = 0, \quad \omega^1 + \omega^2 = 0,$$

respectively. Let us find the connection form of the three-web under consideration, i.e. a 1-form ω such that

$$d\omega^1 = \omega \wedge \omega^1, \quad d\omega^2 = \omega \wedge \omega^2$$

(see Section 1.4). Direct calculations lead to the following result:

$$\begin{aligned}\omega = \ & (R^1_x\partial_1\lambda^1 + R^2_x\partial_2\lambda^2 + R^3_x\partial_3\lambda^3)dt + \partial_1 \ln(\lambda^1 - \lambda^2)(\lambda^1 - \lambda^3)dR^1 \\ & + \partial_2 \ln(\lambda^2 - \lambda^3)(\lambda^2 - \lambda^1)dR^2 + \partial_3 \ln(\lambda^3 - \lambda^1)(\lambda^3 - \lambda^2)dR^3.\end{aligned} \tag{A.3}$$

The hexagonality condition can be written in the form: $d\omega = 0$. Theorem 3 follows from the fact that the hexagonality condition must be satisfied identically by means of equations (A.2), i.e. the three-web of characteristics must be hexagonal on any solution of system (A.2). ■

As we see, it turns out that the hexagonality condition is connected with such notions as the impossibility of the gradient catastrophe and the existence of infinitely many conservation laws.

Remark A.4 *For systems not possessing to Riemann invariants, a theorem similar to Theorem* 3 *was proved in* [Fer 88] *and* [Fer 89]. *It is shown in these papers that any weakly non-linear semi-Hamiltonian system* (A.2) *consisting of three equations can be linearized by an appropriate transformation of the Backlund type* (*see* [V 87]).

To solve the second problem listed above, we must associate with equation (A.2) the differential condition $d\omega = 0$, where ω is given by formula (A.3), and investigate whether the overdetermined system obtained from this, is consistent. We will illustrate below this procedure considering the example of the equations of chromatography. Note that for the equations of chromatography (as well as for some other examples), integration of the corresponding overdetermined system is also based on one of the classical results of the theory of webs, namely on the Graf–Sauer theorem (see Preface to this book).

In Riemann invariants, the three-component system of the equations of chromatography has the following form (see [Ts 90]):

$$\begin{aligned} R^1_t + R^1(R^1R^2R^3)R^1_x = 0, \\ R^2_t + R^2(R^1R^2R^3)R^3_x = 0, \\ R^3_t + R^3(R^1R^2R^3)R^3_x = 0. \end{aligned} \tag{A.4}$$

It is easy to see that this system is not weakly non-linear, although it is semi-Hamiltonian. Following [Fer 91], we will prove that the solution of system (A.4) carrying a hexagonal three-web of characteristics, exists, depends on 9 arbitrary constants and can be parameterized by algebraic curves of third degree. Uncomplicated computations show that the condition $d\omega = 0$ for system (A.4) leads to the differential condition:

$$\frac{\partial^2}{\partial x \ \partial t} \ln(R^1R^2R^3) - 0,$$

which must be combined with equations (A.4). Thus, the product $R^1R^2R^3$ can be written in the form: $R^1R^2R^3 = \frac{\varphi'(t)}{f'(x)}$, where $\varphi(t)$ and $f(x)$ are functions of one variable, and the prime denotes differentiation with respect to the corresponding variable. Substituting the expression for $R^1R^2R^3$ into (A.4), we get:

$$\frac{R^i_t}{\varphi'(t)} + \frac{R^iR^i_x}{f'(x)} = 0.$$

We now introduce new independent variables: $T = \varphi(t)$ and $X = f(x)$. As a result, we obtain three independent equations:

$$R^i_T + R^iR^i_X = 0,$$

whose general solution has the form:

$$g^i(R^i) = X - R^iT, \tag{A.5}$$

where $g^i(R^i)$ are three arbitrary functions of one variable. Thus, in the variables T and X the characteristics $R^i = \text{const}$ are straight lines. Since we are interested in the solutions with a hexagonal three-web of characteristics, we can apply the Graf–Sauer theorem. According to the latter, every rectilinear hexagonal three-web is formed by straight lines belonging to an algebraic curve of third class. In other words, there exists a third degree polynomial $P_3(u, v)$ in two variables such that the coefficients of the straight lines (A.5) forming the web, satisfy:

$$P_3(R^i, g^i(R^i)) = 0.$$

Recalling that $g^i(R^i) = X - R^iT$, we find that R^i are the roots of the third degree equation:

$$P_3(R, X - RT) = 0.$$

In the old coordinates t and x, we get

$$P_3(R, f(x) - R\varphi(t)) = 0. \tag{A.6}$$

Thus, the solutions of system (A.4) with a hexagonal three-web of characteristics are the roots of the third degree polynomial (A.6) in R. The functions $\varphi(t)$ and $f(x)$ can be found from the concordance condition $R^1R^2R^3 = \frac{\varphi'(t)}{f'(x)}$. In fact, by Vieta's theorem, from (A.6) we obtain the expression:

$$R^1R^2R^3 = \frac{f^3 + af^2 + bf + c}{\varphi^3 + \alpha\varphi^2 + \beta\varphi + \gamma},$$

where the constants a, b, c, α, β and γ can be expressed in terms of the coefficients of the polynomial $P_3(u, v)$. The functions $\varphi(t)$ and $f(x)$ can be found from this after the obvious separation of variables.

If $n > 3$, the n-component system of the chromatography equations has the form

$$R_t^i + R^i(\prod_{k=1}^{n} R^k)R_x^i = 0.$$

A procedure similar to that which was just described allows us to construct the exact solutions of this system which are distinguished by the differential condition:

$$\frac{\partial^2}{\partial x \partial t} \ln(\prod_{k=1}^{n} R^k) = 0.$$

This condition also has a definite geometric meaning. Namely, it means that on the corresponding solutions, the n-web of characteristics has the maximal rank $r = \frac{(n-1)(n-2)}{2}$ [Fer 91] (see also Chapter 8 of this book).

A.2 Integration of Weakly Non-Linear Semi-Hamiltonian Systems of Hydrodynamic Type by the Methods of the Theory of Webs

The problem of study of the geometry of characteristics for systems of hydrodynamic type with more than threef equations is more interesting than that with $n = 3$ equations since if $n > 3$, an n-web in the plane possesses a new non-trivial topological invariant, namely, the rank of the web, whose maximal value does not exceed $\frac{(n-1)(n-2)}{2}$. If $n = 3$, the notion of the rank does not bring anything new since webs of rank one are nothing but hexagonal webs. However, starting from $n = 4$, the application of the notion of the rank leads to non-trivial results. For example, in [Fer 90] this gave the opportunity to integrate and get an explicit solution of all weakly non-linear semi-Hamiltonian systems with any number of equations.

We remind the reader that we consider only the diagonal systems $R_t^i + \lambda^i(R)R_x^i = 0$,
$i = 1, \dots, n$, whose eigenvalues satisfy the conditions:

1. $\partial_i \lambda^i = 0$ for any i (weak non-linearity), and

2. $\partial_k\left(\frac{\partial_j \lambda^i}{\lambda^j - \lambda^i}\right) = \partial_j\left(\frac{\partial_k \lambda^i}{\lambda^k - \lambda^i}\right)$ for any triple $i \neq j \neq k \neq i$ (semi-Hamiltonianity).

If $n = 2$ and $\lambda^1, \lambda^2 \neq \text{const}$, any weakly non-linear semi-Hamiltonian system can be reduced to the form:

$$R_t^1 + R^2 R_x^1 = 0, \quad R_t^2 + R^1 R_x^2 = 0. \tag{A.7}$$

Systems similar to system (A.7) arise in differential geometry (in the theory of minimal surfaces of a three-dimensional Euclidean space), in gas dynamics (so-called Chaplygin's gas) etc. The general solution of system (A.7) can be written in the form:

$$\int^{R^1} \frac{\xi d\xi}{f_1(\xi)} + \int^{R^2} \frac{\xi d\xi}{f_2(\xi)} = x + \text{const}, \quad \int^{R^1} \frac{d\xi}{f_1(\xi)} + \int^{R^2} \frac{d\xi}{f_2(\xi)} = t + \text{const}, \tag{A.8}$$

where the functions $f_1(\xi)), f_2(\xi)$ and constants in the right-hand sides of equations (A.8) are arbitrary. Expressing the R^1 and R^2 from these formulas as functions of x and t, we arrive at the general solution of system (A.7). This can be checked by the direct differentiation of (A.8) with respect to x and t. If we take

$$f_1(\xi) = f_2(\xi) = 2i\sqrt{\prod_{s=1}^{5}(\xi - E^s)}, \quad i^2 = -1, \quad E^s = \text{const},$$

then the problem of finding the quantities R^1 and R^2 as functions of x and t becomes the classical inversion Jacobi problem arising, for example, when one constructs finite-zone solutions of Korteweg-de Vries (KDV) equation. The function $U(x,t)$ defined by the formula:

$$U(x,t) = \sum_{s=1}^{5} E^s - 2\big(R^1(x,t) + R^2(x,t)\big),$$

satisfies the non-linear equation:

$$4U_t + U_{xxx} - 6UU_x + 2\sum_s E^s U_x = 0,$$

which up to a normalization, coincides with a KDV equation. The solutions of a KDV equation obtained in this manner, are called "two-zone" solutions (see, for example, [Alb 79]).

For construction of an n-zone solution of a KDV equation, we consider an n-component generalization of system (A.7):

$$R^i_t + \Big(\sum_{k=1}^{n} R^k - R^i\Big) R^i_x = 0, \quad i = 1, \dots, n. \tag{A.9}$$

It is easy to check that this system is both weakly non-linear and semi-Hamiltonian. Its general solution can be given by the formula similar to formula (A.8) (see [Fer 90]):

$$\begin{aligned} &\int^{R^1} \tfrac{\xi^{n-1} d\xi}{f_1(\xi)} + \dots + \int^{R^n} \tfrac{\xi^{n-1} d\xi}{f_n(\xi)} = x + \text{const}, \\ &\int^{R^1} \tfrac{\xi^{n-2} d\xi}{f_1(\xi)} + \dots + \int^{R^n} \tfrac{\xi^{n-2} d\xi}{f_n(\xi)} = t + \text{const}, \\ &\int^{R^1} \tfrac{\xi^{j} d\xi}{f_1(\xi)} + \dots + \int^{R^n} \tfrac{\xi^{j} d\xi}{f_n(\xi)} = \text{const}, \quad j = n-3, \dots, 0, \end{aligned} \tag{A.10}$$

where the functions $f_1(\xi),\ldots,f_n(\xi)$ as well as the constants on the right-hand sides of system (A.10) are arbitrary. Taking

$$f_1(\xi)=\ldots=f_n(\xi)=2i\sqrt{\prod_{s=1}^{2n+1}(\xi-E^s)},$$

we again come to the Jacobi inversion problem and to the following solution of the KDV equation:

$$U(x,t)=\sum_{s=1}^{2n+1}E^s-2\sum_{k=1}R^k(x,t).$$

This is the so-called n-zone solution. Since in the theory of KDV equation the variables R^i have the meaning of zeros of the eigenfunction ψ of the associated Schrödinger operator $\left(-\frac{d^2}{dx^2}+U\right)\psi=E\psi$, weakly non-linear system (A.9) under consideration describes the evolution of zeros of the ψ-function for the potential $U(x,t)$ which evolves according to the KDV equation.

The goal of this section is to obtain formulas similar to formulas (A.10) for any weakly non-linear semi-Hamiltonian system. These formulas always have the following general structure:

$$\begin{aligned}\int^{R^1}\frac{\varphi_1^1(\xi)d\xi}{f_1(\xi)}+\ldots+\int^{R^n}\frac{\varphi_1^n(\xi)d\xi}{f_n(\xi)}&=x+\text{const},\\ \int^{R^1}\frac{\varphi_2^1(\xi)d\xi}{f_1(\xi)}+\ldots+\int^{R^n}\frac{\varphi_2^n(\xi)d\xi}{f_n(\xi)}&=t+\text{const},\\ \int^{R^1}\frac{\varphi_j^1(\xi)d\xi}{f_1(\xi)}+\ldots+\int^{R^n}\frac{\varphi_j^n(\xi)d\xi}{f_n(\xi)}&=\text{const},\quad j=3,\ldots,n,\end{aligned}\tag{A.11}$$

where the functions $f_1(\xi),\ldots,f_n(\xi)$ as well as the constants on the right-hand sides of system (A.11) are arbitrary, while the functions $\varphi_j^i(\xi)$ are determined by the system under consideration and are fixed. In the example considered above, we had $\varphi_j^i(\xi)=\xi^{n-j}$. Note that since the functions $f_1(\xi),\ldots,f_n(\xi)$ are arbitrary, we can always get the condition $\varphi_n^i\equiv 1$, to be valid for any $i=1,\ldots,n$. As a result, the last equation of (A.11) takes the form:

$$\int^{R^1}\frac{d\xi}{f_1(\xi)}+\ldots+\int^{R^n}\frac{d\xi}{f_n(\xi)}=\text{const},$$

After this normalization, the remaining $n(n-1)$ functions $\varphi_j^i(\xi)$, $i=1,\ldots,n$; $j=1,\ldots,n-1$, of one variable are uniquely determined by the system under consideration.

The complete separation of variables which we observed in system (A.11), can be explained by certain simple properties of the web of characteristics. We will now study these properties. For our further goals, it is convenient to extend the n-web of characteristics $dx-\lambda^i dt=0$ (or $R^i=\text{const}$) to the $(n+2)$-web by adding two one-parameter families of curves $x=\text{const}$ and $t=\text{const}$ – the level curves of the independent variables. Let us calculate the rank of this web, i.e. the maximal number of linearly independent abelian equations of the form:

$$dA^1 + \ldots + dA^n + df(t) + dg(x) = 0, \tag{A.12}$$

where the functions A^i are constant along the characteristics of the ith family and the functions $f(t)$ and $g(x)$ are constant along the lines t = const and x = const, respectively. The weakly non-linear semi-Hamiltonian systems possess a natural class of functions A^i which are constant along the characteristics: these are so-called generalized flow functions.

Definition A.5 Let $B^i(R)(dx - \lambda^i dt)$ be an integral of the system under consideration which vanishes along the characteristics of the ith family (recall that an integral is a 1-form that is closed on any solution of the system). Then the function A^i, defined by the relation

$$dA^i = B^i(R)(dx^i - \lambda^i dt),$$

is constant along the characteristics of the ith family and is called the *generalized flow function.*

Of course, the functions A^i depend on the solution we have chosen.

It is easy to show that the integrals $B^i(R)(dx^i - \lambda^i dt)$ vanishing along the characteristics, exist only if the system is both weakly non-linear and semi-Hamiltonian. Moreover, the functions B^i are solutions of the following overdetermined system:

$$dB^i = B^i \sum_{j \neq i} \frac{\partial_j \lambda^i}{\lambda^j - \lambda^i} dR^j + C^i dR^i, \tag{A.13}$$

where C^i are undetermined parameters. It is easy to check that this system is consistent. Note that along with $B^i(R)(dx^i - \lambda^i dt)$, any expression of the form $\varphi^i(R^i)B^i(R)(dx^i - \lambda^i dt)$, where $\varphi^i(R^i)$ is an arbitrary function of the variable R^i, is an integral.

We will try to find as best we can abelian equations of form (A.12), taking as A^i generalized flow functions and assuming that the functions $f(t)$ and $g(x)$ are linear: $f(t) = at$ and $g(x) = bx$. Substituting $dA^i = B^i(R)(dx^i - \lambda^i dt)$ into (A.12), we obtain:

$$B^1(dx - \lambda^1 dt) + \ldots + B^n(dx - \lambda^n dt) + adt + bdx = 0,$$

from which it follows that $\sum_i B^i = -b$ and $\sum_i B^i \lambda^i = a$. Differentiating these relations and using (A.13), we obtain the explicit expression for the coefficients C^i:

$$C^i = -\sum_{j \neq i} B^j \frac{\partial_i \lambda^j}{\lambda^i - \lambda^j}.$$

As a result, we arrive at the following closed linear system:

$$\begin{aligned} dA^i &= B^i(dx - \lambda^i dt), \ i = 1, \dots, n, \\ dB^i &= B^i \textstyle\sum_{j\neq i} \frac{\partial_j \lambda^i}{\lambda^j - \lambda^i} dR^j - \left(\textstyle\sum_{j\neq i} B^j \frac{\partial_i \lambda^j}{\lambda^i - \lambda^j}\right) dR^i. \end{aligned} \tag{A.14}$$

Uncomplicated calculations show that equations (A.14) are consistent for any weakly non-linear and semi-Hamiltonian system since exterior differentiation does not lead to new relations. Therefore, equations (A.14) determine n linearly independent generalized flow functions connected by the relations: $\sum_i B^i = \text{const}$ and $\sum_i B^i \lambda^i = \text{const}$. This provides n linearly independent abelian equations of the form (A.12). Thus, we have proved the following theorem:

Theorem A.6 ([Fer 90]) *The rank of the $(n+2)$-web formed by the characteristics and the lines $x = const$ and $t = const$ is equal to n on the general solution of any weakly non-linear semi-Hamiltonian system.* ■

Note that speaking precisely, the considerations performed above give only the lower bound for the rank since we could get additional equations of the form (A.12) if we do not take the generalized flow functions as A^i and do not assume that the functions $f(t)$ and $g(x)$ are linear. However, this is not the case, and on the general solution the rank is exactly equal to n although it may "jump" on some special solutions.

To integrate a weakly non-linear semi-Hamiltonian system, we take a basis $\{A^i_j, B^i_j\}, i, j = 1, \dots, n$, consisting of n linearly independent solutions of linear system (A.14) and normalize them in such a way that for the first solution we have $\sum_{i=1}^n B^i_1 = 1$ and $\sum_{i=1}^n B^i_1 \lambda^i = 0$, i.e.

$$dA^1_1 + \dots + dA^n_1 = dx; \tag{A.15}$$

for the second solution we have $\sum_{i=1}^n B^i_2 = 0$ and $\sum_{i=1}^n B^i_2 \lambda^i = -1$, i.e.

$$dA^1_2 + \dots + dA^n_2 = dt, \tag{A.16}$$

and on the remaining $n-2$ solutions we have $\sum_{i=1}^n B^i_j = \sum_{i=1}^n B^i_j \lambda^i = 0, \ j = 3, \dots, n$, i.e.

$$dA^1_j + \dots + dA^n_j = 0. \tag{A.17}$$

Since any two generalized flow functions corresponding to the ith family of characteristics are different by a factor $\varphi^i(R^i)$ depending on the corresponding Riemann invariant R^i alone, we have:

$$dA^i_j = \varphi^i_j(R^i) dA^i_n, \ j = 1, \dots, n-1. \tag{A.18}$$

We emphasize that the functions $\varphi^i_j(R^i)$ are uniquely determined by the system (A.14). Since on solutions the relations

$$dR^i \wedge (dx - \lambda^i dt) = 0, \quad dA^i_n \wedge (dx - \lambda^i dt) = 0,$$

hold, the flow function A^i_n depends only on R^i, i.e. on any solution, dA^i_n can be represented in the form:

$$dA^i_n = \frac{dR^i}{f_i(R^i)},$$

where $f_i(R^i)$ depend on the choice of solution. Relations (A.18) can be written in the form:

$$dA^i_j = \frac{\varphi^i_j(R^i)dR^i}{f_i(R^i)}, \quad j = 1, \ldots, n-1,$$

and after this, integration of equations (A.15)–(A.17) leads to formulas (A.11) for the general solution where according to our construction, $\varphi^i_n = 1$.

Example A.7 For the weakly non-linear semi-Hamiltonian system

$$R^i_t + \Big(\sum_{k=1}^n R^k - R^i\Big) R^i_x = 0, \quad i = 1, \ldots, n,$$

linear system (A.14) has the form:

$$\begin{aligned} dA^i &= B^i(dx - \lambda^i dt), \quad \lambda^i = \textstyle\sum_{k=1}^n R^k - R^i, \\ dB^i &= B^i \textstyle\sum_{j\neq i} \frac{B^j}{R^j - R^i} - \Big(\sum_{j\neq i} \frac{B^j}{R^j - R^i}\Big) dR^i. \end{aligned}$$

1. The solution (A^i_1, B^i_1) of this system, satisfying the conditions $\sum_i B^i_1 = 1$ and $\sum_i B^i_1 \lambda^i = 0$, leads to

$$dA^i_1 = \frac{(R^i)^{n-1}}{\prod_{k\neq i}(R^i - R^k)}(dx - \lambda^i dt).$$

2. The solution (A^i_2, B^i_2) of this system, satisfying the conditions $\sum_i B^i_2 = 0$ and $\sum_i B^i_2 \lambda^i = -1$, leads to

$$dA^i_2 = \frac{(R^i)^{n-2}}{\prod_{k\neq i}(R^i - R^k)}(dx - \lambda^i dt).$$

3. The solution (A^i_j, B^i_j) of this system, satisfying the conditions $\sum_i B^i_j = 0$ and $\sum_i B^i_j \lambda^i = 0$, leads to

$$dA^i_j = \frac{(R^i)^{n-j}}{\prod_{k\neq i}(R^i - R^k)}(dx - \lambda^i dt), \ j = 3, \ldots, n.$$

The proof of these results is based on the identities:

$$\sum_{i=1}^{n} \frac{(R^i)^{s-1}}{\prod_{k\neq i}(R^i - R^k)} = \begin{cases} 0, & \text{if } s = 1, \ldots, n-1; \\ 1, & \text{if } s = n; \\ \sum_k R^k, & \text{if } s = n+1. \end{cases}$$

It is easy to see that in our case

$$dA^i_j = (R^i)^{n-j} dA^i_n, \quad j = 1, \ldots, n-1,$$

i.e. $\varphi^i_j(R^i) = (R^i)^{n-j}$. Substitution of these expressions into (A.11) gives the general solution of the system under consideration.

A.3 On the Rank of a Web and Invariants of Wave Systems

Let us start with simple examples (see [BF 92]).

Example A.8 Let us imagine two one-dimensional waves with impulses k_1, k_2 and laws of dispersion $\omega_1(k_1), \omega_2(k_2)$ which while interacting with one another, become two new waves with impulses k_3, k_4 and laws of dispersion $\omega_3(k_3), \omega_4(k_4)$. The conditions of conservation of impulse and energy have the form:

$$\begin{aligned} k_1 + k_2 &= k_3 + k_4, \\ \omega_1(k_1) + \omega_2(k_2) &= \omega_3(k_3) + \omega_4(k_4) \end{aligned} \tag{A.19}$$

and determine a two-dimensional surface in a four-dimensional space R^4 with the coordinates k_1, k_2, k_3 and k_4. This surface is called the *resonance manifold.* Suppose that on the resonance manifold, the additional condition:

$$\varphi_1(k_1) + \varphi_2(k_2) = \varphi_3(k_3) + \varphi_4(k_4) \tag{A.20}$$

holds, and this relation cannot be represented as a linear combination of equations (A.19), i.e. $\varphi_i(k_i) \neq ak_i + b\omega_i(k_i)$, where a and b are constants. Relations (A.20) can be interpreted as an additional conservation law of the wave system, i.e. as its invariant. The importance of the study of invariants of wave systems was emphasized heavily in [ZS 80] and [ZS 88]. From the point of view of web geometry, the presence of the additional relation (A.20) means that the curvilinear four-web defined on the resonance manifold (A.19) by the equations $k_i = \text{const}$, $i = 1, 2, 3, 4$, is of rank three. Note that since the maximal possible value for the rank of a four-web in the plane is equal to three, the wave system under consideration has at most one additional invariant. Thus, the problem of description of resonance manifolds (A.19) admitting the additional invariant (A.20), is the problem of description of four-webs of maximum rank three. The solution of this problem is well-known: any four-web of maximum rank three is rectifiable and formed by the straight lines belonging to an algebraic curve of fourth class [Bl 55].

Example A.9 Next, we consider two two-dimensional waves with impulses $\vec{k}_1, \vec{k}_2$ ($\vec{k}_1, \vec{k}_2$ are vectors from $\mathbf{R}^2$) and laws of dispersion $\omega_1(\vec{k}_1), \omega_2(\vec{k}_2)$ which while interacting with one another, become one new wave with the impulse $\vec{k}$ and the law of dispersion $\omega(\vec{k})$:

$$\begin{aligned} &\vec{k}_1 + \vec{k}_2 = \vec{k}, \\ &\omega_1(\vec{k}_1) + \omega_2(\vec{k}_2) = \omega(\vec{k}). \end{aligned} \tag{A.21}$$

Equations (A.21) determine in the six-dimensional space $\mathbf{R}^6$ a three-dimensional surface M^3 which is called the *resonance manifold.* Each of the equations $\vec{k}_1 =$ const, $\vec{k}_2 =$ const and $\vec{k} =$ const determines a two-parameter family of curves on the resonance manifold M^3. They form a curvilinear three-web on M^3. Three-webs of this kind were studied in [BW 34], [Na 65] and [B 35]. Note that the web under consideration is automatically of rank three since each of equations (A.21) is an abelian equation.

As we did in Example **A.8**, we will consider a *conservation law*: this is an additional relation of the form:

$$\varphi_1(\vec{k}_1) + \varphi_2(\vec{k}_2) = \varphi(\vec{k}), \tag{A.22}$$

which is identically satisfied on the resonance manifold M^3, but cannot be represented as a linear combination of relations (A.21), i.e. $\varphi_i(\vec{k}_i) \neq (\vec{a}, \vec{k}_i) + b\omega_i(\vec{k}_i)$. From a geometrical point of view, the presence of the additional condition (A.21) means that the rank of the curvilinear three-web we have constructed is greater than three.

The rank problem for curvilinear three-webs in three-dimensional space was investigated in detail in the papers [BW 34] and [Na 65]. The description of three-webs of maximum rank, obtained in these papers, allows us to give the complete classification of resonance manifolds (A.21) admitting additional invariants (A.22) (see [BF 92]).

For a wave system of general type, there are d r-dimensional waves with impulses $\vec{k}_i \in \mathbf{R}^r$, $i = 1, \ldots, d$, and laws of dispersion $\omega_i(\vec{k}_i)$. The resonance manifold can be given by the equations:

$$\begin{aligned} &\vec{k}_1 + \ldots + \vec{k}_d = 0, \\ &\omega_1(\vec{k}_1) + \ldots + \omega_d(\vec{k}_d) = 0, \end{aligned} \tag{A.23}$$

and it determines a $(dr - r - 1)$-dimensional surface in the space $\mathbf{R}^{dr}$ with the coordinates $\vec{k}_1, \ldots, \vec{k}_d$. On the resonance manifold, we consider a d-web whose leaves of the ith family are given by the equations: $\vec{k}_i =$ const. This web automatically has rank $r + 1$, since all of equations (A.23) are abelian. The presence of an additional invariant

$$\varphi_1(\vec{k}_1) + \ldots + \varphi_d(\vec{k}_d) = 0$$

means that the rank of the web we have constructed is greater than $r + 1$. Thus, in the most general setting the problem of description of wave systems with additional invariants can be reduced to the description of d-webs of surfaces of codimension r on the manifold of dimension $dr - r - 1$ whose rank is greater than $r + 1$. Examples **A.8** and **A.9** can be obtained from this by setting $d = 4, r = 1$ and $d = 3, r = 2$, respectively. We emphasize that for other values of d and r and the dimension $dr - r - 1$ for the web manifold, the rank problem has not been studied at all.

Some concrete examples of resonance manifolds with additional invariants as well as a discussion of certain physical applications can be found in [BF 92].

Bibliography

[Ac 65] Aczel, J.: *Quasigroups, nets and nomograms.* Adv. in Math. **1** (1965), no. 3, 383–450. (MR **33** #1395; Zbl. **135**, p. 36.)

[Ac 81] Aczel, J.: *On the Thomsen condition for webs.* J. Geom. **17** (1981), no. 2, 155–160. (MR 83e:20081; Zbl. 482.51004.)

[A 69a] Akivis, M.A.: *The canonical expansions of the equations of a local analytic quasigroup.* (Russian) Dokl. Akad. Nauk SSSR **188** (1969), no. 5, 967–970. English translation: Soviet Math. Dokl. **10** (1969), no. 5, 1200–1203. (MR **41** #7021; Zbl. 205, p. 26.)

[A 69b] Akivis, M.A.: *Three-webs of multidimensional surfaces.* (Russian) Trudy Geometr. Sem. **2** (1969), 7–31. (MR **40** #7967; Zbl. 244.53014.)

[A 73a] Akivis, M.A.: *The local differentiable quasigroups and three-webs that are determined by a triple of hypersurfaces.* (Russian) Sibirsk. Mat. Zh. **14** (1973), no. 3, 467–474. English translation: Siberian Math. J. **14** (1973), no. 3, 319–324. (MR **48** #2911; Zbl. 267.53005 & 281.53002.)

[A 73b] Akivis, M.A.: *Local differentiable quasigroups and three-webs of multidimensional surfaces.* (Russian) Studies in the Theory of Quasigroups and Loops, Shtiintsa, Kishinev, 1973, 3–12. (MR **51** #6618.)

[A 74] Akivis, M.A.: *On isoclinic three-webs and their interpretation in a ruled space of projective connection.* (Russian) Sibirsk. Mat. Zh. **15** (1974), no. 1, 3–15. English translation: Siberian Math. J. **15** (1974), no. 1, 1–9. (MR **50** #3129; Zbl. 288.53021 & 289.53020.)

[A 75a] Akivis, M.A.: *The almost complex structure associated with a three-web of multidimensional surfaces.* (Russian) Trudy Geom. Sem. Kazan. Gos. Univ. Vyp. 8 (1975), 11–15. (MR **54** #13772.)

[A 75b] Akivis, M.A.: *Closed G-structures on a differentiable manifold.* (Russian) Problems in geometry, Vol. 7, 69–79. Akad. Nauk SSSR Vsesoyuz. Inst. Nauchn. i Tekhn. Inform., Moscow, 1975. (MR **57** #17549; Zbl. 548.53032.)

[A 76a] Akivis, M.A.: *The local algebras of a multidimensional three-web.* (Russian) Sibirsk. Mat. Zh. **17** (1976), no. 1, 5–11. English translation: Siberian Math. J. **17** (1976), no. 1, 3–8. (MR **53** #9055; Zbl. 337.53018.)

[A 76b] Akivis, M.A.: *Geometric and algebraic structures associated with a three-web of multidimensional surfaces.* (Russian) All-Union Conf. "150 Years of Lobachevskian Geometry" (Kazan, 1976), Tezisy, Moscow, 1976, p. 5.

[A 77a] Akivis, M.A.: *The integration of the structure equations of a Moufang web of the minimal dimension.* (Russian) Differential Geometry, 3–9, Kalinin. Gos. Univ., Kalinin, 1977. (MR 82j:53027.)

[A 77b] Akivis, M.A.: *Multidimensional Differential Geometry.* (Russian) Kalinin. Gos. Univ., Kalinin, 1977, 99 pp. (MR 82k:53001; Zbl. 459.53001.)

[A 78] Akivis, M.A.: *Geodesic loops and local triple systems in a space with an affine connection.* (Russian) Sibirsk. Mat. Zh. **19** (1978), no. 2, 243–253. English translation: Siberian Math. J. **19** (1978), no. 2, 171–178. (MR **58** #7438; Zbl. 388.53007 & 409.53008.)

[A 80] Akivis, M.A.: *Webs and almost Grassmann structures.* (Russian) Dokl. Akad. Nauk SSSR **252** (1980), no. 2, 267–270. English translation: Soviet Math. Dokl. **21** (1980), no. 3, 707–709. (MR 82a:53016; Zbl. 479.53015.)

[A 81a] Akivis, M.A.: *A class of three-webs that are determined by a triple of hypersurfaces.* (Russian) Sibirsk. Mat. Zh. **22** (1981), no. 1, 3–7. English translation: Siberian Math. J. **22** (1981), no. 1, 1–4. (MR 82c:53014; Zbl. 456.53006 & 472.53028.)

[A 81b] Akivis, M.A.: *A geometric condition of isoclinity of a multidimensional web.* (Russian) Webs and Quasigroups, Kalinin. Gos. Univ., Kalinin, 1981, 3–7. (MR 83e:53010; Zbl. 497.53026.)

[A 82a] Akivis, M.A.: *Differential-geometric structures connected with a three-web.* (Russian) Webs and Quasigroups, Kalinin. Gos. Univ., Kalinin, 1982, 3–6. (MR 83k:53020; Zbl. 499.53012.)

[A 82b] Akivis, M.A.: *Webs and almost Grassmann structures.* (Russian) Sibirsk. Mat. Zh. **23** (1982), no. 6, 6–15. English translation: Siberian Math. J. **23** (1982), no. 6, 763–770. (MR 84b:53018; Zbl. 505.53004 & 516.53013.)

[A 83a] Akivis, M.A.: *Differential geometry of webs.* (Russian) Problems in Geometry, Vol. 15, 187–213, Itogi Nauki i Tekhniki, Akad. Nauk SSSR, Vsesoyuz. Inst. Nauchn. i Tekhn. Informatsii, Moscow, 1983. English translation: J. Soviet Math. **29** (1985), no. 5, 1631–1647. (MR 85i:53019; Zbl. 567.53014.)

[A 83b] Akivis, M.A.: *The local algebraizability condition for a system of submanifolds of a real projective space.* (Russian) Dokl. Akad. Nauk SSSR **272** (1983), no. 6, 1289–1291. English translation: Soviet Math. Dokl. **28** (1983), no. 2, 507–509. (MR 85c:53018; Zbl. 547.53006.)

[A 87] Akivis, M.A.: *The conditions of algebraizability of a triple of curves in a three-dimensional projective space.* (Russian) Webs and Quasigroups, Kalinin. Gos. Univ., Kalinin, 1987, 129–136. (MR 88i:53019; Zbl. 617.53012.)

[A 88] Akivis, M.A.: *Certain relations between the curvature and torsion tensors of a multidimensional three-web.* (Russian) Webs and Quasigroups, Kalinin. Gos. Univ., Kalinin, 1988, 23–32. (MR 89g:53019; Zbl. 659.53021.)

[AA 90] Akivis, M.A.; Apresyan, Yu. A.: *On three-webs $W(n+1, n+1, n)$ on a manifold of dimension $2n + 1$.* (Russian) Webs and Quasigroups, Kalinin. Gos. Univ., Kalinin, 1990, 4–10. (MR 91h:53021; Zbl. 701.53028.)

[AGe 82] Akivis, M.A.; Gerasimenko, S.A.: *Some closure figures on manifolds with symmetry.* (Russian) Webs and Quasigroups, Kalinin. Gos. Univ., Kalinin, 1982, 7–11. (MR 84b:53017; Zbl. 499.53011.)

[AGe 86] Akivis, M.A.; Gerasimenko, S.A.: *Multidimensional Bol webs.* (Russian) Problems in Geometry, Vol. 18, 73-103, Itogi Nauki i Tekhniki, Akad. Nauk SSSR Vsesoyuz. Inst. Nauchn. i Tekhn. Informatsii, Moscow, 1986. English translation: J. Soviet Math. **42** (1988), no. 5, 1920–1943. (MR 88j:53023; Zbl. 617.53020.)

[AG 72] Akivis, M.A.; Goldberg, V.V.: *On multidimensional three-webs formed by surfaces of different dimensions.* (Russian) Dokl. Akad. Nauk SSSR **203** (1972), no. 2, 263–266. English translation: Soviet Math. Dokl. **13** (1972), no. 2, 354–357. (MR **36** #821; Zbl. 273.53015.)

[AG 73] Akivis, M.A.; Goldberg, V.V.: *On multidimensional three-webs formed by surfaces of different dimensions.* (Russian) Trudy Geom. Sem. Inst. Nauchn. Inform., Akad. Nauk SSSR **4** (1973), 179–204. (MR **51** #1653; Zbl. 314.53011.)

[AG 74] Akivis, M.A.; Goldberg, V.V.: *The four-web and the local differentiable ternary quasigroup that are determined by a quadruple of surfaces of codimension two.* (Russian) Izv. Vyssh. Uchebn. Zaved. Mat. **1974**, no. 5(144), 12–24. English translation: Soviet Math. (Iz. VUZ) **18** (1974), no. 5, 9–19. (MR **50** #8321; Zbl. 297.53037.)

[AS 71a] Akivis, M.A.; Shelekhov, A.M.: *The computation of the curvature and torsion tensors of a multidimensional three-web and of the associator of the local quasigroup that is connected with it.* (Russian) Sibirsk. Mat. Zh. **12** (1971), no. 5, 953–960. English translation: Siberian Math. J. **12** (1971), no. 5, 685–689. (MR **44** #5876; Zbl. 206.53005 & 236.53033.)

[AS 71b] Akivis, M.A.; Shelekhov, A.M.: *Local differentiable quasigroups and connections that are associated with a three-web of multidimensional surfaces.* (Russian) Sibirsk. Mat. Zh. **12** (1971), no. 6, 1181–1191. English translation: Siberian Math. J. **12** (1971), no. 6, 845–892. (MR **44** #5877; Zbl. 231.53021 & 243.53018.)

[AS 74] Akivis, M.A.; Shelekhov, A.M.: *The structure of the manifold of isoclinic surfaces of an isoclinic three-web.* (Russian) Collection of Articles on Differential Geometry, pp. 11–20. Kalinin. Gos. Univ., Kalinin, 1974. (MR **54** #1111.)

[AS 81] Akivis, M.A.; Shelekhov, A.M.: *Foundations of the theory of webs.* (Russian) Kalinin. Gos. Univ., Kalinin, 1981, 88 pp. (MR 83h:53001; Zbl. 475.53016.)

[AS 85a] Akivis, M.A.; Shelekhov, A.M.: *Introduction to the theory of three-webs.* (Russian) Kalinin. Gos. Univ., Kalinin, 1985, 85 pp. (MR 87f:53019.)

[AS 85b] Akivis, M.A.; Shelekhov, A.M.: *Subwebs of a multidimensional three-web.* (Russian) Sibirsk. Math. Zh., 1985, 20 pp., bibl. 4 titles, Dep. in VINITI on 10/9/85 under no. 7130–B.

[AS 86] Akivis, M.A.; Shelekhov, A.M.: *On the canonical coordinates in a local analytic loop.* (Russian) Webs and Quasigroups, Kalinin. Gos. Univ., Kalinin, 1986, 120–124. (MR 88i:22004; Zbl. 616.22003.)

[AS 88] Akivis, M.A.; Shelekhov, A.M.: *On 3-subwebs of 3-webs and subalgebras of local W_k-algebras.* Acta Math. Hungar. **52** (1988), no. 3–4, 256–271. (MR 90d:53028; Zbl. 674.53017.)

[AS 89] Akivis, M.A.; Shelekhov, A.M.: *On alternators of the fourth order of a local analytic loop and three-webs of multidimensional surfaces.* (Russian) Izv. Vyssh. Uchebn. Zaved. Mat. **1989**, no. 4 (334), 12–16. English translation: Soviet Math. (Iz. VUZ) **33** (1989), no. 4, 13–18. (MR 90f:53032; Zbl. 701.53027.)

[Alb 79] Al'ber, S.I. *Investigation of equations of Korteweg-de Vries by the method of recurrence relations.* J. London Math. Soc. **19** (1979), no. 2, 467–480. (MR 81b:35084; Zbl. 413.53064.)

[Al 43] Albert, A.A.: *Quasigroups I.* Trans. Amer. Math. Soc. **54** (1943), no. 3, 507–519. (MR **5**, p. 229; Zbl. **63**, p. A6.)

[Al 44] Albert, A.A.: *Quasigroups II.* Trans. Amer. Math. Soc. **55** (1944), no. 3, 401–409. (MR **6**, p. 42; Zbl. **63**, p. A6.)

[An 81a] Andikyan, M.A.: *A transversal distribution on a multidimensional three-web.* (Russian) Izv. Vyssh. Uchebn. Zaved. Mat. **1981**, no. 4, 69–73. English translation: Soviet Math. (Iz. VUZ) **25** (1984), no. 4, 81–87. (MR 83c:53021; Zbl. 472.53027.)

[An 81b] Andikyan, M.A.: *Three-webs in a tangent bundle that are defined by a multidimensional surface of an affine space.* (Russian) Ukrain. Geom. Sb. **24** (1981), 3–12. (MR 82k:53028; Zbl. 468.53013 & 593.53004.)

[An 81c] Andikyan, M.A.: *Three-webs in the tangent bundle of a differentiable manifold.* (Russian) Uchen. Zapiski Erevan. Univ., Estestv. Nauki, 1981, no. 1, 3–12. (Zbl. 593.53004.)

[An 81d] Andikyan, M.A.: *Three-webs that are symmetrically adjoint to a normalized surface of an affine space.* (Russian) Akad. Nauk Armyan. SSR Dokl. **72** (1981), no. 4, 231–237. (MR 83b:53016; Zbl. 473.53011.)

[Ap 77] Apresyan, Ju.A.: *On a class of three-webs on a four-dimensional manifold and corresponding third order differential equations.* (Russian) Izv. Vyssh. Uchebn.

Zaved. Mat. **29** (1985), no. 1, 3–8. English translation: Soviet Math. (Iz. VUZ) **29** (1985), no. 1, 1–7. (MR 86f:53013; Zbl. 569.53008.)

[Ap 84] Apresyan, Ju.A.: *Three-parameter families of diffeomorphisms of a line onto a line containing two linear complexes of one-parameter subfamilies of special type.* (Russian) Webs and Quasigroups, Kalinin. Gos. Univ., Kalinin, 1984, 8–15. (MR 88c:53001; Zbl. 558.53009.)

[Ar 81] Arakelyan, G.S.: *Some classes of multidimensional three-webs for which the surfaces of one family belong to the surfaces of another family.* (Russian) Vestnik Moskov. Univ. Ser. I Mat. Mekh., **1981**, no. 2, 3–7. English translation: Moscow Univ. Math. Bull. **36** (1981), no. 2, 1–5. (MR 82m:53016; Zbl. 459.53013.)

[BF 92] Balk, A.M.; Ferapontov, E.V.: *Invariants of wave systems and web geometry.* Advances of Soviet Mathematics, vol. 14, 1992 (Ed. V.E. Zakharov) (to appear).

[BS 83] Barlotti, A.; Strambach, K.: *The geometry of binary systems.* Adv. Math., **49** (1983), no. 1, 1–105. (MR 84k:51005; Zbl. 518.20064.)

[Be 80] Beauville, A.: *Géométrie des tissus (d'apres S.S. Chern and P.A. Griffiths).* Séminaire Bourbaki (1978/79), Exp. no. 531, 103-119. Lecture Notes in Math. **770**, Springer, Berlin, 1980. (MR 82a:53017; Zbl. 436.57088.)

[Bel 66] Belousov, V.D.: *Balanced identities in quasigroups.* (Russian) Mat. Sb. (N.S.) **70** (1966), no. 1, 55–97. (MR **34** #273; Zbl. **199**, p. 52.)

[Bel 67] Belousov, V.D.: *Foundations of the theory of quasigroups and loops.* (Russian) Izdat. "Nauka", Moscow, 1967, 223 pp. (MR **36** #1569; Zbl. **163**, p. 18.)

[Bel 71] Belousov, V.D.: *Algebraic nets and quasigroups.* (Russian) Izdat. "Shtiintsa", Kishinev, 1971, 165 pp. (MR **49** #5214; Zbl. 245.5005.)

[Bel 72] Belousov, V.D.: *n-ary quasigroups.* (Russian) Izdat. "Shtiintsa", Kishinev, 1972, 227 pp. (MR **50** #7369; Zbl. 282.20061.)

[Bel 79] Belousov, V.D.: *Configurations in algebraic nets.* (Russian) Izdat. "Stiintsa", Kishinev, 1979, 142 pp. (MR 80k:20074; Zbl. 447.94058.)

[Bel 88] Belousov, V.D.: *Autopies and anti-autopies in quasigroups.* (Russian) Mat. Issled. No. 102 Kvazigruppy (1988), 3–25. (Zbl. 651.20074.)

[BR 66] Belousov, V.D.; Ryzhkov, V.V.: *On a method of obtaining closure figures.* (Russian) Mat. Issled. 1 (1966), Vyp. 2, 140–150. (MR **36** #6526; Zbl. 226.20074.)

[Bes 58] Beskin, L.N.: *Pseudo-Kählerian systems.* (Russian) Nauchn. Dokl. Vyssh. Shkoly **1958**, no. 5, 21–28.

[BiS 87] Billig, V.A.; Shelekhov, A.M.: *Classification of identities with one variable in a smooth local loop.* (Russian) Webs and Quasigroups, Kalinin. Gos. Univ., Kalinin, 1987, 24–32. (MR 88i:20104; Zbl. 617.20042.)

[BiS 90] Billig, V.A.; Shelekhov, A.M.: *A classification of the identities of length* 12 *and order* 4 *with one variable in a local analytic loop.* (Russian) Webs and Quasigroups, Kalinin. Gos. Univ., Kalinin, 1990, 49–55. (Zbl. 701.22002.)

[Bla 52] Blank, Ya.P.: *On the problem of N.G. Chebotarev concerning generalized translation surfaces.* (Russian) Harkov Gos. Univ. Uch. Zap. 40 = Zap. Mat. Otd.' Fiz.-Mat. Fak. i Harkov Mat. Obshch. (4) **23** (1952), 103–112. (MR **18**, p. 64.)

[Bl 28] Blaschke, W.: *Thomsens Sechseckgewebe. Zueinander diagonale Netze.* Math. Z. **28** (1928), 150–157.

[Bl 33] Blaschke, W.: *Abzählungen für Kurvengewebe und Flächengewebe.* Abh. Math. Sem. Univ. Hamburg **9** (1933), 239–312. (Zbl. **7**, p. 78.)

[Bl 55] Blaschke, W.: *Einführung in die Geometrie der Waben.* Birkhäuser-Verlag, Basel-Stuttgart, 1955, 108 pp. (MR **17**, p. 780; Zbl. **68**, p. 365.)

[BB 38] Blaschke, W.; Bol, G.: *Geometrie der Gewebe.* Springer-Verlag, Berlin, 1938, viii+339 pp. (Zbl. **20**, p. 67.)

[BW 34] Blaschke, W.; Walberer, P.: *Die Kurven* 3*-Gewebe höchsten Ranges im* R^3. Abh. Math. Sem. Univ. Hamburg **10** (1934), 180–200. (Zbl. **9**, p. 378.)

[B 35] Bol, G.: *Über* 3*-Gewebe in vierdimensionalen Raum.* Math. Ann. **110** (1935), 431–463. (Zbl. **10**, p. 222.)

[B 36] Bol, G.: *Über ein bemerkenswertes* 5*-Gewebe in der Ebene.* Abh. Math. Sem. Univ. Hamburg **11** (1936), 387–393. (Zbl. **14**, pp. 230–231.)

[B 37] Bol, G.: *Gewebe und Gruppen.* Math. Ann. 114 (1937), 414–431. (Zbl. **16**, p. 226.)

[Bo 82] Bolodurin, V.S.: *On the invariant theory of point correspondences of three projective spaces.* (Russian) Izv. Vyssh. Uchebn. Zaved. Mat. **1982**, no. 5 (240), 9–15. English translation: Soviet Math. (Iz. VUZ) **26** (1982), no. 5, 8–16. (MR 84b:53010; Zbl. 509.53008.)

[Bot 74] Botsu, V.P.: *A direct proof of the generalized theorem of Graf–Sauer.* (Russian) Collection of Articles on Differential Geometry, Kalinin. Gos. Univ., Kalinin, 1974, 36–51. (MR **58** #30792.)

[Bot 75] Botsu, V.P.: *On a class of four-dimensional hexagonal three-webs.* (Russian) Ukrain. Geom. Sb. **18** (1975), 27–37. (MR **53** #3912; Zbl. 433.53008.)

[Bot 84] Botsu, V.P.: *Isoclinity of four-dimensional hexagonal three-webs.* (Russian) Moscow Hydromeliorative Inst., Moscow, 1984, 27 pp., bibl. 8 titles. Dep. in VINITI on 8/14/84 under no. 5824-84. (Zbl. 560.02205.)

[Br 51] Bruck, R. H.: *Loops with transitive automorphism groups.* Pac. J. Math. **1** (1951), no. 1, 481–483. (MR **13** # 620; Zbl. **44**, p. 11.)

[Br 71] Bruck, R. H.: *A survey of binary systems.* Springer-Verlag, Berlin, 1971, 3rd printing, viii+185 pp. (MR **20** #76; Zbl. **81**, p. 17 & **141**, p. 14.)

[BP 56] Bruck, R. H.; Paige, L.J.: *Loops whose inner mappings are automorphisms.* Ann. Math. (2) **63** (1956), 308–323. (MR **17**, p. 943; Zbl. **74**, p. 17.)

[By 80] Bychek, V.I.: *Coordinate three-webs on submanifolds of a generalized Appell space.* (Russian) Moskov. Gos. Ped. Inst., Moscow, 1980, 34 pp., bibl. 4 titles. Dep. in VINITI on 10/13/80 under no. 4335-80 Dep.

[Ca 08] Cartan, /'E.: *Les sous-groupes des groupes continus de transformations.* Ann. Sci. École Norm. (3) **25** (1908), 57–194. (See also É. Cartan, *Œuvres Completes*, Centre National de la Researche Scientifique, Paris, 1984, Partie II, 719–856.)

[Ca 28] Cartan, É.: *Leçons sur la géométrie des espaces de Riemann.* Gauthier-Villars, Paris, 1928. English translation of 2nd reviewed and augmented edition of 1951: *Geometry of Riemannian spaces.* Math. Sci. Press, Brookline, MA, 1983, xiv+506 pp. (MR 85m:53001; Zbl. **44**, p. 184.)

[Ca 37] Cartan, É.: *La théorie de groupes finis et continus et la géométrie différentielle traitées par la méthode di repère mobile.* Gauthier-Villars, Paris, 1937, vi+269 pp. (Zbl. **18**, p. 298.)

[Ca 60] Cartan, É.: *Riemannian geometry in orthogonal frame.* (Russian) Izdat. Moscow. State Univ., Moscow, 1960, 307 pp. (MR **23** #A1316; Zbl. **99**, p. 372.)

[Ca 62] Cartan, É.: *Spaces with an affine, projective and conformal connection.* (Russian) Izdat. Kazan State Univ., Kazan, 1962, 214 pp. (MR **36** #2080.)

[Ca 71] Cartan, É.: *Les systèmes différentiels extérieurs et leurs applications géometrques.* 2nd ed. Hermann, Paris, 1971, 210 pp. (Zbl. **211**, p. 127.)

[CaH 67] Cartan, H.: *Formes différentielles. Application élémentaires au calcul des variations at à la théorie des courbes et surfaces.* Hermann, Paris, 1967, 186 pp. (MR **37** #6358; Zbl. **184**, p. 127.) English translation: *Differential forms.* H. Mifflin Co., Boston, 1970, 166 pp. (MR **42** #2379; Zbl. **213**, p. 370.)

[CPS 90] Chein, O.; Pflugfelder, H.; Smith, J.D.H. (eds): *Quasigroups and loops: theory and applications.* Heldermann-Verlag, Berlin, 1990, xii+568 pp. (Zbl. 704.00017.)

[Ch 24] Chebotarev, N.G.: *On surfaces of translation.* (Russian) Mat. Sb. **31** (1924), 434–445.

[Ch 27] Chebotarev, N.G.: *Surfaces of translation in a multidimensional space.* (Russian) Trudy Vsesoyuzn. Mat. S'ezda, Moscow, 1927, 232–234.

[C 36a] Chern, S.S.: *Eine Invariantentheorie der Dreigewebe aus r-dimensionalen Mannigfaltigkeiten in* $\mathbf{R}_{2r}$. Abh. Math. Sem. Univ. Hamburg **11** (1936), no. 1–2, 333–358. (Zbl. **13**, p. 418.)

[C 36b] Chern, S.S.: *Abzählungen für Gewebe.* Abh. Math. Sem. Univ. Hamburg **11** (1936), no. 1–2, 163–170. (Zbl. **11**, p. 132.)

[C 82] Chern, S.S.: *Web geometry.* Bull. Amer. Math. Soc. (N.S.) **6** (1982), no. 1, 1–8. (MR 84g:53024; Zbl. 483.53012.)

[CG 77] Chern, S.S.; Griffiths, P.A.: *Linearization of webs of codimension one and maximum rank.* Proc. Intern. Symp. on Algebraic Geometry, 1977, Kyoto, Japan, 85–91. (MR 81k:53010; Zbl. 406.14003.)

[CG 78a] Chern, S.S.; Griffiths, P.A.: *Abel's theorem and webs.* Jahresber. Deutsch. Math.-Verein. **80** (1978), no. 1–2, 13–110. (MR 80b:53008; Zbl. 386.14002.)

[CG 78b] Chern, S.S.; Griffiths, P.A.: *An inequality for the rank of a web and webs of maximum rank.* Ann. Scuola Norm. Sup. Pisa Cl. Sci. (4) **5** (1978), no. 3, 539–557. (MR 80b:53009.; Zbl. 402.57001.)

[CG 81] Chern, S.S.; Griffiths, P.A.: *Corrections and addenda to our paper "Abel's theorem and webs".* Jahresber. Deutsch. Math.-Verein. **83** (1981), 78–83. (MR 82k:53030; Zbl. 474.14003.)

[D 80] Damiano, D.B.: *Webs, abelian equations, and characteristic classes.* Ph.D. Thesis, Brown University, 1980, 98 pp.

[D 83] Damiano, D.B.: *Webs and characteristic forms on Grassmann manifolds.* Amer. J. Math. **105** (1983), 1325–1345. (MR 85g:53014; Zbl. 528.53022.)

[Dr 82] Dragunov, V.K.: *A coordinate three-web on a surface in a space of cubic metric.* (Russian) Webs and Quasigroups, Kalinin. Gos. Univ., Kalinin, 1982, 86–93. (MR 84a:53014; Zbl. 499.53015.)

[DJ 85] Dufour, J.P.; Jean, P.: *Rigidity of webs and families of hypersurfaces.* Singularities and Dynamical Systems (Iráklion, 1983), North-Holland Math. Stud., 103, North-Holland, Amsterdam/New York, 1985, 271–283. (MR 87b:53023; Zbl. 583.57015.)

[ELOS 79] Evtushik, L.E.; Lumiste, Yu.G.; Ostianu, N.M.; Shirokov, A.P.: *Differential-geometric structures on manifolds.* (Russian) Problems in Geometry, Vol. 9, 248 pp, Akad. Nauk SSSR, Vsesoyuz. Inst. Nauchn. i Tekhn. Informatsii, Moscow, 1979. English translation: J. Soviet Math. **14** (1980), no. 4, 1573–1719. (MR 82k:53054b; Zbl. 455.58002.)

[F 76] Fedorova, V.I. (Khasina, V.I.) *On three-webs with partially skew-symmetric curvature tensor.* (Russian) Izv. Vyssh. Uchebn. Zaved. Mat. **1976**, no. 11, 114–117. English translation: Soviet Math. (Iz. VUZ) **20** (1976), no. 11, 101-104. (MR **58** #12790; Zbl. 353.53012.)

[F 77] Fedorova, V.I. (Khasina, V.I.) *On a class of three-webs W_6 with partially skew-symmetric curvature tensor.* (Russian) Ukrain. Geom. Sb. **20** (1977), 115-124. (MR **58** #24090; Zbl. 439.53020 & 499:53014.)

[F 78] Fedorova, V.I.: *A condition defining multidimensional Bol's three-webs.* (Russian) Sibirsk. Mat. Zh. **19** (1978), no. 4, 922–928. English translation: Siberian Math. J. **19** (1978), no. 4, 657–661. (MR **58** #24036; Zbl. 398.53007 & 409.53007.)

[F 81] Fedorova, V.I.: *Six-dimensional Bol three-webs with the symmetric tensor* a^i_j. (Russian) Webs and Quasigroups, Kalinin. Gosud. Univ., Kalinin, 1981, 110–123. (MR 83m:53021; Zbl. 497.53020.)

[F 82] Fedorova, V.I.: *An interpretation of a six-dimensional Bol web in three-dimensional projective space.* (Russian) Webs and Quasigroups, Kalinin. Gos. Univ., Kalinin, 1982, 142–148. (MR 84a:53015; Zbl. 499.53017.)

[F 84] Fedorova, V.I.: *On the classification of six-dimensional Bol webs.* (Russian) Webs and Quasigroups, Kalinin. Gos. Univ., Kalinin, 1984, 124–132. (MR 88c:53001; Zbl. 558.53014.)

[Fe 68] Fenyves, F.: *Extra loops I.* Publ. Math. Debrecen **15** (1968), 235–238. (MR **38** #5970; Zbl. **172**, p. 24.)

[Fe 69] Fenyves, F.: *Extra loops II: On loops with identities of Bol–Moufang type.* Publ. Math. Debrecen **16** (1969), 187–192. (1970) (MR **41** #7017; Zbl. 221.20097.)

[Fer 88] Ferapontov, E.V.: *Weakly non-linear semi-Hamiltonian systems of differential equations from the point of view of web geometry.* Tartu Riikl. Ül. Toimetised Vih. **803** (1988), 103–114. (MR 90a:53027; Zbl. 714.53020.)

[Fer 89] Ferapontov, E.V.: *Systems of three differential equations of hydrodynamical type with a hexagonal three-web of characteristics.* (Russian) Funktsional. Anal. i Prilozhen. **23** (1989), no. 2, 79–80. English translation: Functional Analysis and Its Appl. **23** (1989), no. 3, 151–153. (MR 90k:35157; Zbl. 714.35070.)

[Fer 90] Ferapontov, E.V.: *Integration of weakly non-linear semi-Hamiltonian systems of hydrodynamic type by the methods of web geometry.* Mat. Sb. **181** (1990), no. 9, 1220–1235.

[Fer 91] Ferapontov, E.V.: *Equations of hydrodynamic type from the point of view of web geometry.* Mat. Zametki **50** (1991), no. 5, 97–108.

[Ga 88] Galkin, V.M.: *Quasigroups.* Algebra, Topology, Geometry, Vol. 26, 3–44. Itogi Nauki i Tekhniki, Akad. Nauk SSSR, Vsesoyuz. Inst. Nauchn. i Tekhn. Inform., Moscow, 1988. English translation: J. Soviet Math. **49** (1990), no. 3, 941–967. (MR 89k:20103; Zbl. 675.20057 & 697.20050.)

[Gl 61] Glagolev, N.A.: *Course of nomography.* (Russian) Vyssh. Shkola, Moscow, 1961, 268 pp.

[G 73] Goldberg, V.V.: *$(n+1)$-webs of multidimensional surfaces.* (Russian) Dokl. Akad. Nauk SSSR **210** (1973), no. 4, 756–759. English translation: Soviet Math. Dokl. **14** (1973), no. 3, 795–799. (MR **48** #2919; Zbl. 304.53017.)

[G 74a] Goldberg, V.V.: *$(n+1)$-webs of multidimensional surfaces.* (Russian) Bulgar. Akad. Nauk Izv. Mat. Inst. **15** (1974), 405–424. (MR **51** #13889; Zbl. 346.53010.)

[G 74b] Goldberg, V.V.: *Isoclinic $(n+1)$-webs of multidimensional surfaces.* (Russian) Dokl. Akad. Nauk SSSR **218** (1974), no. 5, 1005–1008. English translation: Soviet Math. Dokl. **15** (1974), no. 5, 1437–1441. (MR **52** #11763; Zbl. 314.53012.)

[G 74c] Goldberg, V.V.: *Transversally geodesic three-webs, hexagonal three-webs and group three-webs formed by surfaces of different dimensions.* (Russian) Collection of Articles on Differential Geometry, Kalinin. Gos. Univ., Kalinin, 1974, 52–69. (MR **54** #5985.)

[G 75a] Goldberg, V.V.: *An invariant characterization of certain closure conditions in ternary quasigroups.* (Russian) Sibirsk. Mat. Zh. **16** (1975), no. 1, 29–43. English translation: Siberian Math. J. **16** (1975), no. 1, 23–34. (MR **51** #6619; Zbl. 304.20046 & 318.20047.)

[G 75b] Goldberg, V.V.: *Local ternary quasigroups that are connected with a four-web of multidimensional surfaces.* (Russian) Sibirsk. Mat. Zh. **16** (1975), no. 2, 247–263. English translation: Siberian Math. J. **16** (1975), no. 2, 190–202. (MR **51** #8318; Zbl. 315.20056 & 349.20026.)

[G 75c] Goldberg, V.V.: *The $(n+1)$-webs defined by $n+1$ surfaces of codimension $n-1$.* (Russian) Problems in Geometry, Vol. 7, 173–195. Akad. Nauk SSSR Vsesoyuz. Inst. Nauchn. i Tekhn. Informatsii, Moscow, 1975. (MR **57** #17537; Zbl. 548.53013.)

[G 75d] Goldberg, V.V.: *The almost Grassmann manifold that is connected with an $(n+1)$-web of multidimensional surfaces.* (Russian) Izv. Vyssh. Uchebn. Zaved. Mat. **1975**, no. 8(159), 29–38. English translation: Soviet Math. (Iz. VUZ) **19** (1975), no. 8, 23–31. (MR **54** #11226; Zbl. 323.53034.)

[G 75e] Goldberg, V.V.: *The diagonal four-web formed by four pencils of multidimensional planes in a projective space.* (Russian) Problems in Geometry, Vol. 7, 197–213. Akad. Nauk SSSR Vsesoyuz. Inst. Nauchn. i Tekhn. Inform., Moscow, 1975. (MR **58** #7422; Zbl. 548.53014.)

[G 75f] Goldberg, V.V.: *A certain property of webs with zero curvature.* (Russian) Izv. Vyssh. Uchebn. Zaved. Mat. **1975**, no. 9(160), 10–13. English translation: Soviet Math. (Iz. VUZ) **19** (1975), no. 9, 7–10. (MR **54** #6007; Zbl. 321.53017.)

[G 76] Goldberg, V.V.: *Reducible, group, and $(2n+2)$-hedral $(n+1)$-webs of multidimensional surfaces.* (Russian) Sibirsk. Mat. Zh. **17** (1976), no. 1, 44–57. English translation: Siberian Math. J. **17** (1976), no. 1, 34–44. (MR **54** #5986; Zbl. 331.53005 & 334.53009.)

[G 77] Goldberg, V.V.: *On the theory of four-webs of multidimensional surfaces on a differentiable manifold X_{2r}.* (Russian) Izv. Vyssh. Uchebn. Zaved. Mat. **1977**, no. 11(186), 15–22. English translation: Soviet Math. (Iz. VUZ) **21** (1977), no. 11, 97–100. (MR **58** #30859; Zbl. 398.53009 & 453.53010.)

[G 80] Goldberg, V.V.: *On the theory of four-webs of multidimensional surfaces on a differentiable manifold* X_{2r}. (Russian) Serdica **6** (1980), no. 2, 105–119. (MR 82f:53023.)

[G 82a] Goldberg, V.V.: *Multidimensional four-webs on which the Desargues and triangle figures are closed.* Geom. Dedicata **12** (1982), no. 3, 267–285. (MR 83i:53031; Zbl. 488.53009.)

[G 82b] Goldberg, V.V.: *A classification of six-dimensional group four-webs of multidimensional surfaces.* Tensor (N.S.) **36** (1982), no. 1, 1–8. (MR 87b:53024; Zbl. 478.53009.)

[G 82c] Goldberg, V.V.: *The solutions of the Grassmannization and algebraization problems for* $(n+1)$*-webs of codimension* r *on a differentiable manifold of dimension* nr. Tensor (N.S.) **36** (1982), no. 1, 9–21. (MR 87a:53027; Zbl. 479.53014.)

[G 82d] Goldberg, V.V.: *Grassmann and algebraic four-webs in a projective space.* Tensor (N.S.) **38** (1982), 179–197. (MR 87e:53024; Zbl. 513.53009.)

[G 83] Goldberg, V.V.: *Tissus de codimension* r *et de* r*-rang maximum.* C. R. Acad. Sci. Paris Sér. I Math. **297** (1983), no. 6, 339–342. (MR 85f:53020; Zbl. 539.53012.)

[G 84] Goldberg, V.V.: *An inequality for the* 1*-rank of a scalar web* $SW(d,2,r)$ *and scalar webs of maximum* 1*-rank.* Geom. Dedicata **17** (1984), no. 2, 109–129. (MR 86f:53014; Zbl. 554.53014.)

[G 85] Goldberg, V.V.: 4*-tissus isoclines exceptionnels de codimension deux et de* 2*-rang maximal.* C.R. Acad. Sci. Paris Sér. I Math. **301** (1985), no. 11, 593–596. (MR 87b:53025; Zbl. 579. 53015.)

[G 86] Goldberg, V.V.: *Isoclinic webs* $W(4,2,2)$ *of maximum* 2*-rank.* Differential Geometry (Peniscola, 1985), 168–183, Lecture Notes in Math. **1209**, Springer-Verlag, Berlin/New York, 1986. (MR 88h:53021; Zbl. 607.53008.)

[G 87] Goldberg, V.V.: *Nonisoclinic* 2*-codimensional* 4*-webs of maximum* 2*-rank.* Proc. Amer. Math. Soc. **100** (1987), no. 4, 701–708. (MR 88i:53037; Zbl. 628.53018.)

[G 88] Goldberg, V.V.: *Theory of Multicodimensional* $(n+1)$*-Webs.* Kluwer Academic Publishers, Dordrecht/Boston/London, 1988, xxii+466 pp. (MR 90h:53021; Zbl. 668:53001.)

[G 89] Goldberg, V.V.: *On a linearizability condition for a three-web on a two-dimensional manifold.* Differential Geometry, (Peniscola, 1988), 223–239, Lecture Notes in Math. **1410**, Springer-Verlag, Berlin/New York, 1989. (MR 91a:53032; Zbl. 689.53008.)

[G 90a] Goldberg, V.V.: *On the Chern-Griffiths formulas for an upper bound for the rank of a web.* Differential Geometry and Its Applications (Brno, 1989), 54–57, World Scientific Publishing, Teaneck, NJ, 1990.

[G 90b] Goldberg, V.V.: *Local differentiable quasigroups and webs.* Chapter X in [CPS 90], 263–311.

[G t.a.] Goldberg, V.V.: *Rank problems for webs* $W(d,2,r)$. Differential Geometry (Eger, 1989), Colloq. Math. Soc. Janos Bolyai, North Holland, Amsterdam/New York (to appear).

[GR 88] Goldberg, V.V.; Rosca, R.: *Geometry of exceptional webs* $EW(4,2,2)$ *of maximum 2-rank.* Note Mat. **VIII** (1988), no. 1, 141–153. (Zbl. 707.53017.)

[Go 71] Golovko, I.A.: *Closure conditions in quasigroups.* (Russian) Izv. Akad. Nauk Moldav. SSR Ser. Fiz.-Tekhn. Mat. Nauk **3** (1971), 17–21. (MR **44** #355; Zbl. 279.20056.)

[Gr 76] Griffiths, P.A.: *Variations on a theorem of Abel.* Invent. Math. **35** (1976), 321–390. (MR **55** #8036; Zbl. 339.14003.)

[Gr 77] Griffiths, P.A.: *On Abel's differential equations.* Algebraic Geometry, J.J. Sylvester Sympos., Johns Hopkins Univ., Baltimore, Md., 1976, 26–51. Johns Hopkins Univ. Press, Baltimore, Md, 1977. (MR **58** #655.)

[Gv 81a] Gvozdovich, N.V.: *On three-webs of maximal mobility.* (Russian) Kaliningrad. Gos. Univ. Differentsial'naya Geom. Mnogoobraz. Figur, no. 12 (1981), 13–17. (MR 83a:53015; Zbl. 522.53011.)

[Gv 81b] Gvozdovich, N.V.: *Infinitesimal automorphisms of Bol and Moufang three-webs.* (Russian) Webs and Quasigroups, Kalinin. Gos. Univ., Kalinin, 1981, 83–91. (MR 83k:53032; Zbl. 497.53023.)

[Gv 82] Gvozdovich, N.V.: *Infinitesimal automorphisms of multidimensional three-webs.* (Russian) Izv. Vyssh. Uchebn. Zaved. Mat. **1982**, no. 5, 73–75. English translation: Soviet Math. (Iz. VUZ) **26** (1982), no. 5, 94–97. (MR 83j:53008; Zbl. 509.53014.)

[Gv 85] Gvozdovich, N.V.: *Some problems of the general theory of infinitesimal automorphisms of multidimensional webs.* (Russian) Problems in the Theory of Webs and Quasigroups, Kalinin. Gos. Univ., Kalinin, 1985, 49–54. (MR 87m:53019; Zbl. 572.53018.)

[Ha 66] Hangan, Th.: *Géométrie différentielle grassmannienne.* Rev. Roumaine Math. Pures Appl. **11** (1966), no. 5, 519–531. (MR **34** #744; Zbl. **163**, p. 434.)

[H 77] Hartshorne, R.: *Algebraic Geometry.* Springer-Verlag, New York/ Heidelberg/ Berlin, 1977, xvi+496 pp. (MR **57** #3116; Zbl. 367.14001.)

[HP 47] Hodge, W.V.D.; Pedoe, D.: *Methods of algebraic geometry*, Vol. 1. Cambridge Univ. Press, Cambridge & MacMillan Co., Cambridge, 1947, viii+440 pp. (MR **10**, p. 396.)

[HP 52] Hodge, W.V.D.; Pedoe, D.: *Methods of algebraic geometry*, Vol. 2. Cambridge Univ. Press, Cambridge, 1952, x+394 pp. (MR **13**, p. 972; Zbl. **48**, p. 145.)

[Ho 58a] Hofmann, K.H.: *Topologische loops.* Math. Z. **70** (1958), 13–37. (MR **21** #1362; Zbl. **95**, p. 27.)

[Ho 58b] Hofmann, K.H.: *Topologische loops mit schwachen Assoziativitätorderungen.* Math. Z. **70** (1958), 125–155. (MR **21** #1363; Zbl. **95**, p. 27.)

[Ho 59] Hofmann, K.H.: *Topologische distributive Doppelloops.* Math. Z. **71** (1959), 36–68. (MR **22** #742; Zbl. **95**, p. 27.)

[HS 85] Hofmann, K.H.; Strambach, K.: *Topological and analytical loops.* Preprint No. 869, 96 pp. Fachbereich Mathematik, Technische Hochschule, Darmstadt, 1985. (Zbl. 560.00274.)

[HS 86a] Hofmann, K.H.; Strambach, K.: *Lie's fundamental theorems for local analytical loops.* Pacific J. Math. **123** (1986), no. 2, 301–327. (MR 87k:17002; Zbl. 596.22002.)

[HS 86b] Hofmann, K.H.; Strambach, K.: *The Akivis algebra of a homogeneous loop.* Mathematika **33** (1986), no. 1, 87–95. (MR 88d:17003; Zbl. 601.22002.)

[HS 86b] Hofmann, K.H.; Strambach, K.: *Topologogical and analytic loops.* Chapter IX in [CPS 90], 205–262.

[HS 91] Hofmann, K.H.; Strambach, K.: *Torsion and curvature in smooth loops.* Publ. Math. Debrecen **38** (1991), no. 3-4, 189–214.

[HSa 80] Holmes, J.P.; Sagle, A.A.: *Analytic H-spaces, Campbell-Hausdorff formula, and alternative algebras.* Pacific J. Math. **91** (1980), no. 1, 105–134. (MR 82d:17005; Zbl. 448.17016.)

[I 76] Indrupskaya, E.I. (Verba, E.I.): *Three-webs of curves satisfying one Pfaffian equation.* (Russian) Geometry of Homogeneous Spaces, Moskov. Gos. Ped. Inst., Moscow, 1976, 82–88.

[I 78] Indrupskaya, E.I. (Verba, E.I.): *Nonholonomic three-webs.* (Russian) Geometry of Imbedded Manifolds, Moskov. Gos. Ped. Inst., Moscow, 1978, 18–25. (MR 82a:53018; Zbl. 445.53011.)

[I 79] Indrupskaya, E.I. (Verba, E.I.): *Non-holonomic three-webs of maximum rank.* (Russian) Geometry of Imbedded Manifolds, Moskov. Gos. Ped. Inst., Moscow, 1979, 57–61. (MR 82f:53024; Zbl. 485.53015.)

[I 81] Indrupskaya, E.I. (Verba, E.I.): *Four-dimensional non-holonomic three-webs of maximum rank.* (Russian) Geometry of Imbedded Manifolds, Moskov. Gos. Ped. Inst., Moscow, 1981, 45–58.

[Iv 73] Ivanov, A.D.: *The interpretation of four-dimensional Bol webs in a three-dimensional projective space.* (Russian) Geometry of Homogeneous Spaces, Moskov. Gos. Ped. Inst., Moscow, 1973, 42–57. (MR **54** #5994.)

[Iv 74] Ivanov, A.D.: *Closed form equations of four-dimensional Bol webs.* (Russian) Collection of Articles on Differential Geometry, Kalinin. Gos. Univ., Kalinin, 1974, 70–78. (MR **54** #13736.)

[Iv 75] Ivanov, A.D.: *Four-dimensional Bol webs of elliptic and hyperbolic types.* (Russian) Izv. Vyssh. Uchebn. Zaved. Mat. **19** (1975), no. 9, 25–34. English translation: Soviet Math. (Iz. VUZ) **19** (1975), no. 9, 20–27. (MR **54** #5995; Zbl. 341.53012.)

[Iv 76] Ivanov, A.D.: *On four-dimensional Bol webs of parabolic type.* (Russian) Izv. Vyssh. Uchebn. Zaved. Mat. **1976**, no. 1, 42–47. English translation: Soviet Math. (Iz. VUZ) **20** (1976), no. 1, 34–38. (MR **55** #8984; Zbl. 337.53020.)

[Iv 79] Ivanov, A.D.: *Mutually polar Bol three-webs of hyperbolic type.* (Russian) Kaliningrad. Gos. Univ. Differentsial'naya Geom. Mnogoobraz. Figur, no. 10 (1979), 30–35. (MR 81f:53012; Zbl. 453.53009.)

[Ke 79] Kerdman, F.S.: *Analytic Moufang loops in the large.* (Russian) Algebra i Logika **18** (1979), no. 5, 523–555. English translation: Algebra and Logic **18** (1980), no. 5, 325–347. (MR 82c:22006; Zbl. 457.22002.)

[Ki 85] Kikkawa, M.: *Canonical connections of homogeneous Lie loops and 3-webs.* Mem. Fac. Sci. Shimane Univ. **19** (1985), 37–55. (MR 87j:53077; Zbl. 588.53014.)

[Kil 71] Kilp, H.O.: *Quasilinear systems of first order partial differential equations with m unknown functions, of two independent variables and with noncoinciding characteristics (geometric theory).* Tartu Riikl. Ül. Toimetised Vih. **281** (1971), 63–85. (MR **47** #1958; Zbl. 341.53020.)

[Kil 75] Kilp, H.O.: *Two quasilinear systems of type* $S^1_{32(1)}$ *from mechanics with a hexagonal web of characteristics.* Tartu Riikl. Ül. Toimetised Vih. **374** (1975), 63–78. (MR **54** #1108; Zbl. 389.35008.)

[Kl 81a] Klekovkin, G.A.: *A pencil of Weyl connections and a normal conformal connection on a manifold with relatively invariant quadratic form.* (Russian) Webs and Quasigroups, Kalinin. Gos. Univ., Kalinin, 1981, 47–55. (MR 83h:53035; Zbl. 497.53028.)

[Kl 81b] Klekovkin, G.A.: *A pencil of Weyl connections associated with a four-dimensional three-web.* (Russian) Geometry of Imbedded Manifolds, Moskov. Gos. Ped. Inst., Moscow, 1981, 59–62.

[Kl 81b] Klekovkin, G.A.: *Examples of four-dimensional isoclinic non-transversally geodesic three-webs.* (Russian) Kirov. Gos. Ped. Inst., Kirov, 1982, 19 pp., bibl. 14 titles. Dep. in VINITI on 10/26/82, no. 5335–82.

[Kl 83] Klekovkin, G.A.: *Weyl geometries generated by a four-dimensional three-web.* (Russian) Ukrain. Geom. Sb. **26** (1983), 56–63. (MR 85h:53017; Zbl. 525.53020.)

[Kl 84] Klekovkin, G.A.: *Four-dimensional three-webs with a covariantly constant curvature tensor.* (Russian) Webs and Quasigroups, Kalinin. Gos. Univ., Kalinin, 1984, 56–63. (MR 88c:53001; Zbl. 558.53010.)

[KT 84] Klekovkin, G.A.; Timoshenko, V.V.: *Real realizations of three-websover two-dimensional algebras.* (Russian) Webs and Quasigroups, Kalinin. Gos. Univ., Kalinin, 1984, 63–69. (MR 88c:53001; Zbl. 558.53011.)

[Kn 32] Kneser, H.: *Gewebe und Gruppen.* Abh. Math. Sem. Univ. Hamburg **9** (1932), 147–151. (Zbl. **6**, p. 33.)

[KN 63] Kobayashi, S.; Nomizu, K.: *Foundations of differential geometry*, 2 vols., Wiley–Interscience, New York/London/Sydney, 1963, xi+329 pp., (MR **27** #2945; Zbl. **119**, p. 375.), Vol. 2, 1969, xv+470 pp. (MR **38** #6501; Zbl. **175**, p. 465.)

[Ku 68] Kuz'min, E.N.: *Mal'cev algebras and their representations.* (Russian) Algebra i Logika **7** (1968), no. 4, 48–69. English translation: Algebra and Logic **7** (1968), no. 4, 233–244. (MR **40** #5688; Zbl. **204**, p. 361.)

[Ku 70] Kuz'min, E.N.: *Mal'cev algebras of dimension five over a field of characteristic zero.* (Russian) Algebra i Logika **9** (1970), no. 5, 691–700. English translation: Algebra and Logic **9** (1970), no. 5, 416–421. (MR **44** #266; Zbl. 244.17018 & 249.17017.)

[Ku 71] Kuz'min, E.N.: *The connection between Mal'cev algebras and analytic Moufang loops.* (Russian) Algebra i Logika **10** (1971), no. 1, 3–22. English translation: Algebra and Logic **10** (1971), no. 1, 1–14. (MR **45** #6968; Zbl. 244.17019 & 248.17001.)

[La 53] Laptev, G.F.: *Differential geometry of imbedded manifolds. Group-theoretic method of differential geometry investigations.* (Russian) Trudy Moskov. Mat. Obshch. **2** (1953), 275–383. (MR **15**, p. 64; Zbl. **53**, p. 428.)

[La 66] Laptev, G.F.: *Fundamental differential structures of higher orders on a smooth manifold.* (Russian) Trudy Geom. Sem. Inst. Nauchn. Inform., Akad. Nauk SSSR **1** (1966), 139–190. (MR **34** #6681; Zbl. **171**, p. 423.)

[La 69] Laptev, G.F.: *The structure equations of the principal fibre bundle.* (Russian) T rudy Geom. Sem. Inst. Nauchn. Inform., Akad. Nauk SSSR **2** (1969), 161–178. (MR **40** #8074; Zbl. 253.53032.)

[L 79] Lazareva, V.B.: *Three-webs on a two-dimensional surface in a triaxial space.* (Russian) Kaliningrad. Gos. Univ. Differentsial'naya Geom. Mnogoobraz. Figur, no. 10 (1979), 54–59. (MR 81m:53012; Zbl. 444.53015.)

[L 81] Lazareva, V.B.: *Three-webs generated by three r-planes in a projective space of dimension* $2r + 1$. (Russian) Webs and Quasigroups, Kalinin. Gos. Univ., Kalinin, 1981, 56–68. (MR 84a:53016; Zbl. 497.53022.)

[L 87] Lazareva, V.B.: *On geometry of an n-axial space.* (Russian) Webs and Quasigroups, Kalinin. Gos. Univ., Kalinin, 1987, 68–75. (MR 88i:53023; Zbl. 617.53017.)

[LS 84] Lazareva, V.B.; Shelekhov, A.M.: *Geometric interpretation of invariant rigging of a point correspondence of three straight lines.* (Russian) Izv. Vyssh. Uchebn. Zaved. Mat. **1984**, no. 9, 43–47. English translation: Soviet Math. (Iz. VUZ) **28** (1984), no. 9, 57–62. (MR 86c:53007; Zbl. 578.53014.)

[LS 89] Lazareva, V.B.; Shelekhov, A.M.: *Examples of G-webs of dimension 4, 6, 8, 10 with different tangent algebras.* (Russian) Kalinin. Gos. Univ., Kalinin, 1989, 16 pp., bibl. 15 titles. Dep. in VINITI on 7/26/89 under no. 5030-B89.

[Li 55] Lichnerowicz, A.: *Théorie globale des connections et des groupes d'holonomie.* Edizioni cremonese, Rome, 1955, xv+282 pp. (MR **19**, p. 453; Zbl. **116**, p. 391.) English translation: *Global theory of connections and holonomy groups.* Noordhof Intern. Pub., 1976, xiv+250 pp. (Zbl. 337.53031.)

[Lie 37] Lie, S.: *Die Theorie der Translationflächen und das Abelesche Theorem.* Gesammelte Abhandlungen, Bd. **2**, Abt. 2, Teubner, Leipzig & Ascheoug & Co., Oslo, 1937, 526–579. (Zbl. **17**, p. 190.)

[Lit 89] Little, J.B.: *On webs of maximum rank.* Geom. Dedicata **31** (1986), no. 19–35. (MR 90g:53023; Zbl. 677.53017.)

[Lo 69] Loos, O.: *Symmetric spaces*, vol. 1, 2, Benjamin, New York, 1969. (MR **39** (1970) #365a,b; Zbl. **175**, p. 486.)

[Ma 55] Mal'cev, A.I.: *Analytical loops.* . (Russian) Mat. Sb. **36** (1955), 569–576. (MR **16**, p. 997; Zbl. **65**, p. 7.)

[M 78] Mikhailov, Ju.I.: *On the structure of almost Grassmann manifolds.* (Russian) Izv. Vyssh. Uchebn. Zaved. Mat. **1978**, no. 2(160), 62–72. English translation: Soviet Math. (Iz. VUZ) **22** (1978), no. 2, 54–63. (MR 81e:53031; Zbl. 398.53006.)

[Mi 86] Mikheev, P.O.: *On a G-property of a local analytic Bol loop.* (Russian) Webs and Quasigroups, Kalinin. Gos. Univ., Kalinin, 1986, 54–59. (MR 88i:20005; Zbl. 677.22002.)

[MiS 88] Mikheev, P.O.; Sabinin, L.V.: *Smooth quasigroups and geometry.* (Russian) Problems in geometry, Vol. 20, 75–110, Itogi Nauki i Tekhniki, Akad. Nauk

SSSR, Vsesoyuz. Inst. Nauchn. i Tekhn. Inform., Moscow, 1988. English Translation: J. Soviet Math. **20** (1990), 2642–2666. (MR 90b:53063.)

[MiS 90] Mikheev, P.O.; Sabinin, L.V.: *Quasigroups and differential geometry.* Chapter X in [CPS 90], 357–430. (Zbl. 721.53018.)

[MS 90] Mishchenko, S.G.; Shelekhov, A.M.: *Subwebs and subloops.* Matem. Issled. No. 113 Kvazigruppy i Ikh Sistemy (1990), 66–71.

[Mo 35] Moufang, R.: *Zur Struktur von Alternativ Körpern.* Math. Ann. **110** (1935), 416–430. (Zbl. **10**, p. 4.)

[Na 87] Nakai, I.: *Topology of complex webs of codimension one and geometry of projective space curves.* Topology **26** (1987, no. 4, 475–504. (MR 89b:14010; Zbl. 582.53020.)

[N 85] Nagy, P.: *On the canonical connection of a three-web.* Publ. Math. Debrecen **32** (1985), no. 1–2, 93–99. (MR 87a:53028; Zbl. 586.53005.)

[N 88] Nagy, P.: *Invariant tensorfields and the canonical connection of a 3-web.* Aequationes Math. **35** (1988), no. 1, 31–44. (MR 89h:53052; Zbl. 644.53013.)

[N 89] Nagy, P.: *On complete group* 3*-webs and* 3*-nets.* Arch. Math. (Basel) **53** (1989), no. 4, 411–413. (MR 90g:53024; Zbl. 696.53008.)

[Na 65] Nazirov, T.K.: *On the maximal rank of a curvilinear three-web in the space.* Vest. Mosk. Univ. Ser. Mat.-Mekh. **1965**, no. 5, 27–34. (MR **32** #9054; Zbl. **148**, p. 155.)

[O 76a] Orlova, V. G.: *On a class of multidimensional three-webs of maximal rank.* (Russian) Geometry of Imbedded Manifolds, Moskov. Gos. Ped. Inst., Moscow, 1976, 89–93.

[O 76b] Orlova, V. G.: *Multidimensional three-webs of maximal rank.* (Russian) Izv. Vyssh. Uchebn. Zaved. Mat. **1976**, no. 1, 45–52. English translation: Soviet Math. (Iz. VUZ) **20** (1976), no. 1, 55–63. (MR **57** #7427; Zbl. 337.53019.)

[Pf 90] Pflugfelder, H.O.: *Quasigroups and loops: introduction.* Heldermann-Verlag, Berlin, 1990, iii+147 pp. (Zbl. 715.20043.)

[Pi 75] Pickert, G.: *Projective Ebenen*, 2nd edition. Springer-Verlag, Berlin/Heidelberg/New York, 1975, ix+371 pp. (MR **51** #6577; Zbl. 307.50001.)

[Po 82] Postnikov, M.M.: *Lectures in geometry, V: Lie groups and algebras.* (Russian) "Nauka", Moscow, 1982, 447 pp. (MR 85b:22001; Zbl. 597.22001.)

[P 01] Poincaré, H.: *Les surfaces de translation et les fonctions abéliennes.* Bull. Soc. Math. France **29** (1901), 61–86.

[R 28] Reidemeister, K.: *Gewebe und Gruppen.* Math. Z. 29 (1928), 427–435.

[R 30] Reidemeister, K.: *Grundlagen der Geometrie.* Springer-Verlag, Berlin, 1930, 147 pp. (Jbuch. **56**, p. 483.)

[RS 67] Rozhdestvenskii, B.L.; Sidorenko, A.D.: *On the impossibility of "gradient catastrophe" for weakly non-linear syatems.* Zh. Vychisl. Mat. i Mat. Fiz. **7** (1967), no. 5, 1176–1179. (MR **36** #1824; Zbl. 526.76001.)

[Sa 72] Sabinin, L.V.: *Geometry of loops.* (Russian) Mat. Zametki **12** (1972), no. 5, 605-616. English translation: Math. Notes **12** (1972), no. 5, 799–805. (MR 49 #5216; Zbl. 258.20006.)

[Sa 88] Sabinin, L.V.: *Non-linear geometric algebra.* (Russian) Webs and Quasigroups, Kalinin. Gos. Univ., Kalinin, 1988, 32–36. (MR 89e:20121; Zbl. 674.53018)

[SM 82a] Sabinin, L.V.; Mikheev, P.O.: *A symmetric connection in the space of an analytic Moufang loop.* (Russian) Dokl. Akad. Nauk SSSR **262** (1982), no. 4, 807-809. English translation: Soviet Math. Dokl. **25** (1982), no. 1, 136–138. (MR 84e:53027; Zbl. 495.53048.)

[SM 82b] Sabinin, L.V.; Mikheev, P.O.: *Analytical Bol loops.* (Russian) Webs and Quasigroups, Kalinin. Gos. Univ., Kalinin, 1982, 102–109. (MR 84c:22007; Zbl. 499.20044.)

[SM 84] Sabinin, L.V.; Mikheev, P.O.: *On the geometry of smooth Bol loops.* (Russian) Webs and Quasigroups, Kalinin. Gos. Univ., Kalinin, 1984, 144–154. (MR 88c:53001; Zbl. 568.53009.)

[SM 85] Sabinin, L.V.; Mikheev, P.O.: *The theory of smooth Bol loops.* Univ. Druzhby Narodov, Moscow, 1985, 81 pp. (MR 87j:22030; Zbl. 584.53001.)

[SM 87] Sabinin, L.V.; Mikheev, P.O.: *Infinitesimal theory of local analytic loops.* (Russian) Dokl. Akad. Nauk SSSR **297** (1987), no. 4, 801–804. English translation: Soviet Math. Dokl. **36** (1988), no. 3, 545–548. (MR 89g:22003; Zbl. 659.53018.)

[Sag 62] Sagle, A.A.: *Mal'cev algebras.* Trans. Amer. Math. Soc. **101** (1961), no. 3, 426–458. (MR **26** #1343; Zbl. **101**, p. 23.)

[Sc 04] Scheffers, G.: *Das Abel'sche Theorem und das Lie'sche Theorem über Translationflächen.* Acta Math. **28** (1904), 65–91.

[Sch 70] Schwartz, L.: *Analyse: deuxème partie; topologie générale et analyse fonctionnelle.* Hermann, Paris, 1970, 433 pp. (MR **57** #7087; Zbl. **206**, p. 63.)

[Se 81] Sedov, L.I.: *Similarity and dimensions methods in mekhanics.* 9th ed., "Nauka', Moscow, 1981, 448 pp. (MR 83k:00015; Zbl. 526.76001.)

[S 80] Shelekhov, A.M.: *Local algebras of a cyclic three-web.* (Russian) Kaliningrad. Gos. Univ. Differentsial'naya Geom. Mnogoobraz. Figur, no. 11 (1980), 115–122. (MR 82k:53031. Zbl. 525.53021.)

[S 81a] Shelekhov, A.M.: *Eight-dimensional cyclic three-webs.* (Russian) Webs and Quasigroups, Kalinin. Gos. Univ., Kalinin, 1981, 124–135. (MR 83j:53010; Zbl. 497.53024.)

[S 81b] Shelekhov, A.M.: *Three-webs with partially symmetric curvature tensor.* (Russian) Sibirsk. Mat. Zh. **22** (1981), no. 1, 210–219. English translation: Siberian Math. J. **22** (1981), no. 1, 156–163. (MR 82k:53032; Zbl. 483.53014 & 471.53019.)

[S 82] Shelekhov, A.M.: *An algebraic eight-dimensional three-web on the manifold of pencils of second-order curves.* (Russian) Webs and Quasigroups, Kalinin. Gos. Univ., Kalinin, 1982, 133–142. (MR 84e:53021; Zbl. 499.53013.)

[S 84] Shelekhov, A.M.: *On a characteristic property of Grassmann three-webs defined by a cubic symmetroid.* (Russian) Webs and quasigroups, Kalinin. Gosud. Univ., Kalinin, 1984, 112–117. (MR 88c:53001; Zbl. 558.53013.)

[S 85a] Shelekhov, A.M.: *Identities with one variable in loops, which are equivalent to the mono-associativity.* (Russian) Problems in the Theory of Webs and Quasigroups, Kalinin. Gos. Univ., Kalinin, 1985, 89–93. (MR 88d:20099; Zbl. 572.20055.)

[S 85b] Shelekhov, A.M.: *On closed g-structures defined by multidimensional three-webs.* (Russian) Kalinin. Gos. Univ., Kalinin, 1985, 49 pp., bibl. 24 titles. Dep. in VINITI on 12/25/85 under no. 8815–B.

[S 85c] Shelekhov, A.M.: *On closure figures and identities defined in a differential neighborhood of fourth order of a multidimensional three-web.* (Russian) Kalinin. Gos. Univ., Kalinin, 1987, 7 pp., bibl. 7 titles. Dep. in VINITI on 11/25/85 under no. 8300–B87.

[S 86a] Shelekhov, A.M.: *Three-webs determined by a determinantal surface.* (Russian) Izv. Vyssh. Uchebn. Zaved. Mat. **1986**, no. 3(286), 84–86. English translation: Soviet Math. (Iz. VUZ) **30** (1986), no. 3, 119–122. (MR 87m:53020; Zbl. 602.53011 & 613:53006.)

[S 86b] Shelekhov, A.M.: *Computation of the covariant derivatives of the curvature tensor of a multidimensional three-web.* (Russian) Webs and Quasigroups, Kalinin. Gos. Univ., Kalinin, 1986, 96–103. (MR 88m:53037; Zbl. 617.53028.)

[S 87a] Shelekhov, A.M.: *On the higher order differential-geometric objects of a multidimensional three-web.* (Russian) Problems in Geometry, Vol. 19, 101–154, Itogi Nauki i Tekhniki, Akad. Nauk SSSR Vsesoyuz. Inst. Nauchn. i Tekhn. Informatsii, Moscow, 1987. English translation: J. Soviet Math. 44 (1989), no. 2, 153–190. (MR 89e:53021; Zbl. 711.53013.)

[S 87b] Shelekhov, A.M.: *On three-webs with elastic coordinate loops.* (Russian) Kalinin. Gos. Univ., Kalinin, 1987, 7 pp., bibl. 6 titles. Dep. in VINITI on 12/2/87 under no. 8465–B87.

[S 87d] Shelekhov, A.M.: *On automorphisms of local analytic loops.* (Russian) Kalinin. Gos. Univ., Kalinin, 1987, 9 pp., bibl. 14 titles. Dep. in VINITI on 12/2/87 under no. 8466–B87.

[S 88a] Shelekhov, A.M.: *The tangent W_4-algebra of a multidimensional three-web.* (Russian) Webs and Quasigroups, Kalinin. Gos. Univ., Kalinin, 1988, 4–16. (MR 89f:53027; Zbl. 659.53019.)

[S 88b] Shelekhov, A.M.: *Three-webs and quasigroups determined by a determinantal hypercubic.* (Russian) Tomsk. Geom. Sb. Vyp.29, 1988, 55–65.

[S 89a] Shelekhov, A.M.: *On integration of closed g_W-structures.* (Russian) Kalinin. Gos. Univ., Kalinin, 1989, 21 pp., bibl. 11 titles. Dep. in VINITI on 7/26/89 under no. 5031–B89.

[S 89b] Shelekhov, A.M.: *A classification of multidimensional three-webs according to closure conditions.* (Russian) Problems in Geometry, Vol. 21, 109–154, Itogi Nauki i Tekhniki, Akad. Nauk SSSR Vsesoyuz. Inst. Nauchn. i Tekhn. Informatsii, Moscow, 1989. English translation: J. Soviet Math. **55** (1991), no. 6, 2140–2167. (MR 91b:53018; Zbl. 705.53014.)

[S 89c] Shelekhov, A.M.: *The g-structure associated with a multidimensional hexagonal 3-web, is closed.* J. Geom. 35 (1989), 167–176. (MR 90h:53022; Zbl. 699.53025.)

[S 90a] Shelekhov, A.M.: *Autotopies of three-webs and geometric G-webs.* (Russian) Izv. Vyssh. Uchebn. Zaved. Mat. **1990**, no. 5, 75–77. English translation: Soviet Math. (Iz. VUZ) **34** (1990), no. 5, 87–90. (MR 91i:53034; Zbl. 714.53017.)

[S 90b] Shelekhov, A.M.: *Computation of the second covariant derivatives of the curvature tensor of a multidimensional three-web.* (Russian) Webs and Quasigroups, Kalinin. Gos. Univ., Kalinin, 1990, 10–18. (MR 91h:53022; Zbl. 701.53026.)

[S 91a] Shelekhov, A.M.: *New closure conditions and some problems in loop theory.* Aequationes Math. **41** (1991), no. 1, 79–84. (Zbl. 719.20037.)

[S 91b] Shelekhov, A.M.: *On the theory of G-webs and G-loops.* Global Differential Geometry and Global Analysis (Berlin, 1990), 265–271, Lecture Notes in Math. **1481**, Springer-Verlag, Berlin et al, 1991.

[S t.a.] Shelekhov, A.M.: *On analytic solutions of functional equations.* (Russian) Mat. Zametki (to appear).

[SD 81] Shelekhov, A.M.; Demidova, L.A.: *Some figures of closure on three-webs.* (Russian) Webs and Quasigroups, Kalinin. Gos. Univ., Kalinin, 1981, 42–46. (MR 83j:53009; Zbl. 497.53021.)

[SS 84] Shelekhov, A.M.; Shestakova, M.A.: *A geometric proof of the universality of some identities in loops.* (Russian) Webs and Quasigroups, Kalinin. Gos. Univ., Kalinin, 1984, 118–124. (Zbl. 558.53016.)

[SS 85] Shelekhov, A.M.; Shestakova, M.A.: *Identities in loops with a weak associativity.* (Russian) Problems in the Theory of Webs and Quasigroups, Kalinin. Gos. Univ., Kalinin, 1985, 115–121. (MR 88d:20100; Zbl. 572.20054.)

[Sh 88] Shestakova, M.A.: *Structure equations of a six-dimensional hexagonal three-web.* (Russian) Webs and Quasigroups, Kalinin. Gos. Univ., Kalinin, 1988, 140–145. (MR 89f:53028; Zbl. 658.53013.)

[Sh 90] Shestakova, M.A.: *An example of a hexagonal three-web with a partially symmetric curvature tensor.* (Russian) Webs and Quasigroups, Kalinin. Gos. Univ., Kalinin, 1990, 22–29. (MR 91h:53024; Zbl. 701.53025.)

[St 83] Sternberg, S.: *Lectures on differential geometry*, 2nd edition, Chelsea Publishing Co., New York, 1983, xv+442 pp. (MR 88f:58001; Zbl. 518.53001.)

[T 27] Thomsen, G.: *Un teoreme topologico sulle schiere di curve e una caratterizzazione geometrica delle superficie isotermo-asintotiche.* Boll. Un. Mat. Ital. Bologna **6** (1927), 80–85.

[Ti 75a] Timoshenko, V.V.: *Three-webs over commutative associative algebras.* (Russian) Ukrain. Geom. Sb. **18** (1975), 136–151. (MR **54** #8516; Zbl. 455.53012.)

[Ti 75b] Timoshenko, V.V.: *On three-webs over commutative associative algebras.* (Russian) Izv. Vyssh. Uchebn. Zaved. Mat. **1975**, no. 11 (133), 109–112. English translation: Soviet Math. (Iz. VUZ) **16** (1975), no. 11, 93–96. (MR **54** #8516; Zbl. 343.53001.)

[Ti 76] Timoshenko, V.V.: *Subwebs defined by divisors of zero of a commutative associative algebra.* (Russian) Geometry of Imbedded Manifolds, Moskov. Gos. Ped. Inst., Moscow, 1976, 102–115.

[Ti 77a] Timoshenko, V.V.: *On three-webs over an algebra the curvature of which being a zero's divisor.* (Russian) Ukrain. Geom. Sb. **20** (1977), 102–114. (MR 81f:53013; Zbl. 449.14001.)

[Ti 77b] Timoshenko, V.V.: *Three-webs over an algebra the curvature of which is a divisor of zero.* (Russian) Izv. Vyssh. Uchebn. Zaved. Mat. **1977**, no. 3 (149), 116–118. English translation: Soviet Math. (Iz. VUZ) **21** (1977), no. 3, 92–94. (MR **57** #17527; Zbl. 381.53010.)

[Ti 78] Timoshenko, V.V.: *Three-webs over some classes of commutative associative algebras.* (Russian) Geometry of Imbedded Manifolds, Moskov. Gos. Ped. Inst., Moscow, 1978, 104–111. (MR 81j:53022; Zbl. 455.53013.)

[Ti 79] Timoshenko, V.V.: *On the structure of a multidimensional three-web which is the real realization of a three-web over a commutative associative algebra.* (Russian) Geometry of Imbedded Manifolds, Moskov. Gos. Ped. Inst., Moscow, 1979, 93–100. (MR 82m:53018.)

[Ti 81] Timoshenko, V.V.: *Four-webs over commutative associative algebras.* (Russian) Geometry of Imbedded Manifolds, Moskov. Gos. Ped. Inst. Moscow, 1981, 104–111.

[To 81] Tolstikhina, G.A.: *Four-dimensional webs with a symmetric curvature tensor.* (Russian) Webs and Quasigroups, Kalinin. Gos. Univ., Kalinin, 1981, 12–22. (MR 83i:53036; Zbl. 497.53025.)

[To 82] Tolstikhina, G.A.: *Invariant transversal distributions of four-dimensional three-webs W_s.* (Russian) Webs and Quasigroups, Kalinin. Gos. Univ., Kalinin, 1982, 115–120. (MR 84a:53017; Zbl. 499.53014.)

[To 85] Tolstikhina, G.A.: *On a property of a 4-web carrying a group three-subweb.* (Russian) Problems of the theory of webs and quasigroups, 121–128, Kalinin. Gos. Univ., Kalinin, 1985. (MR 88b:53019; Zbl. 572.53019.)

[To 87] Tolstikhina, G.A.: *Principal directions on an r-dimensional surface that are defined by a $(2r)$-dimensional three-web.* (Russian) Webs and Quasigroups, Kalinin. Gos. Univ., Kalinin, 1987, 99–105. (MR 88i:53042; Zbl. 617.53026.)

[To 88] Tolstikhina, G.A.: *An affine connection induced by an idempotent quasigroup on a smooth manifold of a multidimensional three-web.* (Russian) Webs and Quasigroups, Kalinin. Gos. Univ., Kalinin, 1988, 16–23. (MR 89f:53029; Zbl. 659.53020.)

[To 90] Tolstikhina, G.A.: *The core of a coordinate quasigroup of a hexagonal Bol three-web.* (Russian) Webs and Quasigroups, Kalinin. Gos. Univ., Kalinin, 1990, 18–22. (MR 91h:53023; Zbl. 701.53024.)

[Tr 89] Trofimov, V.V.: *Introduction to geometry of manifolds with symmetries.* (Russian) Izdat. Mosc. Gos. Univ., Moscow, 1989, 360 pp. (Zbl. 705.53002.)

[Ts 85] Tsarev, S.P.: *On Poisson brackets and one-dimensional Hamiltonian systems of hydronamic type.* (Russian) Dokl. Akad. Nauk SSSR **282** (1985), no. 3, 534–537. English translation: Soviet Math. Dokl. **31** (1985), no. 3, 488-491. (MR 87b:58030; Zbl. 605.53075.)

[Ts 90] Tsarev, S.P.: *Geometry of Hamiltonian systems of hydronamic type. The generalized method of hodograph.* (Russian) Izv. Akad. Nauk SSSR Ser. Mat. **54** (1990), no. 5, 1048–1058. English translation: Math. USSR-Izv. **54**, no. 5, 397–419.

[U 86] Utkin, A.A.: *A three-web defined on a surface by a normcurve.* (Russian) Webs and Quasigroups, Kalinin. Gos. Univ., Kalinin, 1986, 71–77. (MR 88i:53045; Zbl. 617.53022.)

[V 87] Vasiljev, A.M.: *Theory of differential-geometric structures.* (Russian) Moskov. Gos. Univ., Moscow, 1987, 192 pp. (MR 89m:58001; Zbl. 656.53001.)

[Ve 75] Vechtomov, V. E.: *Closure figures for a certain class of universal identities.* (Russian) Mat. Issled. **10** (1975), no. 2, 36–63. (MR **53** #5799; Zbl. 355.20078.)

[Ve 81] Vechtomov, V.E.: *On the question of universal reducible identities.* (Russian) Webs and Quasigroups, Kalinin. Gos. Univ., Kalinin, 1981, 30–37. (MR 83a:20096; Zbl. 539.20048.)

[Ves 86] Veselyaeva, T.Yu.: *A three-web in a projective space R_nP_2 over the matrix algebra R_n.* (Russian) Kalinin. Gos. Univ., Kalinin, 1986, 16–20. (MR 88j:53025; Zbl. 617.53021.)

[Vo 85] Voskanyan, V.K.: *A conformal structure associated with a curvilinear $(n+1)$-web.* (Russian) Problems in the Theory of Webs and Quasigroups, Kalinin. Gos. Univ., Kalinin, 1985, 33–38. (MR 85e:53026; Zbl. 573.53009.)

[Vo 86] Voskanyan, V.K.: *A curvilinear $(n+1)$-web on a hypersurface of the projective space P_{n+1}.* (Russian) Webs and Quasigroups, Kalinin. Gos. Univ., Kalinin, 1986, 21–28. (MR 88i:53049; Zbl. 617.53029.)

[W 82] Wood, J.A.: *An algebraization theorem for local hypersurfaces in projective space.* Ph.D. Thesis, University of California, Berkeley, 1982, 87 pp.

[W 84] Wood, J.A.: *A simple criterion for local hypersurfaces to be algebraic.* Duke Math. J. **51** (1984), no. 1, 235-237. (MR 85d:14069; Zbl. 584.14021.)

[ZS 80] Zakharov, V.E.; Schulman, E.I.: *Degenerative dispersion laws, motion invariants and kinetic equations.* Phys. D **1** (1980), 192–202. (MR 81j:35017.)

[ZS 88] Zakharov, V.E.; Schulman, E.I.: *On additional motion invariants of classical hamiltonian wave systems.* Phys. D **29** (1988), 283–320. (MR 89k:58137; Zbl. 651.35080.)

[Z 78] Zhogova, T.B.: *On the focal three-web of a two-parameter family of two- dimensional planes in P_5.* (Russian) Geometry of Imbedded Manifolds, Moskov. Gos. Ped. Inst., Moscow, 1978, 40–46. (MR 82a:53010; Zbl. 444.53009.)

[Z 79] Zhogova, T.B.: *On a class of two-parameter families of two-dimensional planes in P_5 with a hexagonal focal three-web.* (Russian) Geometry of Imbedded Manifolds, Moscow, 1979, 44–50. (MR 82f:53026; Zbl. 484.53004.)

[Z 80] Zhogova, T.B.: *On a quasigroup generated by a certain class of two-parameter families of two-dimensional planes in P_5.* (Russian) Izv. Vyssh. Uchebn. Zaved. Mat. **1980**, no. 2, 63–66. English translation: Soviet Math. (Iz. VUZ) **24** (1980), no. 2, 65–69. (MR 81f:53009; Zbl. 449.53038.)

SYMBOLS FREQUENTLY USED

The list below contains many of the symbols whose meaning is usually fixed throughout the book.

$A = (A_1, A_2, A_3)$: autotopy of a three-web or a quasigroup, 49, 216
$AG(r,n)$: almost Grassmann structure, 123, 283
$AGW(d,n,r)$: almost Grassmannizable d-web, 293
A^n: affine space of dimension n, 5
Aut l_p: group of automorphisms of a loop l_p, 220
AW: algebraizable web, 5, 296
AW^2: two-dimensional three-web over an algebra A, 131
AX^2: two-dimensional manifold over an algebra A, 131
$a = (a^i_{jk})$: torsion tensor of a three-web or a quasigroup, 12, 23, 27, 82
$\underset{uv}{a}{}^i_{jk}$: torsion tensor of a web $W(n+1,n,r)$, 273
$\underset{\alpha\beta}{a}{}^i_{jk}$: torsion tensor of a connection $\Gamma_{\alpha\beta}$ associated with a three-web, 30
$\mathcal{A}$: group of autotopies of a three-web or a quasigroup, 217, 221
α_l, α_r: left and right commutators of a loop, 75, 78
$[\alpha,\beta,\gamma]$: three-subweb of a d-web, 275, 292
B_l: a) left Bol figure, or b) class of left Bol webs, 56, 62, 140
B_m: a) middle Bol figure, or b) class of middle Bol webs, 56, 62, 138
B_r: a) right Bol figure, or b) class of right Bol webs, 56, 62, 140
b^i_{jkl}: curvature tensor of a three-web, 14, 24, 82
$\underset{\alpha\beta}{b}{}^i_{jkl}$, $\overset{*}{b}{}^i_{jkl}$: curvature tensors of the connections $\underset{\alpha\beta}{\Gamma}$ and $\overset{*}{\Gamma}$, 31–36
$\underset{uv}{b}{}^i_{jkl}$: curvature tensor of a web $W(n+1,n,r)$, 273
β_l, β_r: left and right associators of a loop, 75, 78
$C(r,n)$: Segre cone, 117
$c^\alpha_{\beta\gamma}$: structure tensor of a Lie group (algebra), 2
Der A: algebra of differentiations of an algebra A, 177
δ: symbol of differentiation with respect to secondary parameters, 2, 15
E: a) figure defined by the elasticity identity, or b) class of webs on which all figures E are closed, 57, 62
e: unit of a loop, 49, 73
$\underset{\alpha}{e}{}^i$: vectors tangent to a leaf of the foliation λ_α, 8
ϵ^{ijk}, ϵ_{ijk}: discriminant tensors, 154
$\mathcal{F}_\alpha$: leaf of a web belonging to the foliation λ_α, 7
G: Lie group, 2, 19
GL(n): general linear group, 2, 123, 185, 236
$G(r,n)$: Grassmannian of r dimensional subspaces in a projective space P^n, 5, 276
GW: Grassmann web, 5, 118, 277
G_W: G-structure associated with a three-web W, 9
Γ: Chern connection of a three-web, 27
$\Gamma_{\alpha\beta}$, $\widetilde{\Gamma}_{\alpha\beta}$, $\overset{*}{\Gamma}$: invariant affine connections associated with a three-web W, 30–36
$\gamma(W)$: pencil of affine connections associated with a three-web W, 35
H: a) hexagonal figure, or b) class of hexagonal three-webs, 54
$H^{2\rho}$: transversal (2ρ)-vector of a three-web, 37
$J = (J_1, J_2, J_3)$: isotopy of three-webs or quasigroups, 7, 48

J_l, J_r, J_m: left, right and middle kernels of the tangent W-algebra of a three-web, 232
$J(\xi, \eta, \zeta)$: Jacobian of elements ξ, η and ζ, 179, 183
L_a: left translation, 49
$LW(d, n, r)$: linearizable d-web, 295
$l(a, b)$: LP-isotope (coordinate loop) of a quasigroup , 59
l_p: coordinate loop of a three-web associated with a point p, 64, 70, 219
$\Lambda = (\lambda^i_j)$: basis affinor of a four-web, 291
$\underset{s}{\Lambda}$, $\underset{p,q}{\Lambda}$: terms of the Taylor expansion of the product $x \cdot y$ in a loop, 84, 85
λ_α: foliation of a three-web, 4
M: a) Moufang identity, or b) class of Moufang three-webs 58, 70
N_l, N_r, N_m: left, right and middle kernels of a loop, 217
∇, $\overset{\alpha\beta}{\tilde{\nabla}}$, $\overset{*}{\nabla}$, $\overset{\alpha\beta}{\nabla}$: operators of covariant differentiation with respect to the invariant connections, 13, 30, 32, 33
∇_δ: operators of differentiation with respect to secondary parameters, 15
$O(n)$: orthogonal group, 2, 27
$\omega = (\omega^i_j)$: forms of the Chern connection, 27
$\underset{\alpha}{\omega} = (\underset{\alpha}{\omega}^i)$: basis forms of the foliation λ_α of a web, 8, 273
$\underset{\alpha\beta}{\omega}{}^i_j$: forms of the connection $\underset{\alpha\beta}{\Gamma}$, 30
P^n: projective space of dimension n, 5, 117, 276
$\varphi_{\alpha\beta}$: mapping defined by leaves of the foliation λ_γ, $\gamma \neq \alpha$, $\gamma \neq \beta$, 9, 37
$\pi(d, n)$: Castelnuovo bound, 302
$\pi(d, n, r)$: least upper bound for the rank of a web $W(d, n, r)$, 304
π^i_j: invariant forms of the general linear group, 2, 12, 15
Q: loop, quasigroup, groupoid, 47, 75
$q(\cdot)$: quasigroup, 47, 73
$q = (\cdot, X_\alpha)$: three-base quasigroup, 47, 73
R: a) Reidemeister figure, or b) class of group three-webs, 55, 61
R_a: right translation, 49
$\mathcal{R}(\mathcal{X})$: frame bundle of a manifold X, 1, 185
$\mathcal{R}(\mathcal{W})$:bundle of adapted frames of a three-web W, 9
$\mathbf{SL}(n)$: special linear group, 123, 283
$S(r, n)$: Segre variety, 11, 282, 305
T: a) Thomsen figure, or b) class of parallelizable three-webs, 55, 61
T_e: tangent space to a loop at the unit e, 77
$T_p(X)$: tangent space to a manifold X at a point p, 1
$W(d, n, r)$: d-web of codimension r on a manifold of dimension nr, 272
$W = (X, \lambda_\alpha)$: three-web W formed on a manifold X by foliations λ_α, 4
W_0: three-web formed by three families of parallel planes, 5, 18
X: differentiable manifold, 1
(X, B, π): fibration, 3
X_G: G-structure on a manifold X, 2, 185
X_α: a) base of the foliation λ_α, or b) hypersurfaces defining a Grassmann three-web, 5, 276
$\wedge$: symbol of exterior multiplication, 1
[,]: commutator, 78, 150
(, ,): associator, 79, 150
< , , >: ternary operation, 144
/, \: right and left inverse operations in a quasigroup, 62

Index

GPSR Compliance

The European Union's (EU) General Product Safety Regulation (GPSR) is a set of rules that requires consumer products to be safe and our obligations to ensure this.

If you have any concerns about our products, you can contact us on ProductSafety@springernature.com

In case Publisher is established outside the EU, the EU authorized representative is:

Springer Nature Customer Service Center GmbH
Europaplatz 3
69115 Heidelberg, Germany

Batch number: 10584568

Printed by Printforce, the Netherlands